Hadron Form Factors

From Basic Phenomenology to
QCD Sum Rules

Hadron Form Factors

From Basic Phenomenology to QCD Sum Rules

Alexander Khodjamirian

CRC Press
Taylor & Francis Group
Boca Raton London New York

CRC Press is an imprint of the
Taylor & Francis Group, an **informa** business

CRC Press
Taylor & Francis Group
6000 Broken Sound Parkway NW, Suite 300
Boca Raton, FL 33487-2742

First issued in paperback 2021

© 2020 by Taylor & Francis Group, LLC
CRC Press is an imprint of Taylor & Francis Group, an Informa business

No claim to original U.S. Government works

ISBN-13: 978-1-138-30675-2 (hbk)
ISBN-13: 978-1-03-217303-0 (pbk)
DOI: 10.1201/9781315142005

Publisher's Note
The publisher has gone to great lengths to ensure the quality of this reprint but points out that some imperfections in the original copies may be apparent.

Visit the Taylor & Francis Web site at
http://www.taylorandfrancis.com

and the CRC Press Web site at
http://www.crcpress.com

To my family

Contents

Preface

Modern particle physics is a unique interplay of theory, phenomenology, and experiment. The underlying Standard Model (SM) of particles is expected to have a New Physics extension, to be revealed through current and future experiments. A crucial role in establishing the SM and in searching for effects beyond SM is played by studies of electromagnetic and weak (electroweak) transitions of leptons and quarks, mediated by photons and weak bosons. The rules of computing electroweak transitions at the lepton and quark level are well defined in SM in terms of perturbative expansion expressed via Feynman diagrams. However, transitions are observed between hadrons, the bound states in which quarks are permanently confined; hence, in order to compare predictions of theory with measured observables, it is crucial to identify and compute all relevant hadronic effects accompanying the quark-level transition and to separate these effects from the short-distance part of the transition amplitude. This is the realm of Quantum Chromodynamics (QCD), the part of SM describing quark-gluon interactions. This task, however, is highly complicated by the fact that the bulk of hadronic effects in electroweak transitions are generated by the quark-gluon interactions at long distances, for which perturbative QCD – that is, the expansion in terms of Feynman diagram and in powers of the quark-gluon coupling – is not applicable.

The effects of QCD interactions in the electromagnetic and weak transitions of quarks are described by **hadron form factors**, which are the main subject of this book. The terminology is inherited from atomic physics, where the form factors calculated in quantum mechanics take into account binding effects of atomic electrons, e.g., in the processes of electron scattering on an atomic target. However, one should be aware that the similarity between atoms and hadrons is limited: the quarks confined inside hadrons cannot be described by simple quantum-mechanical wave functions.

In strictly theoretical terms, hadron form factors are defined as Lorentz-invariant functions parametrizing the matrix elements in which quark-transition operators are sandwiched between the initial and final hadronic states. At first sight, hadron form factors are strikingly non-universal objects, depending on the flavors of quarks involved in the electromagnetic or weak transition and on the quantum numbers of the initial and final hadrons. A closer look reveals that, in fact, form factors have many common features, stemming from the symmetries of SM in general and of QCD in particular. One of the underlying reasons is that the quark-gluon interaction is flavor-independent. Hadron form factors, being defined as amplitudes in quantum field theory, also obey the general principles of analyticity and unitarity; hence, it certainly makes sense to analyze the phenomenology of various hadron form factors in a uniform way. Furthermore, the available QCD-based methods used to evaluate hadron form factors are also, to a large extent, universal. These universalities were one of the main motivations for the author to collect the most essential information on the phenomenology and theory of hadron form factors within one text.

The book consists of two parts. The first part (Chapters 1–6) contains a brief introduction to QCD and is mainly devoted to the phenomenology of hadron form factors. A broad variety of examples is discussed, starting from the pion and nucleon electromagnetic form factors, which were studied already at the dawn of particle physics, and ending with the form factors of the B meson decays which are currently investigated at the Large Hadron Collider (LHC) or in experiments at flavor factories, such as Belle-II or BESIII.

In the second part (Chapters 7–10), the QCD-based calculations of hadronic matrix elements are discussed and explained. We focus here on the continuum QCD methods, including conventional QCD sum rules and light-cone sum rules (LCSRs). In the last decades, these techniques successfully complemented lattice QCD computations of hadron form factors. In this book, both sum rule methods are presented at an introductory level, starting from the underlying ideas of the local and light-cone operator-product expansion, and showing derivation steps. Several important applications of LCSRs to hadron form factors are presented and discussed.

This book was planned to a large extent as a textbook, aimed predominantly at master's degree and PhD students who specialize in particle physics, both phenomenological and experimental. In fact, a large part of the material presented below was prepared by the author for his special course of lectures "QCD and Hadrons" at the University of Siegen. A basic knowledge of advanced quantum mechanics and quantum field theory and a familiarity with the SM electroweak theory are desirable.

The main emphasis here is on learning the analytical methods in particle phenomenology. Working through the book will help the reader develop practical skills in this field. They should be able to follow along with detailed derivations presented in the text and reproduce analytical results themselves. The appendices, containing many useful formulas, alongside the bibliography which contains a selection of papers, reviews, and books, should help to fulfill this task. The material presented in this book can then be used by the reader as a starting point for their own research projects. Smaller scale projects could involve a numerical analysis of the analytical results, combining and comparing the obtained plots and tables with experimental and/or lattice QCD results. A very useful source of information to use in parallel are the tables and minireviews by the Particle Data Group available at [1]. The reason this book itself features very few numerical results and shows no plots is that experiments and lattice QCD calculations are continuously advancing, and the current quoted data may soon become outdated. In this sense, what is presented below should not be considered as a comprehensive and up-to-date review of all existing methods and results in the field of hadron form factors. Nevertheless, this book, especially the first part of it, where essential information on various hadron form factors is collected in one place, aims to serve as a sort of handbook for researchers in both phenomenology and experiment.

Finally, a disclaimer is in order. Many interesting and historically important models of hadron form factors – e.g., those based on the concept of constituent quarks – are not considered, as these models seem to have no direct connection to QCD. Due to space limitations – but more importantly, the author's lack of expertise – this book does not contain studies of hadron form factors in lattice QCD. In addition, the application of QCD-based effective theories (chiral perturbation theory, heavy-quark effective theory (HQET) and soft-collinear effective theory) remain beyond our scope. Only the basics of the heavy-quark mass limit, chiral limit, and factorization in QCD are presented. References in the bibliography partially fill this gap. Naturally, the author's selection of the original papers and useful reviews may still be incomplete and somewhat subjective.

When I started this project, an aspirational example for me was the book *Weak Interactions of Elementary Particles*, written by Lev Okun in 1962 [2], in which, by the way, the word "hadron" was coined. The transparent and explanatory style of this book, written long before the era of the Standard Model[1], has almost no precedence in the literature on particle phenomenology. Visiting in the late 1970s the famous ITEP (Institute for Theoretical and Experimental Physics) in Moscow, I had the great privilege to prepare my PhD thesis under

---

[1]Nowadays, a more recent book [3] is better known.

the supervision of Lev Okun. There I also witnessed QCD in the making, and, in particular, the development of QCD sum rules by Mikhail Shifman, Arkady Vainshtein, and Valentin Zakharov, so that I had a unique opportunity to learn the new method directly from its authors. Many of the subtle aspects of QCD and hadron phenomenology I also grasped while working with Boris Ioffe. I am grateful to Vladimir Braun who introduced to me the powerful technique of the QCD light-cone sum rules. Several applications of QCD to quark flavor physics presented in this book originate from my collaboration with Reinhold Rückl. More recent results in this field are based on our works with Thomas Mannel, who taught me nontrivial aspects of the heavy quark expansion.

My special thanks are to Aleksey Rusov for a very careful reading of the manuscript and checking the formulas.

Writing this book, I greatly benefited from a stimulating research environment in the Theoretical Particle Physics group at the University of Siegen. Yet, undoubtedly, the encouragement I received from my family played the greatest role in completing this book.

Introduction

In particle physics, the term **form factor** is inherited from the theory of atomic scattering, where it denotes a function entering the scattering amplitude and describing the momentum distribution of electrons in an atom. The atomic form factors are fully described within the nonrelativistic quantum mechanics. In this theory, the particle number in a scattering process is fixed, the particle momenta are nonrelativistic and an interaction potential is unambiguously introduced. As a result, the initial and final states are described in terms of the quantum-mechanical wave functions, thus making it possible to calculate the form factors.

In this book, we discuss the physics of hadrons, the bound states of strongly interacting quarks. The **hadron form factors** that we are interested in have a broad meaning. Any transition between an initial and final hadronic state, triggered by an electromagnetic (e.m.) or weak interaction of quarks, is described by a form factor. This quantity enters the probability amplitude and quantifies the effects of strong interactions in a transition process. Being a function of the momentum transfer, the form factor essentially depends on how the momenta of hadrons are distributed among their constituent quarks.

Strong interactions are described on the basis of Quantum Chromodynamics (QCD), a quantum field theory of quarks, antiquarks, and gluons. In QCD, the number of these particles participating in a given hadronic process remains to a large extent uncertain, and their energy-momenta are in most cases relativistic. Moreover, the strong interactions that bind quarks inside hadrons are neither described by a potential, nor calculated within a perturbation theory. Therefore, in most of the processes, hadrons cannot be represented as quantum-mechanical states with wave functions and a straightforward analytical calculation of hadron form factors in QCD remains a challenging problem, being tackled with approximate methods and numerical simulations.

Despite these drastic differences between atomic and hadronic physics, there are certain universal features of atomic form factors that are replicated in their hadronic counterparts; hence, for those who are familiar with the basics of quantum mechanics, in this introduction we briefly recollect the origin of atomic form factors. Doing that, we use the formalism maximally adapted to the one used in hadron phenomenology. We will concentrate on general aspects, leaving many details aside and making some simplifications[2].

Suppose an electron with a three-dimensional momentum $\vec{k}$ interacts with an atom in the ground state $|A_0\rangle$. We use hereafter the Dirac $\langle bra|$ and $|ket\rangle$ notation for the states. This particular process is denoted as

$$e^-(\vec{k}) + A_0 \to e^-(\vec{k}') + A_n\,, \qquad (1)$$

where $n = 0, 1, 2, ...$ is the index numerating the energy states of the atomic electrons starting from the ground state $n = 0$. The scattering is elastic if $n = 0$, in which case the electron energy does not change (we neglect the tiny recoil of a very heavy atom), or inelastic if $n \neq 0$, so that the energy difference between the final and initial electrons is equal to the atom excitation energy. We assume that the external electron momentum is

---

[2]More details can be found – e.g., in the textbook [4]. An introductory review on form factors is in [5].

much larger than the average momenta of atomic electrons. On the other hand, the velocity of the electron

$$v = \frac{|\vec{k}|}{m_e} \ll 1, \tag{2}$$

is still small; hence, the nonrelativistic quantum mechanics is fully applicable. Note that here we adopt the same system of physical units as in the particle theory, in which the Planck constant and the velocity of light are dimensionless and equal to one:

$$\hbar = c = 1.$$

The initial and final quantum states of the process (1) are described, respectively, by the state vectors

$$|\vec{k}; 0\rangle \quad \text{and} \quad \langle \vec{k}'; n|, \tag{3}$$

where the states of the scattered electron with a definite 3-momentum are combined with a certain atomic state. Since the initial-state and final-state interactions between the electron and atom – i.e., the interactions before and after scattering – are neglected, the above states can be represented as direct products of the electron and atomic state vectors:

$$|\vec{k}; 0\rangle = |\vec{k}\rangle \otimes |0\rangle, \quad \langle \vec{k}'; n| = \langle \vec{k}'| \otimes \langle n|. \tag{4}$$

The probability amplitude of the process (1) is then obtained, applying the well-known quantum-mechanical formalism of time-independent perturbation theory. The Coulomb force between the external electron and the atom (including the nucleus and atomic electrons) is well described by a potential depending on the distances between the electric charges and is proportional to the square of the electric charge e, which is usually replaced by the electromagnetic (e.m.) coupling

$$\alpha_{em} = \frac{e^2}{4\pi}. \tag{5}$$

The smallness of $\alpha_{em} \simeq 1/137$ justifies using the perturbation theory at $O(\alpha_{em})$.

The amplitude of (1) can be written in a general form of a matrix element

$$\mathcal{M}_{n0} = \langle \vec{k}'; n|\hat{V}|\vec{k}; 0\rangle, \tag{6}$$

where $\hat{V}$ is the operator of Coulomb interaction between the scattered electron and the atom. To proceed, we make use of the coordinate representation of the states and operators in quantum mechanics. The atomic nucleus is supposed to be at the origin $\vec{x} = 0$, and, since we neglect its recoil after scattering, it is treated as a fixed static electric charge eZ. The coordinates of the Z atomic electrons and of the scattered electron are denoted, respectively, as

$$\vec{x}_1, ..., \vec{x_Z} \equiv \{\vec{x}_i\} \ (i = 1, 2, ..., Z) \text{ and } \vec{y}.$$

Apart from using the eigenstates (3) with a definite momentum of the electron and definite energy of atomic electrons, we introduce the total set of eigenstates of the coordinate operators: $\{\hat{\vec{X}}_i\}$ $(i = 1, 2, ..., Z)$ and $\hat{\vec{Y}}$, so that

$$\hat{\vec{X}}_i|\{\vec{x}_i\}; \vec{y}\rangle = \vec{x}_i|\{\vec{x}_i\}; \vec{y}\rangle, \quad \hat{\vec{Y}}|\{\vec{x}_i\}; \vec{y}\rangle = \vec{y}|\{\vec{x}_i\}; \vec{y}\rangle. \tag{7}$$

The operator $\hat{V}$ satisfies the eigenvalue equation

$$\hat{V}|\{\vec{x}_i\}; \vec{y}\rangle = V(\{\vec{x}_i\}, \vec{y}) |\{\vec{x}_i\}; \vec{y}\rangle, \tag{8}$$

where the function on the r.h.s. is the coordinate-dependent Coulomb potential on which the electron is scattered:

$$V(\{\vec{x}_i\}, \vec{y}) = -\frac{Ze^2}{|\vec{y}|} + \sum_i \frac{e^2}{|\vec{y} - \vec{x}_i|} \,. \tag{9}$$

The two parts of this potential correspond to the interaction with the nucleus and with atomic electrons.

The overlaps of the states $|\vec{k}, 0\rangle$ and $\langle \vec{k}'; n|$ with the eigenstates given in (7) are nothing but the wave functions of the electrons in the coordinate space:

$$\langle \{\vec{x}_i; \vec{y}\}|\vec{k}; 0\rangle = \langle \vec{y}|\vec{k}\rangle \langle \{\vec{x}_i\}|0\rangle = e^{i\vec{k}\cdot\vec{y}}\psi_0(\{\vec{x}_i\}) \,,$$
$$\langle \vec{k}'; n|\{\vec{x}_i\}, \vec{y}\rangle = \langle \vec{k}'|\vec{y}\rangle \langle n|\{\vec{x}_i\}\rangle = e^{-i\vec{k}'\cdot\vec{y}}\psi_n^*(\{\vec{x}_i\}) \,, \tag{10}$$

where we explicitly use the factorization of the state vectors. Note that the scattered electron is described by a plane wave and the function $\psi_n(\{\vec{x}_i\})$ represents a generic solution of the Schrödinger equation for atomic electrons in the Coulomb potential of the nucleus:

$$U(\{\vec{x}_i\}) = -Ze^2 \sum_{i=1}^{Z} \frac{1}{|\vec{x}_i|} \,. \tag{11}$$

Furthermore, we also need the representation of the unit operator in the coordinate basis:

$$\mathbb{1} = \int d\vec{y} \int \prod_{i=1}^{Z} d\vec{x}_i |\{\vec{x}_i\}; \vec{y}\rangle \langle \{\vec{x}_i\}; \vec{y}| \,. \tag{12}$$

Returning to the matrix element (6), we insert the above condition into it and use subsequently (8) and (10):

$$\mathcal{M}_{n0} = \langle \vec{k}'; n|\hat{V} \otimes \mathbb{1}|\vec{k}; 0\rangle = \int d\vec{y} \int \prod_{i=1}^{Z} d\vec{x}_i \langle \vec{k}'; n|\hat{V}|\{\vec{x}_i\}; \vec{y}\rangle \langle \{\vec{x}_i\}; \vec{y}|\vec{k}; 0\rangle$$

$$= \int d\vec{y} \int \prod_{i=1}^{Z} d\vec{x}_i V(\{\vec{x}_i\}, \vec{y}) \langle \vec{k}'|\vec{y}\rangle \langle n|\{\vec{x}_i\}\rangle \langle \vec{y}|\vec{k}\rangle \langle \{\vec{x}_i\}|0\rangle$$

$$= \int d\vec{y}\, e^{i(\vec{k}-\vec{k}')\cdot\vec{y}} \int \prod_{i=1}^{Z} d\vec{x}_i V(\{\vec{x}_i\}, \vec{y}) \psi_n^*(\{\vec{x}_i\})\psi_0(\{\vec{x}_i\}) \,. \tag{13}$$

Introducing a notation $\vec{q} = \vec{k} - \vec{k}'$ for the momentum transfer and substituting the potential (9), we obtain:

$$\mathcal{M}_{n0} = \int d\vec{y}\, e^{i\vec{q}\cdot\vec{y}} \int \prod_{i=1}^{Z} d\vec{x}_i \left(-\frac{Ze^2}{|\vec{y}|} + \sum_{i=1}^{Z} \frac{e^2}{|\vec{y} - \vec{x}_i|} \right) \psi_n^*(\{\vec{x}_i\})\psi_0(\{\vec{x}_i\}) \,. \tag{14}$$

The $\vec{y}$-integration is performed using the three-dimensional Fourier-integral:

$$\int d\vec{y}\, e^{i\vec{q}\cdot\vec{y}} \frac{1}{|\vec{y} - \vec{a}|} = \frac{4\pi}{q^2} e^{i\vec{q}\cdot\vec{a}} \,. \tag{15}$$

The result is expressed in a compact form:

$$\mathcal{M}_{n0} = -\frac{4\pi Ze^2}{q^2}\delta_{n0} + \frac{4\pi e^2}{q^2}\mathcal{F}_{n0}(q^2) \,, \tag{16}$$

where in the first term the orthogonality of the wave functions is taken into account and in the second term we introduce a new function

$$\mathcal{F}_{n0}(\vec{q}) = \int \prod_{i=1}^{Z} d\vec{x}_i \, \psi_n^*(\{\vec{x}_i\}) \left(\sum_{i=1}^{Z} e^{i\vec{q}\cdot\vec{x}_i} \right) \psi_0(\{\vec{x}_i\}) , \qquad (17)$$

which is called the *form factor* of atomic electrons and has an equivalent form of a matrix element:

$$\mathcal{F}_{n0}(\vec{q}) = \langle n| \left(\sum_{i=1}^{Z} e^{i\vec{q}\cdot\hat{\vec{X}}_i} \right) |0\rangle . \qquad (18)$$

The formula (16) reflects a pointlike charge in the origin, combined with a smeared electron cloud. Only the latter contributes to the form factor, whereas the former is nonvanishing only for the elastic scattering ($n = 0$). In the limit $\vec{q} \to 0$, the elastic transition form factor is equal to the total charge Z of the atomic electrons, whereas in the case of inelastic transitions, the form factor vanishes at zero momentum.

To simplify the further discussion, we consider the one-electron atom and elastic scattering, so that $\{\vec{x}_i\} = \vec{x}$ and $n = 0$. The form factor

$$\mathcal{F}_{00}(\vec{q}) \equiv f(\vec{q})$$

can be represented as a Fourier transform,

$$f(\vec{q}) = \int d\vec{x} \, e^{i\vec{q}\cdot\vec{x}} \rho(\vec{x}) , \qquad (19)$$

of the spatial probability density (or average charge density) of the atomic electron in a ground state:

$$\rho(\vec{x}) = |\psi_0(\vec{x})|^2.$$

Another important property of the form factor (19) is that in the momentum space it reproduces the symmetry of the spatial density. If the latter is spherically symmetric,

$$\rho(\vec{x}) = \rho(r) , \quad r \equiv \sqrt{|\vec{x}|^2} ,$$

then also the form factor does not depend on the direction of $\vec{q}$, but only on its absolute value $q \equiv |\vec{q}|$. To see this, we choose spherical coordinates with $\vec{q}$ parallel to the z-axis and perform the angular integration in (19):

$$\int d\vec{x} \, e^{i\vec{q}\cdot\vec{x}} \rho(r) = \int_0^\infty dr \, r^2 \rho(r) 2\pi \int_{-1}^{1} d\cos\theta [\cos(qr\cos\theta) + i\sin(qr\cos\theta)]$$

$$= \frac{4\pi}{q} \int_0^\infty dr \, r\rho(r) \sin(qr) \equiv f(q^2) . \qquad (20)$$

Apart from the spherical symmetry, several important features of the form factor follow from this equation. Firstly, $f(q^2)$ is a real valued function of q^2. Secondly, at a given large value of q, the region of integration in (20) is dominated by $r \sim 1/q$, because the integrand at large r is suppressed by the strongly oscillating function $\sin(qr)$. Furthermore, the form factor determines the spatial charge distribution, as can be seen by taking the inverse

Fourier transform of (19). For a spherically symmetric form factor we obtain, after angular integration in $\vec{q}$-space,

$$\rho(r) = \frac{1}{2\pi^2 r} \int\limits_0^\infty dq\, q f(q^2) \sin(qr)\,. \tag{21}$$

Furthermore, defining the Laplace transform of the charge distribution as

$$r\rho(r) = \frac{1}{2\pi^2} \int\limits_0^\infty d\alpha\, \alpha \ell(\alpha^2) e^{-\alpha r}\,, \tag{22}$$

and substituting it into the integral representation (20) of the form factor, we obtain, after integration over r, the so-called spectral representation of the form factor

$$f(q^2) = \frac{1}{\pi} \int\limits_0^\infty d\alpha^2 \frac{\ell(\alpha^2)}{\alpha^2 + q^2}\,, \tag{23}$$

where the function $\ell(\alpha^2)$ is called the spectral density.

Finally, defining the moments of charge distribution

$$Q_n \equiv \int d\vec{x}\, r^{2n} \rho(r)\,, \quad n \geq 0\,, \tag{24}$$

where $Q_0 \equiv Q$ is the total charge, we can easily relate these characteristics with the form factor and its derivatives at $q^2 = 0$. From the definition of the form factor, it directly follows that

$$Q = f(0) = \frac{1}{\pi} \int\limits_0^\infty d\alpha^2 \frac{\ell(\alpha^2)}{\alpha^2}\,. \tag{25}$$

In the course of this book, we will gradually reveal that, to a large extent, the hadron form factors are similar to the atomic ones discussed above. First of all, analogous to (18), hadron form factors are defined as transition matrix elements of certain operators. However, a representation in terms of wave functions similar to (17) is generally not applicable in QCD. Another analogy is with (16): in the scattering amplitude, a hadron form factor is separated (factorized) with respect to other factors in the amplitude, related to the projectile leptons and exchanged photons or W-bosons. Each hadron form factor is also a function of the transferred momentum, but, in this case, of the four-dimensional momentum squared, in order to obey the Lorentz invariance. Furthermore, the dominance of a short-distance region at a large momentum transfer is valid also for the hadron form factors. Spectral representations similar to (23) are in fact prototypes of dispersion relations used for the hadron form factors. Finally, the normalization condition (25) is valid also for the hadron e.m. form factor which, at zero momentum transfer, is equal to the electric charge of that hadron.

Throughout this book the following conventions are used:
- The system of physical units is $\hbar = c = 1$;
- Complex conjugated quantities are denoted by an asterisk $*$;
- A four- (three-) dimensional momentum is named *momentum (three-momentum)*;
- A dot in a scalar product of four-vectors is only shown if necessary;
- In diagrams, the transition from initial to final state is directed from left to right;
- The abbreviations are: e.m. – electromagnetic; l.h.s. (r.h.s.) – left- (right-) hand side.

CHAPTER 1

QCD, QUARK CURRENTS, AND HADRONS

1.1 BASIC ELEMENTS OF QCD

1.1.1 QCD Lagrangian

Within the Standard Model (SM) of elementary particles, *quantum chromodynamics (QCD)* describes the interactions of *quarks* and *antiquarks* with *gluons*. QCD is a quantum field theory that bears certain similarities with quantum electrodynamics (QED), the theory of electromagnetic interactions of electrons and positrons with photons. Quarks and antiquarks, similar to the electron and positron, have spin 1/2 and are described by Dirac equation. Gluons, similar to the photon, are massless, have spin 1 and two polarization states. The *color* charge in quark-gluon interactions is similar to the electric charge in the electromagnetic interactions.

But there are also striking differences between QCD and QED. The most important one is the self-interaction of gluons which has far reaching consequences and essentially shapes QCD. This phenomenon originates due to the specific properties of the color charge. Each quark field in QCD has $N_c = 3$ different states of the color charge. An antiquark field, correspondingly, has three states of the conjugated color charge. The quark-gluon interaction alters the color state of a quark or antiquark. Due to color-charge conservation, the emitted or absorbed gluon has to compensate for this change and, hence, itself carries one of the possible $N_c^2 - 1 = 8$ combinations of color charges. Therefore, a gluon can emit and absorb other gluons.

Interactions of the quark and gluon fields are encoded in the *QCD Lagrangian* density which has the following compact form:

$$
\begin{aligned}
\mathcal{L}_{QCD}(x) &= -\frac{1}{4}G^a_{\mu\nu}(x)G^{a\,\mu\nu}(x) \\
&+ \sum_{q=u,d,s,c,b,t} \overline{q}_i(x)\left(iD^i_{\mu j}\gamma^\mu - m_q\delta^i_j\right)q^j(x)\,,
\end{aligned}
\tag{1.1}
$$

with a summation over the color indices $a = 1, 2, ...8$; $i, j = 1, 2, 3$ and Lorentz-indices μ, ν. The first term of $\mathcal{L}_{QCD}$ contains a product of the gluon field strength tensors,

$$
G^a_{\mu\nu}(x) = \partial_\mu A^a_\nu(x) - \partial_\nu A^a_\mu(x) + g_s f^{abc} A^b_\mu(x)A^c_\nu(x)\,,
\tag{1.2}
$$

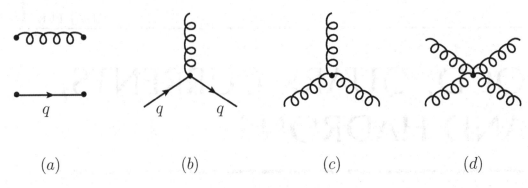

Figure 1.1 Basic elements of QCD diagrams: (a) the propagators of gluon (curly line) and quark with the flavor q (arrow line); (b) the quark-gluon, (c) three-gluon and (d) four-gluon interaction vertices.

constructed from the color-charged gluon field $A_\mu^a(x)$ and its derivatives[1]. This part of the QCD Lagrangian describes *gluodynamics*: the propagation of gluons and their self-interactions.

The second line in (1.1) describes the propagation and interactions of quarks and antiquarks. This part of $\mathcal{L}_{QCD}$ is built from the color-charged quark fields $q^i(x)$ and their Dirac conjugates $\bar{q}_i(x)$ with all possible flavor quantum numbers,

$$q = u, d, s, c, b, t, \tag{1.3}$$

Note that all three color states of a quark with a given flavor q have exactly the same mass m_q. The covariant derivative of the quark field in (1.1) is defined as

$$D_{\mu j}^i = \delta_j^i \partial_\mu - ig_s \frac{(\lambda^a)_j^i}{2} A_\mu^a(x), \tag{1.4}$$

so that the term with gluon field generates the quark-gluon interaction. Further details of the Lagrangian (1.1), in particular, the conventions for the Dirac matrices γ_μ, Gell-Mann matrices λ^a and antisymmetric tensors f_{ijk} are presented in Appendix A.

Summation over the color indices in $\mathcal{L}_{QCD}$ and universality of the coupling g_s in (1.2) and (1.4) ensure the conservation of color charges, which is a manifestation of a more general symmetry of QCD. The Lagrangian (1.1) is invariant with respect to the local gauge transformations of the quark and gluon fields, forming the group $SU(N_c) = SU(3)$. The quark, antiquark and gluon fields transform, respectively, as fundamental, conjugated and adjoint representations of this group, corresponding to the triplet, antitriplet and octet of their color states. Several important issues concerning the quantum field theory aspects of QCD remain beyond our brief description. These are: detailed definitions of the local gauge transformations, gauge fixing conditions, quantization of the quark and gluon fields and auxiliary ghost fields. These topics are described in modern textbooks on quantum field theory and QCD (see, e.g., [6, 7, 8]).

Quantization of QCD leads to the set of "ready-to-use" elements of Feynman diagrams. These are propagators and vertices depicted in Figure 1.1. In Appendix A, their explicit expressions, together with other useful formulae of QCD, are presented.

1.1.2 Perturbative QCD, the running coupling, and quark mass

Having at our disposal the elements of QCD Feynman diagrams, it is possible to calculate the amplitudes of elementary processes involving quarks and gluons. Let us consider as an

---

[1]We use the shorthand notation $\partial_\mu \equiv \partial/\partial x^\mu$.

example the u-quark scattering on a d-quark. The diagram of this process at the lowest possible order, $O(g_s^2)$, in the quark-gluon coupling is shown in Figure 1.2(a). The scattering amplitude is calculated, starting from S-matrix and using definitions and conventions presented in Appendix B. The initial and final states are assumed to be separate quarks[2] with definite momenta and colors:

$$u^i(p_1)d^k(p_2) \to u^j(p_3)d^l(p_4) \,. \tag{1.5}$$

Combining two vertices of quark-gluon interactions with a gluon propagator (taken in the Feynman gauge), we obtain the following expression for the scattering amplitude corresponding to this diagram:

$$\mathcal{A} = \frac{g_s^2}{k^2} \frac{(\lambda^a)^{ij}(\lambda^a)^{kl}}{4} \bar{u}_u(p_3)\gamma_\mu u_u(p_1)\bar{u}_d(p_4)\gamma^\mu u_d(p_2) \,, \tag{1.6}$$

where u_q ($\bar{u}_q$) are the bispinors of initial (final) quarks with flavor $q = u, d$ and

$$k = p_3 - p_1 = p_2 - p_4 \,, \tag{1.7}$$

is the momentum of the intermediate gluon. The factor with λ-matrices in (1.6) originates from the color structure of the diagram. In a scattering process, the kinematical invariant k^2 is spacelike: $k^2 \le 0$. Denoting $k^2 \equiv -\mu^2$, we notice that the scale μ is a measure of the gluon virtuality, that is, of the deviation from the massless gluon with $k^2 = 0$.

Hereafter, we use the conventional definition of the quark-gluon coupling:

$$\alpha_s = \frac{g_s^2}{4\pi} \,,$$

keeping in mind that a gluon exchange always brings two powers of g_s. Note that the dimensionless coupling α_s is analogous to $\alpha_{em} = e^2/(4\pi)$ used in QED. The numerical smallness of $\alpha_{em} \sim 0.01$ makes the perturbation theory in α_{em} almost perfect. E.g., the leading-order diagram of the electron-electron scattering with a photon exchange (the QED analog of the Figure 1.2(a)) is already a sufficiently accurate approximation.

In QCD, the corrections of higher orders in α_s are essential. At the next-to-leading order in quark-gluon coupling, that is at $O(\alpha_s^2)$, one encounters a set of diagrams in which virtual quarks and gluons form loops. Some of these diagrams are depicted in Figures 1.2(b)–(d). A detailed treatment and calculation of loop diagrams remains beyond our scope and can be found in the textbooks on QCD such as [7]. Complications related to the loop diagrams in a quantum field theory arise, because the momenta flowing in the internal quark and gluon lines of the loops can in principle reach infinitely large values. Note that virtual particles propagating in the loops represent purely quantum effects, hence unlimited fluctuations of their energy-momentum are allowed by uncertainty relations for sufficiently short distances and time intervals of their propagation. Thus, a full account of the loop diagram contributions implies integration up to infinite loop momenta which cause divergences of the four-dimensional integrals. The appearance of such *ultraviolet* divergences in the higher-order loop diagrams was known in quantum field theories, such as QED, long before QCD was discovered.

Remarkably, QCD, similar to QED, belongs to an exceptional category of *renormalizable* quantum field theories. All ultraviolet divergences of the loop diagrams can be effectively

---

[2]In reality, quarks or antiquarks, being inseparable parts of bound states, hadrons, cannot form initial or final asymptotic states. Nevertheless, the process of quark-quark scattering is not completely hypothetical. Under certain kinematical conditions, it contributes, e.g., to the production of hadronic jets in the high-energy proton-proton collisions. The u and d quarks from the initial-state protons produce quarks which then transform into the final-state jets.

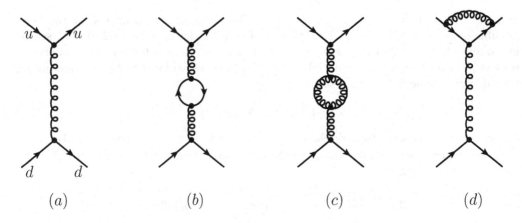

Figure 1.2 Diagrams of the quark-quark scattering process: (a) the leading $O(\alpha_s)$ diagram; (b)–(d) some of the $O(\alpha_s^2)$ diagrams.

removed by a redefinition of the fields $A_\mu^a(x)$ and $q^i(x)$ in the Lagrangian, simultaneously redefining the coupling g_s and quark masses m_q. The residual finite contributions of the loop diagrams introduce additional logarithmic dependence on k^2, hence on the scale μ, with respect to the lowest-order answer (1.6).

Adding loops to the lowest-order diagram amounts to replacing the initially constant parameter α_s by an effective, scale-dependent coupling $\alpha_s(\mu)$. The outcome is the renowned Gross-Wilczek-Politzer formula, relating $\alpha_s(\mu)$ at two different scales:

$$\alpha_s(\mu_1) = \frac{\alpha_s(\mu_0)}{1 + \frac{\alpha_s(\mu_0)}{2\pi}\beta_0 \log \frac{\mu_1}{\mu_0}}. \qquad (1.8)$$

Note that in this relation all effects of $O((\alpha_s \log[\mu])^n)$ are taken into account. The coefficient

$$\beta_0 = 11 - \frac{2}{3}n_f > 0, \qquad (1.9)$$

is one of the fundamental characteristics of QCD, where n_f is the number of quark flavors accounted for in the quark loops and having masses smaller than the scales $\mu_{0,1}$ involved in (1.8). Importantly, β_0 is positive in the SM, because $n_f \leq 6$ and the first term in β_0 dominates over the negative second term. The main contribution to (1.9) originates from the gluonic self-interacting loops, such as the diagram in Figure 1.2(c).

The formula (1.8) reveals that the effective quark-gluon coupling logarithmically decreases when the energy-momentum scale μ increases. Thus, $\alpha_s(\mu)$ asymptotically vanishes at $\mu \to \infty$, manifesting the phenomenon of "asymptotic freedom." Note that only the scale-dependence of the effective coupling (the so-called "running") is predicted in QCD; hence, one needs to fix the normalization of α_s at a certain scale μ_0. This can only be done using experimental data, as will be discussed below.

Not only the coupling but also the quark mass is scale dependent or "running" in QCD. This phenomenon appears as a result of gluon corrections to the quark propagator. The Feynman diagram of the $O(\alpha_s)$ one-loop contribution is shown in Figure 1.3. In the same, as in (1.8), approximation:

$$m_q(\mu_1) = m_q(\mu_0) \left(\frac{\alpha_s(\mu_1)}{\alpha_s(\mu_0)}\right)^{\gamma_0/\beta_0}, \qquad (1.10)$$

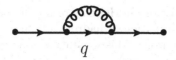

q

Figure 1.3 Diagram of the gluon radiative correction to the quark propagator in $O(\alpha_s)$, contributing to the running of the quark mass.

where $\gamma_0 = 4$ is the so-called anomalous dimension, another universal characteristic of QCD. Note that, according to (1.10), the effective quark mass logarithmically decreases with the scale. Importantly, the running of the quark mass is independent of the quark flavor.

1.1.3 Nonperturbative regime, confinement, and hadrons

The effective coupling (1.8) can be easily inverted towards lower energy-momentum scales. Suppose we probe quark-gluon interactions at scales $\mu < \mu_0$, so that the logarithm in (1.8) becomes negative. If a value $\mu = \Lambda_{QCD}$ is reached, at which the condition

$$\frac{\alpha_s(\mu_0)}{2\pi} \beta_0 \log \frac{\mu_0}{\Lambda_{QCD}} = 1 \qquad (1.11)$$

is fulfilled, the denominator in (1.8) vanishes and, consequently,

$$\lim_{\mu \to \Lambda_{QCD}} \alpha_s(\mu) = \infty. \qquad (1.12)$$

Note that the above divergence and the asymptotic freedom $\alpha_s(\mu \to \infty) = 0$ are intrinsically correlated. Both phenomena are caused by the existence of the gluon self-interactions, leading to positive β_0. The actual value of Λ_{QCD} is obtained solving (1.11), where $\alpha_s(\mu_0)$ is determined from experiment. From (1.12), it follows that if a quark-gluon interaction takes place at sufficiently low energy-momentum scales, in the vicinity of Λ_{QCD}, the value of α_s becomes too large to serve as an expansion parameter. The QCD perturbative expansion in terms of diagrams consisting of vertices, propagators and loops ceases to exist.

Transition of QCD into the *nonperturbative* regime of strong interactions at $\mu \sim \Lambda_{QCD}$ is manifested in a form of *confinement* of quarks and gluons. The color-charged elementary particles of QCD form color-neutral bound states, the *hadrons*. Moreover, separate quarks, antiquarks, gluons or any color-charged combinations of them do not exist as initial or final states in observable processes. Only hadrons are detected in particle physics experiments, together with leptons, photons and weak bosons[3].

To formalize the above statement, we remind that the probability amplitude of a certain process involving particle scattering or decay is determined by an initial-to-final element of the S-matrix:

$$\mathcal{A}(i \to f) \sim \langle f | \hat{S} | i \rangle, \qquad (1.13)$$

(see Appendix B for more details). The part of this matrix containing QCD interactions is

$$\hat{S}_{QCD} = T \left\{ \exp \left(i \int d^4 x \mathcal{L}_{QCD}^{int}(x) \right) \right\}. \qquad (1.14)$$

The notation $\mathcal{L}_{QCD}^{int}$ indicates that only the interaction vertices from the full QCD Lagrangian (1.1) are included. In the above expression the quark and gluon fields entering

---

[3]Note that the superheavy t-quark is exceptional in this sense. The very short, $\tau \ll 1/\Lambda_{QCD}$ lifetime of its weak decay makes the formation of hadrons containing t-quark impossible.

$\mathcal{L}_{QCD}^{int}$ are operators (in the sense of the Heisenberg formalism). On the other hand, due to quark-gluon confinement and formation of hadrons, the initial state $|i\rangle$ and the final state $\langle f|$ in (1.13) can only involve hadrons. This separation of quark-gluon and hadronic degrees of freedom is essentially different from the situation in QED, where electrons, positrons and photons not only enter $\mathcal{L}_{QED}^{int}$ but also form the initial and final states in the S-matrix elements.

Apart from hadronic states, the total sets of initial and final states include a special state of QCD vacuum, denoted as $|i\rangle = |0\rangle$ or $\langle f| = \langle 0|$ and normalized with $\langle 0|0\rangle = 1$. It is defined as a state with a minimal energy with respect to the hadronic states. Any hadron contributes with a positive energy to a state $|i\rangle$ or $\langle f|$, hence, the vacuum state is hadronless.

The absence of hadrons does not necessarily mean that the vacuum in QCD represents an "empty space." It is rather a kind of medium filled with fluctuating quark, antiquark and gluon fields with nonvanishing average densities. This spectacular property of QCD is presumably interrelated with the confinement. All combinations of quark and gluon fields with quantum numbers of the vacuum are allowed, so that QCD vacuum fluctuations are color-neutral, flavorless, have spin-parity $J^P = 0^+$ and vanishing electroweak charges. The most important examples are the local composite fields

$$\bar{q}_i(x)q^i(x), \quad G_{\mu\nu}^a(x)G^{a\mu\nu}(x) \,. \tag{1.15}$$

A full theory of vacuum fields derived from the QCD Lagrangian does not exist in analytic form. Still, the evidence for these fields comes from various sources. An important contribution to the vacuum average of the gluon-field composite in (1.15) is provided by the instantons, which are special solutions of the nonlinear QCD equations of motion (for an introductory review, see, e.g., [9]). Furthermore, the existence of the quark-antiquark vacuum fields in (1.15) is supported by an observable phenomenon of chiral symmetry breaking in QCD to be discussed below in this chapter. Independent information on the vacuum state is also provided by numerical simulations of QCD on the space-time lattices. Characteristic sizes of the quark-antiquark and gluon field fluctuations in the QCD vacuum are of $O(1/\Lambda_{QCD})$, clearly indicating that the state $|0\rangle$ is genuinely nonperturbative.

In many cases the largely unknown space-time structure of the vacuum fields, i.e., the dependence of the composite fields in (1.15) on x, is not relevant and only the average values are essential. To parameterize these averages, one introduces vacuum expectation values of the local densities, briefly called *vacuum condensates*. For the vacuum fields (1.15) the quark and gluon condensates are defined, respectively, as:

$$\langle 0|\bar{q}_i q^i|0\rangle, \quad \langle 0|G_{\mu\nu}^a G^{a\mu\nu}|0\rangle \,, \tag{1.16}$$

where we do not need to specify the coordinate, because of the translational invariance. In the QCD-based methods presented in the last chapters of this book, these vacuum averages play an important role. Together with the effective coupling and quark masses, vacuum condensates belong to the universal parameters of QCD.

Summarizing, the nonperturbative regime of QCD emerging at energy-momentum scales $\mu \sim O(\Lambda_{QCD})$ (at space-time intervals of $O(1/\Lambda_{QCD})$) involves three deeply interrelated phenomena: confinement, QCD vacuum and formation of hadrons. Currently, the only available method to derive these phenomena starting from the first principles, is the numerical simulation of QCD on the space-time lattice. An introduction to lattice QCD can be found, e.g., in [10, 11, 12].

1.1.4   Quark-gluon coupling and quark masses

To normalize the effective quark-gluon coupling α_s at a certain scale, one needs experimental data. An accurate determination of α_s is possible in a process with a large energy-momentum transfer. The Z-boson decay to hadrons is one example. At leading order in the electroweak coupling of Z to quarks this process is described by the two-body decay $Z \to \bar{q}q$, where $q = u, d, s, c, b$ contribute, whereas $q = t$ is kinematically forbidden. The Z-boson mass $m_Z \simeq 91$ GeV is much larger than the quark masses $m_{u,d,s,c,b}$ listed below in (1.20). In the Z rest frame the emitted quark and antiquark have energies of $O(m_Z/2)$; hence, m_Z determines the characteristic scale in this process. The products of Z decay observed in a detector are multiple hadrons. They emerge after the primary quark and antiquark copiously emit gluons and additional quark-antiquark pairs. The most spectacular feature of this hadronization is the dominance of the events with two back-to-back jets of hadrons in the Z rest frame. Without going into much detail, the jet is defined as a bunch of hadrons whose 3-momentum vectors are aligned forming a certain direction, so that the average momentum components orthogonal to that direction are small. The development of a hadronic jet can be traced back to a single energetic quark or gluon; hence, the observation of two-jet dominance tells us that the probability of having just one back-to-back quark pair in Z-decay is larger than a probability of emitting an additional energetic gluon which after developing its own jet would lead to a three-jet event. A primary gluon emission corresponds to an elementary three-body decay $Z \to \bar{q}qg$ and the corresponding probability contains an extra factor of α_s. Therefore, the ratio of the observed three-jet events to the two-jet events provides a measure of the quark-gluon effective coupling α_s at the scale m_Z[4]. The resulting value of $\alpha_s(m_Z)$ determined from this $Z \to jets$ analysis is surprisingly precise on one hand and quite small on the other hand, justifying the use of perturbative QCD at sufficiently large energy-momentum scales. There are several other independent methods to determine α_s at various scales from $O(m_Z)$ up to $O(1\,\text{GeV})$ (see the minireview on QCD in [1]). All the results nicely agree with the predicted behavior (1.8), providing a decisive and nontrivial test of QCD.

Presented in the current edition of [1], the quark-gluon coupling averaged over different determinations is

$$\alpha_s(m_Z) = 0.1181 \pm 0.0011\,. \tag{1.17}$$

Using this value as an input, the values of α_s at any scale can be calculated. For practitioners, a useful computer code is available in [13], with a far more accurate version of the running than (1.8), including higher-order corrections and taking into account the appropriate number n_f of "active" quark flavors in the loop diagrams. In particular, $n_f = 3$ ($n_f = 4$) is adopted for the interval of scales $1.0\,\text{GeV} < \mu < 1.3\,\text{GeV}$ ($1.3\,\text{GeV} < \mu < 4.2\,\text{GeV}$) and $n_f = 5$ for the scales larger than $\mu = 4.2\,\text{GeV}$. Using this code, we obtain

$$\alpha_s(\mu = 1.0\,\text{GeV}) = 0.4620 \pm 0.0043\,, \quad \alpha_s(\mu = 4.2\,\text{GeV}) = 0.2247 \pm 0.0021\,. \tag{1.18}$$

Furthermore, adopting the above values as an input in (1.11) we reproduce the corresponding values of Λ_{QCD}

$$\Lambda_{QCD}^{(n_f=3)} = (332 \pm 17)\,\text{MeV}\,, \quad \Lambda_{QCD}^{(n_f=4)} = (292 \pm 16)\,\text{MeV}\,. \tag{1.19}$$

The basic parameters of the QCD Lagrangian (1.1) are, apart from the quark-gluon coupling, the masses of quarks. In SM, the mass of a quark originates from the Higgs mechanism and is seemingly fixed by the Higgs-boson-quark coupling. However, as we already

---

[4]Note that the gluon loop corrections should be included in addition to the gluon emission. In practice, jet calculus is a quite involved procedure combining perturbative QCD calculations with statistical methods and models of the hadronization dynamics.

know, due to QCD interactions quarks are not directly observable, and their masses become effective and scale-dependent parameters. The quark mass values for different flavors are markedly different, spanning from a few MeV for the u, d quarks to about 170 GeV for the t quark. The methods of quark mass determination also vary depending on the flavor. Generally, each mass $m_q(\mu)$ is fixed at a certain energy-momentum scale μ relating it to a hadron mass or to other observable quantities. Moreover, there are several alternative schemes defining a quark mass in QCD. Hereafter we will adopt the standard $\overline{MS}$ scheme for the running quark mass, omitting the usual bar notation of this scheme, so that $m_q \equiv \overline{m}_q(\mu)$. According to the convention adopted in [1], for the flavors $q = u, d, s$ a sufficiently large scale $\mu = 2\,\text{GeV}$ is chosen. For the c and b quark masses one uses instead the definition

$$m_c \equiv \overline{m}_c(\overline{m}_c), \quad m_b \equiv \overline{m}_b(\overline{m}_b).$$

In the current edition of [1], the values of quark masses normalized at these scales are:

$$m_u = 2.16^{+0.49}_{-0.26}\,\text{MeV}, \quad m_d = 4.67^{+0.48}_{-0.17}\,\text{MeV}, \quad m_s = 93^{+11}_{-5}\,\text{MeV},$$

$$m_c = 1.27 \pm 0.02\,\text{GeV}, \quad m_b = 4.18^{+0.03}_{-0.02}\,\text{GeV}, \quad m_t = 172.9 \pm 0.4\,\text{GeV}. \tag{1.20}$$

To obtain the quark masses at other scales one has to use the relation (1.10) or its more accurate versions presented in the review on quark masses in [1]. A computer code based for running quark masses can be found in [13]. Note that the ratio of quark masses $m_q(\mu)/m_{q'}(\mu)$ for different flavors q, q' remains scale-independent.

QCD has its intrinsic energy scale $\Lambda_{QCD} \sim 300$ MeV, hence the quarks of different flavors are divided into two categories:

- The light quarks $q = u, d, s$ with $m_{u,d} \ll m_s \lesssim \Lambda_{QCD}$,

- The heavy quarks $Q = c, b, t$ with $m_Q \gg \Lambda_{QCD}$.

One should mention that in certain phenomenological models of hadrons, which remain beyond our scope, the concept of "constituent" or "dressed" quarks is used, adding an energy of $O(\Lambda_{QCD})$ to the quark masses in QCD.

1.2 QUARK ELECTROMAGNETIC CURRENT

In this section we consider the electromagnetic (e.m.) interaction of quarks. To switch it on, it is sufficient to extend the covariant derivative in the QCD Lagrangian (1.1):

$$D_\mu \to D'_\mu = \partial_\mu - ig_s \frac{\lambda^a}{2} A^a_\mu(x) - ieQ_q A^{em}_\mu(x), \tag{1.21}$$

where A^{em}_μ is the four-vector of the photon field and

$$Q_u = Q_c = Q_t = +2/3, \quad Q_d = Q_s = Q_b = -1/3$$

are the electric charges of quarks in units of the e.m. coupling $e = \sqrt{4\pi\alpha_{em}}$. The extension to D'_μ generates the e.m. interaction of quarks with the photon:

$$ej^{em}_\mu(x)A^\mu_{em}(x), \tag{1.22}$$

where we introduce the local density of the quark e.m. current:

$$j^{em}_\mu(x) = \sum_{q=u,d,s,c,b,t} Q_q \bar{q}_i(x)\gamma_\mu q^i(x). \tag{1.23}$$

The summation over quark colors implies that the density of the current is a color-neutral operator in QCD. For the sake of brevity we hereafter suppress the quark color indices in the above definition and omit the word "density."

It is convenient to combine (1.22) with the lepton-photon e.m. interaction, introducing the QED interaction Lagrangian:

$$\mathcal{L}_{QED}^{int}(x) = e\Big(j_\mu^{em}(x) - \sum_{\ell=e,\mu,\tau} \bar{\ell}(x)\gamma_\mu\ell(x)\Big)A_{em}^\mu(x)\,, \qquad (1.24)$$

where $\bar{\ell}(x)\gamma_\mu\ell(x)$ is the lepton e.m. current containing the charged lepton field ℓ and its Dirac conjugate $\bar{\ell}$ with the lepton flavors $\ell = e, \mu, \tau$. The combined Lagrangian of QCD and QED for quarks and charged leptons:

$$\mathcal{L}_{QCD\oplus QED}(x) = \mathcal{L}_{QCD}(x) + \mathcal{L}_{QED}^{int}(x)$$
$$-\frac{1}{4}F_{\mu\nu}(x)F^{\mu\nu}(x) + \sum_{\ell=e,\mu,\tau} \bar{\ell}(x)(i\partial_\mu\gamma^\mu - m_\ell)\ell(x)\,, \qquad (1.25)$$

contains also the terms describing propagation of photons and leptons added in the second line, where $F^{\mu\nu} = \partial^\mu A_{em}^\nu - \partial^\nu A_{em}^\mu$ is the photon field-strength tensor. The Lagrangian (1.25), in addition to the gauge symmetry $SU(3)_c$ of QCD, possesses the QED gauge symmetry $U(1)$. We again stress the crucial difference between the two interactions contained in (1.25): in contrast to gluons, a self-interaction of the photons is absent at the Lagrangian level.

The transformations that leave the Lagrangian (1.25) invariant include the phase rotations of quark or lepton fields with coordinate-independent (global) and flavor-dependent phases:

$$q(x) \to q'(x) = \exp\big(-i\chi_q\big)q(x)\,, \quad \bar{q}(x) \to \bar{q}'(x) = \bar{q}(x)\exp\big(i\chi_q\big)\,,$$
$$\ell(x) \to \ell'(x) = \exp\big(-i\chi_\ell\big)\ell(x)\,, \quad \bar{\ell}(x) \to \bar{\ell}'(x) = \bar{\ell}(x)\exp\big(i\chi_\ell\big)\,, \qquad (1.26)$$

where $q = u, d, s, c, b, t$ and $\ell = e, \mu, \tau$. This symmetry is related to the conservation of all individual quark and lepton flavor quantum numbers, and correspondingly, to the vanishing divergences of the vector currents:

$$\partial^\mu[\bar{q}(x)\gamma_\mu q(x)] = 0\,, \quad \partial^\mu[\bar{\ell}(x)\gamma_\mu\ell(x)] = 0\,. \qquad (1.27)$$

Note that the invariance with respect to (1.26) with a uniform global phase $\chi_q = \chi_\ell = \chi$, is related to the gauge symmetry of QED yielding the electric charge conservation and the vanishing divergence of the e.m. current:

$$\partial^\mu j_\mu^{em} = 0\,, \qquad (1.28)$$

Finally, the case $\chi_q = \chi$, when the phase rotations of all quark fields are flavor-uniform but differ from the phases of lepton fields, can be interpreted in terms of the baryon number $B = 1/3$ ($B = -1/3$) conservation for quarks (antiquarks).

1.3   QUARK WEAK CURRENTS

Weak interactions of quarks in SM are driven by the quark currents emitting or absorbing a W-boson. In this book we confine ourselves by the weak leptonic and semileptonic transitions of quarks. These transitions occur when a leptonic weak current couples to the quark current via intermediate virtual W-boson. Its mass, $m_W \simeq 80$ GeV, is much larger than the typical energy-momentum scales in weak decays determined by the quark masses[5]. Therefore, it is possible to apply an effective quark-lepton interaction described by the Hamiltonian[6]:

$$\mathcal{H}_W(x) = \frac{G_F}{\sqrt{2}} \Big\{ j_\mu^W(x) \sum_{\ell=e,\mu,\tau} \bar{\ell}(x)\gamma^\mu(1-\gamma_5)\nu_\ell(x) $$
$$+ j_\mu^{W\dagger}(x) \sum_{\ell=e,\mu,\tau} \bar{\nu}_\ell(x)\gamma^\mu(1-\gamma_5)\ell(x) \Big\}, \tag{1.29}$$

where $G_F \simeq 10^{-5}$ GeV^2 is the Fermi constant which effectively replaces the virtual W-boson propagator. The quark weak current

$$j_\mu^W(x) = \sum_{\substack{q=u,c,t \\ q'=d,s,b}} V_{qq'}^{CKM} \bar{q}(x)\gamma_\mu(1-\gamma_5)q'(x) \tag{1.30}$$

and its Hermitian conjugate

$$j_\mu^{W\dagger}(x) = \sum_{\substack{q=u,c,t \\ q'=d,s,b}} (V_{qq'}^{CKM})^* \bar{q'}(x)\gamma_\mu(1-\gamma_5)q(x) \tag{1.31}$$

contain all possible flavor changing combinations of up-quarks ($q = u, c, t$) and down-quarks ($q' = d, s, b$) in SM, multiplied by the respective elements of the Cabibbo-Kobayashi-Maskawa (CKM) unitary matrix:

$$V^{CKM} = \begin{pmatrix} V_{ud} & V_{us} & V_{ub} \\ V_{cd} & V_{cs} & V_{cb} \\ V_{td} & V_{ts} & V_{tb} \end{pmatrix}, \tag{1.32}$$

which parametrizes the quark-flavor mixing in SM. This matrix depends on the three real parameters and one complex phase, hence the coefficients of the currents in (1.30) differ from their complex-conjugates in (1.31). For completeness, we quote the most convenient Wolfenstein parameterization of the CKM matrix

$$V^{CKM} = \begin{pmatrix} 1-\lambda^2/2 & \lambda & A\lambda^3(\rho-i\eta) \\ -\lambda & 1-\lambda^2/2 & A\lambda^2 \\ A\lambda^3(1-\rho-i\eta) & -A\lambda^2 & 1 \end{pmatrix} + O(\lambda^4), \tag{1.33}$$

where the small parameter $\lambda = 0.22$ sets up the hierarchy. More detailed information on the CKM matrix can be found in the dedicated minireview in [1].

Due to the specific symmetry structure of the electroweak interactions in SM, the quark and lepton weak currents contain superpositions of vector and axial-vector currents, denoted as $V - A$. This structure has a physical interpretation. Each Dirac field can be decomposed into *left-handed* and *right-handed* components (helicities) corresponding, respectively, to the negative and positive projections of its spin on the three-momentum direction. For the quark field we have,

$$q(x) = \frac{1-\gamma_5}{2}q(x) + \frac{1+\gamma_5}{2}q(x) \equiv q_L(x) + q_R(x),$$
$$\bar{q}(x) = \bar{q}(x)\frac{1+\gamma_5}{2} + \bar{q}(x)\frac{1-\gamma_5}{2} \equiv \bar{q}_L(x) + \bar{q}_R(x), \tag{1.34}$$

---

[5]We do not consider t-quark decays.

[6]Traditionally, one uses the weak interaction Hamiltonian instead of an interaction Lagrangian, their relation is simple: $\mathcal{H}^{int} = -\mathcal{L}^{int}$.

leading to

$$\bar{q}(x)\gamma_\mu(1 - \gamma_5)q'(x) = 2\bar{q}_L(x)\gamma_\mu q'_L(x) \,. \tag{1.35}$$

Hence, only transitions between the left-handed quarks are present in the weak current. The form on the r.h.s. of (1.35) is frequently used instead of the $V - A$ form on the l.h.s.

To complete the list of quark currents in the electroweak part of SM, let us mention the flavor-neutral axial-vector current

$$j_{\mu 5}(x) = \bar{q}(x)\gamma_\mu\gamma_5 q(x) \,, \tag{1.36}$$

emerging in a linear combination with the vector current $\bar{q}\gamma_\mu q$ in the interaction of quarks with Z-boson. In this interaction the couplings of axial-vector and vector quark currents with Z differ from each other by a certain factor, and not simply by a sign as in (1.30), so that both left-handed and right-handed quark fields are present.

Finally, in SM there exist also scalar quark currents with spin-parity $J^{PC} = 0^{++}$

$$j(x) = \bar{q}(x)q(x) \,, \tag{1.37}$$

which enter the Yukawa-type interactions of quarks with the Higgs boson.

1.4 EFFECTIVE CURRENTS

Within QCD, it is possible to introduce any local and color-neutral quark-antiquark operator

$$j_\Gamma(x) = \bar{q}_1(x)\Gamma q_2(x) \,, \tag{1.38}$$

with arbitrary flavors q_1, q_2 and a generic combination Γ of γ-matrices. Note that in this way we generate new currents that are not present in the electroweak part of SM Lagrangian.

Currents composed from gluon fields are also possible. The two simplest examples are:

$$j_G(x) = G^a_{\mu\nu}(x)G^{a\mu\nu}(x), \quad j_{G5} = G^a_{\mu\nu}(x)\widetilde{G}^{a\mu\nu}(x) \,, \tag{1.39}$$

where in the latter we use the standard notation (A.43). The above currents are evidently flavorless and color-neutral, preserving also the local gauge symmetry of QCD, because they depend solely on the gluon-field strength. The current j_G has spin-parity $J^{PC} = 0^{++}$, the same as the scalar quark current (1.37). Respectively, the current j_{5G} has the same spin-parity $J^{PC} = 0^{-+}$ as the pseudoscalar quark current

$$j_5(x) = \bar{q}(x)im_q\gamma_5 q(x) \,. \tag{1.40}$$

The latter corresponds to the choice $q_1 = q_2 = q$, and $\Gamma = im_q\gamma_5$ in (1.38). Note that the extra factor i renders this current Hermitian, and the reason it is convenient to multiply $i\gamma_5$ by the quark mass is explained at the end of this section.

Another example of local currents is the tensor current with $J^P = 2^{++}$

$$j_{\mu\nu}(x) = \frac{i}{2}\bar{q}(x)\left(\gamma_\mu\overleftrightarrow{D}_\nu + \gamma_\nu\overleftrightarrow{D}_\mu\right)q(x) \,, \tag{1.41}$$

where $\overleftrightarrow{D}_\nu = (1/2)(\overrightarrow{D}_\nu - \overleftarrow{D}_\nu)$ (see (A.41) for definition of covariant derivative), and its gluonic counterpart:

$$j^G_{\mu\nu} = G^a_{\mu\alpha}(x)G^{a\alpha}_\nu(x) - \frac{1}{4}g_{\mu\nu}G^a_{\alpha\beta}(x)G^{a\alpha\beta}(x) \,. \tag{1.42}$$

Local operators composed from the quark and gluon fields can also emerge effectively, originating from higher orders of the electroweak interactions in SM. The most prominent example is the phenomenon of flavor-changing neutral currents (FCNC). In SM, the loop diagrams, involving virtual gauge bosons (W, Z, γ) and t-quark, generate transitions between quarks of different flavors and the same electric charges. A detailed description of the FCNC decays is beyond our scope (see, e.g., the review [14]). Dominance of the short-distance scales of $O(1/m_W, 1/m_Z, 1/m_t)$ in the loop diagrams allow one to replace these diagrams by pointlike effective operators. For example, the $b \to s$ FCNC transitions accompanied by the lepton pair or a photon are driven by the following quark-lepton and quark-photon operators:

$$
\begin{aligned}
O_9(x) &= \frac{\alpha_{em}}{2\pi} \bar{s}(x)\gamma_\rho(1 - \gamma_5)b(x)\bar{\ell}(x)\gamma^\rho\ell(x)\,, \\
O_{10}(x) &= \frac{\alpha_{em}}{2\pi} \bar{s}(x)\gamma_\rho(1 - \gamma_5)b(x)\bar{\ell}(x)\gamma^\rho\gamma_5\ell(x)\,, \\
O_{7\gamma}(x) &= -\frac{e}{8\pi^2} \bar{s}(x)\sigma_{\mu\nu}\big[m_b(1 + \gamma_5) + m_s(1 - \gamma_5)\big]b(x)F^{\mu\nu}(x)\,.
\end{aligned} \tag{1.43}
$$

Their specific nomenclature and form corresponds to the effective Hamiltonian defined as:

$$
\mathcal{H}_{eff}^{b\to s}(x) = -\frac{G_F}{\sqrt{2}}V_{tb}V_{ts}^* \sum_{i=9,10,7\gamma} C_i O_i(x) + \dots\,, \tag{1.44}
$$

where less important operators are denoted by ellipsis. In the above, the numerical coefficients C_i obtained from the calculation of the loop diagrams (known as *Wilson coefficients*) are normalized to the same Fermi constant as the weak Hamiltonian (1.29) and the CKM factors reflect the dominant virtual t-quark contribution. Important for our discussion is that the quark parts of these operators represent new FCNC currents that are absent in the tree-level vertices in SM. Furthermore, one of the subdominant FCNC operators entering the Hamiltonian (1.44) represents a quark-antiquark-gluon effective current:

$$
O_{8g}(x) = -\frac{g_s m_b}{8\pi^2} \bar{s}(x)\sigma_{\mu\nu}(1 + \gamma_5)\frac{\lambda^a}{2}b(x)G^{a\,\mu\nu}(x)\,, \tag{1.45}
$$

where we neglect the suppressed $\sim m_s$ part.

Performing calculations with quark currents in QCD, one should take into account their dependence on the energy-momentum scale (running) caused by the perturbative gluon exchanges. The relevant one-loop diagrams are shown in Figure 1.4. The logarithmic parts of these diagrams have to be resummed yielding the scale-dependence which, similar to the running (1.10) of the quark mass, is expressed in terms of the quark-gluon coupling:

$$
j_\Gamma\big|_{\mu_1} = j_\Gamma\big|_{\mu_0} \left(\frac{\alpha_s(\mu_1)}{\alpha_s(\mu_0)}\right)^{\gamma^\Gamma/\beta_0}\,. \tag{1.46}
$$

The parameter γ^Γ is the anomalous dimension of the current and depends only on the Dirac structure Γ. Importantly, for the currents related to physical quantities, such as the e.m. current or $V - A$ weak current, the anomalous dimensions vanish. For the scalar current (1.37) one encounters the inverse anomalous dimension of the quark mass (cf. (1.10)):

$$
\bar{q}q\big|_{\mu_1} = \bar{q}q\big|_{\mu_0} \left(\frac{\alpha_s(\mu_1)}{\alpha_s(\mu_0)}\right)^{-\gamma_0/\beta_0}\,. \tag{1.47}
$$

This is anticipated, since the scale-independent product $m_q\bar{q}q$ is a physical quantity – the part of the QCD energy-momentum tensor. The pseudoscalar current $\bar{q}i\gamma_5 q$ has the anomalous dimension $-\gamma_0$, so that being multiplied by the quark mass in (1.40), it remains scale-independent. As we shall see in the next chapter, this current is in fact equal to the divergence of the axial-vector current.

Figure 1.4 Diagrams generating the running of the quark current (1.38). The wavy line represents the external source of the current.

1.5 ISOSPIN AND $SU(3)$-FLAVOR SYMMETRIES

In addition to the exact $SU(3)_c$ gauge symmetry related to the color charge, the QCD Lagrangian (1.1) possesses approximate *flavor symmetries*. Their origin lies in the flavor-independence of the quark-gluon interaction combined with the "accidental" pattern of the quark masses in $\mathcal{L}_{QCD}$. The extent of a flavor symmetry breaking depends on the relations between quark masses with respect to each other and to the scale Λ_{QCD}.

The most accurate among flavor symmetries is the *isospin symmetry*. It emerges due to a very small mass difference between the d-quark and u-quark. Indeed, as follows from (1.20):

$$(m_d - m_u)|_{\mu=2 \text{ GeV}} \simeq 2.5 \text{ MeV} \ll \Lambda_{QCD} \,. \tag{1.48}$$

Neglecting this difference and introducing the average mass

$$m_{ud} \equiv \frac{1}{2}(m_u + m_d) \simeq m_u \simeq m_d \,, \tag{1.49}$$

we rewrite the QCD Lagrangian in the following form:

$$\begin{aligned} \mathcal{L}_{QCD}(x) &= \bar{u}(x)\,(iD_\mu\gamma^\mu - m_{ud})\,u(x) + \bar{d}(x)\,(iD_\mu\gamma^\mu - m_{ud})\,d(x) \\ &\quad + \mathcal{L}_G(x) + \mathcal{L}_{s,c,b,t}(x) \,, \end{aligned} \tag{1.50}$$

where the color indices are not shown and a shorthand notation is used for the terms containing gluons and other quark flavors. An equivalent form of (1.50) is obtained if one unites the u-quark and d-quark in one isospin doublet and its conjugate:

$$\Psi(x) = \begin{pmatrix} u(x) \\ d(x) \end{pmatrix}, \quad \overline{\Psi}(x) = \big(\bar{u}(x)\ \bar{d}(x)\big) \,, \tag{1.51}$$

so that

$$\begin{aligned} \mathcal{L}_{QCD}(x) &= \overline{\Psi}(x)\,(iD_\mu\gamma^\mu - m_{ud})\,\Psi(x) \\ &\quad + \mathcal{L}_G(x) + \mathcal{L}_{s,c,b,t}(x) \,. \end{aligned} \tag{1.52}$$

The above expression is invariant under the following transformation:

$$\Psi(x) \to \Psi'(x) = \exp\left(-i\sum_{a=1}^{3} \chi^a \frac{\sigma^a}{2}\right) \Psi(x) \,,$$

$$\overline{\Psi}(x) \to \overline{\Psi}'(x) = \overline{\Psi}(x) \exp\left(i\sum_{a=1}^{3} \chi^a \frac{\sigma^a}{2}\right) \,, \tag{1.53}$$

where σ^a are the 2×2 Pauli matrices, and the three real parameters χ^a are independent of the coordinate, hence the symmetry is global.

The above transformations represent a sort of rotation in the specific (u, d)-flavor (isospin) space. They form the group $SU(2)_I$ which is mathematically equivalent to the group of transformations of the spin $1/2$ wave functions under three-coordinate rotations. In particular, the isospin $I = 1/2$ is attributed to the doublet (1.51) and the third projections $I_3 = +1/2$ and $I_3 = -1/2$ to its components u and d, respectively. Any state that contains u, d quarks and antiquarks belongs to one of the representations (multiplets) of the $SU(2)_I$ group. The quarks of all other flavors and gluons are singlets with respect to the isospin symmetry.

The e.m. current (1.23) violates isospin symmetry. This simply follows from the fact that the u-quark and d-quark have different electric charges. Therefore, assessing the accuracy of isospin symmetry, one has to take into account also the effects of e.m. interactions. Counting the isospin carried by the e.m. current is similar to adding together two $1/2$ spins in quantum mechanics. The result is a combination of the $I_3 = 0$ member of isospin triplet (*isovector*) with an isospin singlet (*isoscalar*). The following decomposition of the current:

$$j_\mu^{em} = j_\mu^{I=1} + j_\mu^{I=0} \tag{1.54}$$

explicitly separates the isospin components:

$$j_\mu^{I=1} = \frac{1}{2} \left(\bar{u}\gamma_\mu u - \bar{d}\gamma_\mu d \right) , \quad j_\mu^{I=0} = \frac{1}{6} \left(\bar{u}\gamma_\mu u + \bar{d}\gamma_\mu d \right) + \sum_{q=s,c,b,t} Q_q \bar{q}\gamma_\mu q . \tag{1.55}$$

The components of the weak current (1.30) corresponding to the $d \to u$ transition and its conjugate:

$$j_\mu^{W(d \to u)}(x) = V_{ud} \, \bar{u}\gamma_\mu(1 - \gamma_5)d , \quad j_\mu^{W(u \to d)}(x) = V_{ud}^* \, \bar{d}\gamma_\mu(1 - \gamma_5)u , \tag{1.56}$$

transform as isovector components with $I_3 = -1$ and $I_3 = +1$, respectively. Note that the vector parts of the two currents (1.56) form an isospin triplet together with the $I = 1$, $I_3 = 0$ part of the e.m. current (1.55). Furthermore, the parts of the weak current (1.30) containing a single u or d quark, e.g., the $s \to u$, $d \to s$ currents belong to isospin doublets.

A broader but less precise flavor symmetry emerges if one neglects the mass difference between s-quark and u, d-quarks, assuming that

$$m_s \simeq m_{ud} \simeq \bar{m} . \tag{1.57}$$

In full analogy with the isospin case, the QCD Lagrangian can then be rewritten via the quark triplet and its conjugate:

$$\Psi_{\{3\}}(x) = \begin{pmatrix} u(x) \\ d(x) \\ s(x) \end{pmatrix}, \quad \overline{\Psi}_{\{3\}}(x) = \left(\bar{u}(x) \ \bar{d}(x) \ \bar{s}(x) \right) , \tag{1.58}$$

so that

$$\begin{aligned} \mathcal{L}_{QCD}(x) &= \overline{\Psi}_{\{3\}}(x) \left(iD_\mu\gamma^\mu - \bar{m} \right) \Psi_{\{3\}}(x) \\ &+ \mathcal{L}_G(x) + \mathcal{L}_{c,b,t}(x), \end{aligned} \tag{1.59}$$

and the global transformation

$$\Psi_{\{3\}}(x) \to \Psi'_{\{3\}}(x) = \exp\left(-i \sum_{a=1}^{8} \chi^a \frac{\lambda^a}{2} \right) \Psi_{\{3\}}(x) ,$$

$$\overline{\Psi}_{\{3\}}(x) \to \overline{\Psi}'_{\{3\}}(x) = \overline{\Psi}_{\{3\}}(x) \exp\left(i \sum_{a=1}^{8} \chi^a \frac{\lambda^a}{2} \right) , \tag{1.60}$$

leaves this Lagrangian invariant. In this case the symmetry group is $SU(3)$, with eight $(3^2 - 1)$ generators represented by the 3×3 Gell-Mann matrices λ^a. Although formally it is the same symmetry group as $SU(3)_c$, the underlying physics is of course fundamentally different. We denote the approximate three-flavor symmetry as $SU(3)_{fl}$. The parameter of its violation can reach the level of $\sim 30\%$ which roughly corresponds to the ratio $O(m_s/\Lambda_{QCD})$. Despite that, $SU(3)_{fl}$ symmetry is a useful organizing tool for hadron spectroscopy and has plenty of phenomenological applications. The group $SU(3)_{fl}$ has three $SU(2)$ subgroups. One of them corresponds to the isospin symmetry and the two others, which unite (d, s) and (u, s) in doublets, are called U-spin and V-spin, respectively. The U-spin has an apparent advantage: due to the equal electric charges of d and s quarks, this symmetry is preserved by the e.m. interaction[7].

Throughout this book, we shall adopt the isospin symmetry limit in most cases, but try to take into account the $m_s - m_{ud}$ difference explicitly.

1.6   CHIRAL SYMMETRY AND ITS VIOLATION

The u-quark and d-quark masses, having a small difference (1.48), are also themselves small:

$$m_u \lesssim m_d \ll \Lambda_{QCD} \,. \tag{1.61}$$

Therefore, it is conceivable to consider the limit of massless u, d quarks. The isospin-invariant QCD Lagrangian (1.52) becomes:

$$\mathcal{L}_{QCD}(x)\big|_{m_{u,d} \to 0} = \overline{\Psi}(x) i D_\mu \gamma^\mu \Psi(x) + \mathcal{L}_G(x) + \mathcal{L}_{s,c,b,t}(x) \,. \tag{1.62}$$

Splitting the quark doublet into left- and right-handed components similar to (1.34):

$$\Psi(x) = \frac{1 - \gamma_5}{2} \Psi(x) + \frac{1 + \gamma_5}{2} \Psi(x) \equiv \Psi_L(x) + \Psi_R(x) \,,$$

$$\bar{\Psi}(x) = \bar{\Psi}(x) \frac{1 + \gamma_5}{2} + \bar{\Psi} \frac{1 - \gamma_5}{2} \equiv \bar{\Psi}_L(x) + \bar{\Psi}_R(x) \,, \tag{1.63}$$

and substituting this decomposition in (1.62), we obtain:

$$\mathcal{L}_{QCD}(x)\big|_{m_{u,d} \to 0} = \overline{\Psi}_L(x) i D_\mu \gamma^\mu \Psi_L(x) + \overline{\Psi}_R(x) i D_\mu \gamma^\mu \Psi_R(x)$$

$$+ \mathcal{L}_G(x) + \mathcal{L}_{s,c,b,t}(x) \,. \tag{1.64}$$

In this expression the terms with left-handed and right-handed u, d quarks decouple from each other and the quark-gluon vertex splits into two parts:

$$g_s \overline{\Psi}_L(x) \gamma_\mu A^{a\mu}(x) \frac{\lambda^a}{2} \Psi_L(x) \quad \text{and} \quad g_s \overline{\Psi}_R(x) \gamma_\mu A^{a\mu}(x) \frac{\lambda^a}{2} \Psi_R(x) \,,$$

revealing the conservation of left- and right-handedness (helicity) of quarks in the processes with gluon emission or absorption. Note that e.m. interaction also conserves helicity.

Moreover, the Lagrangian (1.64) is invariant with respect to the two separate isospin transformations:

$$\Psi_L(x) \to \Psi_L'(x) = \exp\left(-i \sum_{a=1}^{3} \chi_L^a \frac{\sigma^a}{2}\right) \Psi_L(x) \,,$$

$$\Psi_R(x) \to \Psi_R'(x) = \exp\left(-i \sum_{a=1}^{3} \chi_R^a \frac{\sigma^a}{2}\right) \Psi_R(x) \,, \tag{1.65}$$

---

[7]Contrary to a naive expectation, employing the U or V subgroup instead of the whole $SU(3)_{fl}$, does not improve the symmetry violation caused by the s-quark mass.

with two sets of independent parameters χ_L^a and χ_R^a. The corresponding transformations of the conjugated doublets are not shown for brevity; hence, the massless limit (1.64) possesses a *chiral symmetry* described by the direct product of two groups: $SU(2)_L \times SU(2)_R$.

The mass term in the Lagrangian (1.52), rewritten in terms of the left- and right-handed components:

$$m_{ud}\bar{\Psi}\Psi = m_{ud}(\bar{\Psi}_L\Psi_R + \bar{\Psi}_R\Psi_L)\,, \qquad (1.66)$$

is not invariant with respect to the chiral symmetry transformation (1.65), since in general $\chi_L^a \neq \chi_R^a$.

Naively, one expects the chiral symmetry breaking to be determined by the small parameters $m_{u,d}$. In reality, this symmetry is broken in the nonperturbative regime of QCD to a much larger extent, characterized by a scale of $O(\Lambda_{QCD})$. The underlying sources of symmetry breaking are vacuum fields, in particular, the quark-antiquark fluctuations averaged in a form of the condensate density in (1.16). The quark condensate, similar to the mass term (1.66), violates chiral symmetry:

$$\langle 0|\bar{q}q|0\rangle = \langle 0|(\bar{q}_L q_R + \bar{q}_R q_L)|0\rangle\,, \quad (q = u, d)\,. \qquad (1.67)$$

A propagating quark flips its helicity due to interactions with the vacuum fields. This property of QCD vacuum represents a prominent example of *spontaneous symmetry breaking*, a situation when the ground state in a quantum field theory violates a certain symmetry of the Lagrangian.

According to the Noether theorem, to every continuous symmetry transformation, such as (1.65), there corresponds a four-vector current with vanishing divergence. The currents associated with the left-handed $SU(2)_L$ and right-handed $SU(2)_R$ symmetries (1.65) are:

$$\bar{\Psi}_L(x)\gamma_\mu\frac{\sigma^a}{2}\Psi_L(x) = \frac{1}{2}\bar{\Psi}(x)\gamma_\mu(1-\gamma_5)\frac{\sigma^a}{2}\Psi(x)\,,$$
$$\bar{\Psi}_R(x)\gamma_\mu\frac{\sigma^a}{2}\Psi_R(x) = \frac{1}{2}\bar{\Psi}(x)\gamma_\mu(1+\gamma_5)\frac{\sigma^a}{2}\Psi(x)\,, \qquad (1.68)$$

forming orthogonal linear combinations of the isotriplet vector and axial currents:

$$J_\mu^a(x) = \bar{\Psi}(x)\gamma_\mu\frac{\sigma^a}{2}\Psi(x)\,, \quad J_{\mu 5}^a(x) = \bar{\Psi}(x)\gamma_\mu\gamma_5\frac{\sigma^a}{2}\Psi(x)\,, \quad (a = 1, 2, 3)\,. \qquad (1.69)$$

At $m_{u,d} = 0$, in the chiral symmetry limit, the divergences of the currents (1.68) vanish; hence, for the vector and axial currents we also have:

$$\partial^\mu J_\mu^a(x) = 0\,, \quad \partial^\mu J_{\mu 5}^a(x) = 0. \qquad (1.70)$$

These equalities are confirmed by taking the derivatives of the quark fields and their conjugates and applying the Dirac equation for massless fermions. Similar to the well-known property of the e.m. current, the vanishing divergences (1.70) imply conservation of the corresponding charges, defined as three-dimensional integrals over the time components of the currents (1.69):

$$\hat{Q}^a = \int d\vec{x}\, J_0^a(x_0, \vec{x})\,, \quad \hat{Q}_5^a = \int d\vec{x}\, J_{05}^a(x_0, \vec{x})\,, \quad (a = 1, 2, 3)\,, \qquad (1.71)$$

so that these charges commute with the energy operator $\hat{P}_0 = \hat{H}$ (Hamiltonian):

$$[\hat{H}, \hat{Q}^a] = 0\,, \quad [\hat{H}, \hat{Q}_5^a] = 0\,. \qquad (1.72)$$

In QCD, spontaneous symmetry breaking is associated only with the axial isotriplet current and, accordingly, with $\hat{Q}_5^a$. As a result, the vacuum state in this theory has different properties with respect to the two operators in (1.71). The action of the charges associated with the vector isotriplet currents produces zero:

$$\hat{Q}^a|0\rangle = 0\,, \tag{1.73}$$

reflecting the zero isospin of the vacuum, whereas the axial isotriplet charges transform the vacuum state to other physical states:

$$\hat{Q}_5^a|0\rangle = |r^a\rangle\,, \tag{1.74}$$

Using the second commutation relation in (1.72) we derive that the energy of these states vanishes:

$$\hat{H}|r^a\rangle = \hat{H}\hat{Q}_5^a|0\rangle = \hat{Q}_5^a\hat{H}|0\rangle = 0\,, \tag{1.75}$$

hence, the states $|r^a\rangle$ are massless. This represents a realization of the general Goldstone theorem predicting a massless spin-zero state (boson) for each component of the charge operator associated with the spontaneously broken symmetry.

The operators $\hat{Q}_5^a$, having the same quantum numbers as the time component of the axial four-vector, are pseudoscalars. Therefore, in the limit $m_{u,d} \to 0$ there exists an isospin triplet of massless pseudoscalar states $|r^a\rangle$ playing the role of the Goldstone bosons. Since the only physical states in QCD are hadrons, we expect the emergence of such states in the spectrum of hadrons, to be discussed below in this chapter. Extending the massless limit to the strange quark, and, correspondingly, enlarging the isospin symmetry to $SU(3)_{fl}$, we obtain, in the limit $m_{u,d,s} \to 0$, a spontaneously broken $SU(3)_L \times SU(3)_L$ chiral symmetry, predicting the existence of eight massless pseudoscalar hadrons.

As an important addition to the issue of chiral symmetry in QCD, we also mention the peculiarity of the $SU(3)_{fl}$-singlet axial-vector current. Contrary to naive expectation, this current is not conserved in the massless quark limit. The reason is that its divergence

$$\partial^\mu \left(\sum_{q=u,d,s} \bar{q}\gamma_\mu\gamma_5 q \right) = \sum_{q=u,d,s} 2m_q\bar{q}i\gamma_5 q + \frac{3\alpha_s}{4\pi} G_{\mu\nu}^a \tilde{G}^{a\mu\nu}\,, \tag{1.76}$$

(with the dual gluon field-strength tensor defined in (A.43)), contains, apart from the usual quark mass term, the additional gluonic piece, generated by the triangle loop diagrams formed by the current vertex and two quark-gluon vertices. This purely quantum effect of chiral symmetry violation is known as the *triangle or chiral anomaly* (for an introductory review see, e.g., [15]). A similar anomaly will be discussed in detail in Chapter 6.

A more detailed description of the chiral symmetry in QCD and of the related effective Chiral Perturbation Theory (ChPT) applicable to the low energy hadronic phenomena, can be found in dedicated reviews (see, e.g., [16, 17, 18]).

1.7 HADRONS AND THEIR SPECTROSCOPY

As already discussed in Section 1.1.3, hadrons are the observable color-neutral bound states of quarks, antiquarks and gluons. These states emerge nonperturbatively and are characterized by the energy-momentum scales of $O(\Lambda_{QCD})$. In this section we discuss in more detail the most important properties of hadronic states.

1.7.1 Mesons and baryons

Only three basic combinations of quarks and antiquarks fulfil the condition of color-neutrality in QCD or, in other words, are invariant with respect to the $SU(3)_c$-symmetry transformations:

$$|\text{Meson}\rangle = \frac{1}{\sqrt{N_c}}|q_1^i \bar{q}_{2i}\rangle \, ,$$

$$|\text{Baryon}\rangle = \frac{1}{\sqrt{2N_c}}\epsilon_{ijk}|q_1^i q_2^j q_3^k\rangle \, , \quad |\text{Antibaryon}\rangle = \frac{1}{\sqrt{2N_c}}\epsilon^{ijk}|\bar{q}_{1i}\bar{q}_{2j}\bar{q}_{3k}\rangle, \qquad (1.77)$$

where $N_c = 3$ and the summation over color indices is shown explicitly. Due to flavor universality of the quark-gluon interactions, any combination of flavors, $q_{1,2,3} = u, d, s, c, b$, is possible in these states. For the mesons, the full variety of flavors is experimentally established [1]. For the baryons only a few flavor combinations are still awaiting their discovery, mainly those with two or three heavy quarks.

The quark-antiquark, three-quark and three-antiquark states in (1.77) specify the *valence state* of a meson, baryon and antibaryon, respectively, defined as the state with a minimal possible set of constituents determining the flavor content and other quantum numbers of a given hadron.

1.7.2 Hadronic state in QCD

The valence state represents only one of the possible states of a hadron, albeit an important one. A bound state in a quantum field theory, hence also in QCD, is a superposition of all possible states of the constituents without specifying their amount, provided each state in this superposition has the same quantum numbers.

To illustrate the above statement, it is instructive to consider a simpler bound state: the hydrogen atom. In nonrelativistic quantum mechanics it definitely has two constituents: the electron and proton. At the precision level of QED, however, this "valence" state is incomplete. Quantum corrections, that is, loop diagrams with virtual photons and electron-positron pairs emitted and absorbed during short time intervals have to be taken into account; hence, states with additional constituents are also possible, so that, schematically, one has a superposition

$$|\text{Hydrogen}\rangle = |e^- p\rangle \oplus |e^- p\, \gamma^*\rangle \oplus |e^- p\, e^+ e^-\rangle + \ldots, \qquad (1.78)$$

where dots indicate other states containing virtual photons and electron-positron pairs and having the same quantum numbers. Due to smallness of the e.m. coupling α_{em}, the mass $2m_e$ of the $e^+ e^-$ pair is much larger than the hydrogen binding energy of $O(\alpha_e m_e)$; hence, quantum field fluctuations in the hydrogen atom are suppressed and reveal themselves only in very subtle effects such as the Lamb shift.

In nonperturbative QCD the situation is qualitatively different: the quark-gluon coupling is strong and the average energy of virtual gluons and light quark-antiquark pairs is comparable to the hadronic scale Λ_{QCD}; hence, in the full decomposition of a hadronic state the role of quantum fluctuations involving additional gluons and quark-antiquark pairs is substantial.

To elaborate on this point in more detail, let us consider the positively charged pion, a hadron with the valence state $|u\bar{d}\rangle$. Note that pion will be frequently used throughout this book as an important study case. This hadron is quite stable. Its lifetime determined by the weak decay is many orders of magnitude larger than the typical QCD timescale $1/\Lambda_{QCD}$. We assign a definite momentum $p = (p_0, \vec{p})$ to the pion, so that $p^2 = m_\pi^2$, i.e., the pion is on the mass shell. The superposition of states forming the pion in QCD can, schematically, be represented as:

$$|\pi^+(p)\rangle = \int dP_{[2]}\Phi_{[u\bar{d}]}(p_1, p_2)|u(p_1)\bar{d}(p_2)\rangle$$

$$\oplus \int dP_{[3]}\Phi_{[u\bar{d}G]}(p_1, p_2, p_3)|u(p_1)\bar{d}(p_2)G(p_3)\rangle$$

$$\oplus \sum_{q=u,d,s,..} \int dP_{[4]}\Phi_{[u\bar{d}q\bar{q}]}(p_1, p_2, p_3, p_4)|u(p_1)\bar{d}(p_2)q(p_3)\bar{q}(p_4)\rangle \oplus ..., \quad (1.79)$$

where the integration over momenta with

$$dP_{[N]} \equiv \prod_{i=1}^{N} d^4 p_i \delta^{(4)}\left(\sum_{i=1}^{N} p_i - p\right),$$

obeys the total momentum conservation, and all contributing states are color-neutral and have the same spin-parity $J^P = 0^-$. In (1.79), the coefficient functions $\Phi_{[u\bar{d}]}$ or $\Phi_{[u\bar{d}G]}$, $\Phi_{[u\bar{d}q\bar{q}]}$,... denote probability amplitudes to find the pion in the valence state or in the state with three, four,... constituents, respectively, with the pion momentum distributed among them[8].

We emphasize that none of the states on the r.h.s. of (1.79) belong to the set of observable asymptotic states in physical processes[9], but only their coherent superposition $|\pi^+(p)\rangle$; hence, considering the amplitudes of physical processes in a form of S-matrix elements (1.13), we do not attempt to resolve the initial and final hadronic states in terms of superpositions similar to (1.79). These states in the simplest case of a single hadron are

$$|i\rangle = |h(p, r)\rangle, \quad \langle f| = \langle h'(p', r')|, \quad (1.80)$$

where the notation h, h' has to be replaced by the letters adopted in [1] for the hadrons with a given flavor content and spin-parity. Each hadron is characterized by its on-shell momentum and mass:

$$p = (p_0, \vec{p}), \quad p^2 = m_h^2; \quad p' = (p_0', \vec{p}'), \quad p'^2 = m_{h'}^2,$$

and other quantum numbers denoted as r, r' in (1.80). A usual orthogonality and normalization condition is adopted:

$$\langle h(p', r')|h(p, r)\rangle = 2p_0 (2\pi)^3 \delta^{(3)}(\vec{p}' - \vec{p})\delta_{r'r}, \quad (1.81)$$

which is relativistically invariant. It is usually assumed that (1.80) are stable asymptotic states. This is a reasonable approximation for weakly or electromagnetically decaying hadrons, but not always for the hadrons which are unstable with respect to the quark-gluon interactions and have strong decays. We will discuss stable and unstable hadrons below in this section.

1.7.3   Exotic hadrons and hadronic molecules

To complete our discussion of the bound states in QCD, let as mention that color-neutral states containing "valence" gluons are also possible. Some examples are:

$$|\text{Glueball}\rangle = |G^a G^a\rangle, \quad |\text{Hybrid}\rangle = |q_1^i \bar{q}_{2k} (\lambda^a)_i^k G^a\rangle, \quad (1.82)$$

---

[8]This description resembles the Fock states in quantum mechanics.

[9]In fact, although a general decomposition such as (1.79) remains largely illustrative, under certain asymptotic conditions, it is possible to approximate the pion state with a valence component in the initial and final states, as will be discussed in Chapter 7.

where the nomenclature adopted in the literature is used and, for simplicity, only the color structure is shown. The problem with identification of these so-called "exotic" hadrons is that they can have the same quantum numbers as certain mesons with the quark-antiquark valence state. As a result, transitions of glueball and hybrid states to the quark-antiquark ones and vice versa via strong nonperturbative interactions are unavoidable, leading to mesons with a mixed valence content. Exceptions are the exotic mesons that possess combinations of quantum numbers which are not possible for the quark-antiquark states. One example will be given in the next section.

Furthermore, "molecular" color-neutral bound states are also formed, consisting of the basic "quark atoms" (1.77) and having a valence state with more than three quarks. The commonly known examples are atomic nuclei. Deuteron, being a proton-neutron bound state with a very small binding energy, is nothing but a six-quark state. Ultimately, the potential that binds protons and neutrons inside a nuclei, also originates from nonperturbative QCD interactions. The hadrons composed from two mesons or from a meson and baryon are also being observed [1]. Considered as multiquark states they are dubbed tetraquarks and pentaquarks, respectively. A clear distinction between their molecular and multiquark nature still remains a model-dependent issue.

1.7.4 Basics of hadron spectroscopy

In this book, we only consider mesons and baryons with the basic valence-quark content (1.77). Their main characteristics are listed in the summary tables in [1]. These tables are subdivided in sections devoted to the mesons and baryons with different combinations of flavor quantum numbers. The lightest hadron in each section serves as a ground state in the spectrum of hadronic states with a given flavor content.

To overview the main elements of hadron spectroscopy we first consider the *light unflavored* mesons in [1] with the valence content:

$$\underbrace{|u\bar{d}\rangle\,, \quad \frac{|u\bar{u} - d\bar{d}\rangle}{\sqrt{2}}\,, \quad |d\bar{u}\rangle\,,}_{I=1} \quad \underbrace{\underbrace{\frac{|u\bar{u} + d\bar{d}\rangle}{\sqrt{2}}\,, \quad |s\bar{s}\rangle\,,}_{I=0}}_{\text{2 mixed states}} \tag{1.83}$$

where the color summation is not shown for simplicity. Here isospin symmetry plays a crucial role, e.g., a meson consisting only of $u\bar{u}$ would violate this symmetry. Allowed are the mutually orthogonal linear combinations of the light quark-antiquark states. One of the neutral mesons is the $I_3 = 0$ component of an isovector and the other one is an isoscalar, as shown in (1.83). The flavor-neutral $s\bar{s}$ state is by default isoscalar, hence it has the same quantum numbers as the state $u\bar{u} + d\bar{d}$, and the observed isoscalar mesons represent a mixture of both components. This phenomenon will be discussed in more detail below, in Section 1.7.7.

The ground states of light unflavored mesons with $I = 1$ are the pions:

$$|\pi^+\rangle = |u\bar{d}\rangle\,, \quad |\pi^0\rangle = \frac{|u\bar{u} - d\bar{d}\rangle}{\sqrt{2}}\,, \quad |\pi^-\rangle = |d\bar{u}\rangle\,, \tag{1.84}$$

with a very small mass difference [1]:

$$m_{\pi^\pm} - m_{\pi^0} \simeq 139.57 \text{ MeV} - 134.98 \text{ MeV} = 4.59 \text{ MeV}\,, \tag{1.85}$$

characterizing the isospin symmetry violation. The corresponding isoscalar mesons are $\eta(548)$ and $\eta'(958)$, with larger masses (indicated in the parentheses in the units of MeV) and with a mixed valence content:

$$|\eta\rangle = c_1 \frac{|u\bar{u} + d\bar{d}\rangle}{\sqrt{2}} + c_2|s\bar{s}\rangle \,, \quad |\eta'\rangle = c_3 \frac{|u\bar{u} + d\bar{d}\rangle}{\sqrt{2}} + c_4|s\bar{s}\rangle \,. \tag{1.86}$$

The mixing parameters c_i obey the normalization conditions $c_1^2 + c_2^2 = 1$, $c_3^2 + c_4^2 = 1$. These parameters determined from experimental data (see, e.g., the review on quark model in [1] and also [19]) form an approximately orthogonal matrix:

$$\begin{pmatrix} c_1 & c_2 \\ c_3 & c_4 \end{pmatrix} \simeq \begin{pmatrix} \cos\alpha_P & -\sin\alpha_P \\ \sin\alpha_P & \cos\alpha_P \end{pmatrix} \,, \tag{1.87}$$

where $\alpha_P \simeq 40^\circ$.

We proceed, introducing the basic properties of hadrons with respect to the space-time transformations. Each hadron has a definite spin J and space-parity P denoted as J^P in [1]. Accordingly, the hadronic state transforms as a whole under the space rotations and has $2J + 1$ polarization degrees of freedom, equal to the number of spin projections; hence, apart from the energy-momentum and mass, in case $J = 1$ ($J = 1/2, J > 1$) we should associate a polarization vector (spinor, tensor) with a hadronic state.

To derive the spin-parity and other quantum numbers of a hadron it is sufficient to consider its valence quark content, because the states with additional constituents in the superpositions of the type (1.79) have the same quantum numbers. For the valence states with a fixed number of constituents the usual counting rules of quantum mechanics are used, with some modifications caused by the relativistic effects. In particular, the total spin J of a meson is a vector sum

$$\vec{J} = \vec{S} + \vec{L} \,,$$

where $\vec{S}$ is the vector sum of the quark and antiquark spins with the eigenvalues $S = 0, 1$, and $\vec{L}$ is the orbital angular momentum of the quark-antiquark state. Furthermore, the space-parity P of a meson is determined from the relation:

$$P = -(-1)^L \,, \tag{1.88}$$

where the additional minus sign originates from the opposite signs acquired by the quark bispinor and its conjugate after the P-transformation.

From the above counting rules we expect that the ground state of a quark-antiquark system is a *pseudoscalar* meson with $S = L = 0$ and $J^P = 0^-$. Indeed, the pions and η, η'-mesons are pseudoscalars[10].

The next-to-ground state is the *vector* meson with $S = 1$ and $L = 0$, hence with $J^P = 1^-$. In the spectrum of light unflavored mesons the vector state is represented by the isospin triplet of $\rho(770)$-mesons:

$$|\rho^+\rangle = |u\bar{d}\rangle, \quad |\rho^0\rangle = \frac{|u\bar{u} - d\bar{d}\rangle}{\sqrt{2}} \,, \quad |\rho^-\rangle = |d\bar{u}\rangle \,, \tag{1.89}$$

having the same valence content as the pions, and by the two isospin singlets $\omega(782)$ and $\phi(1020)$:

$$|\omega\rangle = \frac{|u\bar{u} + d\bar{d}\rangle}{\sqrt{2}} \,, \quad |\phi\rangle = |s\bar{s}\rangle \,. \tag{1.90}$$

Note that contrary to (1.86), the admixture of the strange quark-antiquark pair in the valence state of ω-meson is very small and can be neglected, whereas ϕ-meson does not contain an appreciable $u\bar{u} + d\bar{d}$ component. We will discuss this issue in a separate section below.

---

[10]The fact that the lightest hadron is a pseudoscalar meson can be derived rigorously using correlation functions of quark currents in QCD, see, e.g., [20].

The set of meson quantum numbers includes also the C-parity, affiliated with the charge conjugation. After this transformation the state of a neutral meson containing a quark and antiquark of the same flavor, (such as π^0, η or ρ^0, ω) transforms into itself, allowing one to assign a certain parity to this meson. Note that the charge conjugation transformation of a meson state is equivalent to an interchange of quark and antiquark positions and spin projections; hence, for the charge parity the following rule is valid:

$$C = (-1)^{L+S}. \tag{1.91}$$

The C-parity together with the spin and space-parity of a meson are denoted as J^{PC} in [1]. Note in passing that the CP parity, an important combination of quantum numbers for a neutral meson is then simply obtained by multiplying the space and charge parities. Since a general CPT theorem is valid for any quantum theory including QCD, the time-reversal parity T for a meson is also then defined unambiguously. One of the consequences of the CPT theorem is the equality of the masses and total widths (lifetimes) of particles and antiparticles, which is also valid for hadrons, yielding e.g., $m_{\pi^+} = m_{\pi^-}$, $\tau_{\pi^+} = \tau_{\pi^-}$, $m_p = m_{\bar{p}}$, etc.

Furthermore, for the light unflavored mesons an additional quantum number – G-parity is introduced based on the isospin symmetry (see Appendix A). To perform the G-transformation, an $I_3 \to -I_3$ flip for all members of an isospin multiplet is followed by the charge conjugation, e.g., for the pion:

$$|\pi^\pm\rangle \overset{\hat{I}_3}{\to} -|\pi^\mp\rangle \overset{\hat{C}}{\to} -|\pi^\pm\rangle, \qquad |\pi^0\rangle \overset{\hat{I}_3}{\to} -|\pi^0\rangle \overset{\hat{C}}{\to} -|\pi^0\rangle, \tag{1.92}$$

so that the counting rule for the G-parity reads:

$$G = (-1)^I C, \tag{1.93}$$

where C is the charge parity of the neutral member of that multiplet. In the section of light unflavored mesons in [1], the G-parity is quoted in a combination I^G with the isospin.

The spectrum of light unflavored mesons with growing masses continues in two directions. One branch of this spectrum is formed by the orbital excitations, starting from $L=1$. The combination of $L = 1$ with $S=1$ and $S=0$ yields the following mesons:

$(L = 1, S = 1)$ J^{PC}	$I^G = 1^-$	$I^G = 0^+$
0^{++}	$a_0(1450)$	$f_0(1370), f_0(1710)$
1^{++}	$a_1(1260)$	$f_1(1285), f_1(1420)$
2^{++}	$a_2(1320)$	$f_2(1270), f_2'(1525)$
$(L = 1, S = 0)$ J^{PC}	$I^G = 1^+$	$I^G = 0^-$
1^{+-}	$b_1(1235)$	$h_1(1170), h_1(1380)$

where C-parity refers to the neutral states. The mesons with spin zero, one and two and positive P-parity are named, respectively, scalar, axial and tensor mesons. In the above table we quote the two isoscalar states for each J^{PC} according to the assignment given in [1] and based on a phenomenological analysis within the quark model. The mixing pattern is so far clarified only for the tensor mesons, being roughly similar to the one for the vector mesons. The spectrum of scalar mesons is more complicated. The reason lies in the observed scalar mesons $a_0(980), f_0(500), f(980)$ that are lighter than the ones quoted in the above table and may have an exotic valence content (dimeson molecules, tetraquarks or a mixture of both). The spectrum of orbital excitations with growing angular momentum $L = 2, 3, ..$ is easy to predict: for each L there are three mesons with $J = L - 1, L, L + 1$ $(S = 1)$ and one with $J = L$ ($S = 0$), with P-parity counted according to (1.88). According to [1], states up to $L = 5$ have already been observed.

The second branch of the meson spectrum involves *radial excitations*[11]. They have the same J^{PC} as the ground-state meson and differ from the latter only by their mass. The first radially excited states with the pion and ρ-meson quantum numbers are $\pi(1300)$ and $\rho(1450)$, respectively [1].

Importantly, certain orbital and radial excitations have the same J^P, leading to their mutual mixing. E.g., one of the mesons with $L = 2, S = 1$ has $J^P = 1^-$, the same as the radial excitation of a vector meson ($L = 0, S = 1$). In fact, not all combinations of spin and parity are realized for the mesons with the basic valence content (1.77). For example, combining (1.88), (1.91) rules out a bound state of a quark and antiquark with $J^{PC} = 1^{-+}$; hence, mesons with this combination of spin and parities are definitely exotic, possibly having a hybrid valence state (1.82).

Continuing our survey of hadrons, we turn to baryons, the bound states with a three-quark valence content. Each baryon is characterized by its spin, space-parity (J^P) and isospin (I), whereas C and G parities are evidently absent, because charge conjugation converts a baryon to antibaryon. The baryons built from the u and d quarks are listed in the sections "N-baryons" and "Δ-baryons" of [1]. The lowest states have zero angular momentum between quarks. Each of these states can be treated as a system of three identical quarks with certain projections of isospin and spin. Consequently, the Fermi-Dirac statistics demand total antisymmetry of the baryonic state with respect to the permutations of quarks. Since the color-neutral state of three quarks in (1.77) is by default totally antisymmetric, the direct product of the quark spin and isospin states is totally symmetric. This condition is manifested for instance, in the valence content of the lightest baryons – the proton and neutron with $J^P = 1/2^+$, $I = 1/2$:

$$|p\rangle = |u(ud)_{S=0,I=0}\rangle \,, \quad |n\rangle = |d(ud)_{S=0,I=0}\rangle \,. \tag{1.94}$$

Here the diquark (ud) has zero spin and isospin, and the third quark, u or d, determines the quantum numbers of the proton and neutron, respectively. In this book we mostly consider stable baryons, the lightest ones with a given flavor content, therefore we do not dwell on further details of the baryon spectrum, which also includes orbital and radial excitations. They can be found, e.g., in the comprehensive review of the quark model in [1].

1.7.5 Varying the quark flavors

The flavor dimension of the hadron spectroscopy is accessed by replacing the u or d valence quarks (antiquarks) in the light unflavored hadrons by the s, c or b quarks (antiquarks). Varying the quark flavors, we do not alter the color charge of the hadron constituents, hence the QCD interactions forming the hadron are not altered too. The flavored hadrons should then replicate the spin-parity spectrum specified above for the light unflavored mesons and baryons. This is however an oversimplified picture. In reality, varying flavors we alter the quark masses, effectively modifying the quark-gluon interactions. The effect is relatively small while replacing the u, d-quarks by the light s quark ($SU(3)_{fl}$-symmetry) but substantial in the case of the heavy c, b quarks.

The lightest pseudoscalar and vector mesons with various flavor contents are displayed in Table 1.1 with the nomenclature adopted in [1]. To determine the valence state of a meson placed in a certain entry of this table, one has to combine the quark and antiquark flavors, respectively, from the column and row, on the crossing of which that entry is located.

---

[11]The name reflects an analogy with the radial quantum number of bound states in a quantum-mechanical potential.

TABLE 1.1 Hadrons with various flavor content. Pseudoscalar ($J^P = 0^-$) and vector ($J^P = 1^-$) mesons occupy, respectively, the upper and lower lines in each entry. The lightest baryons with $J^P = 1/2^+$ occupy the last row. The masses can be found in [1].

	u	d	s	c	b
$\bar{u}$	π^0, η, η' ρ^0, ω	π^- ρ^-	K^- K^{*-}	D^0 D^{*0}	B^- B^{*-}
$\bar{d}$	π^+ ρ^+	π^0, η, η' ρ^0, ω	$\bar{K}^0$ $\bar{K}^{*0}$	D^+ D^{*+}	$\bar{B}^0$ $\bar{B}^{*0}$
$\bar{s}$	K^+ K^{*+}	K^0 K^{*0}	η, η' ϕ	D_s^+ D_s^{*+}	$\bar{B}_s^0$ $\bar{B}_s^{*0}$
$\bar{c}$	$\bar{D}^0$ $\bar{D}^{*0}$	D^- D^{*-}	D_s^- D_s^{*-}	$\eta_c(1S)$ $J/\psi(1S)$	B_c^- B_c^{*-}
$\bar{b}$	B^+ B^{*+}	B^0 B^{*0}	B_s^0 B_s^{*0}	B_c^+ B_c^{*+}	$\eta_b(1S)$ $\Upsilon(1S)$
(ud)	p	n	Λ	Λ_c	Λ_b

In accordance with their valence content (1.84) and (1.89) dictated by isospin symmetry, the π^0 and ρ^0 mesons with $I = 1$ occupy two diagonal positions, $u\bar{u}$ and $d\bar{d}$, in Table 1.1. The valence content of the isoscalar flavor-neutral mesons essentially depends on their spin-parity quantum numbers. The pseudoscalar mesons η and η' are described by the mixed states (1.86), both with a substantial contribution of the $s\bar{s}$ state; hence, the $J^{PC} = 0^{-+}$ mesons occupy three diagonal positions in Table 1.1. The vector mesons ω and ϕ have a markedly different flavor content (1.90), revealing that at $J^{PC} = 1^{--}$ the mixing between the $u\bar{u} + d\bar{d}$ and $s\bar{s}$ states is negligible. We will discuss the mixing issue from the point of view of QCD below in this section.

A more uniform pattern is observed for the mesons consisting of a heavy quark and antiquark of the same flavor and named *heavy quarkonia*. Both pseudoscalar $\eta_c(2984)$ ($\eta_b(9399)$) and vector $J/\psi(3097)$ ($\Upsilon(9460)$) *charmonium (bottomonium)* ground states listed in Table 1.1 have a negligible admixture of lighter quark-antiquark pairs. Since the heavy quark masses m_c, m_b are much larger than Λ_{QCD}, the nonrelativistic quark approximation with a static confining potential is often used to describe the quarkonium spectroscopy; hence, the traditional spectroscopic notation $1S$ for the lightest quarkonium levels is added. Correspondingly, their radial excitations carry the labels $2S, 3S, \ldots$, whereas the orbital excitations of quarkonia with $L = 1, 2, 3, \ldots$ have the labels $P, D, F, \ldots$, respectively. Finally, there is a specific family of bottom-charmed mesons. The B_c and B_c^* presented in Table 1.1 are the lightest states with $J^P = 0^-$ and $J^P = 1^-$, respectively. These mesons are heavy quark-antiquark states, hence, their spectroscopy can also be described with the help of potential models. At the same time, bottom-charmed mesons carry open flavors and differ from conventional heavy quarkonia not only by the pattern of their decays, but also by other properties, e.g., by the absence of C-parity.

The lightest and stable baryons are listed in the last row of the same Table 1.1. Their valence structure is simply obtained by replacing the u quark in the proton state (1.94) by any other flavor, leaving the diquark (ud) in the symmetric $I = S = 0$ state.

1.7.6 Hadrons are not alike

The presence of a universal scale Λ_{QCD} in nonperturbative QCD suggests a simple rule to calculate the mass of a hadron with a given flavor content. If we confine ourselves by the lowest states with a certain spin-parity, taking, e.g., the mesons listed in Table 1.1, their masses can presumably be estimated summing up the valence quark masses – using their values quoted in (1.20) – and adding a "binding energy" of $O(\Lambda_{QCD})$ per each valence constituent. Doing that, we tacitly ignore the fact that quark masses in (1.20) are defined in the framework of perturbative QCD, at sufficiently large renormalization scales. Nevertheless, this simple addition rule roughly reproduces the measured masses of vector mesons and spin-1/2 baryons[12]. However it completely fails for the lightest pseudoscalar mesons. Especially the pion seems to be anomalously light, about five time lighter than the vector ρ-meson. These two hadron states differ only by the total spin of their valence quark and antiquark; hence, the observed mass difference contradicts our experience with the bound states in quantum mechanics or in QED, where the spin-dependent part of a potential is usually suppressed, causing only a minor effect of hyperfine splitting. Albeit to a lesser extent, the kaon and η-meson are also lighter than expected, whereas the η' meson is distinctively heavier than other pseudoscalar mesons.

The smallness of the pion mass is a spectacular manifestation of the vacuum fields in QCD, specifically, of the quark condensate whose presence spontaneously violates the chiral symmetry (see (1.67)). As we already discussed in Section 1.6, this phenomenon implies existence of massless pseudoscalar ($J^P = 0^-$) mesons playing the role of Goldstone bosons in the spontaneously broken $SU(2)_L \times SU(2)_R$ symmetry. Evidently, the triplet π^+, π^0, π^- has to be identified with these states. Similarly, kaons and η-meson play a role of Goldstone bosons in the extended, spontaneously broken $SU(3)_L \times SU(3)_R$ symmetry.

The fact that pions and other pseudoscalar mesons have small nonvanishing masses is due to explicit violation of the chiral symmetry in the QCD Lagrangian by the masses of u, d, s quarks. Moreover, it is possible to derive the Gell-Mann-Oakes-Renner relation [21] between the pion mass squared, quark masses and quark condensate:

$$m_{\pi^\pm}^2 = -\frac{2}{f_\pi^2}(m_u + m_d)\langle 0|\bar{q}q|0\rangle + O(m_u^2, m_d^2)\,, \tag{1.95}$$

where $q = u$ or d[13]. The parameter f_π entering the coefficient in this relation has a dimension of energy and is related to the matrix element of the quark axial current between the pion state and vacuum. This parameter will be introduced in the next chapter. We note in passing that the quark condensate has a negative sign. Analogous relations are valid at the $SU(3)_{fl}$ symmetry level, e.g., for the charged kaon mass squared:

$$m_{K^\pm}^2 = -\frac{2}{f_K^2}(m_u + m_s)\langle 0|\bar{q}q|0\rangle + O(m_s^2, m_u^2)\,, \tag{1.96}$$

where $q = u, d, s$. Since $m_s \gg m_{u,d}$, the latter relation is violated to a larger extent than (1.95). We already noticed that, in contrast to the light pseudoscalar mesons π, K, η, the $\eta'(958)$ meson is unusually heavy. In QCD the η' mass is linked to the gluon anomaly (1.76) and to the related gluonic vacuum fields.

Another manifestation of spontaneous symmetry breaking is the observed large difference between masses of the lightest axial ($J^P = 1^+$) and vector ($J^P = 1^-$) mesons, that is, between $a_1(1260)$ and $\rho(770)$. In chirally symmetric QCD, the spectra of vector and axial mesons would have been degenerate.

---

[12]The notion of "constituent" quarks with masses, e.g., for u and d in the ballpark of 250–300 MeV, is frequently used in phenomenological quark models of hadrons.

[13]A detailed derivation of this relation will be given in Chapter 8.

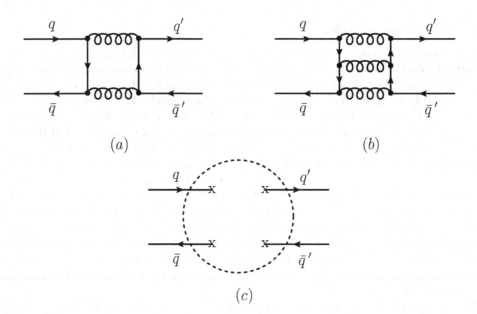

Figure 1.5 Diagrams describing transitions between quark-antiquark states via exchange of (a) two and (b) three gluons in QCD perturbation theory and (c) via QCD vacuum fluctuation depicted by a dashed circle.

Summarizing, the effects of nonperturbative vacuum fields in QCD introduce diversity within hadrons with the same flavor content and different spin-parities. More detailed answers to the question of why hadrons are not alike, can be found in [22].

1.7.7 Mixing of light isoscalar mesons

Development of QCD has also shed light on the observed mixing pattern of isoscalar mesons, explaining, at least qualitatively, why the valence states of $I = 0$ pseudoscalar and vector mesons, presented, respectively, in (1.86) and (1.90), are so different.

In QCD perturbation theory, the transitions

$$q\bar{q} \to q'\bar{q}' \,, \tag{1.97}$$

where q and q' are light quarks with different flavors, are described – at the lowest possible order in α_s – by the diagrams shown in Figure 1.5(a),(b). A color-neutral $q\bar{q}$ pair annihilates in two or three intermediate gluons, and a subsequent fusion of these gluons produces a $q'\bar{q}'$ pair with the same spin-parity. Clearly, a one-gluon intermediate state is not allowed due to the color charge conservation. For a two-gluon state the spin-parities $J^{PC} = 0^{++}, 0^{-+}, 2^{++}$ are possible[14]. Therefore, the quark-antiquark states, for which the transition (1.97) is described by the diagram in Figure 1.5(a), should have one of these combinations of quantum numbers. Note that the gluon field, similar to the photon field, has a negative C-parity; hence, a quark-antiquark state with $J^{PC} = 1^{--}$ cannot couple to a two-gluon state with positive C-parity. The transition (1.97) in this case can be realized with at least three gluons, as shown in Figure 1.5(b).

Based on this observation, it is tempting to explain why mixing between the states $(u\bar{u} + d\bar{d})$ and $s\bar{s}$, being suppressed for vector mesons, is enhanced for the pseudoscalar

---

[14]The corresponding composite two-gluon operators were presented in (1.39) and (1.42).

mesons. Counting the powers of quark-gluon coupling, we see that the $O(\alpha_s^3)$ three-gluon exchange diagram is suppressed with respect to the $O(\alpha_s^2)$ two-gluon exchange diagram, assuming $\alpha_s < 1$. However, we have to keep in mind that the masses of light mesons η, η' and ω, ϕ involved in the mixing lie in the interval 0.5–1.0 GeV; hence, characteristic momenta transferred via gluons in the diagrams in Figure 1.5(a),(b) should then be of $O(\Lambda_{QCD})$. In this energy-momentum region, as we already know, perturbative QCD diagrams consisting of vertices and propagators and having a certain power of α_s cannot be used.

Transitions (1.97) in light mesons are driven by fluctuations of the QCD vacuum which effectively correspond to absorbing a $q\bar{q}$ state and emitting a $q'\bar{q}'$ state. A schematic view of such transition is shown in Figure 1.5(c). This nonperturbative effect was reproduced, employing the instanton solutions of QCD equations of motion (see, e.g., [23]). In this framework, the amplitudes of the transitions

$$\frac{1}{\sqrt{2}}|u\bar{u} + d\bar{d}\rangle \leftrightarrow |s\bar{s}\rangle \tag{1.98}$$

are – in full agreement with experimental observations – comparatively large in the pseudoscalar ($J^P = 0^-$) and scalar ($J^P = 0^+$) mesons and small in the vector ($J^P = 1^-$) and tensor ($J^P = 2^+$) mesons. In the latter case, this is contrary to our expectation based on perturbative diagrams.

To describe the mixing of isoscalar mesons in quantitative terms, we consider the set of two states with a definite flavor content:

$$|f_1\rangle \equiv \frac{|u\bar{u} + d\bar{d}\rangle}{\sqrt{2}},$$
$$|f_2\rangle \equiv |s\bar{s}\rangle. \tag{1.99}$$

Sandwiching the QCD energy operator $\hat{P}_0 = \hat{H}$ between these states we encounter both diagonal and nondiagonal matrix elements, forming a 2×2 mass matrix:

$$\langle f_i|\hat{H}|f_k\rangle \equiv \mathcal{M}_{ik} = \begin{pmatrix} \mu_1 & \Delta \\ \Delta & \mu_2 \end{pmatrix}, \quad (i, k = 1, 2). \tag{1.100}$$

The nondiagonal elements of this matrix originate only from the part of $\hat{H}$ that corresponds to the effective interaction in Figure 1.5(c), whereas the diagonal elements include also the remaining QCD interactions without flavor-state transitions as well as the quark mass terms. The calculation of the matrix elements $\mathcal{M}_{ik}$ demands nonperturbative QCD methods. We only notice that the mass inequality $m_s > m_{u,d}$ implies that $\mu_2 > \mu_1$. Since the matrix is nondiagonal, the states (1.99) cannot represent the observed isoscalar mesons with definite masses. We should find their orthogonal superpositions

$$|M_k\rangle = \mathcal{U}_{k\ell}|f_\ell\rangle, \quad \langle M_i| = \langle f_\ell|\mathcal{U}_{\ell i}^T, \tag{1.101}$$

that diagonalize the mass matrix:

$$\langle M_i|\hat{H}|M_k\rangle = \begin{pmatrix} m_1 & 0 \\ 0 & m_2 \end{pmatrix}, \tag{1.102}$$

so that $|M_1\rangle$ and $|M_2\rangle$ are the meson states with the masses m_1 and m_2. The transformation (1.101) involves a 2×2 orthogonal matrix $\mathcal{U}$, with $\mathcal{U}^T = \mathcal{U}^{-1}$. It can be written as

$$\mathcal{U} = \begin{pmatrix} \cos\alpha & -\sin\alpha \\ \sin\alpha & \cos\alpha \end{pmatrix}, \tag{1.103}$$

with a single real parameter – the mixing angle α. Substituting the states (1.101) in (1.102) and using (1.100), we have:

$$\langle f_n | \mathcal{U}_{ni}^T \hat{H} \mathcal{U}_{k\ell} | f_\ell \rangle = \mathcal{U}_{ni}^T \langle f_n | \hat{H} | f_\ell \rangle \mathcal{U}_{k\ell}$$

$$= \mathcal{U}_{ni}^T \mathcal{M}_{n\ell} \mathcal{U}_{k\ell} = \mathcal{U} \mathcal{M} \mathcal{U}^T = \begin{pmatrix} m_1 & 0 \\ 0 & m_2 \end{pmatrix}. \tag{1.104}$$

Substituting the parameterization (1.103) in this matrix relation, we obtain three equations expressing the masses $m_{1,2}$ and the mixing angle α via matrix elements (1.100):

$$\tan 2\alpha = \frac{2\Delta}{\mu_2 - \mu_1},$$

$$m_{1,2} = \frac{1}{2} \left(\mu_1 + \mu_2 \mp \sqrt{(\mu_2 - \mu_1)^2 + 4\Delta^2} \right). \tag{1.105}$$

The same mixing angle enters the valence content of isoscalar mesons:

$$|M_1\rangle = \cos\alpha \frac{|u\bar{u} + d\bar{d}\rangle}{\sqrt{2}} - \sin\alpha |s\bar{s}\rangle,$$

$$|M_2\rangle = \sin\alpha \frac{|u\bar{u} + d\bar{d}\rangle}{\sqrt{2}} + \cos\alpha |s\bar{s}\rangle. \tag{1.106}$$

For pseudoscalar mesons η and η', the parameter Δ determining the transition between flavor states is larger than the difference $\mu_2 - \mu_1$, explaining why the mixing angle in (1.86) is close to $\alpha = 40°$ [1]. In fact, mixing in the pseudoscalar channel only slightly deviates from prediction of the exact $SU(3)_{fl}$ symmetry, where η is identified with η_8, a member of the octet of pseudoscalar mesons, which includes also the isotriplet of pions and two isodoublets of kaons. Within this approximation, η' coincides with the singlet state η_1:

$$\eta_8 = \frac{|u\bar{u} + d\bar{d} - 2s\bar{s}\rangle}{\sqrt{6}}, \quad \eta_1 = \frac{|u\bar{u} + d\bar{d} + s\bar{s}\rangle}{\sqrt{3}}. \tag{1.107}$$

The opposite situation, typical for the vector mesons ω and ϕ, is when $\Delta \to 0$ and the states in the flavor basis almost do not mix with each other.

1.7.8 Unstable hadrons

The lightest pseudoscalar mesons and baryons with flavor quantum numbers (isospin, strangeness, charm, bottomness or their combinations) are stable with respect to *strong decays* mediated by the flavor conserving quark-gluon interactions. The initial and final states formed by these hadrons serve as the stable asymptotic states (1.80). In reality, only one of these hadronic states is genuinely stable, that is, the proton, with a lifetime exceeding 2.1×10^{21} years [1][15]. All other lightest flavored hadrons decay via weak interactions involving the quark currents (1.31). Nonetheless, considering them stable is a very good approximation because their average lifetimes are much larger than the time scale typical for QCD processes:

$$\left(10^{-12}\,\text{s} \lesssim \tau_{weak} \lesssim 10^{-8}\,\text{s} \right) \gg 1/\Lambda_{QCD} \sim 10^{-24}\,\text{s}, \tag{1.108}$$

where the lower (upper) limit of the above interval of τ_{weak} corresponds to the charmed and bottom hadrons (to the charged pion). Moreover, weakly decaying hadrons – even those

---

[15]Stability of the proton is protected by the baryon number conservation.

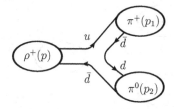

Figure 1.6 Diagram of $\rho \to \pi\pi$ decay.

with the shortest lifetimes – are directly reconstructed in the particle detectors in which either their traces or the paths between their production and decay vertices are observed.

The neutral light-unflavored mesons decay via flavor-conserving e.m. or quark-gluon interactions and have considerably smaller average lifetimes than the interval in (1.108). E.g., the neutral pion decays into two photons with a lifetime $\sim 10^{-16}\,s$. The η-meson and all heavier neutral light-unflavored mesons are already unstable with respect to QCD interactions yielding their strong decays with stable mesons in the final state[16]. Detailed information on these decays can be found in [1]. For *unstable hadrons* it is customary to use the total width $\Gamma_{tot} = 1/\tau$ instead of the lifetime τ. It is already not possible to reconstruct the electromagnetically or strongly decaying hadrons "directly", e.g., using vertex detectors. Still, π^0, η, η' are usually treated as asymptotic states, provided the inequality

$$\Gamma_{tot} \ll \Lambda_{QCD} \sim 300\,\mathrm{MeV} \qquad (1.109)$$

is valid for these mesons. In general, dominant decay modes of unstable hadrons are determined by QCD dynamics, symmetries and by the available phase space, with the values of total widths spreading from $O(100\,\mathrm{keV})$ to $O(\Lambda_{QCD})$. All orbital and radial excitations of lightest hadrons are unstable and decay strongly. Their total width steadily grows with the hadron mass, reflecting the increasing phase space for the final state hadrons.

To have an idea about the mechanism of strong decays, let us consider the light-unflavored vector mesons $\rho(770), \omega(782)$ and $\phi(1020)$. The ρ meson with almost 100 % probability decays into a pair of pions, having quite a large total width $\Gamma_{tot}^{\rho} \simeq 150$ MeV. The diagram of the strong decay mode $\rho^+ \to \pi^+\pi^0$ is shown in Figure 1.6, where the initial and final hadrons are schematically represented by blobs and only the valence quark and antiquark lines are depicted. One can interpret this process as a spontaneous emission of a light quark-antiquark pair from the QCD vacuum which then recombines with the initial quark-antiquark pair and forms the final two-pion state. Note that the angular momentum of the pions is fixed, $L = 1$. The two decay modes $\rho^{\pm} \to \pi^{\pm}\pi^0$ and $\rho^0 \to \pi^+\pi^-$ are related by isospin symmetry, whereas $\rho^0 \to \pi^0\pi^0$ is forbidden by C-parity conservation. Since the characteristic momenta are of $O(m_{\rho}) \sim O(\Lambda_{QCD})$, the $\rho \to 2\pi$ transitions are essentially nonperturbative; hence, no gluon exchange diagrams and/or orders of the quark-gluon coupling α_s can be attributed to the amplitude of this decay. Note that an inverse process is also possible in the partial P-wave of the pion-pion scattering. If the total c.m. energy reaches m_{ρ}, the ρ-meson intermediate state is produced and, during an average time interval of $O(1/\Gamma_{tot}^{\rho})$, it decays back into two pions. In Chapter 5 we will consider the $\rho \to 2\pi$ amplitude in more detail.

The ω meson decays most frequently into three pions and has a much smaller total width, $\Gamma_{tot}^{\omega} \simeq 8.5$ MeV. The reason is twofold: G-parity conservation forbids the decay of an isosinglet vector meson into two pions and, in addition, the phase space of $\omega \to 3\pi$ is substantially smaller than the phase space of $\rho \to 2\pi$.

---

[16]Note that the dominant hadronic decay modes $\eta \to 3\pi$ involve isospin symmetry violation and are therefore suppressed.

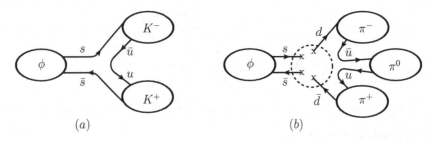

Figure 1.7 Diagrams of (a) $\phi \to K\bar{K}$ and (b) $\phi \to 3\pi$ decays.

Figure 1.8 Gluon annihilation of (a) pseudoscalar and (b) vector charmonia. The corresponding b-quarkonia have the same annihilation diagrams, obtained replacing $\eta_c \to \eta_b$, $J/\psi \to \Upsilon$.

Less simple is the decay pattern of the ϕ meson with a total width $\Gamma^\phi_{tot} \simeq 4.2$ MeV. It predominantly decays into two kaons as shown in Figure 1.7(a). This decay mode has a phase space suppressed with respect to the three-pion decay mode. However, the latter amounts only up to $\sim 15\%$ of the total width. Note that $\phi \to 3\pi$ is not forbidden by any symmetry, but demands a different mechanism of strong decay shown in Figure 1.7(b): The valence $s\bar{s}$ state of ϕ meson transforms via the intermediate gluonic state into a nonstrange quark-antiquark pair which then converts into the three-pion state. This nonperturbative transition is similar to the mixing of isoscalar mesons discussed in the previous subsection and schematically shown in Figure 1.5(c). The $s\bar{s} \leftrightarrow \{u\bar{u}, d\bar{d}\}$ transitions are suppressed in the $J^P = 1^-$ channel, hence it is natural that the amplitudes of ϕ decay to nonstrange final states are also suppressed[17]:

$$\mathcal{A}(\phi \to K\bar{K}) \gg \mathcal{A}(\phi \to 3\pi), \tag{1.110}$$

explaining at least qualitatively the measured decay widths.

Our discussion on unstable hadrons is incomplete without considering the strong decays of $c\bar{c}$ and $b\bar{b}$ mesons (heavy quarkonia). In contrast to their very large masses and plenty of strong decay channels listed in [1], these mesons have quite small total widths. Moreover, the total widths of the lowest vector ($J^{PC} = 1^{--}$) quarkonia, $\Gamma^{J/\psi}_{tot} \simeq 93$ keV and $\Gamma^{\Upsilon(1S)}_{tot} \simeq 54$ keV, are systematically smaller than that of the pseudoscalar ($J^{PC} = 0^{-+}$) quarkonia: $\Gamma^{\eta_c(1S)}_{tot} \simeq 32$ MeV and $\Gamma^{\eta_b(1S)}_{tot} \simeq 10$ MeV, respectively. The small widths of quarkonia have a natural explanation in QCD[18]. A transition of quarkonium into light hadrons consisting of u, d, s quarks takes place via annihilation into gluons. Due to a large mass scale ($2m_c$ or $2m_b$), there are two basic differences between this process and the ϕ-meson decay into

---

[17]This suppression is known in hadron phenomenology as the Okubo-Zweig-Iizuka (OZI) rule.

[18]In fact, their measurement was one of the first manifestations of the asymptotic fall-off of the quark-gluon coupling at large momentum scales.

nonstrange hadrons considered above. First, perturbative QCD is applicable in the quarkonium annihilation process, hence we can use the diagrams shown in Figure 1.8. Second, the gluons emitted after annihilation are very energetic and remain quasifree before eventually converting into one of the multiple hadronic final states with unit probability. In this approximation known as the *parton model*, the total width of the quarkonium is equal to the total width of its purely gluonic decay.

Let us clarify why the heavy quarkonium annihilation takes place in the perturbative domain of QCD. The quarkonium mass is roughly equal to $2m_Q + O(\Lambda_{QCD})$, where $Q = c, b$; hence, to a certain accuracy, we can neglect the $O(\Lambda_{QCD})$ binding energy and approximate quarkonium in the rest frame by a state consisting of a free heavy quark and antiquark with the momenta:

$$p_Q \simeq p_{\bar{Q}} \simeq (m_Q, \vec{0}).$$

Consider as an example the two-gluon annihilation of the pseudoscalar quarkonium shown in Figure 1.8(a). In the rest frame of η_Q, the gluons originating from the annihilation have momenta

$$k_{1,2} \simeq (m_Q, \pm \vec{k}), \quad |\vec{k}| \sim m_Q,$$

where we also neglect small gluon virtualities, $|k^2| \sim O(\Lambda_{QCD}^2)$. Applying momentum conservation in the quark-gluon vertex we obtain (in the adopted approximation) the momentum squared of the virtual heavy quark flowing between the gluon emission points:

$$f^2 = (p_Q - k_1)^2 = m_Q^2 - 2p_Q \cdot k_1 + k_1^2 \simeq m_Q^2 - 2m_Q k_{10} \simeq -m_Q^2. \tag{1.111}$$

Evidently, the virtuality of that quark counted as a deviation from its "mass shell", $f^2 \simeq m_Q^2$, is of order of $-2m_Q^2$, large enough to justify the use of the perturbative diagram with a quark-gluon coupling at a scale of $O(m_Q) \gg \Lambda_{QCD}$. Then also the difference between the total widths of vector and pseudoscalar quarkonia becomes apparent: a vector quarkonium ($J/\psi(1S)$ or Υ) annihilates into three gluons, via the diagram in Figure 1.8(b) which has an extra $\alpha_s(m_Q)$ in the resulting width. In reality, the pseudoscalar quarkonium has a total width larger than the $O(\alpha_s^2)$ value estimated from the two-gluon annihilation. Presumably, in the $J^P = 0^-$ channel certain nonperturbative effects are also partially contributing, or in other words, the parton model is less accurate. Note that while the gluon annihilation widths determine the total widths of heavy quarkonia, their separate (exclusive) strong decay modes represent a complicated interplay of annihilation and subsequent conversion of gluons into a definite hadronic state. A detailed discussion of heavy quarkonium annihilation can be found in the review [24].

Concluding, one has to mention that unstable hadrons with total widths in the ballpark of Λ_{QCD}, such as the ρ meson, cannot, strictly speaking, serve as asymptotic initial or final states in the interaction processes. Indeed, for an unstable hadron, the average time interval of its formation overlaps with the lifetime of its decay into stable hadrons. Experimentally, unstable hadrons manifest themselves as resonance enhancements in the cross sections and differential decay widths. We will return to the origin of hadronic resonances and their total widths in Chapter 5.

1.8 HEAVY QUARK SYMMETRIES AND EFFECTIVE THEORY

The masses of the heavy b and c quarks are considerably larger than the nonperturbative QCD scale:

$$m_b \gg m_c \gg \Lambda_{QCD}. \tag{1.112}$$

To study the consequences of these inequalities, we separate heavy quarks[19] from light quarks and gluons in the QCD Lagrangian:

$$\mathcal{L}_{QCD}(x) = \sum_{Q=b,c} \bar{Q}(x) \left(i D_\mu \gamma^\mu - m_Q \right) Q(x) + \mathcal{L}_{u,d,s}(x) + \mathcal{L}_G(x), \qquad (1.113)$$

where, as usual, the color and Dirac indices are not shown explicitly. The first inequality in (1.112) clearly indicates that no additional flavor symmetry appears in $\mathcal{L}_{QCD}(x)$ if one unifies c and b quarks in a doublet. On the other hand, due to the second inequality in (1.112), it is conceivable to take the limit of QCD in which both quarks are infinitely heavy:

$$m_c \to \infty, \quad m_b \to \infty. \qquad (1.114)$$

As we shall see, in this limit nontrivial symmetries involving b-flavored and charmed hadrons emerge.

To proceed, we somewhat artificially leave only one heavy flavor in the QCD Lagrangian:

$$\mathcal{L}_{QCD}^{(Q)}(x) = \bar{Q}(x) \left(i D_\mu \gamma^\mu - m_Q \right) Q(x) + \mathcal{L}_{u,d,s}(x) + \mathcal{L}_G(x), \qquad (1.115)$$

so that the quark field Q represents either the b or c quark. We also use a generic notation H for the lightest Q-flavored pseudoscalar meson, that is, H is either a $\bar{B}$ or D meson. The key argument is that $\mathcal{L}_{QCD}^{(Q)}$ is sufficient to describe hadrons which contain a single heavy quark Q or antiquark $\bar{Q}$. To see that, let us consider a decomposition of the H-meson state in terms of valence and nonvalence constituents, similar to (1.79):

$$|H\rangle = |Q\bar{q}\rangle \oplus |Q\bar{q}G\rangle \oplus \sum_{q'=u,d,s} |Q\bar{q}q'\bar{q}'\rangle \oplus \ldots, \qquad (1.116)$$

where $q, q' = u, d, s$ and for brevity we do not specify momenta. Due to flavor conservation, an additional heavy flavor Q' can only appear as a nonvalence $Q'\bar{Q}'$-component in the above decomposition. Since Q'-quark is also heavy, $|Q\bar{q}Q'\bar{Q}'\rangle$ is a rare virtual fluctuation within H. More detailed estimates of its contribution to (1.116) reveal that it is suppressed either by a power of $1/(2m_{Q'})$ or by a small quark-gluon coupling $\alpha_s(m_{Q'})$; hence, for example, the $c\bar{c}$ fluctuations inside the B-meson can be neglected to a good approximation.

Our goal is to transform the Lagrangian $\mathcal{L}_{QCD}^{(Q)}$ to its $m_Q \to \infty$ limit, resulting in a QCD-based heavy-quark effective theory (HQET) which involves a systematic expansion of hadronic states and quark currents in the inverse powers of $1/m_Q$. Before doing that, let us explain the physical meaning of the heavy quark limit and related symmetries. To this end we consider an H-meson with the mass m_H and momentum p_H and introduce the corresponding velocity four-vector:

$$v_\mu = \frac{p_{H\mu}}{m_H}. \qquad (1.117)$$

In the rest frame of the H-meson, $v = (1, \vec{0})$ and the heavy quark inside H is also almost at rest. Defining

$$p_H = m_Q v + k, \qquad (1.118)$$

we separate a parametrically large fraction $m_Q v$ of the meson momentum, carried by the heavy quark, from the subdominant fraction k. The latter adds up contributions of the light constituents of H and the small residual momentum of the heavy quark. The light "cloud" surrounding Q inside H includes the valence light antiquark and all nonvalence quark-antiquark and gluon fluctuations indicated in (1.116). Hadronic dynamics is dominated by

---

[19]We omit the superheavy t-quark which does not play a role in the following discussion.

nonperturbative quark-gluon interactions, hence the average values of the k_μ components are functions of Λ_{QCD}. In other words, varying m_Q from one large value $m_Q \gg \Lambda_{QCD}$ to another does not significantly alter k. This is reflected in the scaling relation for the mass of the H meson which follows from (1.118):

$$m_H = \sqrt{p_H^2} = \sqrt{(m_Q v + k)^2} = m_Q + \bar\Lambda + \dots, \qquad (1.119)$$

where the parameter $\bar\Lambda = v \cdot k$ equal to the residual energy in the rest frame of H does not scale with m_Q. On the r.h.s. of this relation we retain only the first term of the expansion in $1/m_Q$, and the rest, starting from terms of $O(\Lambda_{QCD}^2/m_Q)$, is indicated by ellipsis[20].

The fact that the parameter $\bar\Lambda$ does not depend on the heavy quark mass m_Q is first of all caused by the flavor independence of the quark-gluon interactions. The heavy quark in H serves as a static point source of the color charge. Replacing $Q \to Q'$ does not affect the value of $\bar\Lambda$ because the color charge of both heavy quarks is the same. In fact, the situation closely resembles that of atomic electrons. Their energy levels are approximately the same if one compares two atoms in which the nuclei have the same electric charge but different masses, e.g., proton and deuteron.

The approximate equality of binding energies in bottom and charmed hadrons – named *heavy-quark flavor symmetry* – is revealed in a very simple manner. Applying the relation (1.119) to both B and D mesons and taking their masses from [1] we obtain in the first approximation:

$$\bar\Lambda \simeq m_B - m_b = 5.279\,\text{GeV} - 4.67\,\text{GeV} \simeq 0.60\,\text{GeV},$$
$$\bar\Lambda \simeq m_D - m_c = 1.869\,\text{GeV} - 1.03\,\text{GeV} \simeq 0.84\,\text{GeV}, \qquad (1.120)$$

where a simplifying assumption is made, adopting m_b and m_c quoted in (1.20) and normalizing both masses at a uniform scale $\mu = 2.5$ GeV which is roughly an average of the reference scales used in the quark mass determination. The agreement between the two values of the parameter $\bar\Lambda$ is not very convincing, but at least their difference has a correct sign, hinting at the missing inverse heavy-mass correction, which should be larger for the D meson than for the B meson. Indeed, let us add this term to the expansion (1.119):

$$m_H = m_Q + \bar\Lambda + \frac{\bar\Lambda^2}{m_Q}. \qquad (1.121)$$

Considering it as an equation with respect to the parameter $\bar\Lambda$ and solving it first for the B meson and then for the D meson, using the same values of quark masses as in (1.120) as an input, in both cases we obtain the same value $\bar\Lambda \simeq 550$ MeV.

Furthermore, not only changing the mass of a heavy quark but also flipping its spin does not influence the binding energy $\bar\Lambda$. This leads to the *heavy-quark spin symmetry* in the limit $m_Q \to \infty$. This symmetry is manifested e.g., by the smallness of the mass differences between vector and pseudoscalar heavy mesons. Note that these two mesons, H^* and H, differ only by the spin orientation of the heavy quark with respect to the spin of the valence light antiquark. Taking the measured masses [1], we indeed find quite small differences, certainly not related to the b or c mass scales:

$$m_{B^*} - m_B \simeq 45\,\text{MeV}, \quad m_{D^*} - m_D \simeq 140\,\text{MeV}, \qquad (1.122)$$

where again a larger "hyperfine splitting" for charmed mesons can be explained by a larger inverse heavy-mass correction.

---

[20]Note that a practical use of the $1/m_Q$ expansion demands a definition of the scale-dependent heavy quark mass within a certain renormalization scheme. These important details remain out of our scope.

The above discussion reveals a physically motivated way to transform the Lagrangian $\mathcal{L}_{QCD}^{(Q)}$ to the $m_Q \to \infty$ limit. First, one should separate the component of the heavy quark field characterized by the large scale m_Q. The residual components of that field characterized by momenta of $O(\Lambda_{QCD})$, together with the light quarks and gluons in $\mathcal{L}_{QCD}^{(Q)}$, are the essential degrees of freedom for the heavy hadron dynamics. The separation is done assuming that the fields entering $\mathcal{L}_{QCD}^{(Q)}$ are "embedded" within a Q-flavored heavy hadron which has a certain velocity vector v_μ. The starting point is to decompose the heavy quark field in two bispinor components defined with the help of two orthogonal projection operators

$$Q(x) = \frac{1+\slashed{v}}{2}Q(x) + \frac{1-\slashed{v}}{2}Q(x) \equiv \phi_v(x) + \chi_v(x). \tag{1.123}$$

For the Dirac-conjugated field – using the relations for γ-matrices from Appendix A – we obtain:

$$\bar{Q}(x) = \bar{Q}(x)\frac{1+\slashed{v}}{2} + \bar{Q}(x)\frac{1-\slashed{v}}{2} \equiv \bar{\phi}_v(x) + \bar{\chi}_v(x), \tag{1.124}$$

so that both components are eigenstates of the velocity operator in Dirac space

$$\slashed{v}\phi_v = \phi_v, \quad \slashed{v}\chi_v = -\chi_v, \tag{1.125}$$

and the nondiagonal products of these bispinors vanish:

$$\bar{\phi}_v\chi_v = \bar{\chi}_v\phi_v = 0. \tag{1.126}$$

Furthermore, it is convenient to decompose the covariant derivative in the longitudinal and transverse components with respect to the velocity vector:

$$D_\mu = v_\mu(v \cdot D) + (g_{\mu\nu} - v_\mu v_\nu)D^\nu \equiv v_\mu(v \cdot D) + D_\mu^\perp, \tag{1.127}$$

so that $D^\perp \cdot v = 0$. E.g., in the rest frame $v_\mu = (1, \vec{0})$ and $D_\mu^\perp = (0, -\vec{D})$. Multiplying the anticommutator of γ-matrices by both vectors, we obtain:

$$\{\slashed{D}^\perp, \slashed{v}\} = 0. \tag{1.128}$$

To proceed further, we substitute the decompositions (1.123) and (1.124) in the Lagrangian (1.115) and, using the above properties of the ϕ_v and χ_v fields, obtain:

$$\mathcal{L}_{QCD}^{(Q)}(x) = \bar{\phi}_v(x)[i(v \cdot D) - m_Q]\phi_v(x) - \bar{\chi}_v(x)[i(v \cdot D) + m_Q]\chi_v(x)$$
$$+ \bar{\phi}_v(x)i\slashed{D}^\perp\chi_v(x) + \bar{\chi}_v(x)i\slashed{D}^\perp\phi_v(x) + \mathcal{L}_{u,d,s}(x) + \mathcal{L}_G(x). \tag{1.129}$$

Up to this point we only have reparameterized the Lagrangian. A crucial next step involves a redefinition of the fields ϕ_v and χ_v, factoring out a phase involving the large component $m_Q v$ of the heavy quark momentum:

$$\phi_v(x) = e^{-im_Q(v \cdot x)}h_v(x), \quad \chi_v(x) = e^{-im_Q(v \cdot x)}H_v(x). \tag{1.130}$$

Note that the new fields are still eigenstates of $\slashed{v}$:

$$\slashed{v}h_v(x) = h_v(x), \quad \slashed{v}H_v(x) = -H_v(x). \tag{1.131}$$

Inserting the definition (1.130) in (1.129) yields:

$$\mathcal{L}_{QCD}^{(Q)}(x) = \bar{h}_v(x)i(v \cdot D)h_v(x) - \bar{H}_v(x)[i(v \cdot D) + 2m_Q]H_v(x)$$
$$+ \bar{h}_v(x)i\slashed{D}^\perp H_v(x) + \bar{H}_v(x)i\slashed{D}^\perp h_v(x) + \mathcal{L}_{u,d,s}(x) + \mathcal{L}_G(x). \tag{1.132}$$

This formula tells us that h_v describes the massless part of a heavy quark field, whereas H_v involves $\bar{Q}Q$ fluctuations and thereby represents a suppressed part.

The field H_v is related to h_v via equations of motion following from $\mathcal{L}_{QCD}^{(Q)}(x)$. The simplest way to establish this relation is to start from Dirac equation for the heavy quark field:

$$(i\not{D} - m_Q)Q(x) = 0\,, \tag{1.133}$$

which in terms of the new fields reads:

$$i\not{D}h_v(x) + (i\not{D} - 2m_Q)H_v(x) = 0\,. \tag{1.134}$$

Multiplying the above equation with the two projection operators: $(1+\not{v})/2$ and $(1-\not{v})/2$ and using (1.128) and (1.131), we finally obtain a system of two coupled equations:

$$i(v \cdot D)h_v(x) = -i\not{D}^{\perp}H_v(x)\,, \tag{1.135}$$

and

$$(i(v \cdot D) + 2m_Q)H_v(x) = i\not{D}^{\perp}h_v(x)\,. \tag{1.136}$$

The latter one allows us to express the field $H_v(x)$ via derivatives of $h_v(x)$:

$$H_v(x) = \left(\frac{1}{i(v \cdot D) + 2m_Q}\right) i\not{D}^{\perp}h_v(x)\,. \tag{1.137}$$

Note that the operator acting on $h_v(x)$ is nonlocal because it involves an infinite series of derivatives after expanding the denominator in powers of $1/m_Q$. Substituting the above equation in (1.135) we eliminate the field H_v and obtain the equation of motion for the field h_v:

$$i(v \cdot D)h_v(x) = -i\not{D}^{\perp}\left(\frac{1}{i(v \cdot D) + 2m_Q}\right) i\not{D}^{\perp}h_v(x)\,. \tag{1.138}$$

This equation of motion corresponds to the Lagrangian (1.132) in which the same substitution (1.137) is done, resulting in a compact, albeit nonlocal form of the HQET Lagrangian:

$$\mathcal{L}_{HQET}^{(Q)}(x) = \bar{h}_v(x)i(v \cdot D)h_v(x)$$
$$+\bar{h}_v(x)i\not{D}^{\perp}\left(\frac{1}{i(v \cdot D) + 2m_Q}\right) i\not{D}^{\perp}h_v(x) + \mathcal{L}_{u,d,s}(x) + \mathcal{L}_G(x)\,. \tag{1.139}$$

This expression is used to describe the Q-flavored hadrons and their decay processes in terms of the effective spin $1/2$ field $h_v(x)$ together with the light quark and gluon fields. Expanding (1.139) in powers of $1/m_Q$, one generates a set of corrections to the leading, $m_Q \to \infty$ term.

Treated as a quantum field theory, HQET generates Feynman diagrams with propagating effective h_v-fields and vertices of h_v-gluon interactions derived from the Lagrangian (1.139). The propagator of the effective quark field is most easily obtained from the massive free-quark propagator in the momentum space (see (A.49)):

$$S_{(Q)}^{(0)}(p_Q, m_Q) = \frac{\not{p}_Q + m_Q}{p_Q^2 - m_Q^2}\,, \tag{1.140}$$

if we decompose the momentum p_Q of the (virtual) effective quark into the static and residual parts:

$$p_Q^\mu = m_Q v^\mu + \tilde{k}^\mu\,,$$

and take the limit $m_Q \to \infty$:

$$\lim_{m_Q \to \infty} S^{(0)}_{(Q)}(p_Q, m_Q) = \lim_{m_Q \to \infty} \frac{m_Q \not{v} + \tilde{\not{k}} + m_Q}{m_Q^2 + 2m_Q v \cdot \tilde{k} + \tilde{k}^2 - m_Q^2}$$

$$= \left(\frac{1}{v \cdot \tilde{k}}\right) \frac{1 + \not{v}}{2} + O\left(\frac{\tilde{k}}{m_Q}\right). \qquad (1.141)$$

Furthermore, interaction of the field h_v with the gluon field in the HQET Lagrangian (1.139) has a form

$$\bar{h}_v(x) \left(g_s \frac{\lambda^a}{2} v^\mu A^a_\mu(x)\right) h_v(x). \qquad (1.142)$$

Within effective theory, the velocity vector (1.117) serves as a signature of an H-hadron state, replacing the four-momentum p_H. It is then more convenient to use, instead of (1.81), a different normalization of states:

$$\langle H(v')|H(v)\rangle = 2\frac{p_{H0}}{m_H}(2\pi)^3 \delta^{(3)}(\vec{p}_H' - \vec{p}_H), \qquad (1.143)$$

which implies the relation

$$|H(v)\rangle = \frac{1}{\sqrt{m_H}}|H(p_H)\rangle. \qquad (1.144)$$

Considering e.m. or weak transitions between heavy hadrons, we will also need quark currents expressed in terms of effective fields. For example, the heavy quark e.m. current (taken for simplicity at $x = 0$) has the following HQET expression:

$$j^{em(Q)}_\mu = \bar{Q}\gamma_\mu Q \xrightarrow[m_Q \to \infty]{} \bar{h}_v \gamma_\mu h_v, \qquad (1.145)$$

neglecting the $O(1/m_Q)$ corrections and assuming equal velocities of both fields. Using (1.131) for the field h_v and the corresponding equation for the conjugated field

$$\bar{h}_v \not{v} = \bar{h}_v,$$

we derive an alternative form of this current:

$$\bar{h}_v \gamma_\mu h_v = \bar{h}_v \gamma_\mu \not{v} h_v = \bar{h}_v[-\not{v}\gamma_\mu + 2v_\mu]h_v = -\bar{h}_v \gamma_\mu h_v + 2\bar{h}_v v_\mu h_v, \qquad (1.146)$$

hence

$$\bar{h}_v \gamma_\mu h_v = \bar{h}_v v_\mu h_v. \qquad (1.147)$$

Importantly, we can use the heavy-quark flavor symmetry and replace in the above current the flavor $h_v \to h'_v$, yielding the HQET limit of the heavy-to-heavy ($b \to c$) weak vector current:

$$j^{(Q \to Q')}_\mu = \bar{Q}'\gamma_\mu Q \xrightarrow[m_Q \to \infty]{} \bar{h}'_v \gamma_\mu h_v = \bar{h}'_v v_\mu h_v. \qquad (1.148)$$

Both HQET currents taken at equal velocities are conserved:

$$\partial_\mu(\bar{h}_v v^\mu h_v) = \partial_\mu(\bar{h}'_v v^\mu h_v) = 0. \qquad (1.149)$$

To see that, e.g., for the weak current, we convert the simple derivatives into covariant ones and use the equation of motion (1.138) for h_v in $m_Q \to \infty$ limit:

$$\partial_\mu(\bar{h}'_v v^\mu h_v) = \bar{h}'_v v^\mu \partial_\mu h_v + \bar{h}'_v \overleftarrow{\partial}_\mu v^\mu h_v$$

$$= \bar{h}'_v v^\mu (\partial_\mu - ig_s \frac{\lambda^a}{2} A^a_\mu)h_v + \bar{h}'_v(\overleftarrow{\partial}_\mu + ig_s \frac{\lambda^a}{2} A^a_\mu)v^\mu h_v$$

$$= \bar{h}'_v(v \cdot D)h_v + \bar{h}'_v(\overleftarrow{D} \cdot v)h_v = 0. \qquad (1.150)$$

The relations between quark currents in full QCD and HQET discussed above are modified by gluon radiative corrections emerging already at leading order of the heavy mass expansion. In particular, for the heavy-light weak vector current

$$j_\mu^{(Q \to q)} = \bar{q} \gamma_\mu Q \xrightarrow[m_Q \to \infty]{} \bar{q} \gamma_\mu h_v \quad (q = u, d, s), \tag{1.151}$$

radiative corrections modify the effective current:

$$\bar{q} \gamma_\mu h_v \longrightarrow C_\gamma \bar{q} \gamma_\mu h_v + C_v \bar{q} v_\mu h_v \,,$$

inducing the second operator absent at the tree level. The calculational procedure involves Feynman diagrams similar to the ones in Figure 1.4. Their expressions in full QCD are matched to the same diagrams with HQET vertices and propagators. The first-order in α_s results for the coefficients are (see, e.g., [25])

$$C_\gamma = 1 + \frac{\alpha_s}{\pi} \left(\ln \frac{m_Q}{\mu} - \frac{4}{3} \right), \quad C_v = \frac{2\alpha_s}{3\pi} \,. \tag{1.152}$$

Further details on heavy quark symmetries, heavy quark expansion and HQET can be found in the reviews [25, 26, 27, 28] and books [29, 30].

CHAPTER 2

MESON FORM FACTORS

2.1 HADRONIC MATRIX ELEMENT OF THE ELECTRON-PION SCATTERING

We begin with explaining how the pion electromagnetic (e.m.) form factor emerges. It will then serve as a study case throughout this book.

One possible process where the pion form factor contributes to the amplitude is the elastic scattering of an electron on a charged pion[1]

$$e^-(k_1)\pi^+(p_1) \to e^-(k_2)\pi^+(p_2)\,, \tag{2.1}$$

where we assign definite momenta to the initial and final state particles, so that the momentum transfer between electrons and pions is

$$q = k_1 - k_2 = p_2 - p_1\,. \tag{2.2}$$

We also choose a positively charged pion for definiteness. The process (2.1) is a $2 \to 2$ scattering. At the fixed masses $k_{1,2}^2 = m_e^2$ and $p_{1,2}^2 = m_\pi^2$, the kinematics is determined by the two invariant variables:

$$s = (k_1 + p_1)^2 = (k_2 + p_2)^2 \text{ and } t = q^2\,. \tag{2.3}$$

In the electron-pion center-of-mass (c.m.) frame:

$$\vec{p}_1 = -\vec{k_1} \equiv \vec{p}, \quad \vec{p}_2 = -\vec{k_2} \equiv \vec{p}', \quad |\vec{p}| = |\vec{p}'|\,, \tag{2.4}$$

the variable $\sqrt{s}$ is equal to the total energy:

$$\sqrt{s} = \sqrt{m_e^2 + \vec{p}^2} + \sqrt{m_\pi^2 + \vec{p}^2}\,, \tag{2.5}$$

and q^2 is related to the scattering angle:

$$q^2 = -2\vec{p}^2(1 - \cos\theta_{cm})\,. \tag{2.6}$$

Solving (2.5), we express the 3-momentum in c.m. frame via s:

$$|\vec{p}| = \frac{\lambda^{1/2}(s, m_\pi^2, m_e^2)}{2\sqrt{s}}\,, \tag{2.7}$$

where the standard Källen function λ is defined in (B.44).

---

[1]We choose this process because of its relative simplicity. Experimentally it was realized only at small momentum transfers, employing an energetic beam of pions scattered on the atomic electrons, see, e.g., [31].

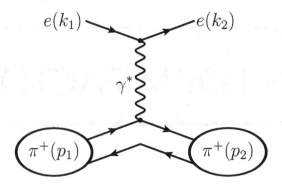

Figure 2.1 Diagram of electron-pion scattering with a virtual one-photon (denoted by γ^*) exchange.

From (2.5)–(2.7), the kinematically allowed region for the scattering process (2.1) is easily read off:

$$s \geq (m_\pi + m_e)^2 \quad \text{and} \quad -\frac{\lambda(s, m_\pi^2, m_e^2)}{s} \leq q^2 \leq 0. \tag{2.8}$$

We see that the squared momentum transfer is negative, approaching zero in the forward scattering limit. In other words, at $\theta_{cm} \neq 0$ the momentum transfer q is a spacelike four-vector. In what follows, we will often use the convenient notation

$$Q^2 = -q^2, \quad Q \equiv \sqrt{Q^2}.$$

The leading contribution to the $e\pi \to e\pi$ scattering amplitude is given by the e.m. interaction with a one-photon exchange between electron and pion. Higher-order corrections caused by the next-to-leading, two-photon exchanges are very small, being suppressed by an extra power of α_{em}. If Q^2 is not too large, the contribution of Z-boson exchange can also be neglected.

The diagram of the electron-pion scattering where the photon interacts with the valence u-quark is shown in Figure 2.1. An additional diagram with the photon coupled to the valence $\bar{d}$ antiquark is not shown. Note that we symbolize initial and final hadrons with ovals, displaying only the valence quark and antiquark lines. One has to emphasize that Figure 2.1 is not a conventional Feynman diagram. The quark-photon interaction vertex, being pointlike, is sandwiched between the initial and final pion states. In these hadronic states, in addition to the valence quark-antiquark component, virtual quark-antiquark pairs and gluons are perpetually and coherently created and annihilated with characteristic momenta of $O(\Lambda_{QCD})$, building up the nonperturbative internal structure of the pions. The quark lines in the quark-photon vertex represent an inseparable part of this structure.

Our next task is to isolate the hadronic part of the $e\pi \to e\pi$ scattering amplitude from the purely e.m. part involving electrons and intermediate photon. We start from the S-matrix element of the process (2.1):

$$S_{fi}^{(e\pi \to e\pi)} = \langle e^-(k_2)\pi^+(p_2)| \hat{S} |e^-(k_1)\pi^+(p_1)\rangle, \tag{2.9}$$

where the S-matrix is:

$$\hat{S} = T\left\{ \exp\left[i \int d^4x \left(\mathcal{L}_{QCD}^{int}(x) + \mathcal{L}_{QED}^{int}(x) \right) \right] \right\}, \tag{2.10}$$

(cf. (B.2)). In the above expression the QCD and QED interaction Lagrangians are taken into account, being the only relevant parts of SM for the process under consideration. We expand the QED part of the S-matrix to the second order in $\mathcal{L}_{QED}^{int}$, that is, to the $O(\alpha_{em})$:

$$
\hat{S} = T\Big\{ \exp\Big[i\int d^4x' \mathcal{L}_{QCD}^{int}(x')\Big]\Big[1 + i\int d^4x\, \mathcal{L}_{QED}^{int}(x)
$$
$$
+ \frac{i^2}{2!}\int d^4x \mathcal{L}_{QED}^{int}(x)\int d^4y \mathcal{L}_{QED}^{int}(y)\Big]\Big\}. \tag{2.11}
$$

Inserting the above expansion in the matrix element (2.9), we notice that the term of the first-order in $\mathcal{L}_{QED}^{int}$ does not contribute because it corresponds to the emission or absorption of a photon and, therefore, does not match the final and initial states in (2.1). The second-order term in $\mathcal{L}_{QED}^{int}$ in (2.11) is the one that generates the one-photon exchange we are interested in. Retaining only that term in (2.11) and applying the expression (1.24) for the interaction Lagrangian in QED, we obtain:

$$
S_{fi}^{(e\pi \to e\pi)} = \frac{i^2 e^2}{2!}\langle e^-(k_2)\pi^+(p_2)| T\Big\{ \exp\Big[i\int d^4x' \mathcal{L}_{QCD}^{int}(x')\Big]
$$
$$
\times \int d^4x \int d^4y \Big[\bar{e}(x)\gamma_\nu e(x)A^\nu(x)A^\mu(y)j_\mu^{em}(y)
$$
$$
+ j_\nu^{em}(x)A^\nu(x)A^\mu(y)\bar{e}(y)\gamma_\mu e(y)\Big]\Big\}|e^-(k_1)\pi^+(p_1)\rangle, \tag{2.12}
$$

where, in order to match the initial and final states, only those terms are retained in which the electron e.m. current is combined with the quark e.m. current j_μ^{em}. The QCD part of S-matrix in (2.12), represented by the exponential factor with $\mathcal{L}_{QCD}^{int}$, implicitly generates all possible quark-gluon interactions accompanying the photon absorption by the quark e.m. current. Hereafter, for brevity this part will not be shown explicitly, however its presence is implied in all matrix elements involving hadronic states.

Returning to (2.12), we notice that the first and second terms on r.h.s are equal, hence we consider only the first term. First of all, to account for the intermediate virtual photon, the T-product of e.m. field operators $A^\mu(x)$ and $A^\nu(y)$ in (2.12) is vacuum averaged, forming the propagator[2]:

$$
\langle 0|T\{A^\nu(x)A^\mu(y)\}|0\rangle = -i\int \frac{d^4k}{(2\pi)^4}\frac{g^{\nu\mu}}{k^2}e^{-ik(x-y)}. \tag{2.13}
$$

Importantly, at the adopted order in the e.m. coupling, we can represent the initial and final states in (2.12) as products of electron and pion states:

$$
|e^-(k_1)\pi^+(p_1)\rangle \rightarrow |e^-(k_1)\rangle \otimes |\pi^+(p_1)\rangle,
$$
$$
\langle e^-(k_2)\pi^+(p_2)| \rightarrow \langle e^-(k_2)| \otimes \langle \pi^+(p_2)|, \tag{2.14}
$$

thus neglecting e.m. interactions beyond the one-photon exchange. As a result, a separate matrix element of the electron e.m. current emerges multiplying the matrix element in which the quark e.m. current is sandwiched between the one-pion states $|\pi^+(p_1)\rangle$ and $\langle\pi^+(p_2)|$. This separation is one of the simplest examples of *factorization*. The purely leptonic matrix element is then easily calculated, employing the expansion of the electron Dirac fields in momentum components with creation and annihilation operators (see (A.21)):

$$
\langle e^-(k_2)|\bar{e}(x)\gamma_\nu e(x)|e^-(k_1)\rangle = \bar{u}_e(k_2)\gamma_\nu u_e(k_1)e^{-i(k_1-k_2)x}, \tag{2.15}
$$

---

[2]For definiteness, we use the Feynman gauge for the e.m. field, although, since both e.m. currents are conserved, the actual choice of gauge is irrelevant.

where $\bar{u}_e$ and u_e are the bispinors of the final and initial electrons. In addition, performing the translation of the quark e.m. current (see (A.78)), we obtain

$$\langle \pi^+(p_2)|j_\mu^{em}(y)|\pi^+(p_1)\rangle = \langle \pi^+(p_2)|j_\mu^{em}(0)|\pi^+(p_1)\rangle e^{-i(p_1-p_2)y} . \tag{2.16}$$

Transforming (2.12) with the help of equations (2.13)–(2.16), we integrate over the coordinates x and y and after that over the momentum q of the virtual photon:

$$\int \frac{d^4k}{(2\pi)^4 k^2} \underbrace{\int d^4x\, e^{-i(k_1-k_2)x-ikx}}_{(2\pi)^4\delta^{(4)}(k+k_1-k_2)} \underbrace{\int d^4y\, e^{-i(p_1-p_2)y+iky}}_{(2\pi)^4\delta^{(4)}(k-p_1+p_2)}$$

$$= \frac{(2\pi)^4}{q^2}\delta^{(4)}(p_1+k_1-p_2-k_2), \tag{2.17}$$

where the resulting δ-function reflects the energy-momentum conservation, fixing the momentum q of the virtual photon according to (2.2). Adding the second part in the square brackets in (2.12), which is equal to the first part, and using the normalization convention (B.20), we finally obtain:

$$S_{fi}^{(e\pi \to e\pi)} = i(2\pi)^4\delta^{(4)}(p_1 + k_1 - p_2 - k_2)\mathcal{A}(e^-\pi^+ \to e^-\pi^+) ,$$

where the $e\pi \to e\pi$ scattering amplitude:

$$\mathcal{A}(e^-\pi^+ \to e^-\pi^+) = \frac{4\pi\alpha_{em}}{q^2}\bar{u}_e(k_2)\gamma^\mu u_e(k_1)\langle \pi^+(p_2)|j_\mu^{em}|\pi^+(p_1)\rangle , \tag{2.18}$$

contains a factorized part describing the e.m. interaction of the electron and the propagation of the virtual photon, whereas all QCD effects are accumulated in the matrix element

$$\langle \pi^+(p_2)|j_\mu^{em}|\pi^+(p_1)\rangle . \tag{2.19}$$

The latter is an example of a *hadronic matrix element*, defined as an amplitude containing a composite operator built from quark and antiquark fields, and sandwiched between hadronic states with definite momenta. In (2.19) we do not specify the coordinate of the current operator putting it by default to $x = 0$. A shift of coordinate only generates a phase factor as in (2.16) which does not play a role in the physical observables (cross sections or decay probabilities), proportional to the modulus squared of the amplitude.

2.2 PION ELECTROMAGNETIC FORM FACTOR

Our next task is to establish the form of the hadronic matrix element (2.19) employing the underlying symmetries. Since the current operator j_μ^{em} is a four-vector, its matrix element can generally be represented as a linear combination of the two independent four-vectors p_1 and p_2[3]:

$$\langle \pi^+(p_2)|j_\mu^{em}|\pi^+(p_1)\rangle = (p_1 + p_2)_\mu f^+ + (p_1 - p_2)_\mu f^- , \tag{2.20}$$

where f^+ and f^- are scalar quantities. Note that although the pion is a pseudoscalar state, the presence of two pion states in the matrix element makes the total P-parity positive. The dependence of $f^\pm$ on $p_{1,2}$ can only emerge via scalar products: p_1^2, p_2^2 and $p_1 p_2$. Since the squares of momenta are fixed,

$$p_1^2 = p_2^2 = m_\pi^2 , \tag{2.21}$$

---

[3] The pion state has spin zero, hence there are no additional Lorentz-covariant quantities related to the spin degrees of freedom.

we are left with one variable $p_1 p_2$. It is more convenient and customary to replace this scalar product by a related variable, the momentum transfer squared:

$$q^2 = (p_2 - p_1)^2 = 2m_\pi^2 - 2p_1 p_2 \,,$$

so that we deal with the functions $f^\pm(q^2)$. Using the relation (A.75) for the commutator of the quark e.m. current together with the current conservation condition (1.27):

$$[j_\mu^{em}, \hat{P}^\mu] = i\partial^\mu j_\mu^{em} = 0 \,, \tag{2.22}$$

we then take a matrix element of the commutator between the pion states and obtain:

$$\langle \pi^+(p_2)|j_\mu^{em}\hat{P}^\mu|\pi^+(p_1)\rangle - \langle \pi^+(p_2)|\hat{P}^\mu j_\mu^{em}|\pi^+(p_1)\rangle$$
$$= (p_1 - p_2)^\mu \langle \pi^+(p_2)|j_\mu^{em}|\pi^+(p_1)\rangle = 0 \,. \tag{2.23}$$

Replacing the matrix element by the parameterization (2.20), yields:

$$(p_1 - p_2)^\mu \left[(p_1 + p_2)_\mu f^+(q^2) + (p_1 - p_2)_\mu f^-(q^2) \right] = q^2 f^-(q^2) = 0 \,, \tag{2.24}$$

where we again use the on-shellness condition (2.21). We conclude that at any $q^2 \neq 0$

$$f^-(q^2) = 0 \,. \tag{2.25}$$

Thus, we are left with a single function of q^2 determining the hadronic matrix element:

$$\langle \pi^+(p_2)|j_\mu^{em}|\pi^+(p_1)\rangle = (p_1 + p_2)_\mu F_\pi(q^2) \,, \tag{2.26}$$

where we replaced $f^+(q^2)$ by the conventional notation $F_\pi(q^2)$ for the *pion electromagnetic form factor*. The definition (2.26) is by default for the positively charged pion. Inserting the charge conjugation transformation $\hat{C}$ (remember that $\hat{C}^2 = \mathbb{1}$), we switch to the π^- form factor:

$$\langle \pi^+(p_2)|j_\mu^{em}|\pi^+(p_1)\rangle = \langle \pi^+(p_2)|\hat{C}\hat{C}j_\mu^{em}\hat{C}\hat{C}|\pi^+(p_1)\rangle$$
$$= \langle \pi^-(p_2)|\big(\hat{C}j_\mu^{em}\hat{C}\big)|\pi^-(p_1)\rangle = \langle \pi^-(p_2)|\big(-j_\mu^{em}\big)|\pi^-(p_1)\rangle \,, \tag{2.27}$$

proving that it has an opposite sign:

$$F_{\pi^-}(q^2) = -F_\pi(q^2) \,. \tag{2.28}$$

Since the neutral pion has a positive C-parity, its e.m. form factor identically vanishes at any q^2, as it formally follows from applying the same transformations as in (2.27) to the hadronic matrix element of e.m. current taken between the π^0 states.

The physical region for $F_\pi(q^2)$ is

$$q^2 \leq 0 \,, \tag{2.29}$$

inferred from the kinematical boundary (2.8) of the $2 \to 2$ elastic scattering. In fact, that boundary also simply follows from the momentum conservation in the hadronic matrix element,

$$p_1 + q = p_2 \,.$$

To demonstrate that, we square both sides of this relation:

$$(p_1 + q)^2 = m_\pi^2 + 2p_1 q + q^2 = p_2^2 = m_\pi^2 \,,$$

hence,

$$2p_1q = -q^2 \, . \tag{2.30}$$

We then choose the rest frame of the initial pion $p_1 = (m_\pi, \vec{0})$, in which $q = (q_0, \vec{q})$ and the final pion has an energy $E_2 = m_\pi + q_0$, hence $q_0 \geq 0$. In this frame (2.30) is reduced to

$$2m_\pi q_0 + q_0^2 = \vec{q}^2 \, , \tag{2.31}$$

yielding

$$|\vec{q}| = \sqrt{2m_\pi q_0 + q_0^2} \geq q_0 \, , \tag{2.32}$$

so that (2.29) is valid. Since q^2 is a Lorentz-invariant variable, this condition holds for an arbitrary frame. The fact that the virtual photon absorbed by a pion has a spacelike momentum, explains why $F_\pi(q^2)$ in (2.26) is frequently named the 'spacelike' form factor. From (2.31) it is evident that the limit $q^2 = 0$ (corresponding to the real photon absorption) is only possible if all components of the photon momentum vanish, $q_0 = |\vec{q}| = 0$, so that the pion remains at rest.

At $q = 0$ and $p_1 = p_2 = (m_\pi, \vec{0})$ the initial decomposition of the hadronic matrix element (2.20) contains a single form factor:

$$\langle \pi^+(p_1)|j_0^{em}|\pi^+(p_1)\rangle = 2m_\pi F_\pi(0) \, . \tag{2.33}$$

The normalization condition for the pion form factor at $q^2 = 0$ follows from the electric charge conservation. In QED, the time component of the e.m. current density $j_0^{em}(x)$, integrated over three-dimensional space, yields the operator of the electric charge:

$$\int d^3x j_0^{em}(x_0, \vec{x}) = \hat{Q} \, , \tag{2.34}$$

which has an eigenvalue $+1$ for a positively charged pion state:

$$\hat{Q}|\pi^+(p)\rangle = |\pi^+(p)\rangle \, . \tag{2.35}$$

Attributing infinitesimal three-momenta to the pions:

$$p_{1,2} = \lim_{\vec{p}_{1,2} \to 0} (m_\pi, \vec{p}_{1,2}), \tag{2.36}$$

and translating the current operator to an arbitrary point x, we rewrite (2.33):

$$\lim_{\vec{p}_{1,2} \to 0} \langle \pi^+(p_2)|j_0^{em}|\pi^+(p_1)\rangle = \lim_{\vec{p}_{1,2} \to 0} \langle \pi^+(p_2)|e^{i\hat{P}x}j_0^{em}(x)e^{-i\hat{P}x}|\pi^+(p_1)\rangle$$

$$= \lim_{\vec{p}_{1,2} \to 0} e^{i(\vec{p}_2 - \vec{p}_1)\vec{x}} \langle \pi^+(p_2)|j_0^{em}(x)|\pi^+(p_1)\rangle = 2m_\pi F_\pi(0) \, . \tag{2.37}$$

Transferring the exponential factor to r.h.s. we integrate both sides over three-dimensional space. Before taking the limit, we obtain on the l.h.s.:

$$\int d^3x \langle \pi^+(p_2)|j_0^{em}(x)|\pi^+(p_1)\rangle = \langle \pi^+(p_2)|\hat{Q}|\pi^+(p_1)\rangle$$

$$= \langle \pi^+(p_2)|\pi^+(p_1)\rangle = (2\pi)^3 \delta^{(3)}(\vec{p}_1 - \vec{p}_2)2E_1 \, ,$$

where (2.34), (2.35) and the normalization condition (1.81) for the pion states were used, whereas on the r.h.s.:

$$\int d^3x e^{-i(\vec{p}_2 - \vec{p}_1)\vec{x}}2m_\pi F_\pi(0) = (2\pi)^3 \delta^{(3)}(\vec{p}_1 - \vec{p}_2)2m_\pi F_\pi(0) \, .$$

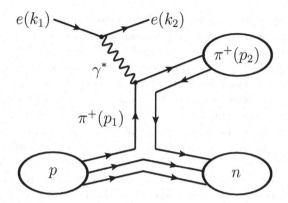

Figure 2.2 Diagram of pion electroproduction on a proton. The quark-antiquark state propagating between nucleons and virtual photon is approximated by a virtual pion.

Taking the limit (2.36) on both sides yields

$$F_\pi(0) = 1 \,, \tag{2.38}$$

fixing the form factor at zero momentum transfer by the π^+ electric charge.

Furthermore, the e.m. current operator is Hermitian:

$$j_\mu^{em\dagger} = j_\mu^{em} \,. \tag{2.39}$$

Hence, the hadronic matrix element of the conjugate operator:

$$\langle \pi^+(p_2)|j_\mu^{em\dagger}|\pi^+(p_1)\rangle = \langle \pi^+(p_1)|j_\mu^{em}|\pi^+(p_2)\rangle^* = (p_2 + p_1)_\mu F_\pi^*(q^2) \,, \tag{2.40}$$

should be equal to (2.26), yielding

$$F_\pi^*(q^2) = F_\pi(q^2) \,, \tag{2.41}$$

which means that the pion e.m. form factor in the spacelike region is a real valued quantity. This important property will be used below, in Chapter 5, when relating the spacelike form factor with its counterpart at positive q^2.

Another important property of the pion e.m. form factor follows from the isospin symmetry. The isospin content of the e.m. current discussed in Chapter 1 is given in (1.54) and (1.55). In fact, only the isovector part $j_\mu^{I=1}$ contributes to the hadronic matrix element (2.26) determining the pion form factor. To explain why the $I = 0$ part is absent, we invoke the G-parity conservation. Remember that a pion has the quantum number $G = -1$. Meanwhile, the isoscalar part (1.55) of the e.m. current is odd with respect to the C-parity transformation, hence it has a negative G-parity. Therefore, acting on the initial pion state, the operator $j_\mu^{I=0}$ generates a state with $G = +1$ that does not match the parity of the final pion state, explicitly violating isospin symmetry.

A standard process used to measure the pion form factor in a spacelike region is the electroproduction of pions on the proton target:

$$e^- p \to e^- \pi^+ n \,. \tag{2.42}$$

The experiment is arranged in such a way that the momentum transfer between the initial electron and proton is small. In this case, a description of this process is possible, introducing the notion of an intermediate pion, propagating between photon and nucleons. A virtual photon vertex with two pions, i.e., the pion form factor, emerges, as shown in Figure 2.2. The dominance of the pion exchange mechanism can be justified by the fact that pion is the lightest intermediate hadronic state and an "effective pion" with $p_1^2 \lesssim 0$ is only slightly off the mass shell $p_1^2 = m_\pi^2$. To extract the form factor $F_\pi(q^2)$ from the measured electroproduction cross section, one still needs to model the strong pion-nucleon interaction as well as to estimate the background to the pion exchange. An elaborated measurement of (2.42), including the pion e.m. form factor determination is presented e.g., in [32]. A calculation of this form factor based on QCD will be discussed in Chapter 10.

In conclusion, we comment on other form factors of the pion. In the context of QCD, any local current has a nonvanishing matrix element between the pion states, provided the quantum numbers of the current match these states. One important example is the *pion scalar form factor* defined as a matrix element of the nonstrange quark-antiquark scalar current:

$$\langle \pi(p_2)| \frac{1}{2}(\bar{u}u + \bar{d}d)|\pi(p_1)\rangle = F_\pi^S(q^2). \tag{2.43}$$

In this case only the isoscalar component of the current contributes as follows from the G-parity conservation. On the other hand, the scalar form factor exists for both charged and neutral pions and from isospin symmetry it follows that:

$$F_{\pi^+}^S(q^2) = F_{\pi^-}^S(q^2) = F_{\pi^0}^S(q^2).$$

2.3   SIMPLER THAN A FORM FACTOR: THE PION DECAY CONSTANT

In Chapter 1 the hadronless QCD-vacuum state $|0\rangle$ was introduced. Let us consider a weak or e.m. process with an initial hadronic state and a final state containing only leptons and/or photons but no hadrons. In this case, after factorizing the non-QCD part of the process amplitude, we encounter a hadronic matrix element where the quark current is sandwiched between the initial hadronic state $|h\rangle$ and the final vacuum state $\langle 0|$.

An important example is provided by *weak leptonic decay* of the charged pion:

$$\pi^-(p) \to \ell^-(p_\ell)\bar{\nu}_\ell(p_\nu), \tag{2.44}$$

where $\ell = e, \mu$, and the charge conjugate is $\pi^+ \to \ell^+\nu_\ell$. At the quark level the pion leptonic decay, shown in Figure 2.3, is a flavor-changing transition

$$d\,\bar{u} \to \ell^-\bar{\nu}_\ell,$$

mediated by the virtual W^--boson.

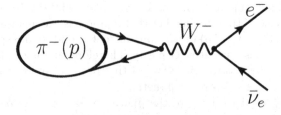

Figure 2.3   Diagram of the $\pi^- \to e^-\bar{\nu}_e$ decay.

The amplitude of (2.44) is given by the S-matrix element of the effective weak Hamiltonian (1.29) in which only the part of the quark weak current (1.30) proportional to

V_{ud} contributes. Neglecting all electroweak higher-order effects which are actually tiny, we expand the S-matrix to the first order in the four-fermion interaction (see (B.4)):

$$S_{fi}^{(\pi \to \ell \nu_\ell)} = \langle \ell^-(p_\ell)\bar{\nu}_\ell(p_\nu)| \exp\left[-i \int d^4x \mathcal{H}_W(x)\right] |\pi^-(p)\rangle$$

$$= \langle \ell^-(p_\ell)\bar{\nu}_\ell(p_\nu)| \left[\mathbb{1} - i \int d^4x \mathcal{H}_W(x)\right] |\pi^-(p)\rangle$$

$$= -i\frac{G_F}{\sqrt{2}}V_{ud}\int d^4x \langle \ell^-(p_\ell)\bar{\nu}_\ell(p_\nu)|\bar{\ell}(x)\Gamma_\mu \nu_\ell(x)\, \bar{u}(x)\Gamma^\mu d(x)|\pi^-(p)\rangle\,, \quad (2.45)$$

where we use a short-hand notation $\Gamma_\mu \equiv \gamma_\mu(1 - \gamma_5)$. Remember that the QCD part is, as usual, contained in the S-matrix but not shown. Using the expansion in the momentum components for the lepton and neutrino field operators, factorizing the leptonic part of the amplitude in terms of bispinors (similar to (2.15)) and performing the quark weak current translation, we obtain:

$$S_{fi}^{(\pi \to \ell \nu_\ell)} = -i\frac{G_F}{\sqrt{2}}V_{ud}\int d^4x\, e^{i(p_\ell x + p_\nu x)}\left[\bar{u}_\ell(p_\ell)\Gamma_\mu v_\nu(p_\nu)\right]\langle 0|\bar{u}(x)\Gamma^\mu d(x)|\pi^-(p)\rangle$$

$$= -i\frac{G_F}{\sqrt{2}}V_{ud}\underbrace{\int d^4x\, e^{i(p_\ell + p_\nu - p)x}}_{(2\pi)^4\delta^{(4)}(p-p_\ell-p_\nu)}\left[\bar{u}_\ell(p_\ell)\Gamma_\mu v_\nu(p_\nu)\right]\langle 0|\bar{u}(0)\Gamma^\mu d(0)|\pi^-(p)\rangle\,, \quad (2.46)$$

so that the amplitude of $\pi \to \ell \nu_\ell$ decay can be written as:

$$\mathcal{A}(\pi^- \to \ell^- \bar{\nu}_\ell) = \frac{G_F}{\sqrt{2}}V_{ud}\,\bar{u}_\ell(p_\ell)\Gamma^\mu v_\nu(p_\nu)\langle 0|j_{\mu 5}|\pi^-(p)\rangle\,. \quad (2.47)$$

It contains the hadronic matrix element of the pion-vacuum transition:

$$\langle 0|j_{\mu 5}|\pi^-(p)\rangle = ip_\mu f_\pi\,, \quad (2.48)$$

where the operator

$$j_{\mu 5} = \bar{u}\gamma_\mu \gamma_5 d \quad (2.49)$$

is the axial-vector part of the (ud) component in the weak current (1.30). This hadronic matrix element depends on a single 4-vector p_μ, with $p^2 = m_\pi^2$ fixed, and is simpler than the expression (2.20) for the pion form factor. Moreover, the Lorentz-invariant parameter f_π is a constant[4] named the pion *decay constant*. The corresponding definition for the inverse matrix element with Hermitian conjugated current $j_{\mu 5}^\dagger = \bar{d}\gamma_\mu \gamma_5 u$ reads:

$$\langle \pi^-(p)|j_{\mu 5}^\dagger|0\rangle = \langle 0|j_{\mu 5}|\pi^-(p)\rangle^* = -ip_\mu f_\pi\,. \quad (2.50)$$

Two questions arise concerning the definition (2.48). First: how can a spin-zero pion state be annihilated by the axial-vector current, which seemingly has spin one, and second: why does the contribution of the vector part of weak current to the vacuum-to-pion matrix element vanish?

Addressing the first question, we notice that, in contrast to the e.m. current, $j_{\mu 5}$ is not conserved, having a nonvanishing divergence. We calculate the latter, converting the simple

---

[4]Since only $|f_\pi|^2$ is observable, the phase convention for the pion state in (2.48) can always be chosen to render f_π real valued.

derivatives to the covariant ones and making use of the QCD equation of motion for the quark field (see (A.40)):

$$\partial^\mu \left(\bar{u}(x)\gamma_\mu\gamma_5 d(x)\right) = \bar{u}(x)\overleftarrow{\partial}\gamma_5 d(x) + \bar{u}(x)\overrightarrow{\partial}\gamma_5 d(x)$$

$$= \bar{u}(x)\left(\overleftarrow{\partial} + ig_s\frac{\lambda^a}{2}A^a(x)\right)\gamma_5 d(x) + \bar{u}(x)\left(\overrightarrow{\partial} - ig_s\frac{\lambda^a}{2}A^a(x)\right)\gamma_5 d(x)$$

$$= \bar{u}(x)\overleftarrow{\slashed{D}}\gamma_5 d(x) - \bar{u}(x)\gamma_5\overrightarrow{\slashed{D}}d(x) = im_u\bar{u}\gamma_5 d(x) + im_d\bar{u}\gamma_5 d(x)\,. \tag{2.51}$$

The divergence is reduced to the pseudoscalar quark current:

$$\partial^\mu j_{\mu 5} = (m_u + m_d)\bar{u}i\gamma_5 d\,. \tag{2.52}$$

It has the same quantum numbers as the charged pion, hence also a nonvanishing matrix element between the pion state and vacuum. In other words, the axial-vector current contains four independent components and a certain combination of them is reduced to a pseudoscalar current. This answers the first question. The second question is clarified applying the same procedure (2.51) to the vector part of the weak current,

$$j_\mu = \bar{u}\gamma_\mu d\,. \tag{2.53}$$

The result for the divergence will be then a scalar current:

$$\partial^\mu j_\mu = i(m_u - m_d)\bar{u}d\,, \tag{2.54}$$

which definitely does not match a pseudoscalar pion state.

Furthermore, applying (A.75) and relating the divergence of the axial-vector current to its commutator with the 4-momentum operator:

$$i\partial^\mu j_{\mu 5} = [j_{\mu 5}, \hat{P}^\mu]\,, \tag{2.55}$$

we take from both sides the matrix element between vacuum and pion state and obtain:

$$\langle 0|i\partial^\mu j_{\mu 5}|\pi^-(p)\rangle = \langle 0|[j_{\mu 5}, \hat{P}^\mu]|\pi^-(p)\rangle$$

$$= \langle 0|\left(j_{\mu 5}\hat{P}^\mu - \hat{P}^\mu j_{\mu 5}\right)|\pi^-(p)\rangle = \langle 0|j_{\mu 5}\hat{P}^\mu|\pi^-(p)\rangle$$

$$= p^\mu\langle 0|j_{\mu 5}|\pi^-(p)\rangle = p^\mu(ip_\mu f_\pi) = im_\pi^2 f_\pi\,, \tag{2.56}$$

where we took into account that the vacuum state has zero momentum. Substituting in l.h.s. of (2.56) the relation (2.52), we arrive at an alternative definition of f_π:

$$\langle 0|(m_d + m_u)\bar{u}i\gamma_5 d|\pi^-(p)\rangle = m_\pi^2 f_\pi\,, \tag{2.57}$$

via the matrix element of the pseudoscalar current. Note that the quark mass factor $(m_d + m_u)$ is kept as a prefactor of the pseudoscalar current rather than as a coefficient at the matrix element. The reason is that quark masses are renormalization scale dependent (running) quantities in QCD (see (1.10)). In fact, the pseudoscalar quark current also has an anomalous dimension, which is inverse to the one of the quark mass; hence, their product entering (2.57) is a scale independent quantity. This is consistent with the fact that the decay constant f_π determines the pion decay width and, being an observable quantity, should be independent of the QCD renormalization scale.

Returning to the decay amplitude (2.47), we insert the parameterization (2.48), replace p by $p_\ell + p_\nu$ and use Dirac equations for the lepton spinors, neglecting the neutrino mass. The amplitude, in terms of the pion decay constant:

$$\mathcal{A}(\pi^- \to \ell^-\bar{\nu}_\ell) = i\frac{G_F}{\sqrt{2}}V_{ud}\, m_\ell \bar{u}_\ell(p_\ell)(1 - \gamma_5)v_\nu(p_\nu)f_\pi\,, \tag{2.58}$$

is proportional to the lepton mass[5], causing dominance of the $\pi \to \mu\bar{\nu}_\mu$ decay versus $\pi \to e\bar{\nu}_e$. The decay width is calculated by squaring the amplitude (2.58) and summing over the lepton polarizations:

$$\overline{|\mathcal{A}(\pi^- \to \ell^- \bar{\nu}_\ell)|^2} = \frac{G_F^2}{2}|V_{ud}|^2 m_\ell^2 f_\pi^2$$
$$\times \text{Tr}\{\overline{u_\ell(p_\ell)\bar{u}_\ell(p_\ell)}(1-\gamma_5)\overline{v_\nu(p_\nu)\bar{v}_\nu(p_\nu)}(1+\gamma_5)\}. \tag{2.59}$$

It remains to compute the trace employing the density matrices for the leptonic spinors:

$$\text{Tr}\{\overline{u_\ell(p_\ell)\bar{u}_\ell(p_\ell)}(1-\gamma_5)\overline{v_\nu(p_\nu)\bar{v}_\nu(p_\nu)}(1+\gamma_5)\}$$
$$= \text{Tr}\{(\not{p}_\ell + m_\ell)(1-\gamma_5)\not{p}_\nu(1+\gamma_5)\}$$
$$= \text{Tr}\{(\not{p}_\ell + m_\ell)\not{p}_\nu(1+\gamma_5)^2\} = 8(p_\ell p_\nu) = 4(m_\pi^2 - m_\ell^2), \tag{2.60}$$

where the last equation follows from $(p_\ell + p_\nu)^2 = m_\pi^2$. Substituting the trace in (2.59) and using the expression (B.42) for a two-body decay width, where in this case

$$\lambda^{1/2}(m_\pi^2, m_\ell^2, 0) = m_\pi^2 - m_\ell^2,$$

we obtain

$$\Gamma(\pi^- \to \ell^- \bar{\nu}_\ell) = \frac{G_F^2 |V_{ud}|^2}{8\pi} m_\ell^2 m_\pi \left(1 - \frac{m_\ell^2}{m_\pi^2}\right)^2 f_\pi^2. \tag{2.61}$$

The charge conjugated decay has an equal width $\Gamma(\pi^+ \to \ell^+ \nu_\ell)$.

The amplitude of the $\tau \to \pi\nu_\tau$ weak decay contains the inverted hadronic matrix element (2.50) and is therefore also determined by the pion decay constant:

$$\mathcal{A}(\tau^- \to \pi^- \nu_\tau) = \frac{G_F}{\sqrt{2}} V_{ud} \bar{v}_\nu(p_\nu)\Gamma^\mu u_\tau(p_\tau)\langle\pi^-(p)|\bar{d}\gamma_\mu\gamma_5 u|0\rangle$$
$$= -i\frac{G_F}{\sqrt{2}} V_{ud} m_\tau \bar{v}_\nu(p_\nu)(1+\gamma_5)u_\nu(p_\tau) f_\pi, \tag{2.62}$$

where $p_\tau = p + p_\nu$. The calculation of the width is very similar to the one described above, with an extra factor $1/2$ in the phase space taking into account the average of the τ-lepton polarizations. The result is:

$$\Gamma(\tau^- \to \pi^- \nu_\tau) = \frac{G_F^2 |V_{ud}|^2}{16\pi} m_\tau^3 \left(1 - \frac{m_\pi^2}{m_\tau^2}\right)^2 f_\pi^2, \tag{2.63}$$

and the same for $\Gamma(\tau^+ \to \pi^+ \bar{\nu}_\tau)$.

Concluding this section, we quote the numerical value of the pion decay constant

$$f_\pi \simeq 130.5 \pm 0.13 \text{ MeV}, \tag{2.64}$$

taken from [1]. This value is obtained equating (2.61) to the measured width which is, up to very small corrections, equal to the inverse mean lifetime of the charged pion. The CKM matrix element $|V_{ud}|$ is determined independently, from certain nuclear β decays.

2.4 THE FORM FACTORS OF WEAK π_{e3} DECAY

The (ud) part of the quark weak current also generates transitions involving a pion in both initial and final states. It is instructive to start from a scattering process

$$\nu_\ell(k_1)\pi^-(p_1) \to \ell(k_2)\pi^0(p_2) \tag{2.65}$$

---

[5]Emergence of the lepton mass factor in the weak leptonic decay amplitude follows from the spin conservation combined with the polarization properties of leptons.

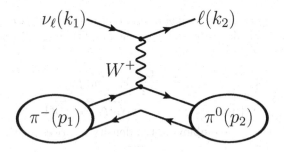

Figure 2.4 Diagram of the hypothetical neutrino-pion inelastic scattering with W-boson exchange.

with a diagram shown in Figure 2.4. This process is almost hypothetical and was actually never observed, due to evident experimental difficulties to scatter a neutrino beam on a pion target. The essential difference between (2.65) and the electron-pion scattering (2.1) is that a W-boson is exchanged between leptons and quarks instead of a virtual photon. In the region of momentum transfer $q^2 \ll m_W^2$, the neutrino-pion scattering is described by the same $\sim V_{ud}$ part of the effective Hamiltonian (1.29) as the pion leptonic decay. To obtain the amplitude of (2.65) we can skip the usual steps with an S-matrix element and start from (2.47), replacing the vacuum state by the pion state and converting the final state neutrino into an initial state one. An important difference has to be taken into account: instead of the axial-vector part of the weak current (2.49) the vector part (2.53) should be retained. We end up with the following scattering amplitude:

$$\mathcal{A}(\nu_\ell \pi^- \to \ell^- \pi^0) = \frac{G_F}{\sqrt{2}} V_{ud} \bar{u}_\ell(k_2) \Gamma^\mu u_\nu(k_1) \langle \pi^0(p_2)|j_\mu|\pi^-(p_1)\rangle. \tag{2.66}$$

Note that the hadronic matrix element emerging in this amplitude is very similar to (2.19), and both contain vector quark currents.

Let us clarify why the axial-vector current does not contribute to the matrix element of the pion-to-pion transition in (2.66). Instead of formal symmetry arguments, we consider a simple model for this matrix element, in which the current operator $j_{\mu 5} = \bar{u}\gamma_\mu \gamma_5 d$ is replaced by a $u\bar{d}$-meson which has the quantum numbers of the current and strongly couples to the initial and final pions. This could be an a_1-meson with $J^P = 1^+$, or a pion which also couples to the axial-vector current, as we already know from (2.48). Thus, we effectively replace the hadronic matrix element $\langle \pi^0(p_2)|j_\mu|\pi^-(p_1)\rangle$ by a three-meson vertex, similar to the $\rho\pi\pi$ diagram shown in Figure 1.6, with a_1 or π instead of ρ. Such a replacement is an admittedly poor approximation from a dynamical point of view. A meson cannot readily replace the current, but can only serve as an intermediate state emitted by the current and absorbed in the three-meson vertex. Moreover, other intermediate hadronic states can also contribute. All these complications are, however, not important if we are only interested in quantum numbers. Indeed, a hadronic matrix element with a given current is certainly forbidden, if the corresponding three-meson vertex violates some of the conservation laws valid in QCD. It is then easy to convince ourselves, that both 3π and $a_1\pi\pi$ strong interaction vertices are forbidden. Considering these vertices as $\pi \to 2\pi$ and $a_1 \to 2\pi$ transition processes[6], we notice that a two-pion state cannot have the same quantum numbers as a single pion or

---

[6]Note that in these processes at least one hadron should be off mass shell (virtual) to obey energy-momentum conservation. But that is allowed, due to the quantum nature of particle processes and within the uncertainty relations. Importantly, a virtuality does not influence symmetry properties, the only ones which interest us here.

a_1 meson. Indeed, counting P-parity of the two-pion state with the orbital momentum L yields:

$$P_{2\pi} = (P_\pi)^2 (-1)^L = +(-1)^L \,,$$

a mismatch with $P_\pi = -1$ ($P_{a_1} = +1$) in which case $L = 0$ ($L = 1$) due to the angular momentum conservation. This argument is independent of the flavor of hadrons. We conclude that a simultaneous conservation of angular momentum and P-parity forbids the matrix element of an axial-vector current with two pseudoscalar mesons. It should be emphasized that although we are considering a weak interaction process, these symmetry arguments only refer to its hadronic part which is mediated by QCD. Therefore, the fact that P-parity is violated by weak interactions, does not play a role here.

Concentrating now on the hadronic matrix element in (2.66):

$$\langle \pi^0(p_2) | j_\mu | \pi^-(p_1) \rangle \,, \tag{2.67}$$

we perform a general decomposition:

$$\langle \pi^0(p_2) | j_\mu | \pi^-(p_1) \rangle = (p_1 + p_2)_\mu \tilde{f}^+(q^2) + (p_1 - p_2)_\mu \tilde{f}^-(q^2), \tag{2.68}$$

similar to (2.20), but with different form factors $\tilde{f}^\pm(q^2)$.

The divergence of the weak vector current, presented in (2.54) vanishes in the isospin-symmetry limit, $m_u = m_d$. Furthermore, applying a chain of equations analogous to (2.22)–(2.24) we convince ourselves that in this limit $\tilde{f}^-(q^2) = 0$. We are left with a single form factor $\tilde{f}^+$ and rename it to F_V:

$$\langle \pi^0(p_2) | j_\mu | \pi^-(p_1) \rangle = (p_1 + p_2)_\mu F_V(q^2) \,. \tag{2.69}$$

Applying the charge conjugation yields for $j_\mu^\dagger = \bar{d} \gamma_\mu u$:

$$-\langle \pi^0(p_2) | j_\mu^\dagger | \pi^+(p_1) \rangle = (p_1 + p_2)_\mu F_V(q^2) \,. \tag{2.70}$$

In reality, isospin symmetry is slightly violated. As a result, the charged pion is heavier than the neutral one[7]. According to [1], $m_{\pi^+} - m_{\pi_0} \simeq 4.6$ MeV. This small mass difference enables a *weak semileptonic* decay

$$\pi^- \to \pi^0 e^- \bar{\nu}_e \,, \quad \text{or} \quad \pi^+ \to \pi^0 e^+ \nu_e \,, \tag{2.71}$$

known also as the π_{e3} decay. Its amplitude:

$$\mathcal{A}(\pi^- \to \pi^0 e^- \bar{\nu}_e) = \frac{G_F}{\sqrt{2}} V_{ud} \bar{u}_e(k_2) \Gamma^\mu v_\nu(k_1) \langle \pi^0(p_2) | j_\mu | \pi^-(p_1) \rangle \,, \tag{2.72}$$

is obtained from (2.66) by transforming the initial-state neutrino to the final-state antineutrino with $k_1 \to -k_1$. Accordingly, instead of (2.2), the momentum transfer between initial and final pion has to be redefined as

$$q = p_1 - p_2 = k_1 + k_2 \,, \tag{2.73}$$

resulting in positive values of $q^2 = (k_1 + k_2)^2$. This variable is equal to the square of lepton-pair invariant mass, with the limits:

$$m_e^2 \leq q^2 \leq (m_{\pi^-} - m_{\pi^0})^2 \,. \tag{2.74}$$

---

[7]This particular isospin violation is mainly due to e.m. effects.

The lower bound is simply the minimal invariant mass of the neutrino-electron pair, neglecting the neutrino mass. To derive the upper bound, we consider the rest frame of decaying π^-, where $\vec{q} = -\vec{p}_2$ and q^2 is determined by the energy of the neutral pion:

$$q^2 = (p_1 - p_2)^2 = m_{\pi^-}^2 - 2m_{\pi^-}E_2 + m_{\pi^0}^2 \,, \tag{2.75}$$

so that the minimal energy $E_2 = m_{\pi^0}$ at $\vec{p}_2 = 0$ (named the *zero recoil point* of the final state pion) corresponds to the maximal value of q^2 in (2.74).

The weak transition form factor F_V and the pion e.m. form factor are related in the isospin limit:

$$F_V(q^2) = \sqrt{2}F_\pi(q^2) \,. \tag{2.76}$$

In particular, since $F_\pi(0) = 1$, this relation predicts the form factor $F_V(0)$, hence, also the width of the pion semileptonic decay with a reasonable accuracy, keeping in mind that $q^2 \simeq 0$ in the whole interval (2.75).

The relation (2.76) follows from comparing the hadronic matrix element (2.26) (where only the isovector part $j_\mu^{I=1}$ of j_μ^{em} contributes), with (2.69) and (2.70). The key observation is that the three vector currents entering these matrix elements form an isovector which we denote as $\mathcal{J}_\mu(I = 1, I_3)$, so that

$$\mathcal{J}_\mu(I = 1, I_3 = +1) = j_\mu = \bar{u}\gamma_\mu d \,,$$
$$\mathcal{J}_\mu(I = 1, I_3 = 0) = -\sqrt{2}j_\mu^{I=1} = -\frac{1}{\sqrt{2}}\left(\bar{u}\gamma_\mu u - \bar{d}\gamma_\mu d\right) \,,$$
$$\mathcal{J}_\mu(I = 1, I_3 = -1) = -j_\mu^\dagger = -\bar{d}\gamma_\mu u \,, \tag{2.77}$$

(cf. isovector components of the pion state in (A.83)). Note that e.g., the current j_μ is assigned by $I_3 = +1$ because, acting on the vacuum, it creates a state with the $u\bar{d}$ content. Using (2.77) one may prove the relation (2.76) invoking the mathematics of isospin $SU(2)$ group: direct products, irreducible representations, Clebsch-Jordan coefficients, Wigner-Eckart theorem and all that.

We will instead consider a more transparent way to substantiate (2.76), attributing to each hadronic matrix element a set of *quark flow diagrams*. Examples of these are e.g., the hadronic parts of diagrams in Figures 2.1 and 2.4, where hadrons are represented by their valence quark content and quark lines are attached to the current vertex in all possible ways. Such a diagrammatic approach allows one to directly relate hadronic matrix elements to each other, employing a certain flavor symmetry.

More specifically, the matrix element (2.26), after replacing the initial and final pions with their valence content, is described by a sum of the two quark flow diagrams, depicted in Figure 2.5(a,b) where in the first (second) diagram the $\bar{u}\gamma_\mu u$ ($\bar{d}\gamma_\mu d$) current couples to the u- (d-) quark line. Each diagram provides a separate contribution to the matrix element (2.26), so that:

$$\begin{aligned}
\langle\pi^+|j_\mu^{em}|\pi^+\rangle &= \langle\pi^+|j_\mu^{(I=1)}|\pi^+\rangle \\
&= \frac{1}{2}\left(\langle u\bar{d}|\,\bar{u}\gamma_\mu u\,|u\bar{d}\rangle - \langle u\bar{d}|\,\bar{d}\gamma_\mu d\,|u\bar{d}\rangle\right) \equiv \frac{1}{2}(A_\mu^{\pi\pi} - \bar{A}_\mu^{\pi\pi}).
\end{aligned} \tag{2.78}$$

where we omit the momentum assignment of states for brevity. Turning to the hadronic matrix element (2.69), we encounter the two diagrams in Figure 2.5(c,d), and, similarly,

$$\langle\pi^0|j_\mu|\pi^-\rangle = \frac{1}{\sqrt{2}}\left(\langle u\bar{u}|\,\bar{u}\gamma_\mu d\,|d\bar{u}\rangle - \langle d\bar{d}|\,\bar{u}\gamma_\mu d\,|d\bar{u}\rangle\right) \,, \tag{2.79}$$

where the two contributions account for the $u\bar{u}$ and $d\bar{d}$ components of the π^0 state. An important feature of a quark flow diagram is that any u- (d-) quark line on it in the

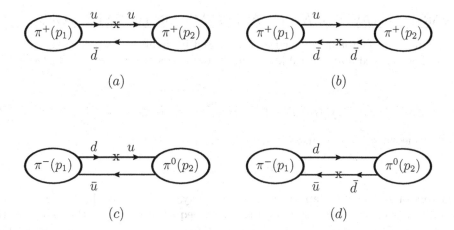

Figure 2.5 Quark flow diagrams corresponding to the hadronic matrix elements of (a),(b) e.m. current and (c),(d) weak vector current. The crosses denote the vertex of the current.

isospin symmetry limit can be replaced by the d- (u)-quark line, leaving the corresponding hadronic matrix element unchanged[8]. QCD interactions in the hadronic transitions simply do not "notice" this replacement, because both u and d quarks have equal masses (in the isospin symmetry limit) and color charges, and additional e.m. or weak interactions that can distinguish between the flavors, are switched off in this approximation. Having that in mind, in the first diagram of the $\pi^- \to \pi^0$ transition (Figure 2.5(c)) we replace the spectator antiquark, $\bar{u} \to \bar{d}$, and the incoming quark in the current, $d \to u$. Analogously, we replace $d \to u$ and $\bar{u} \to \bar{d}$ in the second diagram of that transition (Figure 2.5(d)). These replacements convert the diagrams contributing to (2.79) into the ones of the $\pi^+ \to \pi^+$ transition in (2.78), leading to the equalities:

$$\langle u\bar{u}|\bar{u}\gamma_\mu d|d\bar{u}\rangle = \langle ud|\bar{u}\gamma_\mu u|u\bar{d}\rangle, \quad \langle d\bar{d}|\bar{u}\gamma_\mu d|d\bar{u}\rangle = \langle ud|\bar{d}\gamma_\mu d|u\bar{d}\rangle, \tag{2.80}$$

hence,

$$\langle \pi^0|j_\mu|\pi^-\rangle = \frac{1}{\sqrt{2}}(A^{\pi\pi}_\mu - \bar{A}^{\pi\pi}_\mu), \tag{2.81}$$

Comparing this with (2.78) results in the isospin relation (2.76) between form factors. In fact, the two amplitudes in (2.78) and (2.81) are also related:

$$\bar{A}^{\pi\pi}_\mu = -A^{\pi\pi}_\mu. \tag{2.82}$$

To see that, we notice that the $\pi^+ \to \pi^+$ matrix element with $\bar{d}\gamma_\mu d$ current, does not change after a global replacement $u \leftrightarrow d$:

$$\bar{A}^{\pi\pi}_\mu = \langle u\bar{d}|\,\bar{d}\gamma_\mu d\,|u\bar{d}\rangle = \langle d\bar{u}|\,\bar{u}\gamma_\mu u\,|d\bar{u}\rangle, \tag{2.83}$$

Inserting in the above the charge conjugation transformation, as it was done in (2.27), leads to:

$$\langle d\bar{u}|\,\bar{u}\gamma_\mu u\,|d\bar{u}\rangle = \langle d\bar{u}|\,\hat{C}\hat{C}\bar{u}\gamma_\mu u\hat{C}\hat{C}\,|d\bar{u}\rangle,$$
$$\langle u\bar{d}|\,\hat{C}\bar{u}\gamma_\mu u\hat{C}\,|u\bar{d}\rangle = -\langle u\bar{d}|\,\bar{u}\gamma_\mu u\,|u\bar{d}\rangle = -A^{\pi\pi}_\mu, \tag{2.84}$$

where we essentially use the negative C-parity of the flavor-neutral vector current. Combining the two above equations, we obtain (2.82).

---

[8]Such replacements are not always possible in processes involving baryons, because in the latter the isospin symmetry is correlated with the spin and angular momentum of the three-quark states.

2.5 VARYING FLAVORS: K-, D-, AND B-MESON FORM FACTORS

The pion e.m. form factor introduced and discussed in detail in Section 2.2 has its analog for kaons. The charged and neutral kaon e.m. form factors are defined similar to (2.26):

$$\langle K^+(p_2)|j_\mu^{em}|K^+(p_1)\rangle = (p_1 + p_2)_\mu F_{K^+}(q^2)\,,$$
$$\langle K^0(p_2)|j_\mu^{em}|K^0(p_1)\rangle = (p_1 + p_2)_\mu F_{K^0}(q^2)\,, \qquad (2.85)$$

and their values at $q^2 = 0$ are fixed by electric charges:

$$F_{K^+}(0) = 1\,, \quad F_{K^0}(0) = 0\,. \qquad (2.86)$$

Form factors of K^- and $\bar{K}^0$ are defined, respectively, as charge conjugates of the above. We skip all intermediate steps to obtain the above equations because they are essentially the same as in the pion case. In order to switch from π^+ to K^+ or K^0, one simply has to replace the valence quark flavors in the initial and final pion states entering hadronic matrix elements:

$$(u\bar{d}) \to (u\bar{s}) \quad \text{or} \quad (u\bar{d}) \to (d\bar{s})\,. \qquad (2.87)$$

This replacement, however, brings an essential difference between kaons and pions concerning the isospin. For the kaon form factors both isospin components of the quark e.m. current (1.54) contribute. Only separate isospin components of the kaon form factors, defined as the hadronic matrix elements of $j_\mu^{(I=1)}$ and $j_\mu^{(I=0)}$,

$$F_{K^+(K^0)} = F_{K^+(K^0)}^{(I=1)} + F_{K^+(K^0)}^{(I=0)}\,, \qquad (2.88)$$

are related in the isospin symmetry limit:

$$F_{K^+}^{(I=1)} = -F_{K^0}^{(I=1)}\,, \quad F_{K^+}^{(I=0)} = F_{K^0}^{(I=0)}\,. \qquad (2.89)$$

These relations are easily reproduced making replacements (2.87) in the quark-flow diagrams in Figure 2.5 and comparing them for K^+ and K^0.

The kaon e.m. form factors in the spacelike region $q^2 \leq 0$ can be measured e.g., in the kaon electroproduction:

$$e^- p \to e^- K^+ \Lambda\,, \quad e^- p \to e^- K^0 \Sigma^+\,. \qquad (2.90)$$

The mechanism is the same as in Figure 2.2 where instead of the virtual pion we have a kaon.

For all other stable pseudoscalar mesons, such as D and B, the e.m. form factors can be equally well defined, however they can hardly be measured in the spacelike region, mainly because these mesons are too unstable. The prospects to measure their electroproduction in association with heavy baryons with one of the mesons being in a virtual state sound also not realistic mainly because of their large mass.

On the contrary, the π_{e3} decay (2.71) belongs to a rich variety of flavor-changing semileptonic decays of the type:

$$P_i \to P_f \ell \bar{\nu}_\ell\,, \qquad (2.91)$$

with pseudoscalar mesons P_i and P_f in the initial and final state. One of these decay modes is shown in Figure 2.6. These decays are triggered by the effective weak interaction (1.29), where only the vector part of the quark weak current (1.30),

$$j_\mu = \bar{q}\gamma_\mu q' \quad \text{or} \quad j_\mu^\dagger = \bar{q}'\gamma_\mu q$$

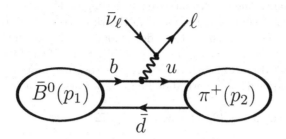

Figure 2.6 Diagram of the semileptonic decay $\bar{B}^0 \to \pi^+\ell^-\nu_\ell$. The wavy line denotes the W-boson.

TABLE 2.1 Weak semileptonic and leptonic decays of pseudoscalar mesons with various flavors and the corresponding quark weak currents. Each decay has a charge conjugate mode which is not shown.

V_{CKM}	j_μ	Semileptonic mode	$f^{+,0}_{P_i P_f}(q^2)$	$j_{\mu 5}$	Leptonic mode	f_P
V_{ud}	$\bar{u}\gamma_\mu d$	$\pi^- \to \pi^0 e^- \bar{\nu}_e$	F_V	$\bar{u}\gamma_\mu\gamma_5 d$	$\pi^- \to \ell^- \bar{\nu}_\ell$	f_π
V_{us}	$\bar{u}\gamma_\mu s$	$\bar{K}^0 \to \pi^+ \ell^- \bar{\nu}_\ell$	$f^{+,0}_{K\pi}$	$\bar{u}\gamma_\mu\gamma_5 s$	$K^- \to \ell^- \bar{\nu}_\ell$	f_K
		$K^- \to \pi^0 \ell^- \bar{\nu}_\ell$	$f^{+,0}_{K\pi}/\sqrt{2}$			
V_{cd}	$\bar{c}\gamma_\mu d$	$\bar{D}^0 \to \pi^+ \ell^- \bar{\nu}_\ell$	$f^{+,0}_{D\pi}$	$\bar{c}\gamma_\mu\gamma_5 d$	$D^- \to \ell^- \bar{\nu}_\ell$	f_D
		$D^- \to \pi^0 \ell^- \bar{\nu}_\ell$	$-f^{+,0}_{D\pi}/\sqrt{2}$			
		$D^- \to \eta^{(')} \ell^- \bar{\nu}_\ell$	$f^{+,0}_{D\eta^{(')}}$			
		$D_s^- \to \bar{K}^0 \ell^- \bar{\nu}_\ell$	$f^{+,0}_{D_s K}$			
V_{cs}	$\bar{c}\gamma_\mu s$	$\bar{D}^0 \to K^+ \ell^- \bar{\nu}_\ell$	$f^{+,0}_{DK}$	$\bar{c}\gamma_\mu\gamma_5 s$	$D_s^- \to \ell^- \bar{\nu}_\ell$	f_{D_s}
		$D^- \to K^0 \ell^- \bar{\nu}_\ell$	$f^{+,0}_{DK}$			
		$\bar{D}_s \to \eta^{(')} \ell^- \bar{\nu}_\ell$	$f^{+,0}_{D_s\eta^{(')}}$			
V_{ub}	$\bar{u}\gamma_\mu b$	$\bar{B}^0 \to \pi^+ \ell^- \bar{\nu}_\ell$	$f^{+,0}_{B\pi}$	$\bar{u}\gamma_\mu\gamma_5 b$	$B^- \to \ell^- \bar{\nu}_\ell$	f_B
		$B^- \to \pi^0 \ell^- \bar{\nu}_\ell$	$f^{+,0}_{B\pi}/\sqrt{2}$			
		$B^- \to \eta^{(')} \ell^- \bar{\nu}_\ell$	$f^{+,0}_{B\eta^{(')}}$			
		$\bar{B}_s^0 \to K^+ \ell^- \bar{\nu}_\ell$	$f^{+,0}_{B_s K}$			
		$B_c^- \to \bar{D}^0 \ell^- \bar{\nu}_\ell$	$f^{+,0}_{B_c D}$			
V_{cb}	$\bar{c}\gamma_\mu b$	$\bar{B}^0 \to D^+ \ell^- \bar{\nu}_\ell$	$f^{+,0}_{BD}$	$\bar{c}\gamma_\mu\gamma_5 b$	$B_c^- \to \ell^- \bar{\nu}_\ell$	f_{B_c}
		$B^- \to D^0 \ell^- \bar{\nu}_\ell$	$f^{+,0}_{BD}$			
		$\bar{B}_s^0 \to D_s^+ \ell^- \bar{\nu}_\ell$	$f^{+,0}_{B_s D_s}$			
		$B_c^- \to \eta_c \ell^- \bar{\nu}_\ell$	$f^{+,0}_{B_c \eta_c}$			

with $q = (u, c)$, $q' = (d, s, b)$ contributes, coupled to the leptonic weak current, with a coefficient proportional to the CKM matrix element $V_{qq'}$ or $V_{qq'}^*$.

All possible decays of that type are listed in the third column of Table 2.1. Note that, due to a sufficient phase space, semileptonic decay modes of kaons and charmed mesons (of bottom mesons) with $\ell = e, \mu$ ($\ell = e, \mu, \tau$) are possible. Some of the form factors of semileptonic transitions included in Table 2.1 are related via isospin symmetry. These relations are easily reproduced, comparing the corresponding quark flow diagrams, as explained in the previous section.

The amplitude of the decay (2.91) generalizes (2.72):

$$\mathcal{A}(P_i \to P_f \ell \bar{\nu}_\ell) = \frac{G_F}{\sqrt{2}} V_{qq'} \bar{u}_\ell(k_2) \Gamma^\mu v_\nu(k_1) \langle P_f(p_2) | j_\mu | P_i(p_1) \rangle \,. \tag{2.92}$$

Denoting the timelike momentum transferred to the lepton pair by $q = k_1 + k_2$, so that $p_1 = p_2 + q$, we keep a nonvanishing charged lepton mass, essential in the case of muon or τ, and, as usual, neglect the neutrino masses, hence $k_1^2 = 0$ and $k_2^2 = m_\ell^2$. In the kinematical region of the decay (2.91) the square of the momentum transfer varies within the limits

$$m_\ell^2 \leq q^2 \leq (m_{P_i} - m_{P_f})^2, \tag{2.93}$$

where m_{P_i, P_f} are the masses of the mesons. The hadronic matrix element in (2.92), similar to (2.68), is decomposed in two independent kinematical structures multiplying two invariant functions of q^2 – the form factors:

$$\langle P_f(p_2) | j_\mu | P_i(p_1) \rangle = (p_1 + p_2)_\mu \, f^+_{P_i P_f}(q^2) + (p_1 - p_2)_\mu \, f^-_{P_i P_f}(q^2) \,. \tag{2.94}$$

Note that quarks in the weak current j_μ have different masses; hence, in general, this current is not conserved:

$$\partial^\mu j_\mu = i(m_q - m_{q'}) \bar{q} q' \,, \tag{2.95}$$

(cf. (2.54)) and the form factor $f^-_{P_i P_f}$ cannot be neglected.

The following redefinition of (2.94) is hereafter used as a default form of this hadronic matrix element:

$$\langle P_f(p_2) | j_\mu | P_i(p_1) \rangle = \left[2p_{2\mu} + \left(1 - \frac{m_{P_i}^2 - m_{P_f}^2}{q^2} \right) q_\mu \right] f^+_{P_i P_f}(q^2)$$

$$+ \frac{m_{P_i}^2 - m_{P_f}^2}{q^2} q_\mu \, f^0_{P_i P_f}(q^2) \,, \tag{2.96}$$

where $f^+_{P_i P_f}$ is the *vector form factor* and $f^0_{P_i P_f}$ is the *scalar form factor*, defined as

$$f^0_{P_i P_f}(q^2) = f^+_{P_i P_f}(q^2) + \frac{q^2}{m_{P_i}^2 - m_{P_f}^2} f^-_{P_i P_f}(q^2) \,, \tag{2.97}$$

so that

$$f^0_{P_i P_f}(0) = f^+_{P_i P_f}(0) \,, \tag{2.98}$$

and there is no divergence at $q^2 = 0$ in (2.96).

The name of the form factor $f^0_{P_i P_f}$ reflects its relation to the hadronic matrix element of the scalar ($J^P = 0^+$) current $\bar{q}' q$. In order to find that relation, we insert the equation (2.95) for the divergence of vector current between the P_i and P_f states:

$$\langle P_f(p_2) | \partial^\mu j_\mu | P_i(p_1) \rangle = i(m_q - m_{q'}) \langle P_f(p_2) | \bar{q} q' | P_i(p_1) \rangle \,. \tag{2.99}$$

The same divergence is then replaced by a commutator of the current with the momentum operator – a relation analogous to (2.55). Repeating similar steps as in (2.56), but this time for the $P_i \to P_f$ matrix element, we substitute the decomposition (2.96) and obtain:

$$\langle P_f(p_2)|i\partial^\mu j_\mu|P_i(p_1)\rangle = q^\mu \langle P_f(p_2)|j_\mu|P_i(p_1)\rangle = (m_{P_i}^2 - m_{P_f}^2)f_{P_iP_f}^0(q^2). \qquad (2.100)$$

Here only $f_{P_iP_f}^0$ contributes, because in (2.96) the vector multiplying $f_{P_iP_f}^+$ is orthogonal to q_μ:

$$(2p_2 q) + q^2 - (m_{P_i}^2 - m_{P_f}^2) = 0, \qquad (2.101)$$

as readily follows from squaring the momentum conservation condition $(p_2 + q)^2 = p_1^2$ and replacing $p_{1,2}^2 = m_{P_i,P_f}^2$. Finally, comparing (2.99) and (2.100) we have the required relation:

$$\langle P_f(p_2)|(m_{q'} - m_q)\bar{q}q'|P_i(p_1)\rangle = (m_{P_i}^2 - m_{P_f}^2)f_{P_iP_f}^0(q^2). \qquad (2.102)$$

Let us now obtain the $P_i \to P_f \ell \bar{\nu}_\ell$ decay width in terms of the form factors. Since these are functions of q^2, the most convenient observable is the differential decay width $d\Gamma/dq^2$. It is experimentally accessible by measuring the decay distribution in the P_f-meson momentum $|\vec{p}_2|$. This variable is directly related to q^2 in the rest frame of the decaying meson, where $p_1 = (m_{P_i}, \vec{0})$ and

$$|\vec{p}_2| = \frac{\lambda^{1/2}(m_{P_i}^2, m_{P_f}^2, q^2)}{2m_{P_i}}, \quad E_2 = \frac{m_{P_i}^2 + m_{P_f}^2 - q^2}{2m_{P_i}}, \qquad (2.103)$$

generalizing the relation (2.75). In particular, the minimal (maximal) value of q^2 specified in (2.93) corresponds to the maximal (zero) recoil of the meson P_f.

A formula for the differential width of a three-body decay, such as $P_i \to P_f \ell \bar{\nu}_\ell$, is given in (B.56). To use it, we need to square the amplitude (2.92), to sum over the lepton polarizations and to represent the result in a form of the polynomial in the second invariant variable, which is in this case

$$(k_2 - k_1)p_2 \equiv rp_2.$$

From the very beginning, it is convenient to substitute in (2.92) the form factor decomposition (2.96). Introducing a short-hand notation:

$$\kappa(q^2) \equiv \frac{m_{P_i}^2 - m_{P_f}^2}{q^2},$$

replacing q_μ by $k_{1\mu} + k_{2\mu}$ and using Dirac equations for the lepton bispinors,

$$\bar{u}_\ell(k_2)\not{k}_2 = m_\ell \bar{u}_\ell(k_2), \quad \not{k}_1 v_\nu(k_1) = 0,$$

we obtain

$$\mathcal{A}(P_i \to P_f \ell \bar{\nu}_\ell) = \frac{G_F}{\sqrt{2}}V_{qq'}\left\{ 2\,\bar{u}_\ell(k_2)\not{p}_2(1-\gamma_5)v_\nu(k_1)f_{P_iP_f}^+(q^2) \right.$$

$$\left. + m_\ell \bar{u}_\ell(k_2)(1-\gamma_5)v_\nu(k_1)\left[f_{P_iP_f}^+(q^2)\left(1 - \kappa(q^2)\right) + f_{P_iP_f}^0(q^2)\kappa(q^2) \right] \right\}. \qquad (2.104)$$

The lepton mass factor is analogous to the one emerging in the leptonic decay amplitudes. We notice that if m_ℓ is neglected, as for example in the semileptonic decays of heavy mesons to muon or electron, only the vector form factor $f_{P_iP_f}^+(q^2)$ contributes to the decay width.

The modulus square of the amplitude (2.104) can be written as

$$|\mathcal{A}(P_i \to P_f \ell^- \bar{\nu}_\ell)|^2 = \frac{G_F^2 |V_{qq'}|^2}{2} \left\{ 2 \left[\bar{u}_\ell \slashed{p}_2 (1 - \gamma_5) v_\nu \right] f_{P_i P_f}^+ \right.$$

$$\left. + m_\ell \left[\bar{u}_\ell (1 - \gamma_5) v_\nu \right] \left((1 - \kappa) f_{P_i P_f}^+ + \kappa f_{P_i P_f}^0 \right) \right\} \left\{ \dots \right\}^* , \tag{2.105}$$

where the second curly bracket contains the complex conjugate of the first bracket, and, to shorten the expressions, we omit the momenta at bispinors and form factors. Note that both form factors are real-valued functions of q^2 in the kinematical region (2.93). This property will be extensively used in Chapter 5.

We transform the complex conjugated products of bispinors using (A.25):

$$\left[\bar{u}_\ell \slashed{p}_2 (1 - \gamma_5) v_\nu \right]^* = \bar{v}_\nu \slashed{p}_2 (1 - \gamma_5) u_\ell , \quad \left[\bar{u}_\ell (1 - \gamma_5) v_\nu \right]^* = \bar{v}_\nu (1 + \gamma_5) u_\ell . \tag{2.106}$$

After that, the sum over lepton polarizations can be taken with the help of (A.26), yielding traces similar to (A.27):

$$\overline{|\mathcal{A}(P_i \to P_f \ell^- \bar{\nu}_\ell)|^2} = G_F^2 |V_{qq'}|^2 \left\{ 4 |f_{P_i P_f}^+|^2 \, \text{Tr} \left[(\slashed{k}_2 + m_\ell) \slashed{p}_2 \slashed{k}_1 \slashed{p}_2 \right] \right.$$

$$+ 2 m_\ell f_{P_i P_f}^+ \left((1 - \kappa) f_{P_i P_f}^+ + \kappa f_{P_i P_f}^0 \right) \left(\text{Tr} \left[(\slashed{k}_2 + m_\ell) \slashed{p}_2 \slashed{k}_1 \right] + \text{Tr} \left[(\slashed{k}_2 + m_\ell) \slashed{k}_1 \slashed{p}_2 \right] \right)$$

$$\left. + m_\ell^2 \left| (1 - \kappa) f_{P_i P_f}^+ + \kappa f_{P_i P_f}^0 \right|^2 \text{Tr} \left[(\slashed{k}_2 + m_\ell) \slashed{k}_1 \right] \right\} . \tag{2.107}$$

In these traces the parts containing γ_5 vanish. Indeed, the only nonvanishing trace with γ_5 that could emerge in the above expression has to contain four independent momentum four-vectors contracted with the totally antisymmetric tensor $\epsilon_{\mu\nu\alpha\beta}$, whereas we only have three independent momenta at our disposal.

To proceed, we consider first the interference of two form factors in the squared and averaged amplitude. Collecting the terms proportional to the product $f_{P_i P_f}^+(q^2) f_{P_i P_f}^0(q^2)$ from the second and third lines in (2.107), and computing the traces, we find that the interference term is equal to

$$8 m_\ell^2 f_{P_i P_f}^+ f_{P_i P_f}^0 \kappa \left[2(k_1 p_2) + (1 - \kappa)(k_1 k_2) \right] . \tag{2.108}$$

Expressing the scalar products in terms of the variables q^2 and rp_2:

$$2 k_1 k_2 = q^2 - m_\ell^2 , \quad 2 k_1 p_2 = qp_2 - rp_2 = -\frac{1}{2} q^2 (1 - \kappa) - rp_2 , \tag{2.109}$$

where (2.101) is used in the last equation, we transform the interference term to

$$8 m_\ell^2 f_{P_i P_f}^+ f_{P_i P_f}^0 \kappa \left[-\frac{1}{2} q^2 (1 - \kappa) - rp_2 + \frac{1}{2} (1 - \kappa)(q^2 - m_\ell^2) \right] . \tag{2.110}$$

In this expression, integration over the phase space will transform only the term proportional to rp_2 (see Appendix B) . According to (B.62), we have to replace

$$rp_2 \to \frac{1}{2} m_\ell^2 (\kappa - 1) , \tag{2.111}$$

hence the interference term (2.110) vanishes. This is also expected from a physical point of view, because the vector and scalar form factors correspond to the contributions of different spin components of the flavor-changing current.

The two terms remaining in (2.107) are proportional to the squares of the form factors. Taking traces, we obtain:

$$
\overline{|\mathcal{A}(P_i \to P_f \ell^- \bar{\nu}_\ell)|^2} = 4G_F^2 |V_{qq'}|^2 \Bigg\{ \Big[8(k_2 p_2)(k_1 p_2) - 4(k_1 k_2) m_{P_f}^2
$$

$$
+ m_\ell^2 (1 - \kappa(q^2))^2 (k_1 k_2) + 4m_\ell^2 (1 - \kappa(q^2))(k_1 p_2) \Big] |f_{P_i P_f}^+ (q^2)|^2
$$

$$
+ m_\ell^2 \kappa(q^2)^2 (k_1 k_2) |f_{P_i P_f}^0 (q^2)|^2 \Bigg\}, \tag{2.112}
$$

which is then reduced to a form depending only on q^2 and $r p_2$. This is done with the help of (2.111) and the additional relation:

$$
4(k_1 p_2)(k_2 p_2) = (q p_2)^2 - (r p_2)^2.
$$

We encounter only terms up to the second power of $r p_2$ for which the integration formulas (B.62) and (B.63) are used. After that the expression for the squared and averaged amplitude becomes:

$$
\overline{|\mathcal{A}(P_i \to P_f \ell^- \bar{\nu}_\ell)|^2} = 2G_F^2 |V_{qq'}|^2 \frac{(q^2 - m_\ell^2)}{(q^2)^2} \Bigg\{ \frac{(2q^2 + m_\ell^2)}{3} \lambda(m_{P_i}^2, m_{P_f}^2, q^2) |f_{P_i P_f}^+ (q^2)|^2
$$

$$
+ m_\ell^2 (m_{P_i}^2 - m_{P_f}^2)^2 |f_{P_i P_f}^0 (q^2)|^2 \Bigg\}. \tag{2.113}
$$

Using (B.56), we obtain the differential width of the semileptonic decay:

$$
\frac{d\Gamma(P_i \to P_f \ell \bar{\nu}_\ell)}{dq^2} = \frac{G_F^2 |V_{qq'}|^2}{24\pi^3} \frac{(q^2 - m_\ell^2)^2}{(q^2)^2 m_{P_i}^2} \Bigg\{ \Big(1 + \frac{m_\ell^2}{2q^2} \Big) m_{P_i}^2 k_f^3 |f_{P_i P_f}^+ (q^2)|^2
$$

$$
+ \frac{3m_\ell^2}{8q^2} (m_{P_i}^2 - m_{P_f}^2)^2 k_f |f_{P_i P_f}^0 (q^2)|^2 \Bigg\}, \tag{2.114}
$$

where $k_f \equiv \lambda^{1/2}(m_{P_i}^2, m_{P_f}^2, q^2)/(2m_{P_i})$ is equal to the absolute value of the three-momentum $|\vec{p}_2|$ of the final P_f meson in the rest frame of the decaying P_i meson.

Another, albeit rarely used observable for the $P_i \to P_f \ell \nu_\ell$ decay is the differential width with respect to the charged lepton energy E_ℓ in the rest frame of P_i. Here we quote the formula for the massless lepton, $m_\ell = 0$, in which case only the vector form factor contributes:

$$
\frac{d\Gamma(P_i \to P_f \ell \nu_\ell)}{dE_\ell} = \frac{G_F^2 |V_{qq'}|^2}{16\pi^3 m_{P_i}} \int_0^{q^2_{max}} dq^2 [2E_\ell (m_{P_i}^2 - m_{P_f}^2 + q^2) - m_{P_i}(q^2 + 4E_\ell^2)] |f_{P_i P_f}^+ (q^2)|^2. \tag{2.115}
$$

The upper limit of the q^2 integration depends on E_ℓ:

$$
q^2_{max} = 2E_\ell \Big(m_{P_i} - \frac{m_{P_f}^2}{m_{P_i} - 2E_\ell} \Big),
$$

and the kinematical region of the lepton energy is

$$0 < E_\ell < \frac{m_{P_i}^2 - m_{P_f}^2}{2m_{P_i}} . \tag{2.116}$$

More details on the kinematics of semileptonic decays can be found in [33].

So far we were considering semileptonic decays (2.91) generated by the flavor-changing weak interaction (1.29) mediated by a virtual W-boson exchange. Another type of semileptonic decays emerges in SM due to the flavor-changing neutral currents (FCNC) stemming from the short-distance loop diagrams. Prominent examples are the rare $B \to K\ell^+\ell^-$ and $B \to \pi\ell^+\ell^-$ decays generated by the effective $b \to s$ operators (1.43) and, respectively, by their $b \to d$ counterparts. Being of different origin, these decays have many similarities with the weak semileptonic decays $B \to \pi\ell\nu_\ell$ or $B_s \to K\ell\nu_\ell$ in what the form factors are concerned. Indeed, the operators $O_{9,10}$ also contain a combination of the vector and axial-vector quark currents, $\bar{s}\gamma_\mu(1-\gamma_5)b$. Accordingly, the contributions of these operators to the $B \to K\ell^+\ell^-$ decay amplitudes (where only the vector part of the $b \to s$ current survives) are parametrized by the two form factors $f_{BK}^+(q^2)$ and $f_{BK}^0(q^2)$, defined as in (2.96) with $P_i = B, P_f = K$. Replacing the s-quark by u-quark, we return to the $B^- \to \pi^0\ell\bar{\nu}_\ell$ form factors. Moreover, the latter form factors are related to the ones in the rare $B \to \pi\ell^+\ell^-$ decays via isospin symmetry. The structure of the FCNC amplitudes will be considered in more detail in Chapter 6. Here we concentrate only on the additional form factor that emerges due to the operator $O_{7\gamma}$ in (1.43). It contributes to $B \to K\ell^+\ell^-$, being coupled, via virtual photon exchange, to the lepton e.m. current. The operator $O_{7\gamma}$ contains an effective tensor quark current

$$\bar{s}\sigma_{\mu\nu}\big[m_b(1+\gamma_5) + m_s(1-\gamma_5)\big]b , \tag{2.117}$$

where, due to P-parity conservation, in the $B \to K$ hadronic matrix element only the part of the current without γ_5 is left. Having at our disposal two independent four-vectors $p_{1\mu}$ and $p_{2\mu}$, we form the single possible antisymmetric Lorentz structure:

$$\langle K(p_2)|\bar{s}\sigma_{\mu\nu}b|\bar{B}(p_1)\rangle = 2\big(p_{2\mu}p_{1\nu} - p_{1\mu}p_{2\nu}\big)\frac{if_{BK}^T(q^2)}{m_B + m_K} , \tag{2.118}$$

where the meson mass factor in the denominator is chosen to render the form factor dimensionless. Note that in the effective operator $O_{7\gamma}$ the tensor quark current is multiplied by the photon field-strength tensor. The latter, for a photon with momentum q, has the form $F_{\mu\nu} = q_\mu A_\nu(q) - q_\nu A_\mu(q)$, hence we need to multiply the above matrix element by q^ν, arriving at the standard parameterization of the tensor form factor:

$$\langle K(p_2)|\bar{s}\sigma_{\mu\nu}q^\nu b|\bar{B}(p_1)\rangle = \Big[2q^2 p_{2\mu} - \big(m_B^2 - m_K^2 - q^2\big)q_\mu\Big]\frac{if_{BK}^T(q^2)}{m_B + m_K} . \tag{2.119}$$

To complete our discussion of hadronic matrix elements relevant for the flavor changing decays of pseudoscalar mesons, we consider their weak leptonic decays

$$P \to \ell\bar{\nu}_\ell ,$$

representing the flavor variations of the $\pi^- \to \ell^-\bar{\nu}_\ell$ decay that we already studied. All possible leptonic decay modes are listed in Table 2.1 with the corresponding quark currents and CKM matrix elements. Note that for heavy pseudoscalar mesons the modes with $\ell = \tau$ are not only kinematically accessible but in fact dominant, due to the lepton mass factor in the amplitude.

A leptonic decay is mediated by the axial-vector part of the quark weak current, $j_{\mu 5} = \bar{q}\gamma_\mu\gamma_5 q'$ ($q = u, c,$ $q' = d, s, b$). The corresponding hadronic matrix element

$$\langle 0|j_{\mu 5}|P(p)\rangle = ip_\mu f_P\,, \tag{2.120}$$

is determined by the decay constant f_P, in full analogy with (2.48). The alternative definition involves the hadronic matrix element of the pseudoscalar quark current (cf. (2.57)):

$$\langle 0|(m_q + m'_q)\bar{q}i\gamma_5 q'|P(p)\rangle = m_P^2 f_P\,. \tag{2.121}$$

Finally, the leptonic decay width

$$\Gamma(P \to \ell\bar{\nu}_\ell) = \frac{G_F^2|V_{qq'}|^2}{8\pi} m_\ell^2 m_P \left(1 - \frac{m_\ell^2}{m_P^2}\right)^2 f_P^2 \tag{2.122}$$

is a straightforward generalization of (2.61).

Numerically, the interval of the kaon decay constant extracted [1] from the measured $K \to \mu\nu_\mu$ width is

$$f_K \simeq 155.72 \pm 0.50 \text{ MeV}\,, \tag{2.123}$$

where the corresponding CKM parameter $|V_{us}|$ is determined independently. Comparing with f_π in (2.64), we reveal a noticeable violation of $SU(3)_{fl}$ symmetry. Being essentially nonperturbative quantities, meson decay constants are calculated using numerical simulation of QCD on the lattice. For the meson decay constants listed in Table 2.1, the results can be found in the periodically updated review [34]. Lattice QCD predictions for f_π and f_K have an impressive agreement with (2.64) and (2.123). Typical values obtained for decay constants of charmed and bottom pseudoscalar mesons are somewhat larger, in the ballpark of 200–250 MeV. In Chapter 8 we will introduce another QCD-based method, enabling to calculate the hadronic decay constants analytically, albeit with larger uncertainties.

A knowledge of the decay constants $f_{D_{(s)}}$ and $f_{B_{(s)}}$ is indispensable for testing SM predictions for the heavy meson leptonic decays, including determination of the corresponding CKM parameters. For instance, the B_s-meson decay constant essentially determines the SM prediction for the rare leptonic decay $B_s \to \ell^+\ell^-$. This decay is mediated by the quark-lepton operator O_{10} entering the effective Hamiltonian (1.44). The decay amplitude

$$\begin{aligned}
\mathcal{A}(\bar{B}_s \to \ell^+\ell^-) &= -\langle \ell^+(p_{\ell+})\ell^-(p_{\ell-})|\mathcal{H}_{eff}^{b\to s}|\bar{B}_s^0(p)\rangle\,, \\
&= \frac{G_F}{\sqrt{2}} V_{tb}V_{ts}^* C_{10}\langle \ell^+(p_{\ell+})\ell^-(p_{\ell-})|O_{10}|\bar{B}_s^0(p)\rangle\,, \tag{2.124}
\end{aligned}$$

after factorizing the leptonic part, retaining only the axial-vector quark current and using the definition (2.120), becomes:

$$\begin{aligned}
\mathcal{A}(\bar{B}_s \to \ell^+\ell^-) &= -\frac{\alpha_{em}G_F}{2\pi\sqrt{2}} V_{tb}V_{ts}^* C_{10}\,\bar{u}_\ell(p_{\ell-})\gamma^\mu\gamma_5 v_\ell(p_{\ell+})\langle 0|\bar{s}\gamma_\mu\gamma_5 b|\bar{B}_s(p)\rangle \\
&= -i\frac{\alpha_{em}G_F}{2\pi\sqrt{2}} V_{tb}V_{ts}^* C_{10}\, 2m_\ell\bar{u}_\ell(p_{\ell-})\gamma_5 v_\ell(p_{\ell+})f_{B_s}\,, \tag{2.125}
\end{aligned}$$

where the Wilson coefficient C_{10} stemming from the short-distance loop diagrams with Z-boson and t-quark, is very precisely calculated in SM (see, e.g., the review [14]). The remaining FCNC operators O_9, $O_{7\gamma}$ in $\mathcal{H}_{eff}^{b\to s}$ do not contribute to this decay amplitude, because the lepton vector current is conserved and vanishes, being multiplied by p_μ. The underlying reason is the angular momentum conservation, forbidding a transition of a pseudoscalar meson to a lepton vector current or to a photon. The width of the B_s leptonic decay is calculated from (2.125) along the same lines as for the weak leptonic decays, with minor differences in the phase space factor, resulting in

$$\Gamma(\bar{B}_s \to \ell^+\ell^-) = \frac{\alpha_{em}^2 G_F^2}{16\pi^3}|V_{tb}|^2|V_{ts}|^2|C_{10}|^2 m_\ell^2 m_{B_s}\sqrt{1 - \frac{4m_\ell^2}{m_{B_s}^2}}f_{B_s}^2. \tag{2.126}$$

In the rest of this section we examine the form factors of weak semileptonic decays with vector mesons ($\rho, \omega, K^*, \phi, D_{(s)}^*$) in the final state:

$$P \to V\ell\bar{\nu}_\ell. \tag{2.127}$$

In fact, only $B_{(s,c)}$ and $D_{(s)}$ mesons are heavy enough to provide sufficient phase space for these decay modes. Since vector mesons are unstable hadrons, the processes (2.127) are observed indirectly, reconstructing a vector meson from the products of its strong decay. We will discuss this issue in Chapter 5. Here we treat vector mesons as asymptotic final states with fixed masses – an assumption that involves a certain approximation.

The hadronic matrix elements of (2.127) differ in two aspects from the ones of $P_i \to P_f\ell\bar{\nu}_\ell$ decays considered so far. First, the complete weak current (1.30), consisting of the axial-vector and vector parts, contributes:

$$j_\mu^W = j_\mu - j_{\mu 5} = \bar{q}\gamma_\mu(1 - \gamma_5)q' \quad \text{or} \quad j_\mu^{W\dagger} = j_\mu^\dagger - j_{\mu 5}^\dagger = \bar{q}'\gamma_\mu(1 - \gamma_5)q.$$

Second, a vector meson has spin $J = 1$ and is therefore characterized, apart from momentum, by its polarization. The three polarization states, two of them transverse and one longitudinal, are described by the unit four-vectors $\varepsilon^\mu(\lambda)$ ($\lambda = +, -, 0$) orthogonal to the vector meson momentum $p_2 = (E_2, \vec{p}_2)$. In the rest frame of the initial P meson, $p_1 = (0, m_P)$, choosing $\vec{p}_2 = (0, 0, |\vec{p}_2|)$ in the z-direction, the polarization vectors are

$$\varepsilon^\mu(\pm) = \mp\frac{1}{\sqrt{2}}(0, 1, \pm i, 0), \quad \varepsilon^\mu(0) = \frac{1}{m_V}(|\vec{p}_2|, 0, 0, E_2), \tag{2.128}$$

so that $\varepsilon(\lambda) \cdot p_2 = 0$. Hereafter, we omit the polarization label λ for brevity.

To construct a Lorentz decomposition of the matrix element $\langle V|j_\mu^W|P\rangle$, we use the three independent four-vectors q, p_2 and ε. Note that by default each structure contains ε^*, representing in the decay amplitude a vector meson in the final state.

We consider first the $P \to V$ matrix element of the weak vector current. Following the line of arguments presented in Section 2.4, we replace the vector current by an intermediate vector meson V' with the flavor content $q\bar{q}'$, and, respectively, the matrix element $\langle V(p_2)|j_\mu|P(p_1)\rangle$ by the three-hadron vertex $P(p_1)V(p_2)V'(q)$. The latter is described by a hadronic matrix element (where, as usual, $\hat{S}_{QCD}$ is implied but not shown) which has a single possible Lorentz structure:

$$\langle V(p_2)V'(q)|P(p_1)\rangle = \varepsilon_{\mu\nu\rho\sigma}\varepsilon'^{*\mu}\varepsilon^{*\nu}q^\rho p_2^\sigma g_{V'VP}, \tag{2.129}$$

where ε' is the polarization vector of V' and $g_{V'VP}$ is the invariant constant named the $V'VP$ *strong coupling*. Contraction of four-vectors with the Levi-Civita pseudotensor provides the only possibility for the two vector mesons to form a state $\langle VV'|$ with the same negative P-parity as the pseudoscalar meson P. In other words, it is implied that on the r.h.s. of (2.129) the pseudoscalar meson is represented by an additional pseudoscalar factor, which is not shown explicitly but flips its sign under P-transformation. All other contractions, e.g., $(\varepsilon'^*p_2)(\varepsilon^*q)$ would then have violated the P-parity conservation. Accordingly, we have a single choice for the Lorentz structure in the hadronic matrix element:

$$\langle V(p_2)|j_\mu|P(p_1)\rangle = \epsilon_{\mu\nu\rho\sigma}\varepsilon^{*\nu}q^\rho p_2^\sigma \frac{2V^{PV}(q^2)}{m_P + m_V}, \tag{2.130}$$

where we use a customary notation and normalization factor for the dimensionless form factor $V^{PV}(q^2)$.

The matrix element of the weak axial-vector current has a richer kinematical structure. To define its decomposition in Lorentz structures, we can, by the same token as above, relate this matrix element to the superposition of AVP and $P'VP$ hadronic vertices, containing respectively, the couplings of an axial ($J^P = 1^+$) and pseudoscalar meson ($J^P = 0^-$) to the vector and pseudoscalar meson. The first coupling considered as a hadronic amplitude of the $A \to VP$ decay has two independent partial waves, $L = 0$ and $L = 2$, where L is the orbital momentum between V and P, $L = 1$ is forbidden by P-parity conservation. One more amplitude corresponds to the $P'VP$ vertex, where $L = 1$; hence, altogether the general decomposition of the hadronic matrix element we are interested in consists of three independent Lorentz structures:

$$
\begin{aligned}
\langle V(p_2)|j_{\mu 5}|P(p_1)\rangle &= i\varepsilon_\mu^*(m_P + m_V)A_1^{PV}(q^2) - 2ip_{2\mu}(\varepsilon^*q)\frac{A_2^{PV}(q^2)}{m_P + m_V} \\
&+ iq_\mu(\varepsilon^*q)\bar{A}(q^2),
\end{aligned}
\tag{2.131}
$$

and, correspondingly, is determined by three independent form factors. Two of them, A_1^{PV} and A_2^{PV}, have a conventional normalization, whereas $\bar{A}(q^2)$ is replaced by an additional form factor determining the $P \to V$ matrix element of the pseudoscalar current, analogous to the scalar form factor $f_{P_iP_f}^0$ in the transitions to pseudoscalar mesons.

The defining relation for the additional form factor is:

$$
\begin{aligned}
2m_V A_0^{PV}(q^2)(\varepsilon^*q) &= \langle V(p_2,\varepsilon)|i(m_q + m_{q'})\bar{q}\gamma_5 q'|P(p_1)\rangle \\
&= -iq^\mu\langle V(p_2)|j_{\mu 5}|P(p_1)\rangle,
\end{aligned}
\tag{2.132}
$$

where we also use connection between the pseudoscalar current and the derivative of the axial-vector current (see (2.52)). Replacing the hadronic matrix element on the r.h.s. by decomposition (2.131), we find a relation between A_0^{PV} and the other three form factors. Solving it with respect to the form factor $\bar{A}$ yields:

$$
q^2\bar{A}(q^2) = 2m_V A_0^{PV}(q^2) - (m_P + m_V)A_1^{PV}(q^2) + \frac{2(p_2 q)}{m_P + m_V}A_2^{PV}(q^2).
\tag{2.133}
$$

Transforming the last term in the above,

$$
\frac{2(p_2 q)}{m_P + m_V}A_2^{PV}(q^2) = (m_P - m_V)A_2^{PV}(q^2) - \frac{q^2}{m_P + m_V}A_2^{PV}(q^2),
$$

we introduce a new notation

$$
2m_V A_3^{PV}(q^2) \equiv (m_P + m_V)A_1^{PV}(q^2) - (m_P - m_V)A_2^{PV}(q^2),
\tag{2.134}
$$

for the combination of two form factors, converting (2.133) into:

$$
q^2\bar{A}(q^2) = 2m_V\left(A_0^{PV}(q^2) - A_3^{PV}(q^2)\right) - \frac{q^2}{m_P + m_V}A_2^{PV}(q^2).
\tag{2.135}
$$

From this equation follows an important link between the form factors at $q^2 = 0$:

$$
A_0^{PV}(0) = A_3^{PV}(0).
$$

With the help of (2.135), we then exclude the form factor $\bar{A}$ from (2.131). Adding together the vector and axial-vector parts, we obtain the final form of the $P \to V$ hadronic matrix element of the weak current[9]:

$$\langle V(p_2)|j_\mu^W|P(p_1)\rangle = \epsilon_{\mu\nu\rho\sigma}\varepsilon^{*\nu}q^\rho p_2^\sigma \frac{2V^{PV}(q^2)}{m_P+m_V} - i\varepsilon_\mu^*(m_P+m_V)A_1^{PV}(q^2)$$

$$+i(2p_2+q)_\mu(\varepsilon^*q)\frac{A_2^{PV}(q^2)}{m_P+m_V} + iq_\mu(\varepsilon^*q)\frac{2m_V}{q^2}\left(A_3^{PV}(q^2)-A_0^{PV}(q^2)\right). \qquad (2.136)$$

For the FCNC decays such as $B \to K^*\ell^+\ell^-$, where $P = B$ and $V = K^*$, in addition to the above definition, one also needs to specify the form factors of the quark tensor current (2.117) contained in the effective operator $O_{7\gamma}$. In the $B \to K^*$ transition both parity components of this current contribute (cf. (2.119) for the $B \to K$ transition). In what follows, we neglect the suppressed part of the tensor $b \to s$ current proportional to m_s which can be easily restored at the end by altering the sign at the contributions with γ_5. We also omit the m_b factor multiplying the current, concentrating on the hadronic matrix element. The important point is that the two parity components of the tensor current contain combinations of Dirac matrices that are related to each other. Rewriting the relevant equality in (A.9)):

$$\sigma_{\mu\nu} = -\frac{i}{2}\epsilon_{\mu\nu\alpha\beta}\sigma^{\alpha\beta}\gamma_5\,,$$

we represent the $B \to K^*$ matrix element of the tensor current as

$$\langle K^*(p_2)|\bar{s}\sigma^{\mu\nu}(1+\gamma_5)b|\bar{B}(p_1)\rangle = \langle K^*(p_2)|\bar{s}\sigma_{\alpha\beta}\gamma_5 b|\bar{B}(p_1)\rangle\left(-\frac{i}{2}\epsilon^{\mu\nu\alpha\beta}+g^{\mu\alpha}g^{\nu\beta}\right). \quad (2.137)$$

The matrix element now contains only the pseudotensor current. The next step is to use the antisymmetry of the σ-matrix and express this matrix element as a linear combination of three possible Lorentz structures built from the three independent four-vectors chosen as ε^* (the polarization vector of K^*), q and $p_1 + p_2 = 2p_2 + q$:

$$\langle K^*(p_2)|\bar{s}\sigma_{\alpha\beta}\gamma_5 b|\bar{B}(p_1)\rangle = \left[\varepsilon_\alpha^*(p_1+p_2)_\beta - \varepsilon_\beta^*(p_1+p_2)_\alpha\right]t_1(q^2)$$

$$+\left[\varepsilon_\alpha^* q_\beta - \varepsilon_\beta^* q_\alpha\right]t_2(q^2) + (\varepsilon^*\cdot q)\left[q_\alpha(p_1+p_2)_\beta - q_\beta(p_1+p_2)_\alpha\right]t_3(q^2)\,. \quad (2.138)$$

Hence, there are altogether three independent form factors, expressed temporarily via the functions $t_{1,2,3}(q^2)$. Substituting the above decomposition in (2.137), we multiply both parts of this equation by q_ν, keeping in mind that in the decay amplitude we practically only need the contracted current, as in (2.119). The result is:

$$\langle K^*(p_2)|\bar{s}\sigma_{\mu\nu}q^\nu(1+\gamma_5)b|\bar{B}(p_1)\rangle = i\epsilon_{\mu\nu\alpha\beta}\varepsilon^{*\nu}q^\alpha p_2^\beta 2t_1(q^2)$$

$$+\left[\varepsilon_\mu^*(m_B^2-m_{K^*}^2) - (\varepsilon^*\cdot q)(p_1+p_2)_\mu\right]t_1(q^2) + \left[\varepsilon_\mu^* q^2 - (\varepsilon^*\cdot q)q_\mu\right]t_2(q^2)$$

$$+(\varepsilon^*\cdot q)\left[q_\mu(m_B^2-m_{K^*}^2) - q^2(p_1+p_2)_\mu\right]t_3(q^2)\,, \quad (2.139)$$

where we used that $(p_1+p_2)q = m_B^2 - m_{K^*}^2$. Renaming the invariant functions and their linear combinations:

$$t_1(q^2) = T_1^{BK^*}(q^2)$$

$$t_1(q^2) + \frac{q^2}{m_B^2-m_{K^*}^2}t_2(q^2) = T_2^{BK^*}(q^2)\,,$$

$$(m_B^2-m_{K^*}^2)t_3(q^2) - t_2(q^2) = T_3^{BK^*}(q^2)\,, \quad (2.140)$$

---

[9]The form factors can be redefined, changing their constant coefficients and global phase, without influencing the observables.

we introduce the three tensor form factors $T_{1,2,3}^{BK^*}(q^2)$ of the $B \to K^*$ transitions. Note that by definition

$$T_1^{BK^*}(0) = T_2^{BK^*}(0). \tag{2.141}$$

After expressing $t_{1,2,3}$ in (2.139) in terms of these form factors, the hadronic matrix element of the tensor current is obtained:

$$\langle K^*(p_2)|\bar{s}\sigma^{\mu\nu}q_\nu(1+\gamma_5)b|\bar{B}(p_1)\rangle = 2i\epsilon^{\mu\nu\alpha\beta}\varepsilon_\nu^* q_\alpha p_{2\beta}\, T_1^{BK^*}(q^2)$$
$$+ \left[(m_B^2 - m_{K^*}^2)\varepsilon^{*\mu} - (\varepsilon^* \cdot q)(2p_2 + q)^\mu\right]T_2^{BK^*}(q^2)$$
$$+ (\varepsilon^* \cdot q)\left[q^\mu - \frac{q^2(2p_2 + q)^\mu}{m_B^2 - m_{K^*}^2}\right]T_3^{BK^*}(q^2). \tag{2.142}$$

Differential width of the $B \to K^*\ell^+\ell^-$ decay and of other semileptonic decays, involving vector meson in the final state, is derived, squaring the decay amplitude and then following basically the same steps as in the calculation of the width (2.114) of semileptonic decays into pseudoscalar meson. Due to the presence of several form factors caused by polarization degrees of freedom of the vector meson, this procedure is technically more involved, hence we will not dwell on it. A detailed derivation including obtaining specific observables (angular distributions and asymmetries), relevant for the uses of FCNC decays in SM phenomenology, can be found e.g., in [35],[36].

2.6   FORM FACTORS IN HQET

For heavy-light hadrons such as a D or B-meson, it is conceivable to redefine or rescale their decay constants and form factors, adapting them to HQET and simplifying the transition to the infinitely heavy quark limit.

We begin with the decay constant of a heavy meson $H = D, B$ defined according to (2.120) as

$$\langle 0|\bar{q}\gamma_\mu\gamma_5 Q|H(p_H)\rangle = ip_\mu f_H, \tag{2.143}$$

where $q = u, d, s$, $Q = c, b$. Introducing the velocity four-vector $v_\mu = p_{H\mu}/m_H$ and transforming the heavy quark field into an effective one:

$$Q(0) \to h_v(0),$$

we also redefine the heavy meson state to the HQET state, changing the normalization according to (1.144). The definition (2.143) transforms to

$$\langle 0|\bar{q}\gamma_\mu\gamma_5 h_v|H(v)\rangle = i\sqrt{m_H}v_\mu f_H, \tag{2.144}$$

so that the hadronic matrix element on the l.h.s. is independent of the heavy mass scale and is parametrized with the only available velocity four-vector:

$$\langle 0|\bar{q}\gamma_\mu\gamma_5 h_v|H(v)\rangle = iv_\mu\hat{f}. \tag{2.145}$$

It is evident that the new parameter $\hat{f}$ does not scale with the heavy mass. It is called the "static" decay constant of the H-meson, keeping in mind the static nature of the infinitely heavy quark. Comparing (2.145) with (2.144), we relate the two definitions[10]:

$$f_H = \frac{\hat{f}}{\sqrt{m_H}}. \tag{2.146}$$

---

[10]Note also that the specific radiative corrections (1.152), generated at the level of quark currents, have to be added to this relation.

At this point we have just redefined the decay constant for a given heavy hadron. But, in fact, (2.146) contains more information: it predicts that in the limit $m_H \to \infty$ there exists a universal static decay constant $\hat{f}$ determined entirely by the nonperturbative QCD dynamics and decoupled from the heavy mass scale. Combining (2.146) for B and D mesons, we also predict the ratio of their decay constants:

$$\frac{f_B}{f_D} = \sqrt{\frac{m_D}{m_B}}. \tag{2.147}$$

These constants for the physical D and B mesons containing c and b quarks with large but finite masses, are calculable in the lattice QCD or using the QCD sum rules, to be discussed in Chapter 8. Alternatively, f_B and f_D can be extracted from the measured leptonic B and D decays, provided the relevant CKM parameters are known independently. A comparison of the results on f_B and f_D with the above scaling relation reveals appreciable numerical corrections caused by an imperfect heavy mass limit; hence, it is more conceivable to include in (2.146) the first-order correction in the inverse heavy mass:

$$f_H = \frac{\hat{f}}{\sqrt{m_H}} \left(1 + \frac{\delta}{m_H}\right). \tag{2.148}$$

Probably, the most important applications of HQET are the heavy-to-heavy meson semileptonic decays $\bar{B}(p_1) \to D^{(*)}(p_2)\ell\bar{\nu}_\ell$. To investigate the heavy-quark limit of their form factors, it is convenient to switch from the momentum transfer squared $q^2 = (p_1 - p_2)^2$ to the dimensionless variable

$$w^{(*)} = v \cdot v' = \frac{m_B^2 + m_{D^{(*)}}^2 - q^2}{2\, m_B\, m_{D^{(*)}}}, \tag{2.149}$$

where

$$v_\mu = \frac{p_{1\mu}}{m_B}, \quad v'_\mu = \frac{p_{2\mu}}{m_{D^*}}$$

are the four-velocities of B and $D^{(*)}$. The kinematically allowed region in a semileptonic decay (neglecting the lepton masses),

$$0 \leq q^2 \leq (m_B - m_{D^{(*)}})^2,$$

converts into the interval

$$w_{max}^{(*)} \equiv \frac{m_B^2 + m_{D^{(*)}}^2}{2\, m_B\, m_{D^{(*)}}} \geq w^{(*)} \geq 1,$$

where, numerically, $w_{max} \simeq 1.589$ ($w_{max}^* \simeq 1.503$), as follows from the measured meson masses [1].

Considering the form factors of $B \to D$ transition defined as in (2.94):

$$\langle D(p_2)|\bar{c}\gamma_\mu b|\bar{B}(p_1)\rangle = (p_1 + p_2)_\mu\, f_{BD}^+(q^2) + (p_1 - p_2)_\mu\, f_{BD}^-(q^2), \tag{2.150}$$

we transform the initial and final states

$$|\bar{B}(p_1)\rangle = |\bar{B}(v)\rangle\sqrt{m_B}, \quad \langle D(p_2)| = \langle D(v')|\sqrt{m_D},$$

and replace both heavy quark fields by the effective ones,

$$b(0) \to h_v(0) \quad c(0) \to h'_{v'}(0),$$

obtaining

$$\langle D(v')|\bar{h}'_{v'}\gamma_\mu h_v|\bar{B}(v)\rangle = \sqrt{\frac{1}{r}}\left[(v+rv')_\mu\, f_{BD}^+(q^2) + (v-rv')_\mu\, f_{BD}^-(q^2)\right], \qquad (2.151)$$

where $r = m_D/m_B$. The hadronic matrix element on the l.h.s. is, similar to (2.145), independent of the heavy mass scales. It can be represented as a superposition of two independent form factors, depending on the invariant variable w:

$$\langle D(v')|\bar{h}'_{v'}\gamma_\mu h_v|\bar{B}(v)\rangle = (v+v')_\mu\, h_+(w) + (v-v')_\mu\, h_-(w)\,. \qquad (2.152)$$

Matching this definition to (2.151) and comparing the coefficients at v_μ and v'_μ we relate the form factors $f^\pm$ to the new ones adapted to HQET:

$$h_\pm(w) = \frac{1}{2\sqrt{r}}\left[(1\pm r)f_{BD}^+(q^2) + (1\mp r)f_{BD}^-(q^2)\right]. \qquad (2.153)$$

Analogous redefinitions are applied to the $B \to D^*$ form factors. The hadronic matrix element, according to (2.136), is written via form factors:

$$\langle D^*(p_2)|j_\mu^W|B(p_1)\rangle = \epsilon_{\mu\nu\rho\sigma}\varepsilon^{*\nu}q^\rho p_2^\sigma \frac{2V^{BD^*}(q^2)}{m_B+m_{D^*}} - i\varepsilon_\mu^*(m_B+m_{D^*})A_1^{BD^*}(q^2)$$

$$+i(2p_2+q)_\mu(\varepsilon^*q)\frac{A_2^{BD^*}(q^2)}{m_B+m_{D^*}} + iq_\mu(\varepsilon^*q)\frac{2m_{D^*}}{q^2}\left(A_3^{BD^*}(q^2) - A_0^{BD^*}(q^2)\right). \qquad (2.154)$$

The heavy-mass independent hadronic matrix element is:

$$\langle D^*(v')|\bar{h}'_{v'}\gamma_\mu(1-\gamma_5)h_v|\bar{B}(v)\rangle = \epsilon_{\mu\nu\alpha\beta}\varepsilon^{*\nu}\,v^\alpha v'^\beta\, h_V(w^*) - i\varepsilon_\mu^*(1+w^*)h_{A_1}(w^*)$$

$$+i(\varepsilon^*\cdot v)\,v_\mu h_{A_2}(w^*) + i(\varepsilon^*\cdot v)\,v'_\mu h_{A_3}(w^*)\,, \qquad (2.155)$$

with the new form factors related to the original ones in (2.154):

$$h_V(w^*) = \frac{2\sqrt{r^*}}{1+r^*}V^{BD^*}(q^2)\,, \qquad h_{A_1}(w^*) = \frac{1+r^*}{\sqrt{r^*}(1+w^*)}A_1^{BD^*}(q^2)\,,$$

$$r^*h_{A_2}(w^*) + h_{A_3}(w^*) = \frac{2\sqrt{r^*}}{1+r^*}A_2^{BD^*}(q^2)\,,$$

$$r^*h_{A_2}(w^*) - h_{A_3}(w^*) = \frac{4r^*\sqrt{r^*}\left[A_3^{BD^*}(q^2) - A_0^{BD^*}(q^2)\right]}{1+r^{*2}-2r^*w^*}\,, \qquad (2.156)$$

where $r^* = m_{D^*}/m_B$.

Using the $B \to D^{(*)}$ transition matrix elements redefined in (2.152) and (2.155), it is possible to obtain relations for the form factors $h_i(w)$ ($i = +,-,V,A_1,A_2,A_3$) valid at $m_Q \to \infty$. To this end, we consider a different hadronic matrix element describing a flavor-diagonal transition in which a heavy pseudoscalar meson, $H = D$ or B, changes its velocity, $H(v) \to H(v')$. The transition is induced by the heavy-quark part of the e.m. current. Note that the HQET e.m. current consists of two fields "labeled" with different velocities

$$j_\mu^{em(Q)} \xrightarrow[m_Q\to\infty]{} \bar{h}_{v'}\gamma_\mu h_v\,.$$

The matrix element has a single possible Lorentz structure:

$$\langle H(v')|\bar{h}_{v'}\gamma_\mu h_v|H(v)\rangle = (v+v')_\mu\xi(w)\,, \qquad (2.157)$$

multiplied by the form factor $\xi(w)$ known as the *Isgur-Wise function*. Note that the structure $(v - v')_\mu$ cannot contribute. To see why, we multiply both parts of the above equation by this four-vector and use the relation (1.131) and its conjugate:

$$\bar{h}_{v'} \gamma_\mu h_v (v - v')^\mu = \bar{h}_{v'} (\slashed{v} h_v) - (\bar{h}_{v'} \slashed{v}') h_v = 0 \,.$$

The r.h.s. of (2.157) also vanishes after multiplying by $(v - v')^\mu$, since $v^2 = v'^2 = 1$, whereas an additional term with $(v - v')^\mu$ would have produced a nonvanishing contribution proportional to $(v - v')^2 = 2(1 - w)$.

Returning to (2.157), we employ the heavy-quark flavor symmetry explained in (1.8) and replace on l.h.s the field $h_{v'}$ by $h'_{v'}$ with the same velocity and a different heavy flavor. This replacement does not alter the long-distance dynamics determining the hadronic matrix element, hence

$$\langle H'(v')| \bar{h}'_{v'} \gamma_\mu h_v | H(v) \rangle \underset{m_Q \to \infty}{=} \langle H(v')| \bar{h}_{v'} \gamma_\mu h_v | H(v) \rangle = (v + v')_\mu \xi(w) \,. \tag{2.158}$$

We conclude that the flavor-changing transition mediated by the vector current is determined by the same Isgur-Wise function as the flavor-diagonal transition. Comparing the relation (2.158) for $H' = D$ and $H = B$ with the definition (2.152), we obtain for the HQET form factors of semileptonic $B \to D$ decays:

$$h_+(w) = \xi(w), \quad h_-(w) = 0 \,. \tag{2.159}$$

Furthermore, using the heavy-quark spin symmetry, it is possible to relate also the $H \to H'^*$ transition form factors with the Isgur-Wise function. Here the main point is that a heavy vector meson H^* differs from a pseudoscalar meson H only by the orientation of the heavy quark spin. Formally, a replacement $H \to H^*$ can be represented as an action of the heavy-quark spin operator on the state H. Using commutation relations between this operator and HQET currents (a detailed derivation can be found e.g., in [25]) it is possible to obtain a representation of the $H(v) \to H^*(v')$ matrix element in terms of the Isgur-Wise function:

$$\langle H^{*'}(v')| \bar{h}'_{v'} \gamma_\mu (1 - \gamma_5) h_v | \bar{H}(v) \rangle \underset{m_Q \to \infty}{=} \epsilon_{\mu\nu\alpha\beta} \varepsilon^{*\nu} v^\alpha v'^\beta \xi(w)$$
$$- \ [i\varepsilon^*_\mu (1 + w) - i(\varepsilon^* \cdot v) v'_\mu] \xi(w) \,, \tag{2.160}$$

where $w^* = w$ in this limit. Comparing this with (2.155), we find the $m_Q \to \infty$ limits for all form factors of $B \to D^*$ transitions:

$$h_V(w) = h_{A_1}(w) = h_{A_3} = \xi(w), \quad h_{A_2}(w) = 0 \,. \tag{2.161}$$

One of the far reaching predictions of HQET is the value

$$\xi(1) = 1 \tag{2.162}$$

of the Isgur-Wise function at the *zero recoil* point, where $v' = v$ and $w = v \cdot v' = 1$. For the defining relation (2.157), the point $w = 1$ corresponds to the zero momentum transfer $q^2 = 0$ and (2.162) is nothing but a normalization condition for the form factor, defining the number of heavy quarks inside H. It is analogous to the pion form factor normalization (2.33) defining the electric charge of the pion. Physically, at $q^2 = 0$ the momentum q transferred to a heavy meson in the rest frame vanishes, so that the final state simply coincides with the initial state and the matrix element indeed reduces to a normalization. For the weak transition matrix element (2.158), $w = 1$ corresponds to $q^2 = (m_B - m_D^{(*)})^2$. In this case, at zero recoil, one heavy quark is promptly replaced by another one with different flavor, so that both initial and final heavy mesons remain at rest; hence, the bound state of light antiquark and other light constituents inside a heavy meson remains intact.

For heavy-to-light transitions, such as $B \to \pi \ell \bar{\nu}_\ell$, the normalization conditions at zero recoil $q^2 \sim (m_B - m_\pi)^2$ cannot be deduced from HQET. Here a different question is addressed: how do the form factors scale with large m_b? At large q^2, near the maximal value $(m_B - m_\pi)^2$, the pion momentum is small and the hadronic matrix element contains a heavy mass scale only in the normalization of the B state. We rewrite the definition (2.96) for $B \to \pi$ transition matrix element, where $p_1 = m_B v$, $p_2 = p_\pi$ and $q = m_B v - p_\pi$, and then switch to HQET:

$$\langle \pi(p_\pi)|\bar{u}\gamma_\mu b|B(m_B v)\rangle = \sqrt{m_B}\langle \pi(p_\pi)|\bar{u}\gamma_\mu h_v|B(v)\rangle$$
$$= \left[2p_{\pi\mu} + \left(1 - \frac{m_B^2 - m_\pi^2}{q^2}\right)q_\mu\right]f_{B\pi}^+(q^2) + \frac{m_B^2 - m_\pi^2}{q^2}q_\mu\, f_{B\pi}^0(q^2). \qquad (2.163)$$

Within the low recoil region of the pion,

$$p_{\pi\mu} \ll m_B v_\mu, \quad q^2 = (m_B - m_\pi)^2 \sim m_B^2(1 - O(m_\pi/m_B)),$$

the definition (2.163) becomes:

$$\sqrt{m_B}\langle \pi(p_\pi)|\bar{u}\gamma_\mu h_v|B(v)\rangle = 2p_{\pi\mu}f_{B\pi}^+(q^2 \sim m_B^2) + m_B v_\mu f_{B\pi}^0(q^2 \sim m_B^2) + \dots, \quad (2.164)$$

where dots indicate small corrections of $O(m_\pi, m_\pi^2/m_B)$. Comparison of the leading powers of m_B on the r.h.s. and l.h.s, yields the following scaling with the heavy mass:

$$f_{B\pi}^+(q^2 \sim m_B^2) \sim \sqrt{m_B}, \quad f_{B\pi}^0(q^2 \sim m_B^2) \sim \frac{1}{\sqrt{m_B}}. \qquad (2.165)$$

More details on the heavy mass scaling of heavy-to-light transition form factors can be found, e.g., in [37]. The situation is much less trivial at a large recoil of the pion, or, equivalently, at $q^2 \ll m_B^2$, in which case the scaling behavior can only be predicted within a certain QCD calculation of the $B \to \pi$ form factors. We will return to this question in Chapter 10.

CHAPTER 3

BARYON FORM FACTORS

3.1 ELECTROMAGNETIC FORM FACTORS OF THE NUCLEON

As already mentioned in Chapter 1, the proton and neutron are the lightest baryons. Having the valence-quark content as shown in (1.94), they form a doublet with respect to the isospin symmetry. The accuracy of this symmetry is characterized by the smallness of their mass difference:

$$m_n - m_p \simeq 1.3 \text{ MeV} \ll m_p \simeq 938.3 \text{ MeV}, \qquad (3.1)$$

determined mainly by a small difference $m_d - m_u$ (see (1.20)). Differing in their electromagnetic characteristics, such as electric charges and magnetic moments, the proton and neutron have very similar properties with respect to hadronic and nuclear interactions, which is another manifestation of isospin symmetry. Accordingly, the proton and neutron states are often considered as components of the *nucleon* state denoted by $|N\rangle$:

$$\left(\begin{array}{c} |N^+\rangle \\ |N^0\rangle \end{array} \right) = \left(\begin{array}{c} |p\rangle \\ |n\rangle \end{array} \right),$$

both having the mass $m_N \simeq m_{p,n}$, neglecting the difference (3.1); hence, in this chapter we consider the form factors of a nucleon, before specifying if it is a proton or neutron.

To introduce the nucleon e.m. form factors, we follow the same way as used in Chapter 2 for the pion e.m. form factor and consider the elastic electron-nucleon scattering

$$e^-(k_1)N(p_1) \to e^-(k_2)N(p_2). \qquad (3.2)$$

The three diagrams of this process at the lowest $O(\alpha_{em})$ in e.m. coupling are similar to the one in Figure 2.1, with nucleons replacing the pions and virtual photon coupled to each valence quark.

Starting from the $S-$matrix element of (3.2), we then factorize out the electron e.m. current and the intermediate photon propagator. The resulting expression for the scattering amplitude:

$$\mathcal{A}(e^-N \to e^-N) = \frac{4\pi\alpha_{em}}{q^2} \bar{u}_e(k_2)\gamma^\mu u_e(k_1)\langle N(p_2)|j_\mu^{em}|N(p_1)\rangle, \qquad (3.3)$$

where $q = p_2 - p_1$, is very similar to (2.18) in the pion case, apart from a different hadronic matrix element

$$\langle N(p_2)|j_\mu^{em}|N(p_1)\rangle, \qquad (3.4)$$

which has a more complicated structure, because nucleons have spin $1/2$. More specifically, the initial and final states in (3.4), in addition to momenta p_1 and p_2 with $p_{1,2}^2 = m_N^2$, are described, respectively, by the four-component bispinor $u_N(p_1)$ and its conjugate $\bar{u}_N(p_2)^1$ satisfying Dirac equations:

$$(\not{p}_1 - m_N)u_N(p_1) = 0, \quad \bar{u}_N(p_2)(\not{p}_2 - m_N) = 0. \tag{3.5}$$

The hadronic matrix element (3.3) is a four-vector; hence, its general decomposition in Lorentz structures is a linear combination of all independent four-vectors constructed from two bispinors, Dirac matrices and 4-momenta p_1, p_2. The form factors we are interested in are the invariant coefficients multiplying Lorentz structures and depending on the invariant variable q^2.

A general form of the four-vector constructed from two bispinors is $\bar{u}_N(p_2)w_\mu u_N(p_1)$, where

$$w_\mu = \gamma_\mu, p_{1\mu}, p_{2\mu}, \ldots, \tag{3.6}$$

and the ellipsis denotes other possible combinations such as $w_\mu = \gamma_\mu \not{p}_1, \gamma_\mu \not{p}_2, p_{1\mu} \not{p}_1, p_{2\mu} \not{p}_1$ etc., a seemingly multiple choice. However, sandwiching the terms containing additional $\not{p}_{1,2}$ between bispinors and using Dirac equations (3.5), allows us to replace these terms one by one with the nucleon mass. As a result, there are only three independent terms in the Lorentz decomposition of the hadronic matrix element (3.4):

$$\langle N(p_2)|j_\mu^{em}|N(p_1)\rangle = \bar{u}_N(p_2)\left[\gamma_\mu \widetilde{F}_1(q^2) + (p_{1\mu} + p_{2\mu})\widetilde{F}_2(q^2) + q_\mu \widetilde{F}_3(q^2)\right]u_N(p_1), \tag{3.7}$$

where, for convenience, we use linear combinations of $p_{1,2}$. Multiplying both parts of the above equation with the momentum transfer vector q_μ leads to a vanishing divergence of the conserved e.m. current on the l.h.s.:

$$q^\mu \langle N(p_2)|j_\mu^{em}|N(p_1)\rangle = -i\langle N(p_2)|\partial^\mu j_\mu^{em}|N(p_1)\rangle = 0, \tag{3.8}$$

whereas r.h.s. becomes:

$$\bar{u}_N(p_2)\not{q}u_N(p_1)\widetilde{F}_1(q^2) + \bar{u}_N(p_2)u_N(p_1)\left[(p_1^2 - p_2^2)\widetilde{F}_2(q^2) + q^2\widetilde{F}_3(q^2)\right]. \tag{3.9}$$

The first term in the above, proportional to $\widetilde{F}_1$, vanishes after replacing $q = p_2 - p_1$ and applying the Dirac equation for nucleon bispinors. The second term proportional to $\widetilde{F}_2$ also vanishes, due to the on-shell condition $p_1^2 = p_2^2 = m_N^2$; hence, from (3.8) we conclude that $\widetilde{F}_3(q^2) = 0$, enforced by the conservation of e.m. current.

Returning to (3.7), it is convenient to represent the second of the two remaining Lorentz structures in a more convenient form containing the momentum transfer q. To this end, we use the product $\sigma_{\mu\nu}q^\nu$, transforming it with the help of anticommutation relations for γ-matrices:

$$\sigma_{\mu\nu}q^\nu = \frac{i}{2}(\gamma_\mu \not{q} - \not{q}\gamma_\mu) = i(p_{2\mu} - \not{p}_2\gamma_\mu + p_{1\mu} - \gamma_\mu \not{p}_1). \tag{3.10}$$

Sandwiched between the bispinors, the above equation yields:

$$-i\bar{u}_N(p_2)\sigma_{\mu\nu}q^\nu u_N(p_1) = \bar{u}_N(p_2)(p_{1\mu} + p_{2\mu} - 2m_N\gamma_\mu)u_N(p_1).$$

Using this relation and redefining in (3.7):

$$\widetilde{F}_1(q^2) + 2m_N\widetilde{F}_2(q^2) = F_1(q^2), \quad \widetilde{F}_2(q^2) = \frac{F_2(q^2)}{2m_N},$$

---

[1] We omit the polarization index for brevity. Note that these bispinors describe the spin 1/2 state of a nucleon as a whole and are not directly related to its quark structure.

we obtain a standard expression for the matrix element (3.4):

$$\langle N(p_2)|j_\mu^{em}|N(p_1)\rangle = \bar{u}_N(p_2)\left[\gamma_\mu F_1(q^2) - i\frac{\sigma_{\mu\nu}q^\nu}{2m_N}F_2(q^2)\right]u_N(p_1), \tag{3.11}$$

determined by the two dimensionless form factors $F_{1,2}(q^2)$ and satisfying the current-conservation condition (3.8). In the electron-nucleon scattering, these form factors are defined in the spacelike region $q^2 \leq 0$, hence we will also use the notation $Q^2 \equiv -q^2$ for the absolute value of the momentum transfer squared.

Using the hermiticity condition (2.39) for the e.m. current, it is possible to prove that the form factors $F_{1,2}(q^2)$ are real valued functions in the spacelike region. One has to use a sequence of equalities similar to (2.40) and take into account the relation (A.25) for complex conjugated products of bispinors. Note also that both isospin components of the e.m. current (1.54) with $I = 1$ and $I = 0$ contribute to the nucleon form factors (3.11), representing linear combinations of the proton and neutron form factors.

In practical applications, the linear combinations

$$
\begin{aligned}
G_E(Q^2) &= F_1(Q^2) - \frac{Q^2}{4m_N^2}F_2(Q^2), \\
G_M(Q^2) &= F_1(Q^2) + F_2(Q^2),
\end{aligned} \tag{3.12}
$$

known as the *electric and magnetic Sachs form factors*, are used. Their normalizations at $Q^2 = 0$ are equal, respectively, to electric charges and magnetic moments of the proton or neutron:

$$G_E^{(p)}(0) = 1, \qquad G_E^{(n)}(0) = 0;$$
$$G_M^{(p)}(0) = \mu_p \simeq 2.7928, \quad G_M^{(n)}(0) = \mu_n \simeq -1.9130.$$

A considerable deviation of magnetic moments from the values expected for pointlike spin 1/2 particles is a striking evidence for a long-distance structure of nucleons, caused by nonperturbative QCD interactions.

Note that, in contrast to the electron-pion scattering, the cross section of electron-nucleon scattering (3.2) expressed via squared form factors $G_{M,E}$ is well measurable, since protons and neutrons (the latter e.g., within a deuteron) represent a stable target. Further details of the nucleon form factor measurements, separation and models can be found in the reviews [38], [39].

3.2 WEAK FORM FACTORS OF THE NUCLEON AND β DECAY

Nucleon form factors are also indispensable for analyzing the neutrino-nucleon scattering

$$\nu_\ell(k_1)\,n(p_1) \to \ell^-(k_2)\,p(p_2), \tag{3.13}$$

and its antineutrino counterpart $\bar{\nu}_\ell\, p \to \ell^+\, n$, where $\ell = e, \mu$. These processes observed in neutrino experiments are mediated by the (ud) part of the weak flavor-changing interaction (1.29). The weak vector current $j_\mu = \bar{u}\gamma_\mu d$ is conserved in the isospin symmetry limit, hence, its matrix element entering the amplitude of (3.13) has a decomposition similar to (3.11):

$$\langle p(p_2)|j_\mu|n(p_1)\rangle = \bar{u}_p(p_2)\left[\gamma_\mu f_1(q^2) - i\frac{\sigma_{\mu\nu}q^\nu}{2m_N}f_2(q^2)\right]u_n(p_1), \tag{3.14}$$

with two form factors $f_{1,2}(q^2)$. The matrix element of the weak axial-vector current $j_{\mu 5} = \bar{u}\gamma_\mu\gamma_5 d$, taken between nucleon states contains three form factors, (cf. (3.7)):

$$\langle p(p_2)|j_{\mu 5}|n(p_1)\rangle = \bar{u}_p(p_2)\left[\gamma_\mu g_1(q^2) - i\frac{\sigma_{\mu\nu}q^\nu}{2m_N}g_2(q^2) - \frac{q_\mu}{2m_N}g_3(q^2)\right]\gamma_5 u_n(p_1), \qquad (3.15)$$

where an extra γ_5 on the r.h.s. warrants the same quantum numbers $J^P = 1^+$ as on the l.h.s.

Isospin symmetry relates the weak vector and e.m. form factors of the nucleons:

$$\langle p|j_\mu|n\rangle = \langle p|j_\mu^{em}|p\rangle - \langle n|j_\mu^{em}|n\rangle \qquad (3.16)$$

leading to the equalities between separate form factors:

$$f_1(Q^2) = F_1^{(p)}(Q^2) - F_1^{(n)}(Q^2), \quad f_2(Q^2) = F_2^{(p)}(Q^2) - F_2^{(n)}(Q^2). \qquad (3.17)$$

They are similar to the relation (2.76) between the pion weak and e.m. form factors, which was substantiated using isospin-symmetric quark flow diagrams instead of a full-scale $SU(2)$ group formalism. Analogous diagrams for the nucleon form factors are in fact more involved, because one has to symmetrize three-quark states with respect to the spin and isospin degrees of freedom. Instead, (3.16) can be proved associating the matrix elements of vector currents with the components of an isospin vector:

$$\bar{\Psi}_N\gamma_\mu\frac{\tau_3}{2}\Psi_N = \frac{1}{2}(\bar{p}\gamma_\mu p - \bar{n}\gamma_\mu n) \sim \langle p|j_\mu^{em}|p\rangle - \langle n|j_\mu^{em}|n\rangle,$$

$$\bar{\Psi}_N\gamma_\mu\frac{\tau_1 + i\tau_2}{2}\Psi_N = (\bar{p}\gamma_\mu n) \sim \langle p|j_\mu|n\rangle, \qquad (3.18)$$

where $\vec{\tau} = \vec{\sigma}$ are Pauli matrices in the isospin space and

$$\Psi_N = \begin{pmatrix} p \\ n \end{pmatrix}, \quad \bar{\Psi}_N = \begin{pmatrix} \bar{p} & \bar{n} \end{pmatrix} \qquad (3.19)$$

are the nucleon isodoublets.

Switching on the $m_d - m_u$ and $m_n - m_p$ mass differences, and drifting from the isospin symmetry limit, we encounter the β-decay of the neutron:

$$n(p_1) \to p(p_2) + e^-(k_1) + \bar{\nu}_e(k_2), \qquad (3.20)$$

which is one of the most important processes in nuclear physics[2]. This semileptonic decay with a low energy-momentum release to the leptons is a nucleon analog of the π_{e3} decay, considered in Chapter 2. The kinematical region of β-decay is characterized by the interval of the momentum transfer squared

$$m_e^2 \leq q^2 \leq (m_n - m_p)^2, \qquad (3.21)$$

For β decay, we cannot ignore isospin symmetry violation, hence also the vector (ud) current is not conserved and the most general decomposition of the hadronic matrix element for both vector and axial-vector currents contains three form factors:

$$\langle p(p_2)|j_\mu|n(p_1)\rangle = \bar{u}_p(p_2)\left[\gamma_\mu f_1(q^2) + i\frac{\sigma_{\mu\nu}q^\nu}{2m_n}f_2(q^2) + \frac{q_\mu}{2m_n}f_3(q^2)\right]u_n(p_1), \qquad (3.22)$$

$$\langle p(p_2)|j_{\mu 5}|n(p_1)\rangle = \bar{u}_p(p_2)\left[\gamma_\mu g_1(q^2) + i\frac{\sigma_{\mu\nu}q^\nu}{2m_n}g_2(q^2) + \frac{q_\mu}{2m_n}g_3(q^2)\right]\gamma_5 u_n(p_1). \qquad (3.23)$$

---

[2]Within nuclei, β-decay of a bounded proton, $p \to n\,e^+\nu_e$ is also possible.

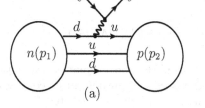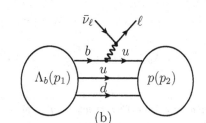

Figure 3.1 Diagrams of the (a) neutron beta-decay and (b) Λ_b-baryon semileptonic decay.

Slightly different (e.g., without nucleon mass factors in the denominator) definitions can be found in the literature. Note also, that here we define $q = p_1 - p_2$ to be the momentum transferred to the lepton pair, hence, with an opposite sign with respect to neutrino-nucleon scattering. Many details on the β-decay form factors can be found in [3].

3.3 VARYING FLAVORS: FORM FACTORS OF Λ, Λ_C, AND Λ_B BARYONS

In addition to the neutron β decay, there are several other semileptonic decays in which the lightest baryon with a given flavor content $s(ud)$, $c(ud)$ or $b(ud)$ and $J^P = 1/2^+$, that is, respectively, Λ, Λ_c or Λ_b, decays into a lighter baryon of the same type and a lepton-neutrino pair. These decay modes are listed in Table 3.1, together with the corresponding flavor-changing weak currents and CKM matrix elements. In Figure 3.1, the diagrams of two of these decays are shown.

TABLE 3.1   Weak semileptonic decays of baryons with various flavors. Each decay has a charge conjugate mode which is not shown.

V_{CKM}	Quark current $j_\mu - j_{\mu 5}$	Semileptonic decay mode
V_{ud}	$\bar{u}\gamma_\mu(1-\gamma_5)d$	$n \to p\,e^-\bar{\nu}_e$
V_{us}	$\bar{u}\gamma_\mu(1-\gamma_5)s$	$\Lambda \to p\,\ell^-\bar{\nu}_\ell$
V_{cd}^*	$\bar{d}\gamma_\mu(1-\gamma_5)c$	$\Lambda_c \to n\,\ell^+\nu_\ell$
V_{cs}^*	$\bar{s}\gamma_\mu(1-\gamma_5)c$	$\Lambda_c \to \Lambda\ell^+\nu_\ell$
V_{ub}	$\bar{u}\gamma_\mu(1-\gamma_5)b$	$\Lambda_b \to p\,\ell^-\bar{\nu}_\ell$
V_{cb}	$\bar{c}\gamma_\mu(1-\gamma_5)b$	$\Lambda_b \to \Lambda_c\ell^-\bar{\nu}_\ell$

For definiteness, let us choose the decay $\Lambda_b \to p\,\ell\nu_\ell$ where modes with $\ell = e, \mu, \tau$ are possible. Form factor decomposition of the hadronic matrix element is simply obtained, replacing quark flavors $d \to b$ in (3.22) and (3.23):

$$\langle p(p_2)|\bar{u}\,\gamma_\mu\,b|\Lambda_b(p_1)\rangle = \bar{u}_p(p_2)\left\{\gamma_\mu f_1(q^2) + i\,\sigma_{\mu\nu}q^\nu\frac{f_2(q^2)}{m_{\Lambda_b}} + q_\mu\frac{f_3(q^2)}{m_{\Lambda_b}}\right\}u_{\Lambda_b}(p_1)\,, \quad (3.24)$$

$$\langle p(p_2)|\bar{u}\,\gamma_\mu\,\gamma_5 b|\Lambda_b(p_1)\rangle = \bar{u}_p(p_2)\left\{\gamma_\mu g_1(q^2) + i\,\sigma_{\mu\nu}q^\nu\frac{g_2(q^2)}{m_{\Lambda_b}} + q_\mu\frac{g_3(q^2)}{m_{\Lambda_b}}\right\}\gamma_5 u_{\Lambda_b}(p_1)\,, \quad (3.25)$$

where the flavor-specific indices at form factors are implicit but omitted for brevity, and the decay kinematical region is

$$m_\ell^2 < q^2 < (m_{\Lambda_b} - m_N)^2\,. \quad (3.26)$$

The decay amplitude, similar to (2.92), reads:

$$\mathcal{A}(\Lambda_b \to p\,\ell\bar{\nu}_\ell) = \frac{G_F}{\sqrt{2}} V_{ub}\bar{u}_\ell(k_2)\Gamma^\mu v_\nu(k_1)\langle p(p_2)|\bar{u}\,\Gamma_\mu\,b|\Lambda_b(p_1)\rangle,\tag{3.27}$$

where $\Gamma_\mu = \gamma_\mu(1-\gamma_5)$. Substituting (3.24) and (3.25) in the above equation, we notice that contributions of the form factors f_3 and g_3 to the decay amplitude, after contracting q_μ with the lepton current, are proportional to m_ℓ and can safely be neglected for the modes with $\ell = e, \mu$.

To obtain the decay observables such as the differential width, one has to square the amplitude (3.27), summing over polarizations of baryons and leptons, and integrate over the three-particle phase space. This is similar to, but slightly more involved than derivation of the meson semileptonic decay width in Chapter 2.5. The only new elements needed here are the sums over polarizations of the baryon bispinors, e.g., for the nucleon:

$$\overline{u_N(p)\bar{u}_N(p)} = \slashed{p} + m_N.$$

For the initial Λ_b we need an average over polarizations which adds a factor $1/2$:

$$\overline{u_{\Lambda_b}(p)\bar{u}_{\Lambda_b}(p)} = \frac{1}{2}\left(\slashed{p} + m_{\Lambda_b}\right).$$

For completeness, we present the resulting expression for the differential width:

$$\begin{aligned}
\frac{d\Gamma}{dq^2}(\Lambda_b \to pl\nu_l) &= \frac{G_F^2 |V_{ub}|^2}{192\pi^3} m_{\Lambda_b}\lambda^{1/2}(m_{\Lambda_b}^2, m_N^2, q^2)\\
&\times \Bigg\{ (c_-^2 - t)\Big[(c_+^2 + 2t)|f_1(q^2)|^2\\
&\quad -6\,c_+\,tf_1(q^2)f_2(q^2) + (2c_+^2 + t)t|f_2(q^2)|^2\Big]\\
&\quad +(c_+^2 - t)\Big[(c_-^2 + 2t)|g_1(q^2)|^2\\
&\quad -6\,c_-\,t\,g_1(q^2)g_2(q^2) + (2c_-^2 + t)t|g_2(q^2)|^2\Big]\Bigg\},
\end{aligned}\tag{3.28}$$

where the following notation is introduced:

$$c_\pm = 1 \pm \frac{m_N}{m_{\Lambda_b}}, \quad t = \frac{q^2}{m_{\Lambda_b}^2}.$$

In order to use the above definitions and equations for the other decay modes listed in the Table 3.1, we have to replace the quark flavors correspondingly, e.g., $b \to c$ and $u \to s$ for $\Lambda_c \to \Lambda\ell\nu$.

Apart from weak semileptonic decays, baryon transition form factors describe FCNC decay channels of which the most important one is:

$$\Lambda_b \to \Lambda\,\ell^+\ell^-,\tag{3.29}$$

triggered by the $b \to s\ell^+\ell^-$ transition and described by the effective interaction (1.44). The dominant part of this decay amplitude is contributed by the effective operators $O_{9,10}$ presented in (1.43). Since the quark parts of these operators have the same $V - A$ structure as the weak currents, the resulting hadronic matrix elements are described by the same form factors as in (3.24) and (3.25) (up to $p \to \Lambda$ replacement). Additional form factors emerge in the hadronic matrix element of the $O_{7\gamma}$ operator containing the tensor $b \to s$ current (2.117). The latter is multiplied by the momentum transfer vector, as in (2.119). For the dominant part of this current proportional to m_b one has the following decomposition (see, e.g., in [40]):

$$\langle \Lambda(p_2)|\bar{s}\, i\sigma_{\mu\nu}q^\nu\, b|\Lambda_b(p_1)\rangle = \bar{u}_\Lambda(p_2)\left\{\gamma_\mu f_1^T(q^2) + i\sigma_{\mu\nu}q^\nu \frac{f_2^T(q^2)}{m_{\Lambda_b}} + q_\mu \frac{f_3^T(q^2)}{m_{\Lambda_b}}\right\}u_{\Lambda_b}(p_1)\,,$$

$$\langle \Lambda(p_2)|\bar{s}\, i\sigma_{\mu\nu}q^\mu\gamma_5 b|\Lambda_b(p_1)\rangle = \bar{u}_\Lambda(p_2)\left\{\gamma_\mu g_1^T(q^2) + i\sigma_{\mu\nu}q^\nu \frac{g_2^T(q^2)}{m_{\Lambda_b}} + q_\mu \frac{g_3^T(q^2)}{m_{\Lambda_b}}\right\}\gamma_5 u_{\Lambda_b}(p_1)\,.$$

$$(3.30)$$

Alternative definitions of the flavor-changing heavy baryon form factors and their HQET limit can be found in [41] where also the so-called helicity form factors are introduced.

3.4   IOFFE CURRENT AND BARYON DECAY CONSTANTS

In Section 1.4 we already discussed that in QCD the choice of quark current operators entering hadronic matrix elements is not limited by the currents originating from the Lagrangian of SM. Any local colorless operator built from quark, antiquark and gluon fields is allowed, and some examples were already given.

Here we consider a specific quark current that has quantum numbers of a baryon. The reason such currents are introduced will be clarified later, in Chapter 10, while discussing QCD-based methods used to calculate baryon form factors. A baryonic current is a local operator composed from three quark fields, with the same flavor content as a given baryon. There are usually several possibilities to combine three quark fields via various combinations of Dirac matrices. We consider as an example a frequently used *Ioffe current* with the proton quantum numbers:

$$\eta_\alpha^{(p)}(x) = \epsilon^{ijk}\left[u_\beta^{Ti}(x)\left(C\gamma_\mu\right)_{\beta\beta'}u_{\beta'}^j(x)\right]\left(\gamma_5\gamma^\mu\right)_{\alpha\delta}d_\delta^k(x)\,, \tag{3.31}$$

where u^T denotes a transposed u-quark field, $C = \gamma_2\gamma_0$ is the charge-conjugation matrix. The color indices i, j, k and Dirac indices $\alpha, \beta, \ldots$ are shown explicitly. The neutron counterpart of this current is

$$\eta_\alpha^{(n)}(x) = \epsilon^{ijk}\left[d_\beta^{Ti}(x)\left(C\gamma_\mu\right)_{\beta\beta'}u_{\beta'}^j(x)\right]\left(\gamma_5\gamma^\mu\right)_{\alpha\delta}d_\delta^k(x)\,. \tag{3.32}$$

From the above definitions it is clear that a baryon current has a net Dirac index, however, due to the presence of the diquark, the total spin of the current can vary. Adopting a free-quark approximation for the quark fields in (3.31) or (3.32), attributing a nonvanishing mass to the quarks, and expanding the bispinors in two-component spinors, it is possible to demonstrate that the total spin of these currents is $J = 1/2$. On the other hand, baryon currents similar to (3.32) do not possess a definite P-parity as opposed to quark-antiquark currents.

Taking the matrix element of the currents (3.31), (3.32) between vacuum and nucleon state, we identify a single possible Dirac structure with the nucleon bispinor

$$\langle 0|\eta^{(N)}|N(p)\rangle = \lambda_N m_N u_N(p)\,, \tag{3.33}$$

where the constant parameter λ_N is the nucleon analog of the pion decay constant. Due to indefinite P-parity, the same current has a nonvanishing matrix element (overlap) with the lightest baryon $N(1520)$ with $J^P = 1/2^-$ listed in [1].

In SM the baryon number is conserved, but many extensions of SM predict violation of this quantum number, e.g., a transformation of baryon into lepton leading to a proton decay with a very large lifetime. This process is mediated by a certain effective pointlike operator (dependent on the particular model beyond the SM) which contains three quark fields and one lepton field; hence, the amplitude of proton decay contains a hadronic matrix element (3.33), and the value of λ_N is needed to estimate the proton lifetime.

Concluding this section, we also present the currents with quantum numbers of a heavy baryon Λ_Q with $Q = c, b$:

$$\eta^{(\Lambda_Q)} = \epsilon^{ijk} \left(u_i \, C \, \Gamma_a \, d_j \right) \widetilde{\Gamma}^a \, Q_k \,. \tag{3.34}$$

There are multiple choices for the (Lorentz-contracted) Dirac structures Γ_a and $\widetilde{\Gamma}^a$ in the above current. To gain a large overlap with the heavy baryon state, that is, to maximize the matrix element

$$\langle 0 | \eta^{(\Lambda_Q)} | \Lambda_Q(p) \rangle = \lambda_{\Lambda_Q} m_{\Lambda_Q} u_{\Lambda_Q}(p) \,, \tag{3.35}$$

one uses the heavy quark limit [42] as a guideline, resulting in the two options for the current (3.34):

$$\Gamma_a = \gamma_5, \ \widetilde{\Gamma}^a = 1; \quad \Gamma_a = \gamma_5 \gamma_\lambda \ \widetilde{\Gamma}^a = \gamma^\lambda \,, \tag{3.36}$$

called, respectively, the pseudoscalar and axial heavy-baryon currents.

CHAPTER 4

HADRONIC RADIATIVE TRANSITIONS

4.1 THE $\rho \to \pi\gamma$ DECAY AND VECTOR DOMINANCE

Form factors considered in Chapters 2 and 3 describe hadronic transitions in which a virtual photon or electroweak boson serves as an external source of quark current, transferring momentum to or from hadrons. In this chapter we discuss processes of *radiative transitions* or radiative decays, in which a real, on-shell photon is emitted off an initial hadron transforming it into a final hadron.

A typical example of a hadronic radiative decay is

$$\rho^{\pm}(p_1) \to \pi^{\pm}(p_2)\gamma(q) , \qquad (4.1)$$

where momenta are indicated in parenthesis and, for definiteness, the charged ρ-meson decay is considered. Momentum conservation $p_1 = p_2 + q$ leaves only two independent four-vectors, and we choose p_2 and q, with

$$p_2^2 = m_\pi^2 , \quad q^2 = 0.$$

The meson $\rho(770)$ is an unstable hadron, hence, before radiative decay takes place, the ρ has to be produced in some other process e.g., in electroproduction:

$$
\begin{array}{l}
e^- p \to e^- \rho^+ n \, . \\
\quad \ \llcorner\, \pi^+\gamma
\end{array}
\qquad (4.2)
$$

The diagram of this process is similar to the one in Figure 2.2, with the final pion replaced by ρ. Another source is the τ-lepton decay

$$
\begin{array}{l}
\tau^- \to \rho^- \nu_\tau \, . \\
\quad \ \llcorner\, \pi^-\gamma
\end{array}
\qquad (4.3)
$$

A neutral ρ-meson can be produced directly in electron-positron collision:

$$e^+ e^- \to \rho^0 \to \pi^0\gamma . \qquad (4.4)$$

In all these processes ρ is observed as a resonance enhancement in the probability distribution over invariant mass of its decay products. The resonance is spread around the mass parameter $m_\rho \simeq 770$ MeV within an interval of the order of total width $\Gamma_\rho^{tot} \simeq 150$ MeV. In Chapter 5 we will discuss the phenomenology of hadronic resonances in more detail. Still, since the characteristics of the ρ-meson, such as its mass and width, are almost universal in all above mentioned processes (see more detailed data in [1]), it is treated as a hadron, similar to a pion or nucleon. Moreover, as already discussed in Chapter 1, the ρ meson, having spin-parity $J^P = 1^-$, from the point of view of a simple quark model differs from the pion only by the spin alignment of its valence quarks; hence, considering the radiative decay (4.1), we will treat ρ as a hadron with a fixed mass m_ρ and neglect its width, keeping

$$p_1^2 = m_\rho^2.$$

In the rest frame of ρ-meson the photon energy E_γ is fixed, being determined from the equation:

$$(p_1 - q)^2 = m_\rho^2 - 2p_1 q = m_\rho^2 - 2m_\rho E_\gamma = m_\pi^2, \tag{4.5}$$

so that

$$E_\gamma = \frac{m_\rho^2 - m_\pi^2}{2m_\rho} \simeq 370 \text{ MeV}. \tag{4.6}$$

Note that all kinematical invariants in this two-body decay are fixed by the on-shell conditions.

The branching fraction of (4.1), according to [1], is at the level of 10^{-4}. Its smallness reflects suppression of the e.m. interaction with respect to QCD interactions inducing the dominant strong decay $\rho \to \pi\pi$. The latter was discussed in Chapter 1.7 and its quark flow diagram is shown in Figure 1.6.

One could attempt to describe the $\rho \to \pi\gamma$ decay as a radiative transition between quark-antiquark bound states, similar to the photon emission off an atomic electron. More specifically, since the valence quark and antiquark are – both in ρ and π – in the S wave, we would have a magnetic-dipole transition, induced by the quark magnetic moment interacting with the magnetic component of the photon field. However, such a quantum-mechanical description is not applicable to light mesons consisting of relativistic quarks and antiquarks; hence, there is no feasible way to model $\rho \to \pi\gamma$ as a magnetic-dipole transition, for the same reason it is impossible to explain a very large mass difference $m_\rho - m_\pi$ in terms of the hyperfine splitting in a potential.

A perturbative description of $\rho \to \pi\gamma$ is also not possible, because in this decay the photon is emitted with an energy in the ballpark of $O(\Lambda_{QCD})$. The perturbative expansion in the e.m. coupling α_{em} is perfectly valid and higher-order corrections in α_{em} are strongly suppressed. This however, does not help to resolve the hadronic matrix element of a radiative decay. Even though the photon is emitted from a pointlike e.m. current, an average distance between the quark and antiquark forming this current is of $O(1/\Lambda_{QCD})$, hence their strong nonperturbative interaction is unavoidable.

Approaching the $\rho \to \pi\gamma$ decay from a phenomenological side, we, as usual, single out the hadronic matrix element in the decay amplitude. Starting from a general expression for the S-matrix element, we retain the first-order term in the e.m. coupling, corresponding to a real photon emission (cf. (2.11)):

$$S_{fi}^{(\rho \to \pi\gamma)} = \langle \pi^+(p_2)\gamma(q,\lambda)|\hat{S}|\rho(p_1)\rangle$$

$$= \int d^4x \langle \pi^+(p_2)\gamma(q,\lambda)|ieA^\mu(x)j_\mu^{em}(x)|\rho(p_1)\rangle. \tag{4.7}$$

Expanding the e.m. field operator $A^\mu(x)$ in components with definite momenta and polarizations:

$$A_\mu(x) = \int \frac{d\vec{k}}{(2\pi)^3\sqrt{2k_0}} \sum_{\lambda=\pm} \left(a(k,\lambda)\varepsilon_\mu^{(\gamma)}(k,\lambda)e^{-ikx} + a^\dagger(k,\lambda)\varepsilon_\mu^{(\gamma)*}(k,\lambda)e^{ikx} \right), \tag{4.8}$$

we factorize out the photon with momentum q and polarization λ from the final state. The photon polarization vector $\varepsilon^{(\gamma)}(q, \lambda)$ has the following properties[1]:

$$\varepsilon^{(\gamma)}(q, \lambda) \cdot q = 0, \quad \sum_{\lambda = \pm 1} \varepsilon_\mu^{(\gamma)}(q, \lambda)\, \varepsilon_\nu^{(\gamma)}(q, \lambda) = -g_{\mu\nu}, \tag{4.9}$$

where the summation goes over two polarization states of a real photon and Feynman gauge for the photon field is adopted.

Performing translation of the e.m. current and transforming the integral over x into a δ-function, we obtain:

$$S_{fi}^{(\rho \to \pi\gamma)} = i(2\pi)^4 \delta^{(4)}(p_1 - p_2 - q)\mathcal{A}(\rho^+ \to \pi^+\gamma), \tag{4.10}$$

where the decay amplitude

$$\mathcal{A}(\rho^+ \to \pi^+\gamma) = \sqrt{4\pi\alpha_{em}}\,\varepsilon^{(\gamma)\mu}(q, \lambda)\langle \pi^+(p_2)|j_\mu^{em}|\rho^+(p_1)\rangle, \tag{4.11}$$

contains the hadronic matrix element of the e.m. current between the ρ and π states.

To parameterize the hadronic matrix element entering (4.11), we use the ρ-meson polarization four-vector $\varepsilon^{(\rho)}(p_1, \lambda)$, where the index $\lambda = \pm 1, 0$ counts the three possible polarization states of a massive vector meson (see also (2.128)). These polarization vectors fulfil the orthogonality and summation conditions:

$$\varepsilon^{(\rho)}(p_1, \lambda) \cdot p_1 = 0, \quad \sum_{\lambda = \pm 1, 0} \varepsilon_\alpha^{(\rho)}(p_1, \lambda)\, \varepsilon_\beta^{(\rho)}(p_1, \lambda) = -g_{\alpha\beta} + \frac{p_{1\alpha} p_{1\beta}}{m_\rho^2}. \tag{4.12}$$

In what follows, we omit the arguments of polarization vectors for brevity.

Furthermore, we take into account that the pion state is a pseudoscalar, hence the hadronic matrix element in (4.11) transforms as an axial vector. With the polarization vector $\varepsilon^{(\rho)}$ and two independent momenta p_2 and q at our disposal, the only possible Lorentz structure, which transforms as an axial vector, is the one with the Levi-Civita tensor:

$$\langle \pi^+(p_2)|j_\mu^{em}|\rho^+(p_2 + q)\rangle = \epsilon_{\mu\nu\alpha\beta} q^\nu \varepsilon^{(\rho)\alpha} p_2^\beta\, g_{\rho\pi\gamma}, \tag{4.13}$$

where the invariant amplitude is reduced to a constant parameter $g_{\rho\pi\gamma}$ accumulating QCD effects in this transition. The charge-conjugated process $\rho^- \to \pi^-\gamma$ has the same constant up to an inessential phase factor.

Note that at spacelike momentum transfer ($q^2 < 0$) the same matrix element (4.13) describes the $\rho \to \pi$ transition via a virtual photon[2]. To parameterize q^2-dependence of this transition, a specific $\rho \to \pi$ form factor, similar to the pion spacelike form factor has to be introduced:

$$\langle \pi^+(p_2)|j_\mu^{em}|\rho^+(p_2 + q)\rangle = \epsilon_{\mu\nu\alpha\beta} q^\nu \varepsilon^{(\rho)\alpha} p_2^\beta\, f_{\rho\pi}(q^2). \tag{4.14}$$

The radiative decay constant (4.13) coincides with the limiting value of this form factor:

$$g_{\rho\pi\gamma} = f_{\rho\pi}(0). \tag{4.15}$$

Note that there are no constraints for this hadronic parameter, in contrast to the pion e.m. form factor at $q^2 = 0$, fixed by the electric charge.

---

[1] For a photon in the final state it should be $\varepsilon^{(\gamma)*}$; we omit the index $*$ hereafter in order not to overload the notation.

[2] The inverse $\pi \to \rho$ transition can in principle be extracted from the ρ-meson electroproduction, $e^- p \to e^- \rho^+ n$. Here, as in the case of pion electroproduction shown in Figure 2.2, the pion exchange dominates.

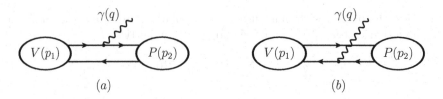

Figure 4.1 Diagrammatic representation of a vector to pseudoscalar meson radiative transition.

Substituting (4.13) in (4.11), we square the decay amplitude $\mathcal{A}(\rho^+ \to \pi^+\gamma)$ and average (sum) over polarizations of the initial ρ (final γ). In doing that, we employ the summation conditions in (4.9) and (4.12), with an extra factor $1/3$ stemming from the average. Finally, taking into account the kinematical factors in two-body decay (see (B.47)), we obtain the width of the radiative transition:

$$\Gamma(\rho^\pm \to \pi^\pm\gamma) = \frac{\alpha_{em}}{24}|g_{\rho\pi\gamma}|^2\left(\frac{m_\rho^2 - m_\pi^2}{m_\rho}\right)^3 = \frac{\alpha_{em}}{3}|g_{\rho\pi\gamma}|^2 E_\gamma^3. \qquad (4.16)$$

Comparing with experimental data from [1] it is possible to extract the $\rho\pi\gamma$ decay constant.

Naively, one would expect that the $\rho^0 \to \pi^0\gamma$ transition is suppressed with respect to its charged counterpart (4.1), because both initial and final hadrons are electrically neutral. In fact, the two hadronic matrix elements are equal in the isospin symmetry limit:

$$\mathcal{A}(\rho^0 \to \pi^0\gamma) \simeq \mathcal{A}(\rho^\pm \to \pi^\pm\gamma), \qquad (4.17)$$

revealing that QCD dynamics dominates in these transitions. This equality owes to the fact that the G-parities of ρ and π are opposite, hence, only $j_\mu^{I=0}$ – the isoscalar part of the e.m. current with negative G-parity – contributes to the hadronic matrix element. The relation (4.17) then simply follows from comparison of the coefficients[3] at the $I = 0$ part of the product of two isovector states.

It is also instructive to confirm (4.17), comparing the quark flow diagrams as it was done in Chapter 2.4 for the pion form factors. The two diagrams for a generic $V \to P\gamma$ transition, shown in Figure 4.1(a) and 4.1(b), correspond, respectively, to a photon emission from the valence quark and antiquark. For $\rho^+ \to \pi^+\gamma$ we have the following decomposition:

$$\begin{aligned}
\langle\pi^+|j_\mu^{(I=0)}|\rho^+\rangle &= \frac{1}{6}\Big(\langle(u\bar{d})_\pi|\,\bar{u}\gamma_\mu u\,|(u\bar{d})_\rho\rangle + \langle(u\bar{d})_\pi|\,\bar{d}\gamma_\mu d\,|(u\bar{d})_\rho\rangle\Big) \\
&= \frac{1}{6}(A_\mu^{\rho\pi} + \bar{A}_\mu^{\rho\pi}),
\end{aligned} \qquad (4.18)$$

(cf. (2.78)) where the $I = 0$ component of the e.m. current is separated into the $\bar{u}u$ and $\bar{d}d$ components[4] and the amplitudes $A_\mu^{\rho\pi}$ and $\bar{A}_\mu^{\rho\pi}$ are identified, respectively, with the diagrams in Figures 4.1(a) and 4.1(b). Counting an analogous diagram for the $\rho^0 \to \pi^0\gamma$

---

[3]They can be found in the tables of Clebsch-Gordan coefficients in [1].

[4]The s, c, b quark components of the isoscalar e.m. current (see (1.55)) only interact with the nonvalence quarks and antiquarks and are irrelevant for the symmetry relations which hold at the level of the valence states.

decay, we obtain:

$$\langle \pi^0 | j_\mu^{(I=0)} | \rho^0 \rangle = \frac{1}{6} \left(\left\langle \frac{1}{\sqrt{2}} (u\bar{u} - d\bar{d})_\pi \left| \bar{u}\gamma_\mu u \right| \frac{1}{\sqrt{2}} (u\bar{u} - d\bar{d})_\rho \right\rangle \right.$$

$$\left. + \left\langle \frac{1}{\sqrt{2}} (u\bar{u} - d\bar{d})_\pi \left| \bar{d}\gamma_\mu d \right| \frac{1}{\sqrt{2}} (u\bar{u} - d\bar{d})_\rho \right\rangle \right)$$

$$= \frac{1}{6} \left(\frac{1}{2} (A_\mu^{\rho\pi} + \bar{A}_\mu^{\rho\pi}) + \frac{1}{2} (A_\mu^{\rho\pi} + \bar{A}_\mu^{\rho\pi}) \right) = \frac{1}{6} (A_\mu^{\rho\pi} + \bar{A}_\mu^{\rho\pi}), \quad (4.19)$$

yielding (4.17).

The advantage of quark-flow diagrams is that one easily reproduces flavor-symmetry relations between $V \to P\gamma$ decay amplitudes even if the initial and final hadrons belong to different isospin multiplets. E.g, we obtain for the hadronic matrix element of $\omega(782) \to \pi\gamma$, where only the $I = 1$ component of the current contributes,

$$\langle \pi^0 | j_\mu^{(I=1)} | \omega^0 \rangle = \frac{1}{2} \left(\left\langle \frac{1}{\sqrt{2}} (u\bar{u} - d\bar{d})_\pi \left| \bar{u}\gamma_\mu u \right| \frac{1}{\sqrt{2}} (u\bar{u} + d\bar{d})_\omega \right\rangle \right.$$

$$\left. - \left\langle \frac{1}{\sqrt{2}} (u\bar{u} - d\bar{d})_\pi \left| \bar{d}\gamma_\mu d \right| \frac{1}{\sqrt{2}} (u\bar{u} + d\bar{d})_\omega \right\rangle \right)$$

$$= \frac{1}{2} \left(\frac{1}{2} (A_\mu^{\rho\pi} + \bar{A}_\mu^{\rho\pi}) + \frac{1}{2} (A_\mu^{\rho\pi} + \bar{A}_\mu^{\rho\pi}) \right) = \frac{1}{2} (A_\mu^{\rho\pi} + \bar{A}_\mu^{\rho\pi}), \quad (4.20)$$

yielding

$$\mathcal{A}(\omega^0 \to \pi^0 \gamma) \simeq 3\mathcal{A}(\rho^\pm \to \pi^\pm \gamma). \quad (4.21)$$

Keeping in mind that the masses of ω and ρ mesons are almost equal, we immediately predict the ratio of the widths:

$$\Gamma(\omega^0 \to \pi^0 \gamma) \simeq 9\Gamma(\rho^\pm \to \pi^\pm \gamma), \quad (4.22)$$

which is confirmed by comparing the measured branching fractions multiplied by total widths [1].

Furthermore, taking into account the s-quark component of the e.m. current, let us assume that hadronic matrix elements of the diagrams in Figure 4.1 do not change if the u or d quark line is replaced by an s-quark line. This assumption leads to amplitude relations equivalent to the $SU(3)_{fl}$-symmetry relations, e.g., between $\rho \to \pi\gamma$ and $K^* \to K\gamma$ amplitudes. More involved are quark diagram relations involving the decay modes with η and η' mesons, such as $\rho^0 \to \eta\gamma$ or $\phi(1020) \to \eta'\gamma$, where mixing between the isoscalar pseudoscalar mesons has to be taken into account. We will not discuss all these relations further. In fact, comparison with experiment manifests appreciable $SU(3)_{fl}$ violation effects which are usually fitted with model-dependent parameters.

It is possible to describe the $\rho \to \pi\gamma$ decay as a two-stage process. First, the ρ meson transforms into pion and ω^0 via isospin conserving strong interaction and, subsequently, the ω meson, via e.m. current, annihilates into a photon. Shown in Figure 4.2 is the corresponding quark-flow diagram. This is one of the examples of the so-called *vector dominance model*. Since the photon is massless, and also $m_\omega > m_\rho - m_\pi$, the intermediate ω meson on this diagram is off mass shell, being exchanged between the strong and interaction vertices. Postponing the definitions of relevant hadronic matrix elements to the next chapter, we only mention that the vector dominance model provides [43] a mutually consistent description

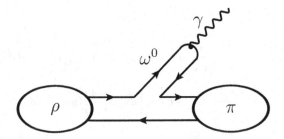

Figure 4.2   Vector dominance model for the $\rho \to \pi\gamma$ decay with intermediate ω^0 meson.

of the $\rho \to \pi\gamma$ radiative decay and the $\omega \to 3\pi$ strong decay if the latter is assumed to proceed in two stages:

$$\omega^0 \to \rho^-\pi^+ .$$
$$ \hookrightarrow \pi^-\pi^0$$

The vector dominance model provides a viable phenomenological description not only for various radiative transitions but also for photoproduction processes where one attributes a hadronic component to the photon.

Still, vector dominance and similar models, where off-shell hadrons are introduced, cannot be directly related to QCD. E.g., considering ω as a virtual intermediate state propagating in $\rho \to \pi\gamma$ we tacitly promote this composite state to a sort of fundamental particle, which evidently contradicts the fact that there are many hadronic states with the same quantum numbers as ω. Indeed, adopting the mechanism in Figure 4.2, we neglect the intermediate three-pion state and radially excited ω mesons. In the next chapter we return to this question and discuss a more general interpretation of the vector dominance model.

4.2 $D^* \to D\gamma$ AND $B^* \to B\gamma$ DECAYS

Flavor-conserving radiative decays of heavy mesons, $D^* \to D\gamma$ and $B^* \to B\gamma$, have the same spin-parity quantum numbers as $\rho \to \pi\gamma$ in both initial and final states; hence, we can easily write down their decay amplitudes, e.g., for $D^{*+} \to D^+\gamma$, replacing in (4.11) and (4.13)

$$u \to c, \ \rho^+ \to D^{*+}, \ \pi^+ \to D^+,$$

so that

$$\mathcal{A}(D^{*+} \to D^+\gamma) = \sqrt{4\pi\alpha_{em}} \, \epsilon_{\mu\nu\alpha\beta} \varepsilon^{(\gamma)\mu} q^\nu \varepsilon^{(D^*)\alpha} p^\beta g_{D^*D\gamma} . \tag{4.23}$$

The constant $g_{D^*D\gamma}$ parameterizes the hadronic matrix element in this decay. It can be extracted from the measured decay width which is, correspondingly, reproduced from (4.16):

$$\Gamma(D^* \to D\gamma) = \frac{\alpha_{em}}{24} \left(\frac{m_{D^*}^2 - m_D^2}{m_{D^*}}\right)^3 |g_{D^*D\gamma}|^2 , \tag{4.24}$$

for both D^{*+} and D^{*-}.

Furthermore, replacing $\bar{d} \to \bar{s}$, we obtain the $D_s^* \to D_s\gamma$ amplitude and width. The analogous formulae for $B^* \to B\gamma$ and $B_s^* \to B_s\gamma$ transitions are easily obtained by replacing

$$c \to b, \ D^{*+} \to \bar{B}^{*0}, \ D^+ \to \bar{B}^0$$

in (4.23) and (4.24).

Examining the data available in [1], we notice that $D_{(s)}^*$ mesons have much larger branching fractions of radiative decays, than light mesons. The reason lies in a very small phase space suppressing the strong decay $D^* \to D\pi$. Another indication for that is the total width of D^* which is in the ballpark of 100 KeV. The strong $B^* \to B\pi$ decays are forbidden by the lack of phase space, and $D_s^* \to D_s\pi$ and $B_s^* \to B_s\pi$ by isospin symmetry, hence $D_{(s)}^* \to D_{(s)}\gamma$ and $B_{(s)}^* \to B_{(s)}\gamma$ are the dominant decay channels.

We can use the same quark flow diagrams in Figure 4.1 for the heavy meson radiative decays, but there are no isospin symmetry relations anymore, e.g., between $D^{*+} \to D^+\gamma$ and $D^{*0} \to D^0\gamma$. Indeed, here the e.m. current contributes with both isospin components, hence there are two independent isospin amplitudes per two decay modes. It is more convenient to parametrize the diagrams in terms of the two amplitudes corresponding to the photon emission off the heavy c-quark and light $\bar{d}, \bar{u}$ antiquark. To this end we represent the relevant part of quark e.m. current as:

$$j_\mu^{em} = Q_c \bar{c}\gamma_\mu c + \sum_{q=u,d} Q_q \bar{q}\gamma_\mu q \qquad (4.25)$$

Then, separating the contributions of the two diagrams we obtain for both radiative decay modes

$$\begin{aligned}
\langle D^+|j_\mu^{em}|D^{*+}\rangle &= Q_c\langle (c\bar{d})_D|\,\bar{c}\gamma_\mu c\,|(c\bar{d})_{D^*}\rangle + Q_d\langle (c\bar{d})_D|\,\bar{d}\gamma_\mu d\,|(c\bar{d})_{D^*}\rangle \\
&\equiv Q_c A_\mu^{D^*D} + Q_d \bar{A}_\mu^{D^*D},
\end{aligned} \qquad (4.26)$$

and

$$\begin{aligned}
\langle D^0|j_\mu^{em}|D^{*0}\rangle &= Q_c\langle (c\bar{u})_D|\,\bar{c}\gamma_\mu c\,|(c\bar{u})_{D^*}\rangle + Q_u\langle (c\bar{u})_D|\,\bar{u}\gamma_\mu u\,|(c\bar{u})_{D^*}\rangle + \\
&\equiv Q_c A_\mu^{D^*D} + Q_u \bar{A}_\mu^{D^*D}.
\end{aligned} \qquad (4.27)$$

Since $Q_d \neq Q_u$ we indeed have two different linear combinations of amplitudes in the above hadronic matrix elements.

It is conceivable that the dominant contribution to the $H^* \to H\gamma$ $(H = D, B)$ transition stems from a photon emission off the light antiquark in the initial H^* meson. QCD-based calculations using various methods definitely support this conjecture. A somewhat schematic presentation of this amplitude suggested by the above decompositions is:

$$g_{H^*H\gamma} = \hat{g}\left(\frac{Q_Q}{m_H^*} + \frac{Q_q}{\mu_q}\right), \qquad (4.28)$$

where $\mu_q \sim \Lambda_{QCD}$ is the energy scale related to the light degrees of freedom in the radiative transition and $\hat{g}$ is a dimensionless constant. Note that this formula is also consistent with the HQET limit $m_Q \to \infty$ of the $H^* \to H$ transition matrix element, so that $\hat{g}$ does not scale with the heavy mass.

4.3   RADIATIVE TRANSITIONS IN HEAVY QUARKONIUM

The $V \to P\gamma$ decays with $c\bar{c}$, $b\bar{b}$ or $c\bar{b}$ valence quark and antiquark belong to the rich spectrum of radiative transitions in heavy quarkonia. The vector-to-pseudoscalar transitions between the lowest levels are

$$J/\psi(1S) \to \eta_c(1S)\gamma, \quad \Upsilon(1S) \to \eta_b(1S)\gamma, \quad \text{or} \quad B_c^{*+} \to B_c\gamma, \qquad (4.29)$$

of which only the charmonium transition has been observed so far [1]. In the following, we omit the notation $(1S)$ at the ground states J/ψ and η_c.

Dynamics of these radiative decays is markedly different from the photon emission off a light or heavy-light meson. As already mentioned in Section 1.7, the $c\bar{c}$, $b\bar{b}$ and $c\bar{b}$ mesons can be approximately treated as nonrelativistic two-body bound states, due merely to the large masses of the valence quark and antiquark, $m_b, m_c \gg \Lambda_{QCD}$. In fact, radiative decays (4.29) resemble transitions between orthopositronium and parapositronium, the spin-one and spin-zero bound states of electron and positron in a Coulomb potential.

Observed heavy quarkonium spectroscopy is reproduced reasonably well, introducing an interquark potential and solving the Schrödinger equation for the energy levels and wave functions. This approach works at least for the lowest quarkonium states, below the open heavy-flavor threshold. Typical QCD-inspired potentials used to fit heavy quarkonia represent a superposition of two components: the confining one that grows at large distances and the short-distance one of the Coulomb type. The latter effectively corresponds to a one-gluon exchange. Note however that, in order to describe radially excited states of quarkonium and their transitions, a nonrelativistic potential is not sufficient, because it does not include the coupling of excited states with open-flavored meson pairs. More advanced models involving coupled channels are used, see, e.g., [44].

Within a nonrelativistic potential model, the hadronic matrix element of a radiative transition is calculable in a form of an integral containing a convolution of initial and final quarkonia wave functions with a nonrelativistic operator stemming from the quark e.m. current. For the charmonium transition in (4.29) taken as an example, we can write, after defining the invariant constant similar to (4.13):

$$\langle \eta_c(p)|j^{em}_\mu|J/\psi(p+q)\rangle = \epsilon_{\mu\nu\alpha\beta} q^\nu \varepsilon^{(J/\psi)\alpha} p^\beta \, g_{J/\psi\,\eta_c\gamma}\,, \qquad (4.30)$$

a schematic expression in the charmonium potential model:

$$g_{J/\psi\,\eta_c\gamma} \sim \int d\vec{r}\, \Psi_{1^1S_0}(\vec{r})\, \hat{O}^{em}\, \Psi_{3^1S_1}(\vec{r})\,, \qquad (4.31)$$

where we use spectroscopic notations $^3S_1$ and $^1S_0$ for the wave functions of J/ψ and η_c, respectively, and $\hat{O}^{em}$ denotes the effective operator emerging from j_{em} in the nonrelativistic limit. This transition matrix element somewhat resembles the atomic form factors described in the Introduction.

The vector-to-pseudoscalar transitions in (4.29) represent, using the terminology of atomic physics, magnetic-dipole transitions, due to the fact that the transition operator is determined by the c-quark magnetic moment. The latter is independent of coordinates and can be factorized in (4.31) reducing it to[5]

$$g_{J/\psi\,\eta_c\gamma} \sim \mu \int_0^\infty dr\, r^2\, R_{1^1S_0}(r) R_{3^1S_1}(r)\,, \qquad (4.32)$$

where $\mu \sim e/m_c$ is the quark magnetic moment in units of e, $R_{1^1S_0}(r)$ and $R_{3^1S_1}(r)$ are the radial wave functions, and a universal normalization-dependent coefficient is not shown. In the nonrelativistic limit, neglecting the hyperfine part of the potential, the radial wave functions of $1S$-levels coincide and the integral reduces to a unity. This brings an approximate equality:

$$g_{J\psi\,\eta_c\gamma} \simeq g_{\psi(2S)\eta_c(2S)\gamma} \qquad (4.33)$$

between the radiative decay constants of $J/\psi \to \eta_c\gamma$ and of the similar transition $\psi(2S) \to \eta_c(2S)\gamma$ between the first radial excitations. The resulting relation between decay widths, corrected by phase space factors, is

$$\frac{\Gamma(\psi(2S) \to \eta_c(2S)\gamma)}{\Gamma(J/\psi \to \eta_c\gamma)} = \frac{(m^2_{\psi(2S)} - m^2_{\eta_c(2S)})^3 m^3_{J/\psi}}{(m^2_{J/\psi} - m^2_{\eta_c})^3 m^3_{\psi(2S)}}\,. \qquad (4.34)$$

---

[5]For more details see, e.g., a comprehensive review in [24].

Using data on branching fractions, total widths and masses from the $c\bar{c}$-meson section of [1], we find that (4.33) is violated by about 30%, which roughly reflects the accuracy of nonrelativistic approximation in these decays. Furthermore, in the same approximation the radiative decay $\eta(2S) \to J/\psi\gamma$ should be forbidden because the overlap of $2S$ and $1S$ wave functions vanishes due to orthogonality. The decay is observed [1], but the extracted decay constant is indeed smaller than for the $1S \to 1S$ transition:

$$g_{\psi(2S)\eta_c(1S)\gamma} \ll g_{J/\psi\eta_c\gamma}.$$

Concluding, we have to admit that a simple nonrelativistic potential is certainly not a very accurate approximation even for the heavier b-quarkonium. A far more elaborated effective theory based on a combination of a heavy quark limit with an expansion in terms of quark-antiquark velocity is obtained on a basis of QCD and applied to calculate the spectra of heavy quarkonia and the transitions between the levels. The basics of this approach are given e.g., in the review [45]. A quite different method to treat the $J/\psi \to \eta_c\gamma$ decay, which does not rely on nonrelativistic approximation and is based on QCD sum rules, will be presented in Chapter 8.

4.4 $B \to K^*\gamma$ DECAY

Flavor-changing radiative decays of hadrons are generated by a weak boson exchange combined with a photon emission. One prominent example is $B \to K^*\gamma$, triggered by the $b \to s\gamma$ FCNC transition. In SM, the most important contribution is due to the effective operator $O_{7\gamma}$ presented in (1.43) and entering the effective Hamiltonian (1.44).

To obtain the decay amplitude, we start, as usual, from the S-matrix element, where in $\mathcal{H}_{eff}^{b \to s}$ we retain only the contribution of $O_{7\gamma}$, where we neglect the suppressed term proportional to m_s:

$$S_{fi}^{(B \to K^*\gamma)} = \langle K^*(p_2)\gamma(q)| - i\int d^4x\, \mathcal{H}_{eff}^{b \to s}(x)|\bar{B}(p_1)\rangle$$

$$= i\int d^4x \langle K^*(p_2)\gamma(q)|\frac{G_F}{\sqrt{2}}V_{tb}V_{ts}^* C_{7\gamma}O_{7\gamma}(x)|\bar{B}(p_1)\rangle = -\frac{G_F}{\sqrt{2}}V_{tb}V_{ts}^*\frac{C_{7\gamma}e}{8\pi^2}$$

$$\times i\int d^4x \langle K^*(p_2)\gamma(q)|m_b\bar{s}(x)\sigma_{\mu\nu}(1+\gamma_5)b(x)F^{\mu\nu}(x)|\bar{B}(p_1)\rangle. \quad (4.35)$$

Factorizing the photon part of the above matrix element and using the momentum decomposition of the photon field, we obtain:

$$\langle \gamma(q)|F_{\mu\nu}(x)|0\rangle = \langle \gamma(q)|\partial^\mu A_{em}^\nu(x) - \partial^\nu A_{em}^\mu(x)|0\rangle$$

$$= \partial^\mu\left(\varepsilon^{(\gamma)\nu*}e^{iqx}\right) - \partial^\nu\left(\varepsilon^{(\gamma)\mu*}e^{iqx}\right) = i\left(q^\mu\varepsilon^{(\gamma)\nu*} - q^\nu\varepsilon^{(\gamma)\mu*}\right)e^{iqx}. \quad (4.36)$$

Using this expression in (4.35), we make a translation of the current and separate the factor $i(2\pi)^4\delta(p_1 - p_2 - q)$, emerging after the x-integration, so that the decay amplitude is:

$$\mathcal{A}(B \to K^*\gamma) = i\frac{G_F}{\sqrt{2}}V_{tb}V_{ts}^*\frac{e}{4\pi^2}C_{7\gamma}m_b\varepsilon^{(\gamma)\nu*}\langle K^*(p_2)|\bar{s}\sigma_{\mu\nu}q^\nu(1+\gamma_5)b|\bar{B}(p_1)\rangle. \quad (4.37)$$

For the $B \to K^*$ hadronic matrix element we can directly use the decomposition (2.142) at $q^2 = 0$. Taking into account the orthogonality condition $q \cdot \varepsilon^{(\gamma)*} = 0$ and the relation (2.141) between the two form factors, we come to the conclusion that the radiative B-decay amplitude contains a single hadronic parameter which is the $B \to K^*$ form factor $T_1^{BK^*}$ at zero momentum transfer:

$$\mathcal{A}(B \to K^*\gamma) = \frac{G_F}{\sqrt{2}} \frac{e}{2\pi^2} V_{tb} V_{ts}^* C_{7\gamma} m_b \Big\{ - \epsilon_{\mu\nu\rho\sigma} \varepsilon^{(\gamma)*\mu} \varepsilon^{*\nu} q^\rho p_2^\sigma$$

$$+ i \big[(\epsilon^{(\gamma)*} \cdot \varepsilon^*)(p_2 \cdot q) - (\varepsilon^{(\gamma)*} \cdot p_2)(\varepsilon^* \cdot q) \big] \Big\} T_1^{BK^*}(0) . \tag{4.38}$$

This amplitude respects gauge invariance of the e.m. interaction: it vanishes if we replace $\varepsilon^* \to q$. By the same token, the radiative decay $B \to K\gamma$ is forbidden. Indeed, the $B \to K$ matrix element (2.119) of the tensor current vanishes at $q^2 = 0$ if we multiply it by the real photon polarization vector, due to orthogonality of the latter to the photon momentum. A transition $B \to K\ell^+\ell^-$ via the photonic $O_{7\gamma}$ operator is, however, allowed at $q^2 \neq 0$, because the virtual photon has an additional, longitudinal polarization.

The decay width of $B \to K^*\gamma$ is easily obtained squaring the amplitude (4.38), using the summation condition (4.9) for the photon polarizations and the one, similar to (4.12), for the K^* polarizations. Finally, we employ the general formula (B.47) for the two-particle decay width. The result is:

$$\Gamma(B \to K^*\gamma) = \frac{\alpha_{em} G_F^2 |V_{tb} V_{ts}^*|^2 |C_{7\gamma}|^2}{32\pi^4} m_B^3 m_b^2 \left(1 - \frac{m_{K^*}^2}{m_B^2} \right)^3 |T_1^{BK^*}(0)|^2 . \tag{4.39}$$

Note that the b-quark mass, being a scale-dependent parameter of QCD, appears in the formula for the physical width. The point is that the effective tensor quark current – in contrast to the vector and axial-vector weak currents – has a compensating scale dependence (anomalous dimension).

CHAPTER 5

FORM FACTORS IN THE TIMELIKE REGION

5.1 CROSSING TRANSFORMATION AND ANALYTICITY

In Chapter 2 we introduced the pion e.m. form factor $F_\pi(q^2)$, parameterizing the hadronic matrix element of the electron-pion scattering:

$$\langle \pi^+(p_2)|j_\mu^{em}|\pi^+(p_1)\rangle = (p_1 + p_2)_\mu F_\pi^{spc}(q^2), \tag{5.1}$$

where we now add the index *spc*, to emphasize that this form factor is defined in the region of spacelike momentum transfers,

$$q^2 = (p_2 - p_1)^2 \le 0.$$

The process of electron-positron annihilation into a pair of pions:

$$e^-(k_1)e^+(k_2) \to \pi^-(p_1)\pi^+(p_2), \tag{5.2}$$

is also mediated by e.m. interaction with an exchange of a virtual photon. Moreover, the diagram of the annihilation process is obtained, according to our convention (initial-to-final transition directed from left to right), rotating the diagram of electron-pion scattering in Figure 2.1 by 90 degrees counterclockwise. Note that the transition from scattering to annihilation implies that the outgoing electron with momentum k_2 transforms to the incoming positron with momentum $-k_2$, and the initial-state π^+ with momentum p_1 becomes the final-state π^- with momentum $-p_1$. The two processes (2.1) and (5.2) are related to each other by a *crossing* transformation, which means interchanging some of the initial-state and final-state particles and reversing the signs of momenta, electric charges and other additive quantum numbers of these particles to maintain their conservation. Apart from (5.2), a kinematically allowed crossing channel of $e^-\pi^+ \to e^-\pi^+$ scattering is $e^-\pi^- \to e^-\pi^-$. If we consider the weak scattering $\nu_e\pi^- \to e^-\pi^0$, not only the annihilation $\nu_e e^+ \to \pi^+\pi^0$ is possible under crossing transformation, but also the decay $\pi^- \to \pi^0 e^- \bar\nu_e$, which emerges due to the mass difference between the charged and neutral pion.

To obtain the amplitude of (5.2), we return to Chapter 2 and repeat the derivation of the amplitude with transformed initial and final states. The result:

$$\mathcal{A}(e^-e^+ \to \pi^-\pi^+) = \frac{4\pi\alpha_{em}}{q^2}\bar{\mathrm{v}}_e(k_2)\gamma^\mu \mathrm{u}_e(k_1)\langle \pi^-(p_1)\pi^+(p_2)|j_\mu^{em}|0\rangle, \tag{5.3}$$

contains a hadronic matrix element, different from the one in (5.1). The momentum transfer changes correspondingly:

$$q = p_2 - p_1 \rightarrow q = p_2 + p_1,$$

and the invariant variable $q^2 = (p_1 + p_2)^2$ becomes timelike and bounded from below. To derive the bound, we choose the c.m. frame of the process (5.2), in which $\vec{p_1} + \vec{p_2} = 0$, $E_1 = E_2$ and $q^2 = 4E_1^2$, hence,

$$q^2 = (p_1 + p_2)^2 \geq 4m_\pi^2. \tag{5.4}$$

Applying the usual symmetry arguments to the hadronic matrix element in (5.3), we define the pion e.m. form factor in the timelike region:

$$\langle \pi^-(p_1)\pi^+(p_2)|j_\mu^{em}|0\rangle = (p_2 - p_1)_\mu F_\pi^{tml}(q^2). \tag{5.5}$$

At first sight, the form factors in (5.1) and (5.5) are not related to each other, describing two markedly different physical processes – scattering and annihilation. Moreover, as we shall see in the next sections, the timelike form factor acquires an imaginary part and involves intermediate hadronic states, whereas the spacelike form factor is a real valued function of q^2. In fact, the crossing transformation leads to the following nontrivial connection between these two different form factors: there exists a single function $F_\pi(q^2)$ which is equal to the spacelike or timelike form factor if q^2 belongs to the respective region:

$$F_\pi(q^2) = \left\{ \begin{array}{ll} F_\pi^{spc}(q^2), & q^2 \leq 0, \\ F_\pi^{tml}(q^2), & q^2 \geq 4m_\pi^2. \end{array} \right. \tag{5.6}$$

Importantly, in this case there is also an intermediate region $0 < q^2 < 4m_\pi^2$, where the function $F_\pi(q^2)$, being continuous, exists but does not correspond to any physical process.

To substantiate (5.6), we start from the following representation of the pion state:

$$|\pi^+(p)\rangle = \frac{2p_0}{m_\pi^2 f_\pi} \int d^3x e^{-ipx} j_5(x)|0\rangle, \tag{5.7}$$

where $p^2 = m_\pi^2$ and

$$j_5(x) = (m_u + m_d)\bar{u}(x)i\gamma_5 d(x) \tag{5.8}$$

is the pseudoscalar current operator with the pion quantum numbers. To prove (5.7), we multiply the ket-vector states on both sides by the bra-vector state of the pion:

$$\langle \pi^+(p')|\pi^+(p)\rangle = \frac{2p_0}{m_\pi^2 f_\pi} \int d^3x e^{-ipx} \langle \pi^+(p')|j_5(x)|0\rangle. \tag{5.9}$$

Applying a translation similar to (A.79) and using the relation (2.57) for the pion decay constant, we end up with the normalization condition (1.81) for the pion state:

$$\langle \pi^+(p')|\pi^+(p)\rangle = \frac{2p_0}{m_\pi^2 f_\pi} \int d^3x e^{-ipx} e^{ip'x} \langle \pi^+(p')|j_5(0)|0\rangle$$

$$= \frac{2p_0}{m_\pi^2 f_\pi}(2\pi)^3 \delta^{(3)}(\vec{p} - \vec{p}')e^{-i(p_0-p_0')x_0} m_\pi^2 f_\pi = 2p_0(2\pi)^3 \delta^{(3)}(\vec{p} - \vec{p}'). \tag{5.10}$$

Note that, adopting (5.7), we implicitly use that at $p^2 = m_\pi^2$ no hadronic states, apart from the single-pion state, are generated by the current j_5. The analog of (5.7) for the bra-vector state is:

$$\langle \pi^-(p)| = \frac{2p_0}{m_\pi^2 f_\pi} \int d^3x e^{+ipx} \langle 0|j_5(x). \tag{5.11}$$

To proceed, we use (5.7) to transform the hadronic matrix element (5.1):

$$\langle \pi^+(p_2)|j_\mu^{em}|\pi^+(p_1)\rangle = \frac{2p_{10}}{m_\pi^2 f_\pi} \int d^3x e^{-ip_1 x} \langle \pi^+(p_2)|j_\mu^{em}(0)j_5(x)|0\rangle$$

$$= \frac{2p_{10}}{m_\pi^2 f_\pi} \int d^3x e^{-ip_1 x} \langle \pi^+(p_2)|[j_\mu^{em}(0), j_5(x)] + j_5(x)j_\mu^{em}(0)|0\rangle, \qquad (5.12)$$

rewriting it identically as a sum of two terms, one of which contains the commutator of two currents. In the second term we then use (5.11):

$$\frac{2p_{10}}{m_\pi^2 f_\pi} \int d^3x e^{-ip_1 x} \langle \pi^+(p_2)|j_5(x)j_\mu^{em}(0)|0\rangle = -\langle \pi^+(p_2)\pi^-(-p_1)|j_\mu^{em}(0)|0\rangle, \qquad (5.13)$$

and notice that the resulting matrix element on the r.h.s. vanishes because at $q^2 = (p_2 - p_1)^2 < 0$ a two-pion state cannot be produced by the e.m. current; hence,

$$\langle \pi^+(p_2)|j_\mu^{em}|\pi^+(p_1)\rangle = (p_2 + p_1)_\mu F_\pi^{spc}(q^2)$$

$$= \frac{2p_{10}}{m_\pi^2 f_\pi} \int d^3x e^{-ip_1 x} \langle \pi^+(p_2)|[j_\mu^{em}(0), j_5(x)]|0\rangle. \qquad (5.14)$$

Performing similar transformations of the hadronic matrix element (5.5), we reduce it to the same form:

$$\langle \pi^-(p_1)\pi^+(p_2)|j_\mu^{em}|0\rangle = (p_2 - p_1)_\mu F_\pi^{tml}(q^2)$$

$$= -\frac{2p_{10}}{m_\pi^2 f_\pi} \int d^3x e^{ip_1 x} \langle \pi^+(p_2)|[j_\mu^{em}(0), j_5(x)]|0\rangle, \qquad (5.15)$$

where at the last step we, correspondingly, use the absence of $\pi^+ \to \pi^+$ scattering at positive $q^2 > 4m_\pi^2$. We notice that r.h.s. of (5.14) and (5.15) both contain a uniform expression defining a function of p_1 and p_2. The hadronic matrix elements of scattering and annihilation are equal to this function in the two separate regions of variables, differing by the sign of the momentum p_1, in accordance with the crossing transformation. In terms of the invariant variable q^2, this function is parameterized via single form factor $F_\pi(q^2)$, coinciding with the spacelike and timelike form factors in the two separate intervals of q^2, as indicated in (5.6). Hereafter, we remove the indices at these form factors, identifying them according to the region of q^2.

Up to now, the momentum transfer squared q^2 was a real number. We now follow a useful method to employ the connection between form factors defined in different regions of q^2 and extend this variable from the real axis to the complex plane,

$$q^2 \to \text{Re}[q^2] + i\,\text{Im}[q^2].$$

Furthermore, following the S-matrix theory, we adopt the conjecture that transition amplitudes are analytic functions of their invariant variables and that the only singularities these functions have are poles and branch points emerging from the unitarity of the S-matrix. These general properties of transition amplitudes are discussed and substantiated in many textbooks on quantum field theory; a more thorough description can be found e.g., in [46]. The principle of *analyticity* in the case under consideration states that the form factor $F_\pi(q^2)$ is an analytic function in the whole complex plane of the variable q^2 apart from isolated singularities – poles and branch points. These singularities, as we shall see in the next section, are generated by the unitarity condition.

5.2 UNITARITY AND DISPERSION RELATION

In the spacelike region the pion form factor is equal to its complex conjugate, as we have shown in (2.41), hence

$$\text{Im} F_\pi(q^2) = 0, \quad \text{Re}[q^2] \leq 0. \tag{5.16}$$

This condition guarantees absence of singularities (poles and/or branch points) of the form factor on the negative real axis of the complex q^2-plane. That is not the case in the timelike region, where the form factor has a nonvanishing imaginary part at $q^2 \geq 4m_\pi^2$. It follows from the unitarity relation (B.11) for the transition amplitude (T-matrix element) derived in Appendix B. Here we will adapt this relation to a specific case of the pion form factor, being a part of the $e^+e^- \to \pi^+\pi^-$ transition amplitude.

Instead of considering the full $e^+e^- \to \pi^+\pi^-$ annihilation process, it suffices to introduce an effective interaction of a virtual photon γ^* with the quark e.m. current:

$$\widetilde{\mathcal{L}}(x, q) = e\tilde{a}^\mu(q)e^{-iqx}j_\mu^{em}(x), \tag{5.17}$$

where the (fictitious) virtual photon field has a definite momentum q ($q^2 \neq 0$) and polarization vector $\tilde{a}^\mu$.

The S-matrix for the $\gamma^* \to \pi^+\pi^-$ process can be represented as

$$\hat{S} = T\left\{ e^{\left(i\int d^4x' \mathcal{L}_{QCD}^{int}(x')\right)}\left[1 + i\int d^4x \widetilde{\mathcal{L}}(x, q)\right]\right\}, \tag{5.18}$$

keeping only the first order in the interaction (5.17). According to our convention, hereafter we do not display the QCD part of the S-matrix. The corresponding S-matrix element is then

$$S_{fi} = \langle \pi^-(p_1)\pi^+(p_2)| \left[i\int d^4x \widetilde{\mathcal{L}}(x, q)\right]|0\rangle$$

$$= ie\tilde{a}^\mu(q)\int d^4x e^{-iqx}\langle \pi^-(p_1)\pi^+(p_2)|j_\mu^{em}(x)|0\rangle. \tag{5.19}$$

Applying translation of the e.m. current: $j_\mu^{em}(x) = e^{i\hat{P}x}j_\mu^{em}(0)e^{-i\hat{P}x}$ (cf. (A.78)), we transform (5.19) to

$$S_{fi} = ie\tilde{a}^\mu(q)\int d^4x e^{-iqx+i(p_1+p_2)x}\langle \pi^-(p_1)\pi^+(p_2)|j_\mu^{em}|0\rangle$$

$$= ie\tilde{a}^\mu(q)(2\pi)^4\delta^{(4)}(q - p_1 - p_2)F_\pi(q^2)(p_2 - p_1)_\mu, \tag{5.20}$$

where the definition (5.5) of the timelike form factor is used; hence, the T-matrix element is

$$T_{fi} = e\tilde{a}^\mu(q)F_\pi(q^2)(p_2 - p_1)_\mu. \tag{5.21}$$

Furthermore, we use the unitarity relation for this matrix element (see (B.11)):

$$2\,\text{Im} T_{fi} = \sum_n (2\pi)^4\delta^{(4)}(p_f - p_n)T_{fn}^* T_{ni}, \tag{5.22}$$

where the sum goes over all possible intermediate states with total momentum p_n and momentum conservation implies $p_n = p_f = p_1 + p_2 = p_i = q$. For consistency, the order in the e.m. coupling e should be the same on both sides of the unitarity relation. This leaves us with a single possible choice for the first matrix element entering the r.h.s. of (5.22),

$$T_{fn}^* = \langle \pi^-(p_1)\pi^+(p_2)|n\rangle^*, \tag{5.23}$$

and describing the pure QCD (strong interaction) process $|n\rangle \to \pi^+\pi^-$, whereas the second matrix element

$$T_{ni} = e\tilde{a}^\mu(q)\langle n|j_\mu^{em}|0\rangle, \tag{5.24}$$

corresponds to a first-order in e transition $\gamma^* \to \langle n|$. The above equation is obtained, writing down the S-matrix element of this transition, and performing transformations similar to the ones leading to (5.20):

$$
\begin{aligned}
S_{ni} &= \langle n|\left[i \int d^4x \widetilde{\mathcal{L}}(x, q)\right]|0\rangle \\
&= ie\tilde{a}^\mu(q) \int d^4x \, e^{-iqx + ip_n x}\langle n|j_\mu^{em}(0)|0\rangle \\
&= ie\tilde{a}^\mu(q)(2\pi)^4 \delta^{(4)}(q - p_n)\langle n|j_\mu^{em}|0\rangle.
\end{aligned} \tag{5.25}
$$

It remains to substitute (5.21), (5.23) and (5.24) in the unitarity relation (5.22). Comparing both sides of the resulting relation, we equate the four-vectors multiplying the same coefficient $e\tilde{a}_\mu$, and finally obtain the unitarity relation for the imaginary part of the pion form factor:

$$2\operatorname{Im}F_\pi(q^2)(p_2 - p_1)_\mu = \sum_n \int d\tau_{N(n)}\langle \pi^-(p_1)\pi^+(p_2)|n\rangle^*\langle n|j_\mu^{em}|0\rangle. \tag{5.26}$$

In this relation, the sum over intermediate states includes the phase-space integration. Assuming that the state $|n\rangle$ consists of $N(n)$ particles we use the phase-space element $d\tau_{N(n)}$ defined in (B.19). Accordingly, all hadronic matrix elements entering the r.h.s. have to be understood as the transition amplitudes in the sense of redefined T-matrix elements in (B.17).

As follows from (5.26), the lightest possible intermediate state with $J^P = 1^-$ and $I = 1$ is $|n\rangle = |\pi^+\pi^-\rangle$; hence, $\operatorname{Im}F_\pi(q^2) \neq 0$ above the kinematical threshold $q^2 = 4m_\pi^2$. At the same time a $\pi^0\pi^0$ state consisting of two identical bosons cannot have such quantum numbers and is forbidden. Furthermore, three-pion intermediate states with negative G-parity do not contribute either. Ordering the terms in the unitarity sum in (5.26) by their thresholds, the next ones after $|\pi^+\pi^-\rangle$ state are the four-pion and kaon-antikaon states. The pattern of these contributions becomes rather complicated at $q^2 > 1$ GeV^2, where the lightest intermediate states overlap with the onset of multimeson states. On top of that there are also intermediate states of vector mesons, most importantly the ρ^0-meson, whose contribution to (5.26) will be examined below in this chapter.

The fact that $F_\pi(q^2)$ acquires an imaginary part at $q^2 \geq 4m_\pi^2$, results in a branch point of this function located at $q^2 = 4m_\pi^2$. To prove that, we apply the Schwarz reflection principle from complex analysis: if an analytic function is real valued on a part of the real axis, then this function is equal to its complex conjugate taken at the point reflected with respect to the real axis. In our case, $F_\pi(q^2)$ is analytic and real-valued function at $q^2 < 4m_\pi^2$, hence

$$F_\pi(q^2 - i\epsilon) = F_\pi^*(q^2 + i\epsilon), \tag{5.27}$$

where ϵ can be infinitesimally small. On the other hand, due to the appearance of the imaginary part,

$$F_\pi(q^2) \neq F_\pi^*(q^2) \quad \text{at} \quad q^2 \geq 4m_\pi^2.$$

Then, as immediately follows from (5.27), the function $F_\pi(q^2)$ has a finite discontinuity related to the imaginary part:

$$
\begin{aligned}
\Delta[F_\pi(q^2)] &\equiv \lim_{\epsilon \to 0}\left(F_\pi(q^2 + i\epsilon) - F_\pi(q^2 - i\epsilon)\right) \\
&= \lim_{\epsilon \to 0}\left(F_\pi(q^2 + i\epsilon) - F_\pi^*(q^2 + i\epsilon)\right) = 2i\operatorname{Im}F_\pi(q^2).
\end{aligned} \tag{5.28}
$$

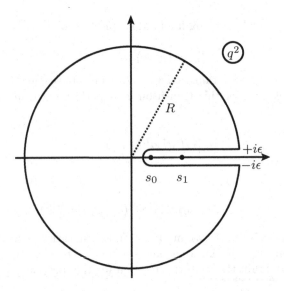

Figure 5.1 Contour in the complex plane of the variable q^2 used to derive the dispersion relation for the pion form factor.

Therefore, $q^2 = 4m_\pi^2$ is a branch point and, correspondingly, we have to insert a cut at $4m_\pi^2 < \mathrm{Re}[q^2] < \infty$, transforming the complex plane into a twofold Riemann surface. As we shall see later, a one-hadron intermediate state in the unitarity relation triggers a singularity of a different type, namely, a simple pole located at the point whose real part is equal to the mass squared of that hadron. Each intermediate state contributing to unitarity relation adds one more pole or branch point on the complex plane, so that the resulting Riemann surface has in fact a complicated structure. This does not preclude employing analyticity of the function $F_\pi(q^2)$ because all these singularities are concentrated on the positive real axis, above $4m_\pi^2$. The rest of the complex q^2-plane is free from singularities, as follows from the analyticity principle formulated at the end of the previous section.

Being an analytic function, the pion form factor obeys the Cauchy formula:

$$F_\pi(q^2) = \frac{1}{2\pi i} \oint_C d\tilde{q}^2 \frac{F_\pi(\tilde{q}^2)}{\tilde{q}^2 - q^2}, \qquad (5.29)$$

where a convenient choice of the integration contour C (counterclockwise directed) is shown in Figure 5.1. The points on the positive real axis mark the positions of the two lowest thresholds, $s_0 = 4m_\pi^2$ and $s_1 = 16m_\pi^2$, of the two-pion and four-pion contributions to the unitarity relation, respectively. The part of the real axis, where these and other singularities of the function $F_\pi(q^2)$ are located, is circumvented. To proceed, we subdivide the contour in the following four parts, parameterizing each of them correspondingly:

(1,2) two direct lines parallel to the real axis, shifted to the complex plane at infinitesimal distances $+\epsilon$ and $-\epsilon$:

$$q^2 = s \pm i\epsilon; \quad 4m_\pi^2 \le s \le R,$$

(3) large circle $\tilde{C}$ with radius R:

$$q^2 = Re^{i\phi}, \quad \epsilon/R < \phi < 2\pi - \epsilon/R,$$

(4) small semicircle $\tilde{c}$ with radius ϵ:

$$q^2 = 4m_\pi^2 + \epsilon e^{i\phi}, \quad -\pi/2 \leq \phi \leq \pi/2.$$

Analyticity of the function $F_\pi(q^2)$ guarantees that the Cauchy formula (5.29) is independent of the values of the two parameters ϵ and R determining the sizes of the contour; hence, the simultaneous limits $R \to \infty$ and $\epsilon \to 0$, which we are aiming at, are allowed. Subdividing the complex integral into three parts, we transform (5.29):

$$F_\pi(q^2) = \frac{1}{2\pi i} \int_{4m_\pi^2}^{R} ds \frac{F(s+i\epsilon) - F(s-i\epsilon)}{s - q^2}$$

$$+ \frac{1}{2\pi i} \oint_{\tilde{C}} d\tilde{q}^2 \frac{F_\pi(\tilde{q}^2)}{\tilde{q}^2 - q^2} + \frac{1}{2\pi i} \oint_{\tilde{c}} d\tilde{q}^2 \frac{F_\pi(\tilde{q}^2)}{\tilde{q}^2 - q^2}. \tag{5.30}$$

Replacing the integration variable $\tilde{q}^2$ in the second and third integrals in the above by the corresponding angular parameters, we convince ourselves that the integral over the small semicircle $\tilde{c}$ vanishes in the limit $\epsilon \to 0$. The $R \to \infty$ limit of the integral over the large circle $\tilde{C}$ depends on the asymptotic behavior of $F_\pi(q^2)$ at $|q^2| \to \infty$. In fact, as will be discussed in the last chapters of this book, QCD predicts a definite asymptotic behavior of the form factor: in the limit of infinitely large spacelike momentum transfer, $Q^2 = -q^2 \to \infty$,

$$F_\pi(Q^2) \sim \frac{\alpha_s(Q^2)}{Q^2}, \tag{5.31}$$

where the constant part of the coefficient is not specified, and, as we know, the Q^2-behavior of QCD coupling is only logarithmical. Due to analyticity of the function $F_\pi(q^2)$, the same asymptotics is valid at any $|q^2| \to \infty$, that is, in all directions of the complex plane, including the positive real axis $q^2 = s \to \infty$; hence, the integral over infinitely large contour also vanishes, and, taking the limit $\epsilon \to 0$, $R \to \infty$ we obtain from (5.30):

$$F_\pi(q^2) = \frac{1}{2\pi i} \int_{4m_\pi^2}^{\infty} ds \frac{\lim_{\epsilon \to 0} \left[F(s+i\epsilon) - F(s-i\epsilon) \right]}{s - q^2}, \tag{5.32}$$

expressing the form factor in terms of its discontinuity which, according to (5.28), is expressed via imaginary part:

$$\lim_{\epsilon \to 0} \left[F_\pi(s+i\epsilon) - F_\pi(s-i\epsilon) \right] = 2i \mathrm{Im} F_\pi(s). \tag{5.33}$$

We finally transform (5.32) to the form of a dispersion relation for the pion form factor:

$$F_\pi(q^2) = \frac{1}{\pi} \int_{4m_\pi^2}^{\infty} ds \frac{\mathrm{Im} F_\pi(s)}{s - q^2 - i\epsilon}, \tag{5.34}$$

where, conventionally, the infinitesimal $i\epsilon$ in the denominator is left to indicate that this relation, when applied at $q^2 > 4m_\pi^2$, yields a complex-valued function $F_\pi(q^2 + i\epsilon)$, defined above the real axis.

It is important that the dispersion relation (5.34) is valid for any value of q^2 in the complex plane; hence, it can be used to match QCD-based predictions available at negative

q^2 to the integral over $\mathrm{Im}F_\pi(s)$ inferred from a certain phenomenological model of the form factor in the timelike region. In particular, the normalization condition (2.38) at $q^2 = 0$ yields a simple sum rule for the integrated imaginary part of the pion form factor:

$$\frac{1}{\pi} \int\limits_{4m_\pi^2}^{\infty} ds \frac{\mathrm{Im}F_\pi(s)}{s} = 1 . \tag{5.35}$$

Dispersion relations similar to (5.34) will also be used for other form factors and hadronic amplitudes considered in this book. Therefore, it is important that the method works even if the asymptotic behavior of the form factor at $|q^2| \to \infty$ is not sufficient to render the contour integral finite. The solution in this case is to perform *subtractions* in the dispersion relation. The idea is simple: subtract from the function of complex variable q^2 its value at a certain point q_0^2 and then apply the Cauchy theorem to the resulting difference, which vanishes at large $|q^2|$ faster than the function itself. In case of the pion form factor, a natural choice is the subtraction point $q_0^2 = 0$ where the form factor is normalized to a unit. The once-subtracted dispersion relation is:

$$F_\pi(q^2) - 1 = \frac{1}{\pi} \int\limits_{4m_\pi^2}^{\infty} ds \, \mathrm{Im}F_\pi(s) \left(\frac{1}{s - q^2} - \frac{1}{s} \right) = \frac{q^2}{\pi} \int\limits_{4m_\pi^2}^{\infty} ds \frac{\mathrm{Im}F_\pi(s)}{s(s - q^2)} . \tag{5.36}$$

To perform n subtractions at some point q_0^2 on the real axis one needs to subtract the n terms of Taylor expansion of the form factor at q_0^2:

$$F_\pi(q^2) - F_\pi(q_0^2) - (q^2 - q_0^2)\frac{d}{dq^2}F_\pi(q^2)\Big|_{q_0^2} - \dots$$

$$-\frac{1}{(n-1)!}(q^2 - q_0^2)^{n-1}\frac{d^{n-1}}{dq^{2(n-1)}}F_\pi(q^2)\Big|_{q_0^2} = \frac{(q^2 - q_0^2)^n}{\pi} \int\limits_{4m_\pi^2}^{\infty} ds \frac{\mathrm{Im}F_\pi(s)}{(s - q_0^2)^n(s - q^2)} . \tag{5.37}$$

Sometimes subtractions are used to make the dispersion integral less dependent on the poorly known imaginary part at $s \to \infty$. This procedure, however, demands the knowledge of the function and its first $n - 1$ derivatives at the subtraction point.

5.3 PION FORM FACTOR IN THE ELASTIC REGION AND THE WATSON THEOREM

The unitarity relation (5.26) is greatly simplified in the so-called elastic region:

$$4m_\pi^2 \leq s < 16m_\pi^2 , \tag{5.38}$$

where only the two-pion intermediate state contributes:

$$2\mathrm{Im}F_\pi(s)(p_2 - p_1)_\mu = \int d\tau_{2\pi}' \langle \pi^+(p_2)\pi^-(p_1)|\pi^+(p_2')\pi^-(p_1')\rangle$$
$$\times \langle \pi^+(p_2')\pi^-(p_1')|j_\mu^{em}|0\rangle^* . \tag{5.39}$$

Here

$$d\tau_{2\pi}' = \frac{d^3p_1'}{(2\pi)^3 2p_{10}'} \frac{d^3p_2'}{(2\pi)^3 2p_{20}'} \delta^{(4)}(p_1 + p_2 - p_1' - p_2') \tag{5.40}$$

is the two-body phase space element and the momentum conservation yields

$$q^2 = s = (p_1 + p_2)^2 = (p'_1 + p'_2)^2.$$

The first hadronic matrix element on the r.h.s. of (5.39) is the amplitude of pion-pion elastic scattering. In the second one we recognize the complex conjugated matrix element (5.5) of the timelike pion form factor. Altogether, (5.39) has a form of an integral equation, relating the imaginary part of the form factor to its convolution with the pion-pion scattering amplitude.

Since pions have zero spin, this amplitude is a plain Lorentz-invariant function:

$$\langle \pi^+(p_2)\pi^-(p_1)|\pi^+(p'_2)\pi^-(p'_1)\rangle \equiv \mathcal{A}_{\pi\pi}(s,t), \tag{5.41}$$

depending on the two Mandelstam variables, s and

$$t = (p'_1 - p_1)^2 = (p_2 - p'_2)^2.$$

Furthermore, we choose a c.m. frame of the two pions, where $\vec{p}_1 + \vec{p}_2 = 0$ and $\vec{p}_2$ is directed along the z-axis, so that

$$p_{1(2)} = \left(\frac{\sqrt{s}}{2}, 0, 0, -(+)p(s)\right).$$

In this frame the momenta of the final-state pions are

$$p'_{1(2)} = \left(\frac{\sqrt{s}}{2}, 0, -(+)p(s)\sin\theta, -(+)p(s)\cos\theta\right),$$

where the angle θ is between $\vec{p}_1$ and $\vec{p}'_1$,

$$p(s) = \beta_\pi(s)\frac{\sqrt{s}}{2} \quad \text{and} \quad \beta_\pi(s) = \sqrt{1 - \frac{4m_\pi^2}{s}}$$

is the velocity of pions in their c.m. frame. The scattering angle θ is related to the invariant variables:

$$\cos\theta = 1 + \frac{2t}{s - 4m_\pi^2}. \tag{5.42}$$

In the absence of spin degrees of freedom, the orbital angular momentum is the relevant quantum number for a state of two pions; hence, we can use the *partial wave expansion* for the scattering amplitude:

$$\mathcal{A}_{\pi\pi}(s,t) = 8\pi\sqrt{s}\sum_{\ell=0}^{\infty}(2\ell+1)f_\ell(s)P_\ell(\cos\theta). \tag{5.43}$$

In this expansion hadronic dynamics is encoded in the coefficient functions (partial wave amplitudes) $f_\ell(s)$, whereas the angular dependence is described by the Legendre polynomials P_ℓ.

Furthermore, we employ the unitarity relation for the elastic scattering amplitude $\mathcal{A}_{\pi\pi}(s,t)$[1]. In the elastic region (5.38) this relation is similar to (5.26) to the extent that only the two-pion intermediate state contributes:

$$2\,\mathrm{Im}\mathcal{A}_{\pi\pi}(s,t) = \int d\tau''_{2\pi}\langle \pi^+(p_2)\pi^-(p_1)|\pi^+(p''_2)\pi^-(p''_1)\rangle$$

$$\times \langle \pi^+(p''_2)\pi^-(p''_1)|\pi^+(p'_2)\pi^-(p'_1)\rangle^* = \int d\tau''_{2\pi}\mathcal{A}_{\pi\pi}(s,t')\mathcal{A}^*_{\pi\pi}(s,t''). \tag{5.44}$$

---

[1]Note that analytical properties of the scattering amplitude with respect to the variable s (at fixed t) are more involved than for the form factor, because an additional cut at $s < 0$ is generated by singularities in the channel of the third Mandelstam variable $u = 4m_\pi^2 - t - s$.

Here the phase space integration goes over the intermediate momenta p_1'' and p_2'' and the momentum conservation yields $(p_1'' + p_2'')^2 = s$, so that all three two-pion states in (5.44) have the same c.m. frame and the invariant variables

$$t' = (p_1'' - p_1)^2 = (p_2 - p_2'')^2, \quad t'' = (p_1'' - p_1')^2 = (p_2' - p_2'')^2, \tag{5.45}$$

are, via equations similar to (5.42), related to the corresponding angles θ', θ'' in this frame. To disentangle the equation (5.44), we use the partial wave expansion (5.43) for each of the amplitudes and obtain

$$16\pi\sqrt{s}\sum_{\ell=0}^{\infty}(2\ell+1)\mathrm{Im}f_\ell(s)P_\ell(\cos\theta) = 64\pi^2 s\int d\tau_{2\pi}''\sum_{\ell'=0}^{\infty}(2\ell'+1)f_{\ell'}(s)P_{\ell'}(\cos\theta')$$

$$\times \sum_{\ell''=0}^{\infty}(2\ell''+1)f_{\ell''}^*(s)P_{\ell''}(\cos\theta''). \tag{5.46}$$

Integrating over the angular part of the phase space,

$$\int d\tau_{2\pi}'' = \frac{\beta_\pi(s)}{32\pi^2}\int_{-1}^{1}d\cos\theta''\int_{0}^{2\pi}d\phi, \tag{5.47}$$

and employing the orthogonality (triangle) relation between Legendre polynomials:

$$\int_{-1}^{+1}d\cos\theta''\int_{0}^{2\pi}d\phi P_{\ell'}(\cos\theta')P_{\ell''}(\cos\theta'') = \delta_{\ell'\ell''}\frac{4\pi}{2\ell'+1}P_{\ell'}(\cos\theta), \tag{5.48}$$

where the trigonometric relation

$$\cos\theta' = \cos\theta\cos\theta'' + \sin\theta\sin\theta''\cos\phi \tag{5.49}$$

between the adjacent angles is used, we obtain the unitarity relation for a separate partial wave:

$$\mathrm{Im}f_\ell(s) = p(s)|f_\ell(s)|^2, \tag{5.50}$$

valid in the kinematical region (5.38). Parameterizing

$$f_\ell(s) = |f_\ell(s)|\exp(i\delta_\ell(s)),$$

we introduce the elastic phase $\delta_\ell(s)$ of the ℓ-th partial wave. Substituting this parameterization in (5.50), we find that

$$|f_\ell(s)| = \frac{\sin\delta_\ell(s)}{p(s)}. \tag{5.51}$$

and

$$f_\ell(s) = \frac{\sin\delta_\ell(s)}{p(s)}\exp(i\delta_\ell(s)). \tag{5.52}$$

We conclude that a partial wave amplitude in the elastic region of pion-pion scattering is entirely determined by its phase.

The ansatz (5.51) helps to simplify the unitarity relation (5.39) for the pion form factor in the elastic region. To this end, we substitute in this relation the partial wave expansion (5.43) of the pion-pion scattering amplitude and the definition of the pion form factor:

$$2\,\mathrm{Im}F_\pi(s)(p_2-p_1)_\mu = \int d\tau_{2\pi}'\mathcal{A}_{\pi\pi}(s,t)(p_2'-p_1')_\mu F_\pi^*(s)$$

$$= \frac{\beta_\pi}{16\pi}\int_{-1}^{1}d\cos\theta\left(8\pi\sqrt{s}\sum_{\ell=0}^{\infty}(2\ell+1)f_\ell(s)P_\ell(\cos\theta)\right)(p_2'-p_1')_\mu F_\pi^*(s). \tag{5.53}$$

After that we multiply both parts of (5.53) by a unit 4-vector $n_\mu = (0, 0, 0, -1)$, so that

$$n \cdot (p_2 - p_1) = 2p(s) \quad \text{and} \quad n \cdot (p_2' - p_1') = 2p(s) \cos \theta.$$

The equation for the imaginary part of F_π becomes:

$$2 \operatorname{Im} F_\pi(s) = p(s) \int_{-1}^{1} d\cos\theta \sum_{\ell=0}^{\infty} (2\ell + 1) f_\ell(s) P_\ell(\cos\theta) \cos\theta \, F_\pi^*(s). \tag{5.54}$$

The factor $\cos\theta$ emerging on the r.h.s. is nothing but the first Legendre polynomial $P_1(\cos\theta)$. This is expected, because the only partial wave that contributes to the form factor is the P-wave. Indeed, after integrating r.h.s. over $\cos\theta$, only the $\ell = 1$ term in the partial wave expansion remains, due to the orthogonality of polynomials:

$$\int_{-1}^{1} dz P_\ell(z) P_{\ell'}(z) = \frac{2}{2\ell + 1} \delta_{\ell\ell'}, \tag{5.55}$$

and we obtain

$$\operatorname{Im} F_\pi(s) = p(s) f_1(s) F_\pi(s)^*. \tag{5.56}$$

Using (5.52) for the P-wave, and rewriting the complex-conjugated form factor on the r.h.s. as $F_\pi(s)^* = |F_\pi(s)| \exp(-i\delta_{F_\pi}(s))$, we obtain the following equation for the imaginary part of the form factor:

$$\operatorname{Im} F_\pi(s) = \sin \delta_1(s) \exp \left[i(\delta_1(s) - \delta_{F_\pi}(s)) \right] |F_\pi(s)|. \tag{5.57}$$

Since, by default, l.h.s. of the above equation is a real-valued function, we conclude that

$$\delta_{F_\pi}(s) = \delta_1(s),$$

telling that the phase of the pion form factor is equal to the phase of the P-wave pion-pion scattering in the elastic region:

$$F_\pi(s) = |F_\pi(s)| \exp(i\delta_1(s)), \quad 4m_\pi^2 \leq s < 16m_\pi^2. \tag{5.58}$$

This statement is known in hadron phenomenology as the *Watson theorem*. An equivalent version of the latter equality reads:

$$\frac{\operatorname{Im} F_\pi(s)}{\operatorname{Re} F_\pi(s)} = \tan \delta_1(s). \tag{5.59}$$

At larger momentum transfers, $s \geq 16m_\pi^2$ other (inelastic) channels contribute to the unitarity relation, such as 4π, $K\bar{K}$,... and make this relation far more complicated.

5.4 OMNES REPRESENTATION

If we neglect inelastic effects at large s, assuming that the unitarity condition (5.58) holds up to $s \to \infty$, a very useful representation of the pion form factor can be derived. To this end, we substitute the elastic unitarity condition in the form (5.59) under the dispersion integral in (5.34):

$$F_\pi(s) = \frac{1}{\pi} \int_{4m_\pi^2}^{\infty} ds' \frac{\tan \delta_1(s') \operatorname{Re} F_\pi(s')}{s' - s - i\epsilon}, \tag{5.60}$$

and define an auxiliary function:

$$\Phi(s) \equiv \frac{1}{2i} F_\pi(s),$$

whose real (imaginary) part is given by $\mathrm{Im} F_\pi$ ($\mathrm{Re} F_\pi$). From (5.60) it follows that the values of this function at $s \pm i\epsilon$ are given by:

$$\Phi(s \pm i\epsilon) = \frac{1}{2\pi i} \int\limits_{4m_\pi^2}^{\infty} ds' \frac{\tan \delta_1(s') \mathrm{Re} F_\pi(s')}{s' - s \mp i\epsilon}. \tag{5.61}$$

Using the formula:

$$\frac{1}{s' - s \mp i\epsilon} = \mathrm{PV} \frac{1}{s' - s} \pm i\pi\delta(s' - s), \tag{5.62}$$

we obtain

$$\Phi(s + i\epsilon) - \Phi(s - i\epsilon) = \int\limits_{4m_\pi^2}^{\infty} ds' \tan \delta_1(s') \mathrm{Re} F_\pi(s')\delta(s' - s)$$

$$= \tan \delta_1(s) \mathrm{Re} F_\pi(s)\theta(s - 4m_\pi^2). \tag{5.63}$$

Furthermore, from the dispersion relation (5.34), we infer that

$$\mathrm{Im} F_\pi(s + i\epsilon) = -\mathrm{Im} F_\pi(s - i\epsilon).$$

Hence,

$$F_\pi(s+i\epsilon) + F_\pi(s-i\epsilon) = 2\mathrm{Re}\, F_\pi(s) = 2i\big[\Phi(s+i\epsilon) + \Phi(s-i\epsilon)\big]. \tag{5.64}$$

Substituting the second equation to (5.63), we obtain a nontrivial relation between the values of the newly defined function taken near the real axis, at $s \pm i\epsilon$:

$$\Phi(s+i\epsilon) - \Phi(s-i\epsilon) = i \tan \delta_1(s)\big[\Phi(s+i\epsilon) + \Phi(s-i\epsilon)\big]\theta(s - 4m_\pi^2). \tag{5.65}$$

This relation can be easily rearranged in the following form:

$$\Phi(s+i\epsilon) = \Phi(s-i\epsilon) \ \text{ at } s < 4m_\pi^2,$$
$$\Phi(s+i\epsilon) \exp\big[-2i\delta_1(s)\big] = \Phi(s-i\epsilon) \ \text{ at } s \geq 4m_\pi^2. \tag{5.66}$$

Taking logarithms of the above equations we obtain:

$$\ln \Phi(s+i\epsilon) = \ln \Phi(s-i\epsilon) \ \text{ at } s < 4m_\pi^2,$$
$$\ln \Phi(s+i\epsilon) - 2i\delta_1(s) = \ln \Phi(s-i\epsilon) \ \text{ at } s \geq 4m_\pi^2. \tag{5.67}$$

We conclude that the function $\ln \Phi(s)$ is real valued everywhere, except the part of the real axis at $s \geq 4m_\pi^2$. Therefore, it is possible to apply the Cauchy formula for this function and derive a dispersion relation, similar to (5.34). Before doing that, it is necessary to perform at least one subtraction, keeping in mind that a logarithmic function is not vanishing at $|s| \to \infty$. For the subtracted function $\ln \Phi(s) - \ln \Phi(0)$ the dispersion relation can be written as

$$\ln \Phi(s) - \ln \Phi(0) = \frac{1}{2\pi i} \int\limits_{4m_\pi^2}^{\infty} ds' \left[\frac{1}{s' - s} - \frac{1}{s'}\right]$$

$$\times \ \Big(\ln \Phi(s' + i\epsilon) - \ln \Phi(s' - i\epsilon)\Big). \tag{5.68}$$

Using on the r.h.s. the second equation in (5.67) and returning to the form factor on the l.h.s., we have:

$$\ln\left[\frac{F_\pi(s)}{F_\pi(0)}\right] = \frac{s}{\pi}\int\limits_{4m_\pi^2}^{\infty} ds'\frac{\delta_1(s')}{s'(s'-s)}. \tag{5.69}$$

Exponentiating both sides of this equation, we obtain the simplest possible *Omnes representation* for the pion e.m. form factor:

$$F_\pi(s) = \exp\left[\frac{s}{\pi}\int\limits_{4m_\pi^2}^{\infty} ds'\frac{\delta_1(s')}{s'(s'-s)}\right], \tag{5.70}$$

consistent with the form factor normalization $F_\pi(0) = 1$. In practical applications, more subtractions are used in this representation, allowing one to suppress the large s part of the integral, thus making it less sensitive to the accuracy of the elastic approximation for the phase.

5.5   RESONANCES IN THE PION FORM FACTOR

The vector meson $\rho(770)$ has a twofold nature. Viewed as a quark-antiquark state, it is a partner of the pseudoscalar pion, differing from the latter only by the spin alignment of constituents; hence, ρ is a part of the spectrum of mesons and, to a certain approximation, is considered as a stable hadron, as in our analysis of the $B \to \rho$ form factors (Chapter 2) and of the $\rho \to \pi\gamma$ radiative decay (Chapter 4). In reality, as already discussed in Chapter 1, all three members of the isospin ρ-triplet, $\rho^\pm$ and ρ^0 are extremely unstable and decay strongly into a two-pion final state as shown in Figure 1.6. Therefore, it is only possible to observe a ρ-meson as an intermediate resonant state in hadronic processes[2]. Here we explain the resonance nature of ρ in more detail.

Let us consider the neutral meson $\rho^0(770)$. Its valence quark content quoted in (1.89) coincides with the isovector component of the quark e.m. current; hence, it is possible to generate a single ρ^0 directly in the electron-positron collision $e^+e^- \to \rho^0$ via virtual timelike photon. At this point we deliberately oversimplify physics, assuming that the two-pion strong decay of ρ is switched off, together with all other secondary decay modes such as $\rho \to \pi\gamma$, and leaving only the e.m. coupling to the lepton pair. In this idealized situation, the produced ρ^0 remains quasistable and decays into an e^+e^- or $\mu^+\mu^-$ pair with a very small width,

$$\Gamma(\rho^0 \to e^+e^-) \sim 0.07 \text{ MeV}, \tag{5.71}$$

as inferred from [1], multiplying the branching fraction by the total width. Note that for our hypothetical stable ρ^0 the total and leptonic widths become equal.

Furthermore, we denote the invariant amplitude of the ρ^0 production as $\mathcal{A}(e^+e^- \to \rho^0)$. In fact, below we will write down in detail the amplitude of an inverse process. Squaring the production amplitude, summing over polarizations and multiplying by the phase space including an additional flux factor, we obtain the cross section, schematically:

$$\sigma(e^+e^- \to \rho^0) \sim \overline{|\mathcal{A}(e^+e^- \to \rho^0)|^2}d\tau_\rho. \tag{5.72}$$

---

[2]Note that the instability of ρ is indirectly caused by the spontaneous violation of chiral symmetry in QCD which suppresses the pion mass, opening up a large phase space for the $\rho \to \pi\pi$ decay.

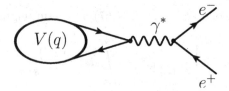

Figure 5.2 Diagram of vector meson leptonic decay $V \to e^+ e^-$.

Essential for our discussion is the one-particle phase space of the final ρ. Using (B.19) at $N = 1$, we have

$$d\tau_\rho = \frac{d\vec{p}_\rho}{(2\pi)^3 2E_\rho}(2\pi)^4 \delta^{(4)}(p_\rho - p_{e^+} - p_{e^-}) = \frac{2\pi}{2m_\rho}\delta(m_\rho - \sqrt{s}) = 2\pi\,\delta(m_\rho^2 - s)\,, \quad (5.73)$$

and, after integrating out the three-momentum δ-function, we choose the c.m. frame of $e^+ e^-$ collision, where $E_{e^+} + E_{e^-} = \sqrt{s}$ and $E_\rho = m_\rho$. Replacing the phase space factor in (5.72) by (5.73), we see that the cross section

$$\sigma(e^+ e^- \to \rho^0) \sim \overline{|\mathcal{A}(e^+ e^- \to \rho^0)|^2}\delta(s - m_\rho^2) \quad (5.74)$$

has a form of an infinitely high and narrow peak. It is located at $\sqrt{s} = m_\rho$ and vanishes at all other energies. Naturally, a stable ρ-resonance is produced at rest only if the collision energy is equal to its mass, $\sqrt{s} = m_\rho$.

After restoring the coupling of ρ to the two-pion state, the $e^+ e^-$ production of ρ is followed by the strong decay:

$$e^+ e^- \to \rho^0 \to \pi^+ \pi^-\,, \quad (5.75)$$

and we expect that a resonance peak at $\sqrt{s} = m_\rho$ remains in the cross section. However, the two-pion production is now possible at any $\sqrt{s} > 4m_\pi^2$ being determined by the timelike pion form factor; hence, we have to clarify how the form factor $F_\pi(s)$ is influenced by an intermediate ρ resonance.

To obtain the ρ-meson contribution to the form factor we need to combine two hadronic matrix elements. The first one describes the first stage of the process (5.75), a transition of e.m. current to ρ^0:

$$\langle \rho^0(q, \lambda)|j_\mu^{em}|0\rangle = \langle \rho^0(q, \lambda)|j_\mu^{(I=1)}|0\rangle = \varepsilon_\mu^{(\rho)*}(q, \lambda)\frac{m_\rho f_\rho}{\sqrt{2}}\,, \quad (5.76)$$

where $\varepsilon_\mu^{(\rho)*}(q, \lambda)$ is the polarization vector with spin projection $\lambda = \pm 1, 0$ specified in (4.12). The above parametrization is consistent with both the e.m. current conservation and the orthogonality $\varepsilon_\mu^{(\rho)*}(q, \lambda) \cdot q = 0$.

The constant parameter f_ρ in (5.76) is the decay constant of ρ analogous to the pion decay constant f_π. It can be estimated from the width of

$$\rho^0(q) \to e^+(k_1)e^-(k_2). $$

The diagram of this decay is shown in Figure 5.2. The decay amplitude can be easily obtained following our standard procedure, e.g., the one used for the electron-pion scattering in Chapter 2.1 (cf. (2.18)). The resulting expression

$$\mathcal{A}(\rho^0 \to e^+ e^-) = \frac{4\pi\alpha_{em}}{q^2}\bar{v}_e(k_2)\gamma^\mu u_e(k_1)\langle 0|j_\mu^{em}|\rho^0(q, \lambda)\rangle\,, \quad (5.77)$$

being of the second order in e.m. interaction, consists of the electron-positron current, the propagator of the timelike photon with $q^2 = m_\rho^2$ and the hadronic matrix element which is the inverse of (5.76):

$$\langle 0|j_\mu^{em}|\rho^0(q,\lambda)\rangle = \langle 0|j_\mu^{(I=1)}|\rho^0(q,\lambda)\rangle = \varepsilon_\mu^{(\rho)}(q,\lambda)\frac{m_\rho f_\rho}{\sqrt{2}}. \tag{5.78}$$

The coefficient $1/\sqrt{2}$ corresponds to f_ρ being defined as the decay constant of the charged ρ meson:

$$\langle 0|\bar{u}\gamma_\mu d|\rho^-(q,\lambda)\rangle = m_\rho f_\rho \varepsilon_\mu^{(\rho)}(q,\lambda), \tag{5.79}$$

and reflects the isospin symmetry relation between the two matrix elements (5.76) and (5.79), similar to the relation between the weak vector and e.m. pion form factors. This relation can be easily confirmed, counting quark flow diagrams:

$$\langle 0|j_\mu^{(I=1)}|\rho^0\rangle = \langle 0|\frac{1}{2}\left(\bar{u}\gamma_\mu u - \bar{d}\gamma_\mu d\right)|\frac{1}{\sqrt{2}}\left(u\bar{u} - d\bar{d}\right)_{\rho^0}\rangle$$

$$= \frac{1}{\sqrt{2}}\langle 0|\bar{u}\gamma_\mu u|(u\bar{u})_{\rho^0}\rangle = \frac{1}{\sqrt{2}}\langle 0|\bar{u}\gamma_\mu d|(d\bar{u})_{\rho^-}\rangle = \frac{1}{\sqrt{2}}\langle 0|\bar{u}\gamma_\mu d|\rho^-\rangle, \tag{5.80}$$

where the replacements of u and d quarks in the diagrams are consistent with isospin symmetry. Note that the charged ρ decay constant is related to a leptonic weak decay $\rho^+ \to \ell^+ \nu_\ell$, which has a vanishingly small branching fraction and is therefore practically unobservable.

The width of the $\rho^0 \to e^+e^-$ decay is easily calculated from the amplitude (5.77). Using (5.78), we have:

$$\mathcal{A}(\rho^0 \to e^+e^-) = \frac{4\pi\alpha_{em}}{\sqrt{2}m_\rho}\bar{v}_e(k_2)\gamma^\mu u_e(k_1)\varepsilon_\mu^{(\rho)}(q,\lambda)f_\rho. \tag{5.81}$$

Squaring this amplitude and summing (averaging) over the lepton (ρ) polarizations (using (4.12)), we obtain:

$$\frac{1}{3}\overline{|\mathcal{A}(\rho^0 \to e^+e^-)^2|} = \frac{8\pi^2\alpha_{em}^2}{3m_\rho^2}\text{Tr}\left[\slashed{k}_2\gamma^\mu\slashed{k}_1\gamma^\nu\right]\left(-g_{\mu\nu} + \frac{q_\mu q_\nu}{m_\rho^2}\right)f_\rho^2$$

$$= \frac{8\pi^2\alpha_{em}^2}{3m_\rho^2}(8k_1k_2)f_\rho^2 = \frac{32\pi^2}{3}\alpha_{em}^2 f_\rho^2, \tag{5.82}$$

neglecting the electron mass. Multiplying this expression by the two-particle phase space $d\tau_2 = 1/(8\pi)$ and dividing by the factor $2m_\rho$ (cf. (B.47) in the limit of two massless particles in the final state), we obtain the width

$$\Gamma(\rho^0 \to e^+e^-) = \frac{2\pi\alpha_{em}^2}{3m_\rho}f_\rho^2. \tag{5.83}$$

The second hadronic matrix element relevant for our discussion describes the strong coupling of ρ to two pions:

$$\langle \pi^+(p_2)\pi^-(p_1)|\rho^0(q)\rangle = (p_1 - p_2)^\nu \varepsilon_\nu^{(\rho)}(q,\lambda)g_{\rho\pi\pi}. \tag{5.84}$$

or the inverse one:

$$\langle \rho^0(q))|\pi^+(p_2)\pi^-(p_1)\rangle = (p_1 - p_2)^\nu \varepsilon_\nu^{(\rho)*}(q,\lambda)g_{\rho\pi\pi}^*. \tag{5.85}$$

The matrix element (5.84) serves as an amplitude of the $\rho^0 \to \pi^+\pi^-$ decay induced by QCD interactions[3] and shown in Figure 1.6. This amplitude is expressed via single Lorentz structure and contains one invariant hadronic parameter – the *strong coupling* $g_{\rho\pi\pi}$. The second possible combination of momenta $p_1 + p_2 = q$ vanishes, being multiplied by the polarization vector of ρ. Squaring and averaging the amplitude (5.84), we use (B.47) and obtain the width:

$$\Gamma(\rho^0 \to \pi^+\pi^-) = \frac{1}{3}\overline{|\mathcal{A}(\rho^0 \to \pi^+\pi^-)|^2} \frac{\lambda^{1/2}(m_\rho^2, m_\pi^2, m_\pi^2)}{16\pi m_\rho^3}$$

$$= \frac{1}{3}(p_1 - p_2)^\mu (p_1 - p_2)^\nu \left(-g_{\mu\nu} + \frac{q_\mu q_\nu}{m_\rho^2}\right) \frac{|g_{\rho\pi\pi}|^2}{16\pi m_\rho} \left(1 - \frac{4m_\pi^2}{m_\rho^2}\right)^{1/2}$$

$$= -(p_1 - p_2)^2 \frac{|g_{\rho\pi\pi}|^2}{48\pi m_\rho} \left(1 - \frac{4m_\pi^2}{m_\rho^2}\right)^{1/2} = \frac{m_\rho}{48\pi}\left(1 - \frac{4m_\pi^2}{m_\rho^2}\right)^{3/2} |g_{\rho\pi\pi}|^2. \quad (5.86)$$

It is instructive to obtain the numerical values of the hadronic parameters characterizing the ρ meson. Equating the leptonic width of ρ^0 (5.83) to its experimental value (5.71) and, respectively, the two-pion width (5.86) to the measured ρ^0 total width [1], (neglecting tiny branching fractions of other decay modes), we obtain:

$$f_\rho \simeq 220 \text{ MeV}, \quad |g_{\rho\pi\pi}| \simeq 5.95. \quad (5.87)$$

Note that f_ρ is in the same ballpark as f_π and both parameters are of the order of Λ_{QCD}.

We are now in a position to calculate the ρ-resonance contribution to the unitarity relation (5.26) for the pion form factor:

$$2\,\text{Im}F_\pi^{(\rho)}(q^2)(p_2 - p_1)_\mu = \int d\tau_\rho \langle \pi^+(p_2)\pi^-(p_1)|\rho^0(q)\rangle\langle\rho^0(q)|j_\mu^{em}|0\rangle, \quad (5.88)$$

denoting by $d\tau_\rho$ the one-particle phase space. Substituting the matrix elements (5.76) and (5.84), we then take the sum over polarizations and integrate over the phase space in the rest frame of intermediate ρ (or, equivalently, in the c.m. frame of two pions). The result is:

$$2\,\text{Im}F_\pi^{(\rho)}(q^2)(p_2 - p_1)_\mu = \int \frac{d^3p}{(2\pi)^3 2p_0}(2\pi)^4 \delta^{(4)}(p - q)$$

$$\times \sum_\lambda \varepsilon_\nu^{(\rho)}(q,\lambda)\varepsilon_\mu^{(\rho)*}(q,\lambda)(p_1 - p_2)^\nu g_{\rho\pi\pi} \frac{m_\rho f_\rho}{\sqrt{2}}$$

$$= (2\pi)\frac{\delta(p_0 - q_0)}{2p_0}\left(-g_{\mu\nu} + \frac{q_\mu q_\nu}{m_\rho}\right)(p_1 - p_2)^\nu \frac{m_\rho f_\rho}{\sqrt{2}} g_{\rho\pi\pi}$$

$$= (2\pi)\delta(p_0^2 - q_0^2)(p_2 - p_1)_\mu \frac{m_\rho f_\rho}{\sqrt{2}} g_{\rho\pi\pi}, \quad (5.89)$$

where $\vec{p} = 0$, $p^2 = p_0^2 = m_\rho^2$, $q_0^2 = q^2$. Comparing the invariant functions on both sides of the resulting equation, we obtain the imaginary part of the ρ-meson contribution at $q^2 = s$:

$$\text{Im}F_\pi^{(\rho)}(s) = \frac{m_\rho f_\rho}{\sqrt{2}} g_{\rho\pi\pi}\pi\delta(s - m_\rho^2). \quad (5.90)$$

Substituting it in the dispersion relation and integrating out the δ-function, we single out the ρ contribution

$$F_\pi^{(\rho)}(q^2) = \frac{1}{\pi}\int_{4m_\pi^2}^\infty ds \frac{\text{Im}F_\pi^{(\rho)}(s)}{s - q^2 - i\epsilon} = \frac{m_\rho f_\rho g_{\rho\pi\pi}}{\sqrt{2}(m_\rho^2 - q^2 - i\epsilon)}, \quad (5.91)$$

---

[3] As usual, we imply but do not show explicitly the QCD part of S-matrix inserted between initial and final hadronic states.

having a form of an infinitely narrow resonance at $q^2 = m_\rho^2$, as anticipated from the previous qualitative discussion of the cross section $e^+e^- \to \rho$.

The expression (5.91) corresponds to a simple pole of the function $F_\pi(q^2)$ on the real axis $q^2 = s$ of the complex plane. Note that, since $m_\rho > 4m_\pi$, this pole is located above the elastic region $4m_\pi^2 < s < 16m_\pi^2$. As we already know, there are branch points on the s-axis for each threshold, starting from the two-pion threshold at $s = 4m_\pi^2$, and there is an opening of a new Riemann sheet for each branch point; hence, the ρ meson pole is located on one of those sheets[4]. On the other hand, the denominator in (5.91) also resembles a free-particle propagator for ρ, which is not surprising, because ρ propagates without interactions between the e.m. current and $\rho\pi\pi$ vertex.

In reality, the resonance formula (5.91) is more complicated and, strictly speaking, can only be written in a closed form under certain assumptions. Indeed, from experiment we know that ρ is a broad resonance and its total width is saturated by the decay into two pions. Meanwhile, calculating the ρ contribution to the pion form factor, we only included a single interaction of this vector meson with the two-pion state. In fact, since the $\rho\pi\pi$ coupling (5.87) is large, multiple sequential transitions between ρ and $\pi\pi$ should be taken into account, so that the full amplitude of $e^+e^- \to \pi^+\pi^-$ via the ρ-meson is a superposition:

$$
\begin{aligned}
\mathcal{A}^{(\rho)}(e^+e^- \to \pi^+\pi^-) &= \mathcal{A}(e^+e^- \to \rho^0 \to \pi^+\pi^-) \\
&+ \mathcal{A}(e^+e^- \to \rho^0 \to \pi^+\pi^- \to \rho^0 \to \pi^+\pi^-) + \cdots, \quad (5.92)
\end{aligned}
$$

in which only the first term was taken into account so far, obtaining (5.91). Let us demonstrate how this infinite sum leads to the width of ρ appearing in the denominator in (5.91), significantly modifying the resonance contribution to the pion form factor.

To this end, we return to the expression in the first line of (5.53) for the imaginary part of the form factor:

$$
2\,\mathrm{Im}F_\pi(s)(p_2 - p_1)_\mu = \int d\tau'_{2\pi} \mathcal{A}_{\pi\pi}(s, t)(p'_2 - p'_1)_\mu F_\pi^*(s), \quad (5.93)
$$

and approximate both amplitudes under the integral by an intermediate ρ-meson contribution. This corresponds to the second term in the series of amplitudes in (5.92). For the form factor we can directly use the expression (5.91). For the pion-pion scattering amplitude we derive an analogous expression, starting from its unitarity relation, where only the ρ contribution is retained:

$$
\begin{aligned}
2\,\mathrm{Im}\mathcal{A}_{\pi\pi}^{(\rho)}(s, t) &= \int d\tau_\rho \langle \pi^+(p_2)\pi^-(p_1)|\rho^0(q)\rangle \langle \rho^0(q)|\pi^+(p'_2)\pi^-(p'_1)\rangle^* \\
&= |g_{\rho\pi\pi}|^2 (p_2 - p_1) \cdot (p'_1 - p'_2)\, 2\pi\delta(m_\rho^2 - s). \quad (5.94)
\end{aligned}
$$

Substituting this imaginary part into the dispersion relation for the pion-pion scattering amplitude in the variable s at fixed t results in

$$
\mathcal{A}_{\pi\pi}^{(\rho)}(s, t) = \frac{1}{\pi} \int ds' \frac{\mathrm{Im}\mathcal{A}_{\pi\pi}^{(\rho)}(s', t)}{s' - s} = \frac{|g_{\rho\pi\pi}|^2 (p_2 - p_1) \cdot (p'_1 - p'_2)}{m_\rho^2 - s - i\epsilon}, \quad (5.95)
$$

Using the above expression together with (5.91) in (5.93), we obtained the next iteration for the imaginary part of the form factor, in which the transition $\rho \to 2\pi \to \rho$ is added to

---

[4]A more detailed discussion of the resonances on the complex plane of momentum transfer can be found e.g., in the dedicated minireview in [1].

the simple ρ propagation

$$
\begin{aligned}
2\,\mathrm{Im}F_\pi^{(\rho\to 2\pi\to\rho)}(s)(p_2 - p_1)_\mu \;=\;& \int d\tau'_{2\pi}\frac{|g_{\rho\pi\pi}|^2(p_2 - p_1)\cdot(p'_1 - p'_2)}{m_\rho^2 - s - i\epsilon}\\
&\times\;(p'_2 - p'_1)_\mu\frac{m_\rho f_\rho g^*_{\rho\pi\pi}}{\sqrt{2}(m_\rho^2 - s + i\epsilon)}\,.
\end{aligned}
\tag{5.96}
$$

The integral over the two-pion phase space can be factorized out and calculated separately:

$$
I_{\alpha\mu}(q) \equiv \int d\tau'_{2\pi}(p'_1 - p'_2)_\alpha(p'_2 - p'_1)_\mu = \left(g_{\alpha\mu} - \frac{q_\alpha q_\mu}{s}\right)\frac{[p(s)]^3}{3\pi\sqrt{s}}\,,
\tag{5.97}
$$

where we introduce a compact notation for the three-momentum of the pions in a c.m. system:

$$
p(s) = \frac{1}{2}(s - 4m_\pi^2)^{1/2}\theta(s - 4m_\pi^2).
$$

The θ-function indicates that $p(s)$ vanishes below the two-pion threshold.

To obtain the integral (5.97), instead of performing an explicit calculation involving angular integration, we start from the most general decomposition:

$$
I_{\alpha\mu}(q) = \left(g_{\alpha\mu} - \frac{q_\alpha q_\mu}{s}\right)I(s)\,,
\tag{5.98}
$$

which is transverse with respect to the conserved total momentum $q = p_1 + p_2 = p'_1 + p'_2$, because $q(p'_1 - p'_2) = 0$. The invariant coefficient $I(s)$ is obtained multiplying both parts of the above equality by $g_{\alpha\mu}$:

$$
g^{\alpha\mu}I_{\alpha\mu}(q) = -\int d\tau'_{2\pi}(p'_2 - p'_1)^2 = \int d\tau'_{2\pi}(-2m_\pi^2 + 2p'_1p'_2) = \frac{(s - 4m_\pi^2)^{3/2}}{8\pi\sqrt{s}} = 3I(s)\,,
$$

yielding, in terms of $p(s)$, the coefficient on the r.h.s. of (5.97).

Finally, we obtain from (5.93):

$$
\mathrm{Im}F_\pi^{(\rho\to 2\pi\to\rho)}(s) = \frac{m_\rho f_\rho}{\sqrt{2}(m_\rho^2 - s - i\epsilon)}\left\{\frac{|g_{\rho\pi\pi}|^2[p(s)]^3}{6\pi\sqrt{s}}\right\}\frac{g^*_{\rho\pi\pi}}{m_\rho^2 - s + i\epsilon}\,.
\tag{5.99}
$$

This expression[5] should be interpreted as an imaginary part of a two-pion loop insertion in the intermediate ρ state, which we write as

$$
F_\pi^{(\rho\to 2\pi\to\rho)}(q^2) = \frac{m_\rho f_\rho}{\sqrt{2}(m_\rho^2 - q^2)}\mathcal{A}^{(2\pi)}(q^2)\frac{g^*_{\rho\pi\pi}}{m_\rho^2 - q^2}\,,
\tag{5.100}
$$

where the imaginary part $\mathrm{Im}\mathcal{A}^{(2\pi)}(s)$ is equal to the expression in curly brackets in (5.99). At $s = m_\rho^2$ it is normalized to the $\rho\to 2\pi$ width (5.86), which is almost equal to the total width:

$$
\mathrm{Im}\mathcal{A}^{(2\pi)}(m_\rho^2) = \frac{|g_{\rho\pi\pi}|^2}{6\pi m_\rho}\left[p(m_\rho^2)\right]^3 = m_\rho\Gamma(\rho^0\to\pi^+\pi^-) \simeq m_\rho\Gamma_{tot}(\rho)\,.
\tag{5.101}
$$

As already mentioned, (5.100) represents the second term of an infinite series describing sequential transitions between ρ and 2π states, whereas the first term is given by (5.91).

---

[5]Note that singularities in denominators at $s = m_\rho^2$ are spurious because this expression represents only one term in the infinite sum corresponding to the expansion of the amplitude (5.75). The sum, as we shall see below, has no singularity, being protected by the resonance width.

Subsequent terms are derived via iterations, resulting altogether in a geometric series. The sum of the series,

$$F_\pi^{(\rho \oplus 2\pi)}(s) = \left(\frac{f_\rho g_{\rho\pi\pi}}{\sqrt{2} m_\rho} \right) \frac{m_\rho^2}{m_\rho^2 - s - \text{Re}\mathcal{A}^{(2\pi)}(s) - i\text{Im}\mathcal{A}^{(2\pi)}(s)}, \qquad (5.102)$$

yields the full contribution of the ρ-resonance to the timelike pion form factor. Expanding the above expression in powers of the ratio $\mathcal{A}^{(2\pi)}(s)/(m_\rho^2 - s)$, we reproduce, at the zeroth and first order in this ratio, the expressions (5.91) and (5.100), respectively. At the same time, the magnitude of $\mathcal{A}^{(2\pi)}(s)$ determines the finite width and height of the resonance peak near $s = m_\rho^2$.

The resonance formula (5.102) is rather general and leaves certain freedom to the choice of the function $\mathcal{A}^{(2\pi)}(s)$. The simplest option is to neglect the s-dependence of $\mathcal{A}^{(2\pi)}(s)$, absorbing the real part in the m_ρ^2 definition and approximating the imaginary part by (5.101). This leads to the *Breit-Wigner (BW)* formula for the ρ resonance with a constant width[6]:

$$F_\pi^{(\rho, BW)}(s) = \left(\frac{f_\rho g_{\rho\pi\pi}}{\sqrt{2} m_\rho} \right) \frac{m_\rho^2}{m_\rho^2 - s - i m_\rho \Gamma_{tot}(\rho)}. \qquad (5.103)$$

This formula is usually applied at the values of s in the vicinity of the resonance. Its main drawback is the unitarity violation. Indeed, continuing (5.102) to the region $q^2 < 4m_\pi^2$, where the form factor is real-valued, is, strictly speaking, not possible because of the remnant imaginary part generated by a constant width.

A more elaborated version of the Breit-Wigner formula (see, e.g., [48]) takes into account the s-dependence of $\text{Im}\mathcal{A}^{(2\pi)}(s)$ in a form of the p-wave two-pion phase space, with the normalization at $s = m_\rho^2$ given by (5.101):

$$\text{Re}\mathcal{A}^{(2\pi)}(s) = 0, \quad \text{Im}\mathcal{A}^{(2\pi)}(s) = \sqrt{s}\Gamma_{\rho \to 2\pi}(s), \qquad (5.104)$$

where the function

$$\Gamma_{\rho \to 2\pi}(s) = \frac{m_\rho^2}{s} \left(\frac{p(s)}{p(m_\rho^2)} \right)^3 \Gamma(\rho \to 2\pi), \qquad (5.105)$$

is sometimes called energy-dependent width. This function vanishes at $s < 4m_\pi^2$, below the two-pion threshold, protecting the resonance contribution from violating the unitarity condition. The modified resonance formula reads:

$$F_\pi^{(\rho, \widetilde{BW})}(s) = \left(\frac{f_\rho g_{\rho\pi\pi}}{\sqrt{2} m_\rho} \right) \frac{m_\rho^2}{m_\rho^2 - s - i\sqrt{s}\Gamma_{\rho \to 2\pi}(s)}. \qquad (5.106)$$

The *Gounaris-Sakurai* representation [49] is another frequently used version of (5.102) with an initial form

$$F_\pi^{(\rho, GS)}(s) = \frac{c}{m_\rho^2 - s - \text{Re}\mathcal{A}^{(2\pi)}(s) - i\text{Im}\mathcal{A}^{(2\pi)}(s)}, \qquad (5.107)$$

in which both the imaginary and real part of $\mathcal{A}^{(2\pi)}(s)$ are nonvanishing and s-dependent, and c is a normalization constant, adjusted to reproduce the same form factor at $q^2 = 0$ as in (5.106):

$$F_\pi^{(\rho, GS)}(0) = F_\pi^{(\rho, \widetilde{BW})}(0). \qquad (5.108)$$

---

[6]The first application of this formula in nuclear physics goes back to [47].

The expression for the imaginary part is fixed by (5.104). To derive the real part, we employ the dispersion relation for $\mathcal{A}^{(2\pi)}(s)$. Taking into account the asymptotics

$$\mathrm{Im}\mathcal{A}^{(2\pi)}(s) \sim s \quad \text{at } s \to \infty,$$

we perform two subtractions at $s = 0$, to render the dispersion integral finite:

$$\mathcal{A}^{(2\pi)}(s) = \mathcal{A}^{(2\pi)}(0) + s\frac{d\mathcal{A}^{(2\pi)}(0)}{ds} + \frac{s^2}{\pi}\int\limits_{4m_\pi^2}^{\infty} ds' \frac{\mathrm{Im}\mathcal{A}^{(2\pi)}(s')}{s'^2(s'-s-i\epsilon)}. \tag{5.109}$$

Furthermore, we decompose the denominator of the integrand

$$\frac{1}{s'-s-i\epsilon} = PV\frac{1}{s'-s} + i\pi\delta(s'-s).$$

The imaginary part is then correctly reproduced and, in order to obtain the real part, we need to take the principal value (PV) of the integral in (5.109). Before that, it is convenient to transform the integration variable, $s' \to v'$ where

$$v' = \sqrt{1 - \frac{4m_\pi^2}{s'}}, \quad \int\limits_{4m_\pi^2}^{\infty} ds' = 4m_\pi^2 \int\limits_0^1 dv' \frac{2v'}{(1-v'^2)^2}.$$

In addition, we use the velocity of pions in the c.m. frame:

$$v(s) = \sqrt{1 - \frac{4m_\pi^2}{s}}.$$

In terms of the new variable, the dispersion integral becomes:

$$\frac{s^2}{\pi}\int\limits_{4m_\pi^2}^{\infty} ds' \frac{\mathrm{Im}\mathcal{A}^{(2\pi)}(s')}{s'^2(s'-s-i\epsilon)} = \left(\frac{m_\rho^2\Gamma(\rho \to 2\pi)}{8\pi[p(m_\rho^2)]^3}\right) s\mathcal{I}(v), \tag{5.110}$$

where the dimensionless integral

$$\mathcal{I}(v) = \frac{1}{v}\int_0^1 dv' v'^4 \left[\frac{1}{v'-v-i\epsilon} - \frac{1}{v'+v}\right]$$

has the following real part:

$$\mathrm{Re}\,\mathcal{I}(v) = \frac{1}{v}\left[PV\left(\int_0^1 dv'\frac{v'^4}{v'-v}\right) - \int_0^1 dv'\frac{v'^4}{v'+v}\right] = \frac{2}{3} + 2v^2 - v^3\ln\frac{1+v}{1-v}. \tag{5.111}$$

Returning to the dispersion relation (5.109) we can now obtain its real part, applying the above equation. Introducing a compact notation

$$-\mathrm{Re}\mathcal{A}^{(2\pi)}(s) \equiv H(s),$$

we have

$$H(s) = H(0) + s\frac{dH(0)}{ds} - \left(\frac{m_\rho^2\Gamma(\rho \to 2\pi)}{2\pi[p(m_\rho^2)]^3}\right)\left[\frac{2}{3}s - 2m_\pi^2 - \left(\frac{s}{4} - m_\pi^2\right)v\ln\frac{1+v}{1-v}\right]. \tag{5.112}$$

Subtraction terms in this relation are fixed by the normalization conditions for the ρ-meson mass and width:

$$H(m_\rho^2) = \frac{d}{ds} H(m_\rho^2) = 0. \tag{5.113}$$

We notice that, without loss of generality, (5.112) can be represented in a form

$$H(s) = h + h's + \hat{H}(s), \tag{5.114}$$

where the function

$$
\begin{aligned}
\hat{H}(s) &= \left(\frac{m_\rho^2 \Gamma(\rho \to 2\pi)}{2\pi [p(m_\rho)]^3} \right) \left(\frac{s}{4} - m_\pi^2 \right) v \ln \frac{1+v}{1-v} \\
&= \frac{2m_\rho^2 \Gamma(\rho \to 2\pi)}{\pi \sqrt{s}} \ln \left(\frac{\sqrt{s} + 2p(s)}{2m_\pi} \right).
\end{aligned} \tag{5.115}
$$

is added to a linear polynomial in s. Note that this function vanishes at $\sqrt{s} < 2m_\pi$. Using (5.113), we fix the coefficients h and h' in (5.114) and obtain

$$H(s) = \hat{H}(s) - \hat{H}(m_\rho^2) - (s - m_\rho^2) \frac{d}{ds} \hat{H}(m_\rho^2). \tag{5.116}$$

Finally, normalizing the Gounaris-Sakurai representation at $s = 0$ according to (5.108), we restore its conventional form:

$$F_\pi^{(\rho, GS)}(s) = \left(\frac{f_\rho g_{\rho\pi\pi}}{\sqrt{2} m_\rho} \right) \frac{m_\rho^2 + H(0)}{m_\rho^2 - s + H(s) - i\sqrt{s}\Gamma_{\rho \to 2\pi}(s)}. \tag{5.117}$$

With the estimates (5.87) of the ρ-meson decay constant and $\rho\pi\pi$ coupling, the numerical value of the ρ contribution to the pion e.m. form factor at $s = q^2 = 0$ is the same for both (modified) Breit-Wigner and Gounaris-Sakurai representations:

$$F_\pi^{(\rho)}(0) = \frac{f_\rho g_{\rho\pi\pi}}{\sqrt{2} m_\rho} \simeq 1.20, \tag{5.118}$$

deviating by $\sim 20\%$ from the unit normalization (2.38). Here we clearly observe the effect of other than ρ-meson contributions to the unitarity relation (5.26) and, therefore, also to the dispersion relation (5.34). We emphasize again that the latter relation is valid at any q^2, far beyond the timelike region. In other words, all intermediate hadronic states with ρ-meson quantum numbers have a virtual impact on the form factor at $q^2 \leq 0$. Moreover, because the imaginary part $\mathrm{Im} F_\pi(s)$ is not positive definite, a destructive interference between different contributions to $F_\pi(0)$ is possible and, in fact, is needed to decrease the ρ-meson part estimated in (5.118).

There is no unique prescription of adding all relevant hadronic states to the ρ. It is anticipated that the dominant part of that infinite sum is provided by the radially excited ρ-mesons. These resonances are identified in various experiments and listed in [1]. The established first and second radial excitations are

$$\rho' \equiv \rho(1450), \quad \rho'' \equiv \rho(1700).$$

Their contributions to the pion form factor are accounted in the same manner as for ρ, that is, introducing the decay constants $f_{\rho'}$, $f_{\rho''}$ and strong couplings $g_{\rho'\pi\pi}$, $g_{\rho''\pi\pi}$ and resumming the transitions between ρ', ρ'' and the two-pion loops.

One ansatz frequently used is a superposition of three resonance contributions taken in the form (5.117):

$$F_\pi(s) = \frac{\sum_{i=\rho,\rho',\rho''} c_i F_\pi^{(i,GS)}(s)}{\sum_{i=\rho,\rho',\rho''} c_i F_\pi^{(i,GS)}(0)}, \qquad (5.119)$$

where the coefficients are adjusted to fulfil the normalization $F_\pi(0) = 1$.

The above resonance model has, however, several deficiencies. One is that the two-pion decay modes of ρ', ρ'' are not the dominant ones. In fact, the four-pion and other modes become equally important. This problem is partially solved, including a total width instead of a two-pion one, more specifically, replacing in the denominator of $F_\pi^{(\rho',GS)}$ the function $\Gamma_{\rho' \to 2\pi}(s)$ by

$$\Gamma_{\rho'}(s) = \frac{m_{\rho'}^2}{s} \left(\frac{p(s)}{p(m_{\rho'}^2)} \right)^3 \Gamma_{tot}(\rho'), \qquad (5.120)$$

and the same for ρ''. In this way, we implicitly include all intermediate hadronic states coupled to ρ', ρ'', but the energy dependence of the width remains the same as for the two-pion mode. It can still be improved with more complicated phase-space factors. The second problem is more difficult and has no model-independent solution. Strictly speaking, one has to take into account strongly coupled channels and mixing between the ρ-resonances, because, e.g., transitions like $\rho \to \pi^+\pi^- \to \rho'$ etc. are possible. For this purpose, models of coupled channels developed in hadron phenomenology are employed for the form factors (e.g., in [50]).

Furthermore, a model of the pion e.m. form factor in the timelike region is not necessarily limited to the three resonance states. In fact, from the point of view of QCD, it is more consistent to use the ansatz with an infinite amount of ρ-resonances (see, e.g., [51]). This approach is inspired by the properties of hadronic amplitudes in QCD in the limit of infinite number of colors. At $N_c \to \infty$, the spectrum of hadronic states represents an infinite series of stable mesons with equidistant masses[7]. A modified version of the $N_c = \infty$ model, where the widths of the first few resonances are installed, can be found in [53].

As already mentioned in the beginning of this section, the pion e.m. form factor in the timelike region is directly measured in the electron-positron annihilation $e^+e^- \to \pi^+\pi^-$, where the total energy of e^+e^- collision in the c.m. frame is equal to the square root of the timelike momentum transfer: $E_{e^+} + E_{e^-} = \sqrt{s}$. If the collision energy is fixed to a certain value, determined by the characteristics of the collider, it is still possible to access smaller values of $\sqrt{s}$ applying the technique of initial state radiation [54], when events with additional photon emission, $e^+e^- \to \pi^+\pi^-\gamma$ are selected, so that a part of the collision energy is carried away by the radiated photon. Note that measuring the cross section, one cannot access the phase of the form factor, but only its absolute value. The phase is known only in the elastic region, where, due to the Watson theorem, it can be determined from the measurements of the pion-pion scattering phase in other hadronic processes.

Concluding this section, let us return to the $\rho \to \pi$ transition form factor $f_{\rho\pi}(q^2)$ introduced in (4.14). At $q^2 = 0$ it coincides with the $\rho \to \pi\gamma$ radiative decay constant. For definiteness, we consider the $\rho^+ \to \pi^+$ transition. Following the procedure we used to obtain the ρ meson contribution to the pion e.m. form factor, we can derive a similar dispersion relation for $f_{\rho\pi}(q^2)$. This form factor, via crossing transformation from spacelike region, is related to the $\gamma^* \to \rho^+\pi^-$ form factor in the timelike region $s > (m_\rho + m_\pi)^2$ where it can be measured in $e^+e^- \to \gamma^* \to \rho^+\pi^-$. The form factor $f_{\rho\pi}(q^2)$ is an analytic function

---

[7]QCD at $N_c = \infty$ and the related $1/N_c$ expansion of hadronic amplitudes are not discussed in this book, for an introductory review see, e.g., [52].

and receives the contribution of intermediate ω resonance at $q^2 = s = m_\omega^2$ as follows from unitarity. Introducing the $\omega\rho\pi$ strong coupling

$$\langle \pi^-(p_2)\rho^+(p_1)|\omega^0(q)\rangle = \epsilon_{\mu\nu\alpha\beta}\varepsilon^{(\omega)\mu}q^\nu\varepsilon^{(\rho)\alpha}p_2^\beta\, g_{\omega\rho\pi}(q^2) \tag{5.121}$$

and the ω decay constant

$$\langle \omega^0(q,\lambda)|j_\mu^{em}|0\rangle = \langle \omega^0(q,\lambda)|j_\mu^{(I=0)}|0\rangle = \varepsilon_\mu^{(\omega)*}(q,\lambda)m_\omega f_\omega\,, \tag{5.122}$$

we obtain the contribution of the ω-pole to the dispersion relation:

$$f_{\rho\pi}(q^2) = \frac{1}{\pi}\int\limits_0^\infty ds\frac{\mathrm{Im}f_{\rho\pi}(s)}{s - q^2 - i\epsilon} = \frac{m_\omega f_\omega g_{\omega\rho\pi}}{m_\omega^2 - q^2 - i\epsilon} + \cdots\,, \tag{5.123}$$

where the terms corresponding to other intermediate states are denoted by ellipsis. An essential difference between this relation and the one for the pion form factor is that ω meson is located below the physical threshold, because $m_\omega < m_\rho + m_\pi$. We will return to this point in the next section. Retaining in (5.123) only the ω contribution, we recover at $q^2 = 0$ the vector dominance model of the $\rho \to \pi\gamma$ transition discussed in qualitative terms in Chapter 4.1 and depicted in Figure 4.2:

$$g_{\rho\pi\gamma} = f_{\rho\pi}(0) = \frac{f_\omega g_{\omega\rho\pi}}{m_\omega} + \cdots\,. \tag{5.124}$$

We see that the vector dominance model is in fact based on a hadronic dispersion relation in the channel of e.m. current, which has the same quantum numbers as vector mesons. The accuracy of this model for $\rho \to \pi\gamma$ is determined by the magnitude of other than ω contributions (e.g., of the $\omega(1420)$ [1]) to the dispersion relation at $q^2 = 0$.

5.6 KAON FORM FACTORS AND τ DECAYS

The charged and neutral kaon e.m. form factors in the spacelike region were discussed in Section 2.5. In the timelike region these form factors are also measured in electron-positron collisions above the threshold $s = 4m_K^2$, being defined as:

$$\langle K^+(p_1)K^-(p_2)|j_\mu^{em}|0\rangle = (p_1 - p_2)_\mu\,F_{K^+}(s)\,,$$
$$\langle K^0(p_1)\bar{K}^0(p_2)|j_\mu^{em}|0\rangle = (p_1 - p_2)_\mu\,F_{K^0}(s)\,, \tag{5.125}$$

with the electric charge normalization given in (2.86). The isospin $I = 1$ and $I = 0$ components of these form factors are analytic continuations of the respective components (2.88) in the spacelike region.

Here we focus on the intermediate vector meson contributions to the kaon form factors. Importantly, all three neutral mesons, ρ^0, ω and ϕ strongly couple to the kaon pair. For these mesons we adopt the valence content as in (1.89) and (1.90):

$$|\rho^0\rangle = \frac{1}{\sqrt{2}}|u\bar{u} - d\bar{d}\rangle,\quad |\omega\rangle = \frac{1}{\sqrt{2}}|u\bar{u} + d\bar{d}\rangle,\quad |\phi\rangle = |s\bar{s}\rangle\,,$$

neglecting small mixing effects. Due to isospin symmetry, the intermediate ρ (ω and ϕ) meson contributes only to the $I = 1$ ($I = 0$) component of the kaon form factor.

Note that the energy threshold $2m_K$ of the kaon pair production is larger than the masses of ρ and ω, and only ϕ is located slightly above that threshold; hence, the strong decays $\rho \to K\bar{K}$ and $\omega \to K\bar{K}$ ($K = K^+, K^0$) are kinematically forbidden. Nevertheless, both ρ

and ω mesons contribute to the dispersion relation for the kaon form factors with the poles located in the unphysical region $0 < s < 4m_K^2$ of the real axis $q^2 = s$. To substantiate that, we notice that, after analytical continuation, the function $F_K(q^2)$ is defined in the whole complex q^2-plane and, consequently, the imaginary part $\mathrm{Im}F_K(s)$ obeys unitarity relation, similar to (5.26), at all s. Only at $s > 4m_K^2$ ($q^2 < 0$) this function coincides with the timelike (spacelike) form factor accessible in physical processes with asymptotic on-shell kaon states, such as $e^+e^- \to K\bar{K}$ ($e^-K \to e^-K$). Here we may repeat for the kaon form factor the derivation of analyticity given in Section 5.1 for the pion form factor. For the latter, the unphysical region is $0 < s < 4m_\pi^2$, which is evidently free from intermediate resonances. In the $K\bar{K}$ channel the situation is different, with ρ or ω located below $4m_K^2$; hence, the unitarity sum for $\mathrm{Im}F_K(s)$ should also contain the ρ and ω contributions proportional to $\delta(s - m_\rho^2)$ and $\delta(s - m_\omega^2)$, respectively, similar to (5.90). Moreover, since the thresholds of the states with two and more pions also lie in the interval $4m_\pi^2 < s < 4m_K^2$, we have to resum their contributions into the widths of ρ and ω, obtaining resonance formulas for $F_K(s)$ similar to (5.106) or (5.117). The above discussion based on analyticity may seem somewhat too formal. The fact that light vector mesons influence, via dispersion relation, the kaon form factors also in the physical region $s > 4m_K^2$, has the following qualitative interpretation. The quark e.m. current can emit any intermediate state including ρ and ω provided the quantum numbers are the same. There is no contradiction to energy conservation, provided these states exist during short time intervals in a form of quantum fluctuations before they are converted, via strong interaction, into the final on-shell $K\bar{K}$ state[8].

Assuming that only the three lightest vector mesons contribute to the kaon form factors, and adopting e.g., a simple Breit-Wigner formula for each resonance, we have

$$F_K(s) = \sum_{V=\rho,\omega,\phi} \frac{\kappa_V f_V g_{VK\bar{K}} m_V}{m_V^2 - s - im_V\Gamma_{tot}(V)} = \sum_{V=\rho,\omega,\phi} \frac{\kappa_V f_V g_{VK\bar{K}}}{m_V} F_V^{(BW)}(s), \qquad (5.126)$$

$(K = K^+, K^0)$ where

$$F_V^{(BW)}(s) = \frac{m_V^2}{m_V^2 - s - im_V\Gamma_{tot}(V)} \qquad (5.127)$$

is the resonance function that only depends on the mass and width of the vector meson V. In (5.126) the vector meson decay constants are defined as in (5.79):

$$\langle V(q,\lambda)|j_\mu^{em}|0\rangle = \kappa_V \varepsilon_\mu^{(V)*}(q,\lambda)m_V f_V, \qquad (5.128)$$

where $\varepsilon^{(V)}$ is the polarization vector of V; the labels of momentum q and spin projection λ, are hereafter omitted for brevity. The normalization coefficients κ_V are adjusted so that in the $SU(3)_{fl}$ symmetry limit

$$f_\rho = f_\omega = f_\phi = f_V.$$

To fix κ_V, we substitute in the hadronic matrix element (5.128) the quark content of V and j_μ^{em} and reduce this matrix element to a single-flavor one defined as

$$\langle q\bar{q}|\bar{q}\gamma_\mu q|0\rangle \equiv \varepsilon_\mu^{(V)*}m_V f_V \qquad (q = u, d, s).$$

---

[8]We can also call that a virtual hadron propagation, which is however a model-dependent concept, not well defined in QCD.

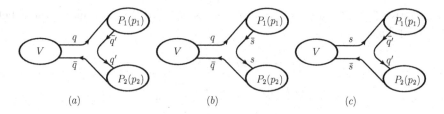

Figure 5.3 Quark flow diagrams for the strong couplings of vector and pseudoscalar mesons; $q, q' = u, d$. The diagrams (a),(b),(c) correspond to the three different amplitudes A_{qq}, A_{qs} and A_{sq}, respectively.

For the three vector mesons we obtain

$$
\langle \rho^0 | j_\mu^{(I=1)} | 0 \rangle = \langle \frac{u\bar{u} - d\bar{d}}{\sqrt{2}} | \frac{\bar{u}\gamma_\mu u - \bar{d}\gamma_\mu d}{2} | 0 \rangle = \frac{1}{\sqrt{2}} \varepsilon_\mu^{(V)*} m_V f_V \,,
$$

$$
\langle \omega^0 | j_\mu^{(I=0)} | 0 \rangle = \langle \frac{u\bar{u} + d\bar{d}}{\sqrt{2}} | \frac{\bar{u}\gamma_\mu u + \bar{d}\gamma_\mu d}{6} | 0 \rangle = \frac{1}{3\sqrt{2}} \varepsilon_\mu^{(V)*} m_V f_V \,,
$$

$$
\langle \phi^0 | j_\mu^{(I=0)} | 0 \rangle = \langle s\bar{s} | -\frac{\bar{s}\gamma_\mu s}{3} | 0 \rangle = -\frac{1}{3} \varepsilon_\mu^{(V)*} m_V f_V \,, \tag{5.129}
$$

hence,

$$
\kappa_\rho = \frac{1}{\sqrt{2}}, \quad \kappa_\omega = \frac{1}{3\sqrt{2}}, \quad \kappa_\phi = -\frac{1}{3}.
$$

We emphasize that (5.128) is just a convenient parameterization. Numerical values of the decay constants f_V are calculated from the measured widths of leptonic e.m. decays available in [1], as it was done for f_ρ using (5.83). For a generic vector meson this width is equal to

$$
\Gamma(V \to e^+ e^-) = \frac{4\pi \alpha_{em}^2}{3m_V} \kappa_V^2 f_V^2 \,. \tag{5.130}
$$

Obtaining f_ω and f_ϕ, we see that their values have rather small, $O(10\%)$ deviations from f_ρ. Keeping in mind that ω and ρ contain only u, d quarks, in what follows we adopt the approximation

$$
f_\omega = f_\rho \,, \tag{5.131}
$$

which leaves only two independent decay constants, f_ρ and f_ϕ in the parameterizations of the form factors via vector mesons.

The strong couplings of kaons to vector mesons are defined as

$$
\langle K(p_1)\bar{K}(p_2)|V(q)\rangle = (p_2 - p_1)^\nu \varepsilon_\nu^{(V)} g_{VK\bar{K}} \,. \tag{5.132}
$$

Applying quark flow diagrams, we can derive relations between separate contributions to the kaon form factors in (5.126). It is instructive to include in this analysis also the pion e.m. form factor, because then we can trace the limit of $SU(3)_{fl}$ symmetry for the kaon and pion form factors. Therefore, we consider a generic $VP\bar{P}$ coupling where $V = \rho^0, \omega, \phi$ and $P = \pi, K$. In the limit of isospin symmetry, the three diagrams shown in Figure 5.3 are distinguished by the presence and position of the s-quark: (a) diagram without s quarks; (b) diagram with $\bar{s}s$ pair in the pseudoscalar mesons $P\bar{P}$ only; (c) diagram with $\bar{s}s$ pair in both V and $P\bar{P}$. The corresponding hadronic invariant amplitudes are denoted as A_{qq}, A_{qs} and A_{sq}. Each quark-flow diagram has its counterpart with charge conjugated quark lines and with an amplitude differing from the initial one by an additional minus sign, in accordance with the negative C-parity of neutral vector mesons. We find the following relations between the VPP couplings:

$$g_{\rho^0\pi^+\pi^-} \equiv g_{\rho\pi\pi} = \sqrt{2}A_{qq},$$

$$g_{\rho^0 K^+ K^-} = g_{\omega K^+ K^-} = \frac{1}{\sqrt{2}}A_{qs},$$

$$g_{\rho^0 K^0 \bar{K}^0} = -g_{\omega K^0 \bar{K}^0} = -\frac{1}{\sqrt{2}}A_{qs},$$

$$g_{\phi K^+ K^-} = g_{\phi K^0 \bar{K}^0} = -A_{sq}. \tag{5.133}$$

The diagram counting also correctly predicts that the isospin forbidden couplings vanish: $g_{\omega\pi^+\pi^-} = g_{\rho\pi^0\pi^0} = 0$. The $SU(3)_{fl}$ symmetry corresponds to the limit of three equal quark-flow diagrams:

$$A_{qq} = A_{qs} = A_{sq}.$$

Finally, we obtain the pion and kaon e.m. form factors in terms of vector meson contributions:

$$F_\pi(s) = \frac{f_\rho A_{qq}}{m_\rho}F_\rho^{(BW)}(s),$$

$$F_{K^+}(s) = \frac{f_\rho A_{qs}}{2m_\rho}F_\rho^{(BW)}(s) + \frac{f_\rho A_{qs}}{6m_\omega}F_\omega^{(BW)}(s) + \frac{f_\phi A_{sq}}{3m_\phi}F_\phi^{(BW)}(s),$$

$$F_{K^0}(s) = -\frac{f_\rho A_{qs}}{2m_\rho}F_\rho^{(BW)}(s) + \frac{f_\rho A_{qs}}{6m_\omega}F_\omega^{(BW)}(s) + \frac{f_\phi A_{sq}}{3m_\phi}F_\phi^{(BW)}(s), \tag{5.134}$$

so that in the $SU(3)_{fl}$ limit:

$$F_\pi(s) = F_{K^+}(s), \quad F_{K^0}(s) = 0.$$

The form factor decompositions can be developed further, choosing a more elaborate resonance formula than a simple Breit-Wigner one, and adding radial excitations of ρ, ω, ϕ (see, e.g., [53]).

Decays of τ lepton with two pseudoscalar mesons in the final state,

$$\tau^-(p_\tau) \to P_1(p_1)P_2(p_2)\nu_\tau(p_\nu), \tag{5.135}$$

and their charge conjugate τ^+-decays provide information on the timelike flavor-changing form factors generated by the weak vector current. The diagram of these decays is shown in Figure 5.4. In Table 5.1 we list all possible modes with various flavor combinations of mesons allowed in the isospin symmetry limit. The weak axial current does not contribute for the same reason as in the semileptonic $P_i \to P_f \ell \bar{\nu}_\ell$ decays.

The τ decays (5.135) are mediated by the part of the weak interaction Hamiltonian (1.29) in which the τ component of leptonic current is combined with the u-quark part of the vector current (1.30),

$$j_\mu^\dagger = \bar{q}'\gamma_\mu u,$$

where $q' = d, s$. The hadronic matrix element of $\tau^- \to P_1 P_2 \nu_\tau$ represents a crossing transform, $p_1 \to -p_1, q \to -q$, of the matrix element (2.96). The initial-state meson P_i becomes the final-state meson P_1, and we rename $P_f \to P_2$ to have a consistent notation. Correspondingly, the momentum transfer to the meson pair is

$$q = p_1 + p_2 = p_\tau - p_\nu.$$

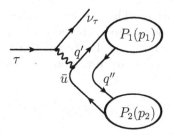

Figure 5.4 Diagram of the $\tau^- \to \nu_\tau P_1 P_2$ decay. The wavy line denotes W-boson exchange. Various flavor combinations of pseudoscalar mesons P_1, P_2 are obtained combining $q' = d, s$ and $q'' = u, d, s$.

TABLE 5.1 Final states with two
pseudoscalar mesons in the τ^- decays.
Notation is specified in Figure 5.4.

q'	V_{CKM}	q''	$\tau^- \to P_1 P_2 \nu_\tau$
d	V_{ud}^*	u, d	$\tau^- \to \pi^- \pi^0 \nu_\tau$
		s	$\tau^- \to K^0 K^- \nu_\tau$
s	V_{us}^*	u	$\tau^- \to K^- \pi^0 \nu_\tau$
		d	$\tau^- \to \bar{K}^0 \pi^- \nu_\tau$
		s	$\tau^- \to \eta^{(')} K^- \nu_\tau$

As a result, the hadronic matrix element of (5.135) is written in terms of two timelike form factors (up to a general minus sign which plays no role):

$$\langle P_1(p_1) P_2(p_2) | \bar{q}' \gamma_\mu u | 0 \rangle = \left[\bar{p}_\mu - \frac{m_{P_1}^2 - m_{P_2}^2}{q^2} q_\mu \right] f_{P_1 P_2}^+(q^2)$$
$$+ \frac{m_{P_1}^2 - m_{P_2}^2}{q^2} q_\mu f_{P_1 P_2}^0(q^2), \qquad (5.136)$$

where $\bar{p} = p_1 - p_2$, and the scalar form factor $f_{P_1 P_2}^0$ is defined via matrix element of the scalar flavor-changing current (cf. (2.102)):

$$\langle P_1(p_1) P_2(p_2) | (m_{q'} - m_q) \bar{q}' q | 0 \rangle = (m_{P_1}^2 - m_{P_2}^2) f_{P_1 P_2}^0(q^2). \qquad (5.137)$$

The physical region of the meson pair invariant mass squared is

$$(m_{P_1} + m_{P_2})^2 < q^2 < m_\tau^2.$$

Using (5.136) and factorizing the leptonic current, we obtain the τ decay amplitude:

$$\mathcal{A}(\tau \to P_1 P_2 \nu_\tau) = \frac{G_F}{\sqrt{2}} V_{uq'} \bar{v}_\nu(p_\nu) \gamma^\mu (1 - \gamma_5) u_\tau(p_\tau)$$
$$\times \left\{ \left[\bar{p}_\mu - \frac{m_{P_1}^2 - m_{P_2}^2}{q^2} q_\mu \right] f_{P_1 P_2}^+(q^2) + \frac{m_{P_1}^2 - m_{P_2}^2}{q^2} q_\mu f_{P_1 P_2}^0(q^2) \right\}. \qquad (5.138)$$

To derive the decay width, we square this amplitude in a very similar way to how it was done in Section 2.5 for the $P_i \to P_f \ell \bar{\nu}_\ell$ decay, hence, we skip the details here. We only mention that one has to multiply the result by an additional factor $1/2$, taking into account the average over the τ polarizations. The interference term between the two form factors vanishes and the whole amplitude squared is reduced to a form depending only on q^2 and $\bar{p} \cdot p_\nu$. After that the integration formulas (B.62) and (B.63) are used, where the momenta p_1, p_2, p_3 in the three-particle final state correspond, respectively, to p_ν, p_1, p_2 in (5.135). The resulting amplitude squared is

$$\frac{1}{2}\overline{|\mathcal{A}(\tau \to P_1 P_2 \nu_\tau)|^2} = 2G_F^2 |V_{uq'}|^2 (m_\tau^2 - q^2)$$

$$\times \left\{ \left(\frac{m_\tau^2 + 2q^2}{3q^4}\right) \lambda(q^2)|f_{P_1 P_2}^+(q^2)|^2 + \frac{(m_{P_1}^2 - m_{P_2}^2)^2 m_\tau^2}{q^4}|f_{P_1 P_2}^0(q^2)|^2 \right\}, \quad (5.139)$$

where $\lambda(q^2) \equiv \lambda(q^2, m_{P_1}^2, m_{P_2}^2)$. Finally, applying the formula (B.56) for the differential width of three-body decay, we obtain the decay distribution in the invariant mass $q^2 = s$ of the meson pair:

$$\frac{d\Gamma(\tau \to P_1 P_2 \nu_\tau)}{ds} = \frac{G_F^2 |V_{uq'}|^2 m_\tau^3}{768\pi^3 s^3} \left(1 - \frac{s}{m_\tau^2}\right)^2 \sqrt{\lambda(s)}$$

$$\times \left[\left(1 + \frac{2s}{m_\tau^2}\right) \lambda(s)|f_{P_1 P_2}^+(s)|^2 + 3(m_{P_1}^2 - m_{P_2}^2)^2 |f_{P_1 P_2}^0(s)|^2 \right]. \quad (5.140)$$

Comparing the diagrams generated from Figure 5.4 with various flavors q', q'' and equating those which differ by $u \leftrightarrow d$ replacement, we obtain isospin relations between the form factors $f_{P_1 P_2}^+$. In the same way, we relate the latter to the e.m. form factors. Note that $f_{\pi^0 \pi^-}^+(s) = F_V(s)$, where $F_V(s)$ is the timelike continuation of the form factor defined in (2.69). Therefore, from the isospin relation (2.76) it follows that

$$f_{\pi^0 \pi^-}^+(s) = \sqrt{2}F_\pi(s). \quad (5.141)$$

One more relation between $f_{K^0 K^-}^+(s)$ and the $I = 1$ part of the kaon e.m. form factor is

$$f_{K^0 K^-}^+(s) = -2F_K^{(I=1)}(s). \quad (5.142)$$

Thus, measuring the pion and kaon e.m. form factors in $e^+ e^-$ annihilation, it is possible to predict the widths of $\tau \to \pi^- \pi^0 \nu_\tau, K^0 K^- \nu_\tau$ decays. Note that, in accordance with the isospin symmetry limit, we neglect the mass differences $m_{\pi^-} - m_{\pi^0}$ and $m_{K^0} - m_{K^-}$; hence, in this approximation, the scalar form factors do not contribute to the amplitudes of these decays.

For $\tau \to K^- \pi^0 \nu_\tau$ and $\tau \to \bar{K}^0 \pi^- \nu_\tau$ the scalar form factors are essential, being multiplied by the factor $m_K^2 - m_\pi^2$ in the amplitude (5.138). The form factors in these decay channels are also related by isospin symmetry:

$$f_{K^- \pi^0}^{+,0}(s) = \frac{1}{\sqrt{2}} f_{\bar{K}^0 \pi^-}^{+,0}(s). \quad (5.143)$$

The resonance contribution to the vector form factors $f_{K^- \pi^0}^+$ and $f_{\bar{K}^0 \pi^-}^+$ is dominated by the vector meson K^{*-}. To obtain that in a form similar to (5.134), we introduce the K^* decay constant,

$$\langle K^{*-}|\bar{s}\gamma_\nu u|0\rangle = \langle \bar{K}^{*0}|\bar{s}\gamma_\nu d|0\rangle = \varepsilon_\mu^{(K)*} m_{K^*} f_{K^*}, \quad (5.144)$$

and the strong coupling of K^* to $K\pi$ (cf. (5.132)):

$$\langle \bar{K}^0(p_1)\pi^-(p_2)|K^{*-}(q)\rangle = (p_2 - p_1)^\nu \varepsilon_\nu^{(K^*)} g_{K^{*-}\bar{K}^0 \pi^-}. \quad (5.145)$$

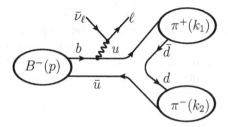

Figure 5.5 Diagram of the semileptonic decay $B^- \to \pi^+\pi^-\ell^-\bar{\nu}$. The intermediate quark-antiquark state, formed after the emission of lepton pair indicates a resonance contribution.

For the latter the following isospin relations are valid:

$$g_{K^{*-}\bar{K}^0\pi^-} = \sqrt{2}g_{K^{*-}K^-\pi^0} = -\sqrt{2}g_{\bar{K}^{*0}\bar{K}^0\pi^0} = g_{\bar{K}^{*0}K^-\pi^+}\,. \tag{5.146}$$

Combining the two above definitions, we obtain the K^* contribution:

$$f^+_{\bar{K}^0\pi^-}(s) = \sqrt{2}f^+_{K^-\pi^0}(s) = \frac{f_{K^*}g_{K^{*-}\bar{K}^0\pi^-}}{m^*_K}F^{(BW)}_{K^*}(s)\,, \tag{5.147}$$

which can be improved further taking into account the energy-dependent width of K^* and adding radial excitations of K^*. Note that, due to its origin from the matrix element of scalar current, the form factor $f^0_{K\pi}$ receives its resonance contributions from scalar ($J^P = 0^+$) mesons with strangeness, such as $K^*_0(700), K^*_0(1430)$ [1]. Finally, in the low momentum-transfer region

$$m^2_\ell \le q^2 \le (m_K - m_\pi)^2,$$

the same form factors $f^{+,0}_{K\pi}(s)$ determine the kaon semileptonic decays $K \to \pi\ell\nu_\ell$ ($\ell = e, \mu$).

5.7 $B \to \pi\pi\ell\nu_\ell$ DECAYS

In Chapter 2 we considered form factors of transitions with a single meson in the final state, such as the pion e.m. or $B \to \pi$ weak form factors. The next in complexity are transitions with two mesons in the final state. An instructive example is the B-meson weak semileptonic decay

$$B^-(p) \to \pi^+(k_1)\,\pi^-(k_2)\,\ell(q_1)\bar{\nu}_\ell(q_2)\,, \tag{5.148}$$

denoted also as $B_{\ell 4}$ decay, with the diagram shown in Figure 5.5. This particular decay mode, generated by the weak $b \to u$ current, represents a rich set of similar processes, among them the FCNC decay $B \to K\pi\ell^+\ell^-$ or the weak semileptonic decay $D \to \pi\pi\ell\bar{\nu}_\ell$. Historically, the first observed and studied decay of this type was $K \to \pi\pi e\nu_e$ (or K_{e4}). This process and its $K \to \pi\pi$ form factors, are reviewed, e.g., in [33]. Small momenta of the final-state pions in K_{e4} allow one to use the chiral perturbation theory with the effective soft-pion interactions. In $B_{\ell 4}$ decay, where the available phase space is much larger, the $B \to \pi\pi$ form factors span over broad kinematical regions and reveal different behavior depending on the region (see, e.g., [55, 56]).

We discuss the $B_{\ell 4}$ decay form factors in one row with the other timelike form factors, because the system of two pions (or *dipion*) in the final state of this decay has a positive invariant mass and develops strong interaction with rescattering phase and resonance contributions; hence, the form factors parameterizing the $B_{\ell 4}$ decay amplitude at spacelike momentum transfers are timelike with respect to the dipion invariant mass. It is customary to isolate the ρ resonance, restricting the invariant mass of the dipion around m_ρ, hence, reducing the $B_{\ell 4}$ decay to a $B \to \rho \ell \nu_\ell$ and neglecting also the width of ρ. However, this approximation provides a limited information on the dynamics of $B \to \pi\pi$ transition. To clarify the role of excited ρ-mesons and to assess the contribution of the dipions with other spin-parities ($J^P = 0^+, 2^+, ...$) one has to study the amplitude of the full process (5.148).

We start with kinematics, introducing the combinations of momenta

$$k = k_1 + k_2 , \quad \bar{k} = k_1 - k_2 , \quad q = q_1 + q_2 ,$$

From the momentum conservation $p = k + q$, combined with $p^2 = m_B^2$, it follows that

$$qk = \frac{1}{2}(m_B^2 - q^2 - k^2) , \quad (qk)^2 - k^2 q^2 = \frac{1}{4}\lambda(m_B^2, q^2, k^2) \equiv \frac{\lambda_B}{4} , \qquad (5.149)$$

where the function λ is defined in (B.44). Since $k_{1,2}^2 = m_\pi^2$, the scalar product $k\bar{k} = 0$.

To focus on the hadronic matrix element of (5.148), we treat the lepton pair as a single particle with a mass $\sqrt{q^2}$. This is justified, because, as usual, we will factorize the leptonic current. For the hadronic matrix element, the set of kinematical variables is the same as for a $2 \to 2$ scattering process or for a three-body decay, that is, there are three Mandelstam variables:

$$s = (p - q)^2 = k^2 , \quad t = (p - k_2)^2 = (k_1 + q)^2 , \quad u = (p - k_1)^2 = (k_2 + q)^2 , \qquad (5.150)$$

obeying the condition

$$s + t + u = m_B^2 + q^2 + 2m_\pi^2 \equiv h ,$$

which leaves only two of these variables independent[9]. As such, we choose $s = k^2$ and the linear combination

$$\frac{t - u}{2} = \frac{1}{2}((p - k_2)^2 - (p - k_1)^2) = p(k_1 - k_2) = q\bar{k} . \qquad (5.151)$$

The variable $q\bar{k}$ is related to the polar angle θ_π of the π^- in the rest frame of the dipion, where $\vec{k} = 0$, $k_0 = \sqrt{k^2}$, $\vec{k_1} = -\vec{k_2}$, hence $\vec{p} = \vec{q}$. The z-axis is chosen in the direction of $\vec{p}$, so that

$$q\bar{k} = \vec{p}(\vec{k_1} - \vec{k_2}) = 2|\vec{p}||\vec{k_1}| \cos \theta_\pi .$$

The absolute values of 3-momenta are obtained, solving the energy conservation conditions:

$$\sqrt{|\vec{p}|^2 + m_B^2} = \sqrt{|\vec{p}|^2 + q^2} + \sqrt{k^2} , \quad 2\sqrt{|\vec{k_1}|^2 + m_\pi^2} = \sqrt{k^2} .$$

Finally, we obtain

$$q\bar{k} = \frac{1}{2}\sqrt{1 - \frac{4m_\pi^2}{k^2}} \sqrt{\lambda_B} \cos \theta_\pi . \qquad (5.152)$$

Together with the momentum transfer q^2, the invariants k^2 and $q\bar{k}$ form a set of three kinematical variables. The interval of the variable q^2 at fixed k^2 is the same as in the semileptonic processes with one hadron in the final state:

$$m_\ell^2 \leq q^2 \leq (m_B - \sqrt{k^2})^2 , \qquad (5.153)$$

and, at fixed q^2, the limits on the dipion mass squared are

$$4m_\pi^2 \leq k^2 \leq (m_B - \sqrt{q^2})^2 , \qquad (5.154)$$

---

[9]Note that h is the height of the triangle on the Mandelstam (s, t, u) plane used to display the allowed kinematical regions in $2 \to 2$ processes.

where the lower limit is given by the two-pion threshold, and the upper limit is simply the reversed upper limit of (5.153). Finally, the allowed region of the angular variable $q\bar{k}$ is determined varying (5.152) within the interval $-1 \leq \cos\theta \leq 1$.

For the amplitude of $B \to \pi\pi\ell\nu_\ell$ decay, we can use an expression similar to (2.92), taking into account that here both vector and axial-vector parts of the weak current contribute:

$$
\begin{aligned}
\mathcal{A}(B^- \to \pi^+\pi^-\ell\bar{\nu}_\ell) &= \frac{G_F}{\sqrt{2}} V_{ub} \bar{u}_\ell(q_1) \Gamma^\mu v_\nu(q_2) \\
&\times \langle \pi^+(k_1)\pi^-(k_2)|\bar{u}\gamma_\mu(1-\gamma_5)b|B^-(p)\rangle .
\end{aligned}
\tag{5.155}
$$

The next step is to represent the $B \to \pi\pi$ matrix element in terms of Lorentz structures multiplied by form factors. There are three independent momentum four-vectors, and we choose k, $\bar{k}$ and q. A rich kinematics in this process is reflected by the fact that form factors are functions of three invariant variables q^2, k^2 and $q\bar{k}$. Furthermore, we should take into account that the two pions have a certain orbital angular momentum L, hence, spin-parity of the dipion will depend on L:

$$
J_{(2\pi)} = L, \quad P_{(2\pi)} = (P_\pi)^2(-1)^L = (-1)^L.
$$

The form factor components with different L form a partial wave expansion, similar to the one we already used for the pion timelike form factor. Since the variable $q\bar{k}$, according to (5.152), depends on the angle θ_π between pions in a certain frame, it is natural to expand each form factor in a series of Legendre polynomials depending on θ_π. The coefficients in this expansion depend on the two remaining variables q^2, k^2 and represent partial form factors describing $B \to (2\pi)_L$ transitions to a dipion with the spin L and mass $\sqrt{k^2}$.

To determine the number of independent form factors in $B \to \pi\pi\ell\nu_\ell$ decay, we use the same qualitative argument as for the $P \to V$ transitions in Chapter 2.5. Taking the part of the hadronic matrix element in (5.155) generated by the quark vector current $\bar{u}\gamma_\mu b$, we replace this current by a B^* meson with $J^P = 1^-$. The hadronic matrix element is then effectively replaced by the $B \to B^*(2\pi)$ strong interaction amplitude, where we treat the dipion as a single hadron with $J^P = L^{P(2\pi)}$. The total angular momentum of B^* and dipion is equal to the vector sum

$$
\vec{J}_{B^*} \oplus \vec{L} \oplus \vec{L}'
$$

where L' is the orbital momenum between B^* and dipion. The sum $\vec{J}_{B^*} \oplus \vec{L}$ at $L \geq 1$ has three possible eigenvalues $L+1$, L and $L-1$, hence L' has to be equal to one of these values to warrant a zero total angular momentum of the final state equal to the spin of the initial B-meson. Simultaneously, the P-parity conservation yields

$$
P_{(2\pi)}P_{B^*}(-1)^{L'} = (-1)^L(-1)(-1)^{L'} = P_B = -1.
$$

Combining spin and parity conservations, we conclude that, for each $L \geq 1$, only one value $L' = L$ is allowed, while $L' = L = 0$ is forbidden. We infer from these arguments that for the vector current there is a single form factor whose partial wave expansion does not contain an S-wave component. The hadronic matrix element of the vector current in terms of this form factor is:

$$
i\langle \pi^+(k_1)\pi^-(k_2)|\bar{u}\gamma_\mu b|B^-(p)\rangle = -\frac{4}{\sqrt{k^2\lambda_B}} i\epsilon_{\mu\alpha\beta\gamma} q^\alpha k^{1\beta} k^{2\gamma} F_\perp(q^2, k^2, q\bar{k}),
\tag{5.156}
$$

where the appearance of the Levi-Civita tensor in the Lorentz structure is again dictated by P-parity conservation. The partial wave expansion of the form factor reads:

$$F_\perp(q^2, k^2, q\bar{k}) = \sum_{\ell=1}^{\infty} \sqrt{2\ell+1} F_\perp^{(\ell)}(q^2, k^2) \frac{P_\ell^{(1)}(\cos\theta_\pi)}{\sin\theta_\pi} = -\sqrt{3} F_\perp^{(\ell=1)}(q^2, k^2) + \ldots, \quad (5.157)$$

where $P_\ell^{(1)}(\cos\theta_\pi)$ are associated Legendre polynomials, and $P_{\ell=1}^{(1)}(\cos\theta_\pi) = -\sin\theta_\pi$. The normalization factors and partial wave expansion are chosen as in [56]. With these conventions, the form factors are reduced to specially chosen helicity amplitudes: the products of hadronic matrix elements and particularly defined polarization vectors associated with leptonic currents. We will not dwell on these details here.

Counting in the same way the form factors of the axial-vector current $\bar{u}\gamma_\mu\gamma_5 b$ in (5.155), we effectively replace this current by the axial B-meson ($J^P = 1^+$) and find that there are two possible solutions for angular momenta that satisfy both spin and parity conservation: $L' = L + 1$ and $L' = L - 1$; hence, there are two independent form factors and one of them has no S-wave ($L = 0$) contribution. The third form factor originates from the corresponding pseudoscalar current $\bar{u}i\gamma_5 b$. A generic decomposition of the hadronic matrix element consistent with the spin and P-parity conservation is simply a linear combination of all three independent momenta:

$$\langle \pi^+(k_1)\pi^-(k_2)|\bar{u}\gamma_\mu\gamma_5 b|B^-(p)\rangle = k_\mu f_1 + \bar{k}_\mu f_2 + q_\mu f_3, \quad (5.158)$$

where, for brevity, we do not show dependence of the form factors $f_{1,2,3}$ on $q^2, k^2, q\bar{k}$. In parallel, we introduce the form factor of the pseudoscalar current related to the divergence of the axial-vector current:

$$\begin{aligned}\bar{f} &\equiv i\langle \pi^+\pi^-|(m_b + m_u)\bar{u}i\gamma_5 b|B^-\rangle = i\langle \pi^+\pi^-|\partial_\mu(\bar{u}\gamma^\mu\gamma_5)b|B^-\rangle \\ &= q_\mu\langle \pi^+\pi^-|\bar{u}\gamma^\mu\gamma_5 b|B^-\rangle. \end{aligned} \quad (5.159)$$

Substituting the decomposition (5.158) in r.h.s. of the above equation, we obtain a relation

$$f_3 = \frac{1}{q^2}\left(\bar{f} - (qk)f_1 - (q\bar{k})f_2\right), \quad (5.160)$$

allowing us to replace the form factor f_3 by a linear combination of the other form factors:

$$\langle \pi^+(k_1)\pi^-(k_2)|\bar{u}\gamma_\mu\gamma_5 b|B^-(p)\rangle = \left(k_\mu - \frac{qk}{q^2}q_\mu\right)f_1 + \left(\bar{k}_\mu - \frac{q\bar{k}}{q^2}q_\mu\right)f_2 + \frac{q_\mu}{q^2}\bar{f}. \quad (5.161)$$

Redefining these form factors to the ones adopted e.g., in [56] (where they were adjusted to helicity amplitudes):

$$f_1 \equiv i\left(\frac{2\sqrt{q^2}}{\lambda_B}F_0 - 4\frac{(qk)(q\bar{k})}{\sqrt{k^2}\lambda_B}F_\|\right), \quad f_2 \equiv \frac{i}{\sqrt{k^2}}F_\|, \quad \bar{f} \equiv i\sqrt{q^2}F_t, \quad (5.162)$$

we obtain the hadronic matrix element of the axial-vector current in terms of three form factors $F_{0,\|,t}$:

$$\begin{aligned}-i\langle \pi^+(k_1)\pi^-(k_2)|\bar{u}\gamma_\mu\gamma_5)b|B^-(p)\rangle &= \frac{2\sqrt{q^2}}{\sqrt{\lambda_B}}\left(k^\mu - \frac{qk}{q^2}q^\mu\right)F_0(q^2, k^2, q\bar{k}) \\ &+ \frac{1}{\sqrt{k^2}}\left(\bar{k}^\mu - \frac{4(qk)(q\bar{k})}{\lambda_B}k^\mu + \frac{4k^2(q\bar{k})}{\lambda_B}q^\mu\right)F_\|(q^2, k^2, q\bar{k}) \\ &+ \frac{q^\mu}{\sqrt{q^2}}F_t(q^2, k^2, q\bar{k}). \end{aligned} \quad (5.163)$$

The partial wave expansion of the form factor $F_\parallel$ is the same as for $F_\perp$ in (5.157), one just has to replace the index $\perp$ to $\parallel$, whereas for the two other form factors it is written as

$$
\begin{aligned}
F_{0,t}(q^2, k^2, q\bar{k}) &= \sum_{\ell=0}^{\infty} \sqrt{2\ell+1} F_{0,t}^{(\ell)}(q^2, k^2) P_\ell^{(0)}(\cos\theta_\pi) \\
&= F_{0,t}^{(0)}(q^2, k^2) + \sqrt{3} F_{0,t}^{(1)}(q^2, k^2) \cos\theta_\pi + \dots,
\end{aligned}
\tag{5.164}
$$

including an S-wave contribution.

So far we were considering the $B^- \to \pi^+\pi^-\ell^-\bar{\nu}_\ell$ decay mode. Via isospin it is related to $B^- \to \pi^0\pi^0\ell^-\bar{\nu}_\ell$ and $\bar{B}^0 \to \pi^+\pi^0\ell^-\bar{\nu}_\ell$. In principle, these relations can be derived comparing quark flow diagrams, i.e., the quark part of the diagram shown in Figure 5.5 with various combinations of u, d quarks. However, in this case the dipion states with odd and even orbital momenta L have different isospins and there is more than one quark flow diagram in the isospin symmetry limit. We therefore use the conventional isospin formalism, expanding direct products of isospin states and forming $B \to \pi\pi$ hadronic matrix elements with definite isospin. All necessary definitions of the pion and B-meson states in terms of isospin multiplets are given in Appendix A (see (A.83) and (A.86)). First, we form direct products of two pion states and, using the tables of Clebsch-Gordan coefficients from [1], expand these products in irreducible representations $\langle I, I_3|$ of the isospin group $SU(2)_I$:

$$
-\langle \pi^+(k_1)\pi^-(k_2)| = \langle 1, +1| \otimes \langle 1, -1| = \frac{1}{\sqrt{6}}\langle 2, 0| + \frac{1}{\sqrt{2}}\langle 1, 0| + \frac{1}{\sqrt{3}}\langle 0, 0|,
$$

$$
\langle \pi^0(k_1)\pi^0(k_2)| = \langle 1, 0| \otimes \langle 1, 0| = \frac{\sqrt{2}}{\sqrt{3}}\langle 2, 0| - \frac{1}{\sqrt{3}}\langle 0, 0|,
$$

$$
-\langle \pi^+(k_1)\pi^0(k_2)| = \langle 1, +1| \otimes \langle 1, 0| = \frac{1}{\sqrt{2}}\langle 2, +1| + \frac{1}{\sqrt{2}}\langle 1, +1|.
\tag{5.165}
$$

The weak current $\bar{u}\gamma_\mu(1-\gamma_5)b$ can also be considered as a component of an isospin doublet. The action of the weak current operator on the initial B-state increases I_3 by $1/2$. The states emerging after that have the following isospin decomposition:

$$
(\bar{u}...b)|\bar{B}^0\rangle = |1/2, +1/2\rangle \otimes |1/2, +1/2\rangle = |1, +1\rangle,
$$

$$
(\bar{u}...b)|B^-\rangle = -|1/2, +1/2\rangle \otimes |1/2, -1/2\rangle = -\frac{1}{\sqrt{2}}|1, 0\rangle - \frac{1}{\sqrt{2}}|0, 0\rangle,
\tag{5.166}
$$

where we omit Lorentz indices for simplicity, retaining only the flavor content of the current. Comparing the states (5.165) and (5.166), we conclude that an isospin conserving transition between these states is only possible with $I = 0$ and $I = 1$. One immediate consequence is that a dipion state with $I = 2$ is forbidden in $B \to \pi\pi\ell\nu_\ell$ decays. Furthermore, we represent hadronic matrix elements of these decays in terms of two independent isospin amplitudes. The latter are defined as matrix elements of an isospin conserving operator $\hat{S}$ of strong interactions (keeping in mind the S-matrix of QCD) sandwiched between the states (5.165) and (5.166), where only the diagonal combinations with $I = 0, 1$ are retained. We obtain:

$$
\langle \pi^+(k_1)\pi^-(k_2)|\bar{u}...b|B^-\rangle = \frac{1}{2}\langle 1, 0|\hat{S}|1, 0\rangle + \frac{1}{\sqrt{6}}\langle 0, 0|\hat{S}|0, 0\rangle
$$

$$
\langle \pi^0(k_1)\pi^0(k_2)|\bar{u}...b|B^-\rangle = \frac{1}{\sqrt{6}}\langle 0, 0|\hat{S}|0, 0\rangle
$$

$$
\langle \pi^+(k_1)\pi^0(k_2)|\bar{u}...b|\bar{B}^0\rangle = -\frac{1}{\sqrt{2}}\langle 1, 1|\hat{S}|1, 1\rangle,
\tag{5.167}
$$

yielding one relation between the hadronic matrix elements, and hence, between the form factors:

$$F_{\perp(\parallel,0,t)}^{(B^-\to\pi^+\pi^-)} + \frac{1}{\sqrt{2}}F_{\perp(\parallel,0,t)}^{(\bar{B}^0\to\pi^+\pi^0)} = F_{\perp(\parallel,0,t)}^{(B^-\to\pi^0\pi^0)} . \tag{5.168}$$

The formation of a resonance in $B^- \to \pi^+\pi^-\ell^-\bar{\nu}_\ell$ is schematically shown in Figure 5.5. An unstable meson, being formed by the u-quark from b-decay and the spectator $\bar{u}$-quark from B meson, strongly decays into two pions. Considering only dipions in the P- and S-waves, we restrict the quantum numbers of resonances, respectively, to $J^P = 1^-$, $I^G = 1^+$ (the ρ meson and its radial excitations) and $J^P = 0^+$, $I^G = 0^+$ (the scalar isoscalar f_0 mesons starting from $f_0(500)$). Masses and widths of all these mesons are listed in [1]. Note that the isospin and spin of a resonance are correlated, due to the fixed final state of the dipion. In particular, the G-parity for a state of two pions with the orbital momentum L and isospin I, according to (1.93), is equal to $(-1)^L(-1)^I$.

To obtain a resonance contribution in the explicit form, we invoke hadronic dispersion relation, considering the $B \to \pi\pi$ hadronic matrix element as an analytic function of the variable k^2, with all other variables kept fixed:

$$\langle\pi^+(k_1)\pi^-(k_2)|\bar{u}\Gamma_\mu b|B^-\rangle = \frac{1}{\pi}\int_{4m_\pi^2}^{\infty} ds\frac{\mathrm{Im}_s\langle\pi^+(k_1)\pi^-(k_2)|\bar{u}\Gamma_\mu b|B^-\rangle}{s - k^2 - i\epsilon}, \tag{5.169}$$

where $\Gamma_\mu = \gamma_\mu(1 - \gamma_5)$ and the imaginary part at $k^2 = s$ is given by the unitarity relation,

$$2\,\mathrm{Im}_{k^2}\langle\pi^+(k_1)\pi^-(k_2)|\bar{u}\Gamma_\mu b|B^-\rangle = \sum_h d\tau_h\langle\pi^+(k_1)\pi^-(k_2)|\overline{h(k)}\rangle\langle h(k)|\bar{u}\Gamma_\mu b|B^-\rangle . \tag{5.170}$$

To isolate the ρ-meson term in the hadronic sum, we use the definition (5.84) of the $\rho\pi\pi$ coupling (adapted to our choice of momenta) and the one-particle phase space (5.73):

$$\mathrm{Im}_s\langle\pi^+(k_1)\pi^-(k_2)|\bar{u}\Gamma_\mu b|B^-\rangle = \pi\delta(s - m_\rho^2)g_{\rho\pi\pi}(k_2 - k_1)^\alpha\overline{\varepsilon_\alpha^{(\rho)}\langle\rho^0(k)|\bar{u}\Gamma_\mu b|B^-\rangle} + \dots . \tag{5.171}$$

where all other contributions are denoted by dots. Substituting the $B \to \rho$ transition matrix element in the form (2.136):

$$\sqrt{2}\langle\rho^0(k)|\bar{u}\Gamma_\mu b|B(p)\rangle = \epsilon_{\mu\nu\rho\sigma}\varepsilon^{(\rho)*\nu}q^\rho k^\sigma\frac{2V^{B\rho}(q^2)}{m_B + m_\rho} - i\varepsilon_\mu^{(\rho)*}(m_B + m_\rho)A_1^{B\rho}(q^2)$$

$$+ i(2k + q)_\mu(\varepsilon^{(\rho)*}q)\frac{A_2^{B\rho}(q^2)}{m_B + m_\rho} + iq_\mu(\varepsilon^{(\rho)*}q)\frac{2m_\rho}{q^2}\left(A_3^{B\rho}(q^2) - A_0^{B\rho}(q^2)\right), \tag{5.172}$$

where the coefficient $\sqrt{2}$ reflects the normalization of the form factors to the final state with $\rho^\pm$, we sum over polarizations of the vector meson:

$$\overline{\varepsilon_\alpha^{(\rho)*}\varepsilon_\beta^{(\rho)*}} = -g_{\alpha\beta} + \frac{k_\alpha k_\beta}{m_\rho^2}.$$

The resulting expression for the imaginary part is inserted in the dispersion integral (5.169). Integrating the δ-function yields a simple pole for the ρ contribution to the $B \to \pi\pi$ hadronic matrix element:

$$\langle\pi^+(k_1)\pi^-(k_2)|\bar{u}\Gamma_\mu b|B^-\rangle^{(\rho)} = \frac{g_{\rho\pi\pi}}{\sqrt{2}(m_\rho^2 - k^2 - i\epsilon)}\left\{\epsilon_{\mu\nu\rho\sigma}\bar{k}^\nu q^\rho k^\sigma\frac{2V^{B\rho}(q^2)}{m_B + m_\rho}\right.$$

$$- i\bar{k}_\mu(m_B + m_\rho)A_1^{B\rho}(q^2) + i(2k + q)_\mu(q\bar{k})\frac{A_2^{B\rho}(q^2)}{m_B + m_\rho}$$

$$\left. + iq_\mu(q\bar{k})\frac{2m_\rho}{q^2}\left(A_3^{B\rho}(q^2) - A_0^{B\rho}(q^2)\right)\right\}. \tag{5.173}$$

Then, it is conceivable to follow the procedure we used for the pion timelike form factor in (5.106), and include the energy-dependent $\rho \to 2\pi$ width in the denominator of the pole factor, effectively taking into account the resummation of all intermediate two-pion loops:

$$\frac{1}{m_\rho^2 - k^2 - i\epsilon} \to \mathcal{D}^{(\rho)}(k^2) \equiv \frac{1}{\left[m_\rho^2 - k^2 - i\sqrt{k^2}\Gamma_{\rho \to 2\pi}(k^2) \right]}, \qquad (5.174)$$

Comparing (5.173) with the definitions (5.156) and (5.163), we obtain the ρ-resonance contributions to the P-wave $B \to \pi\pi$ form factors, e.g.:

$$F_\perp^{(\ell=1|\rho)}(k^2, q^2) = -\frac{\sqrt{k^2}\sqrt{\lambda_B}}{\sqrt{2}\sqrt{3}} g_{\rho\pi\pi} \mathcal{D}^{(\rho)}(k^2) \frac{V^{B\rho}(q^2)}{(m_B + m_\rho)},$$

$$F_\parallel^{(\ell=1|\rho)}(k^2, q^2) = \frac{\sqrt{k^2}}{\sqrt{2}\sqrt{3}} g_{\rho\pi\pi} \mathcal{D}^{(\rho)}(k^2)(m_B + m_\rho)A_1^{B\rho}(q^2). \qquad (5.175)$$

Further relations and details on various observables in the $B \to \pi\pi\ell\bar{\nu}_\ell$ decays in terms of $B \to \pi\pi$ form factors can be found in [56]. Analogous form factors are introduced for the $B \to K\pi\ell\ell$ FCNC decays, including the K^* resonance contributions.

CHAPTER 6

NONLOCAL HADRONIC MATRIX ELEMENTS

6.1 $\pi^0 \to 2\gamma$ DECAY

In this chapter we consider hadronic processes mediated by two local quark currents. In general, the two currents are separated from each other by an average long distance of $O(1/\Lambda_{QCD})$, and therefore the locality, a common feature of all hadron form factors considered so far, is lost. Still, it is possible to parametrize the nonlocal hadronic amplitudes in terms of Lorentz-invariant functions which will play the role of form factors.

The radiative two-photon decay of the neutral pion

$$\pi^0(p) \to \gamma(k_1)\gamma(k_2)$$

is our first example. We start, as usual, from the S-matrix element, retaining the part of the expansion (2.11) which is of the second-order in e.m. interaction (the QCD part is implied but not shown):

$$S_{fi}^{(\pi^0 \to 2\gamma)} = \frac{i^2 e^2}{2!} \langle \gamma(k_1)\gamma(k_2)| T\left\{ \int d^4x A_{em}^\mu(x) j_\mu^{em}(x) \int d^4y A_{em}^\nu(y) j_\nu^{em}(y) \right\} |\pi^0(p_1)\rangle. \quad (6.1)$$

Using the momentum decomposition of the photon field operator in (4.8), we single out the relevant components with momenta k_1, k_2 and polarization vectors ε_1, ε_2 (we omit * for brevity) and factorize the decay amplitude in a form of the two-photon part multiplied by a hadronic matrix element:

$$S_{fi}^{(\pi^0 \to 2\gamma)} = 4\pi i \alpha_{em} \varepsilon_1^\mu \varepsilon_2^\nu i \int d^4x \int d^4y e^{ik_1 x} e^{ik_2 y} \langle 0|T\left\{ j_\mu^{em}(x) j_\nu^{em}(y) \right\}|\pi^0(p)\rangle$$

$$= i(2\pi)^4 \delta^{(4)}(p - k_1 - k_2)\mathcal{A}(\pi^0 \to 2\gamma), \quad (6.2)$$

where

$$\mathcal{A}(\pi^0 \to 2\gamma) = 4\pi \alpha_{em} \varepsilon_1^\mu \varepsilon_2^\nu i \int d^4x e^{ik_1 x} \langle 0|T\left\{ j_\mu^{em}(x) j_\nu^{em}(0) \right\}|\pi^0(p)\rangle. \quad (6.3)$$

In obtaining (6.2), we take into account that it receives two equal contributions (a photon with given momentum can be emitted from both currents), hence, the factor $1/2$ cancels out. Also the translation property of the current is used in order to obtain the δ-function after y-integration.

The hadronic matrix element in this case is a T-product of two quark e.m. currents sandwiched between the neutral pion state and vacuum. The pion is a pseudoscalar state, therefore, there is only a single possible Lorentz structure composed from two independent momenta and consistent with the P-parity conservation:

$$i \int d^4x e^{ik_1 x} \langle 0| T\left\{ j_\mu^{em}(x) j_\nu^{em}(0) \right\} |\pi^0(p)\rangle = \epsilon_{\mu\nu\alpha\beta} k_1^\alpha k_2^\beta \, g_{\pi\gamma\gamma} . \tag{6.4}$$

The constant $g_{\pi\gamma\gamma}$ is the only hadronic parameter in this decay. Since we consider a two-body decay with a fixed mass of the initial and final particles, there are no additional kinematical variables. In Chapter 9 we will consider a related form factor describing a transition of two virtual photons with $k_{1,2}^2 \neq 0$ into π^0, so that $g_{\pi\gamma\gamma}$ is a limiting value of that form factor at $k_{1,2}^2 = 0$.

Using (6.4), we rewrite the amplitude,

$$\mathcal{A}(\pi^0 \to 2\gamma) = 4\pi\alpha_{em} \epsilon^{\mu\nu\alpha\beta} \varepsilon_{1\mu} \varepsilon_{2\nu} k_{1\alpha} k_{2\beta} \, g_{\pi\gamma\gamma} . \tag{6.5}$$

It is gauge invariant with respect to the photon fields and vanishes if we replace a photon polarization vector by its momentum. The decay width is easily obtained squaring the above amplitude, summing over the photon polarizations which is equivalent to a replacement $\varepsilon_{1(2)}^{\mu'} \varepsilon_{1(2)}^{*\mu} \to -g_{\mu\mu'}$ and multiplying by the phase space factors:

$$\Gamma(\pi^0 \to 2\gamma) = \frac{1}{2!} \overline{|\mathcal{A}(\pi^0 \to 2\gamma)|^2} \frac{1}{16\pi m_\pi} = \frac{\pi\alpha_{em}^2}{4} m_\pi^3 |g_{\pi\gamma\gamma}|^2 , \tag{6.6}$$

where an extra statistical factor $1/2!$ accounts for the identical photons in the final state.

Being nonlocal and long-distance dominated, the amplitude of $\pi^0 \to \gamma\gamma$ is unique because it is determined by the chiral anomaly. In a quantum field theory, an anomaly is a common name for the effects taking place when a certain symmetry of the Lagrangian (or, equivalently, a conservation of the corresponding current) is violated for physical amplitudes due to quantum corrections generated by loop diagrams. The gluonic chiral anomaly in QCD was already mentioned in Chapter 1.6 (see (1.76)). The chiral anomaly we are interested in is related to QED. It emerges via triangle loop diagrams formed by an axial-vector current and two e.m. currents. The divergence of this current does not vanish in the limit of massless quarks in the presence of e.m. fields.

Avoiding lengthy calculation of triangle diagrams and their renormalization, let us reproduce the chiral anomaly in a slightly more general way. To this end, we consider a local single-flavored axial-vector current $\bar{q}(x)\gamma_\mu\gamma_5 q(x)$ and temporarily represent it as a limit of a bilocal quark-antiquark operator:

$$\bar{q}(x)\gamma_\mu\gamma_5 q(x) = \lim_{\delta \to 0} j_{\mu 5}(x,\delta), \quad j_{\mu 5}(x,\delta) = \bar{q}(x+\delta)\gamma_\mu\gamma_5 [x+\delta, x-\delta] q(x-\delta), \tag{6.7}$$

where the gauge link

$$[x+\delta, x-\delta] = \exp\left(ieQ_q \int_{x-\delta}^{x+\delta} dy^\alpha A_\alpha^{em}(y) \right) \tag{6.8}$$

restores the local gauge invariance of e.m. interaction. In (6.7), as usual, the sum over quark colors is implied. Furthermore, it is convenient to adopt the fixed $x_0 = 0$ point gauge for the e.m. field, similar to the one for the gluon field. In this gauge, the e.m. field is expressed in terms of its strength tensor (cf. (A.64)):

$$A_\alpha^{em}(y) = \frac{1}{2} y^\nu F_{\nu\alpha}(0) . \tag{6.9}$$

Without loss of generality, we neglect derivatives of the field strength, assuming constant $F_{\mu\nu}$. We then calculate the divergence of the bilocal current (6.7), using the Dirac equation for the quark in e.m. field:

$$(\vec{\not{\partial}} - ieQ_q A^{em}(x))q(x) = -im_q q(x)\,, \quad \bar{q}(x)(\overleftarrow{\not{\partial}} + ieQ_q A^{em}(x)) = im_q \bar{q}(x)\,. \qquad (6.10)$$

Differentiating the quark fields and gauge link, we obtain:

$$\frac{\partial}{\partial x_\mu} j_{\mu 5}(x,\delta) = \Big(-\bar{q}(x+\delta)\gamma_5\big(ieQ_q A^{em}(x-\delta) - im_q\big)q(x-\delta)$$

$$+\bar{q}(x+\delta)\big(-ieQ_q A^{em}(x+\delta) + im_q\big)\gamma_5 q(x-\delta)\Big)\big[x+\delta, x-\delta\big]$$

$$+\bar{q}(x+\delta)\gamma_\mu\gamma_5 q(x-\delta)\frac{\partial}{\partial x_\mu}\big[x+\delta, x-\delta\big]$$

$$= 2im_q\bar{q}(x+\delta)\gamma_5 q(x-\delta)\big[\dot{x}+\delta, x-\delta\big]$$

$$+\bar{q}(x+\delta)\big(ieQ_q A^{em}(x-\delta) - ieQ_q A^{em}(x+\delta)\big)\gamma_5 q(x-\delta)\big[x+\delta, x-\delta\big]$$

$$+\bar{q}(x+\delta)\gamma_\mu\gamma_5 q(x-\delta)\frac{\partial}{\partial x_\mu}\big[x+\delta, x-\delta\big]\,. \qquad (6.11)$$

In the local limit, the first line on the r.h.s. of the above equation converts into a usual divergence term with the quark mass multiplied by the pseudoscalar current. To simplify the further discussion, we take the chiral limit $m_q \to 0$, discarding this term. Our goal is to demonstrate that the sum of remaining terms in the second and third line generates a nonvanishing contribution at $\delta \to 0$. To proceed, we transform the divergence using (6.9) for the e.m. field and collecting together all terms proportional to the field strength:

$$\frac{\partial}{\partial x_\mu} j_{\mu 5}(x,\delta) = \bar{q}(x+\delta)\left(ieQ_q\gamma^\alpha\frac{1}{2}(x-\delta)^\nu F_{\nu\alpha} - ieQ_q\gamma^\alpha\frac{1}{2}(x+\delta)^\nu F_{\nu\alpha}\right)\gamma_5 q(x-\delta)$$

$$+O(\delta^2) + \cdots + \bar{q}(x+\delta)\gamma_\mu\gamma_5 q(x-\delta)\frac{\partial}{\partial x_\mu}\left(1 + ieQ_q\int\limits_{x-\delta}^{x+\delta} dy^\alpha\frac{1}{2}y^\nu F_{\nu\alpha} + \dots\right)$$

$$= -2ieQ_q\delta^\nu\bar{q}(x+\delta)\gamma^\alpha\gamma_5 q(x-\delta)F_{\nu\alpha} + O(\delta^2) + \dots\,, \qquad (6.12)$$

where we retain in the gauge link only terms of the first order in the infinitesimal distance δ between quark fields.

As a next step, we calculate a certain amplitude involving the nonlocal divergence (6.12). One possibility is to take the vacuum average of this operator in the presence of e.m. fields. We choose $x = 0$ for simplicity and obtain:

$$\langle 0|\partial^\mu j_{\mu 5}(0,\delta)|0\rangle_F = -2ieQ_q\delta^\nu F_{\nu\alpha}\langle 0|\bar{q}(\delta)\gamma^\alpha\gamma_5 q(-\delta)|0\rangle_F$$

$$= 2ieQ_q\delta^\nu F_{\nu\alpha}\langle 0|q_\tau^i(-\delta)\bar{q}_{\omega k}(\delta)|0\rangle_F (\gamma^\alpha\gamma_5)_{\tau\omega}\delta_i^k\,, \qquad (6.13)$$

where the index F indicates the presence of e.m. field, Dirac and color indices are shown explicitly and an additional minus sign takes into account the permutation of anticommuting quark fields. In (6.13) the vacuum average of quark and antiquark fields separated by the four-dimensional distance 2δ is identified with a quark propagator in the external e.m. field. Note that we are calculating a physical amplitude, hence a time-ordering should be added. Jumping slightly ahead, we refer to the expression (8.86) for a quark propagator in the external gluon field derived below in Chapter 8. The propagator that we need here is obtained, replacing in (8.86) the gluon field strength by an e.m. one. With our choice of coordinates we have:

$$-i\langle 0|T\{q_\tau^i(-\delta)\bar{q}_{\omega k}(\delta)\}|0\rangle_F = S_{\omega\tau}(-\delta,\delta) = \frac{eQ_q}{16\pi^2}\frac{\delta^\rho}{\delta^2}\widetilde{F}_{\rho\beta}(\gamma^\beta\gamma_5)_{\tau\omega}\delta_k^i \,. \tag{6.14}$$

Substituting this expression in (6.13), and taking Dirac and color traces, we obtain:

$$\langle 0|\partial^\mu j_{\mu 5}(0,\delta)|0\rangle_F = \frac{e^2 Q_q^2 N_c}{2\pi^2}\frac{\delta^\nu\delta^\rho}{\delta^2}F_{\nu\alpha}\widetilde{F}_{\rho\beta}g^{\beta\alpha} \,, \tag{6.15}$$

where $N_c = 3$ results from the color trace. In the case of a lepton current this factor is absent and $Q_q \to 1$. Since the vacuum average cannot depend on the direction of the four-vector δ, we have to average it:

$$\langle 0|\frac{\delta^\nu\delta^\rho}{\delta^2}|0\rangle = \frac{1}{4}g^{\nu\rho} \,.$$

After that the $\delta \to 0$ limit of (6.15) can safely be taken and we obtain the nonvanishing vacuum average of the local divergence:

$$\langle 0|\partial^\mu j_{\mu 5}|0\rangle_F = \frac{e^2 Q_q^2 N_c}{8\pi^2}F_{\mu\nu}\widetilde{F}^{\mu\nu} \,, \tag{6.16}$$

which violates the chiral symmetry of both QCD and QED Lagrangians. The reason we obtain a nonvanishing quantity is nontrivial and is deeply related to the fact that the local products of field operators in QCD, QED and other renormalizable theories are singular functions. Here this singularity, reflected by the $1/\delta^2$ denominator in the propagator, is canceled by the numerator in the vacuum average. A more detailed analysis based on direct calculation of triangle diagrams and their renormalization or using more involved field-theoretical considerations (see, e.g., the review [15]) reveals that (6.16) is valid not only for the vacuum matrix elements but also at the operator level, so that

$$\partial^\mu j_{\mu 5} = \frac{e^2 Q_q^2 N_c}{8\pi^2}F_{\mu\nu}\widetilde{F}^{\mu\nu} \,. \tag{6.17}$$

Returning to the $\pi^0 \to 2\gamma$ decay, we choose the corresponding current with $I = 1$:

$$j_{\mu 5}^{I=1}(x) = \bar{u}(x)\gamma_\mu\gamma_5 u(x) - \bar{d}(x)\gamma_\mu\gamma_5 d(x) \,.$$

The decay constant of the neutral pion parameterizes the matrix element of this current:

$$\langle \pi^0|j_{\mu 5}^{I=1}|0\rangle = -\sqrt{2}iq_\mu f_\pi \,, \tag{6.18}$$

where in the isospin symmetry limit f_π is equal to the charged pion decay constant (cf. (2.50)). For this current the chiral anomaly equation reads:

$$\partial^\mu j_{\mu 5}^{I=1} = \frac{e^2 N_c(Q_u^2 - Q_d^2)}{8\pi^2}F_{\mu\nu}\widetilde{F}^{\mu\nu} \,. \tag{6.19}$$

To obtain the $\pi \to 2\gamma$ amplitude, we consider the vacuum-to-2γ matrix element of the axial-vector current:

$$\langle \gamma(k_1)\gamma(k_2)|j_{\mu 5}^{I=1}|0\rangle = iq_\mu\epsilon^{\rho\lambda\alpha\beta}\varepsilon_{1\rho}\varepsilon_{2\lambda}k_{1\alpha}k_{2\beta}T(q^2) \,. \tag{6.20}$$

Here we assume that photons are on shell but the external momentum q transferred via the current is virtual, $q^2 \neq 0$. Note that r.h.s. of the above equation contains the single possible Lorentz structure. We skip a detailed analysis of this point (see, e.g., [15]) using instead a qualitative argument. The axial-vector ($J^P = 1^+$) intermediate states cannot contribute to this amplitude in the case of real photons[1], hence there are only pseudoscalar states propagating between the current and final state, hence only one invariant amplitude.

Multiplying (6.20) by q^μ, transforms the l.h.s. to the divergence (see (2.56)):

$$\langle\gamma(k_1)\gamma(k_2)|\partial^\mu j_{\mu 5}^{I=1}|0\rangle = q^2\epsilon^{\rho\lambda\alpha\beta}\varepsilon_{1\rho}\varepsilon_{2\lambda}k_{1\alpha}k_{2\beta}T(q^2)\,. \tag{6.21}$$

After that we use the anomaly equation inserting it instead of the derivative on l.h.s:

$$\langle\gamma(k_1)\gamma(k_2)|\partial^\mu j_{\mu 5}^{I=1}|0\rangle = \frac{e^2 N_c(Q_u^2 - Q_d^2)}{8\pi^2}\langle\gamma(k_1)\gamma(k_2)|F_{\mu\nu}\widetilde{F}^{\mu\nu}|0\rangle\,. \tag{6.22}$$

The emerging matrix element is a pure QED object which is easily calculated using the momentum representation (4.8) of the e.m. field:

$$\langle\gamma(k_1)\gamma(k_2)|F_{\mu\nu}\widetilde{F}^{\mu\nu}|0\rangle = 2\epsilon^{\mu\nu\alpha\beta}\langle\gamma(k_1)\gamma(k_2)|\partial_\mu A_\nu \partial_\alpha A_\beta|0\rangle = -4\epsilon^{\mu\nu\alpha\beta}k_{1\mu}\varepsilon_{1\nu}k_{2\alpha}\varepsilon_{2\beta}\,. \tag{6.23}$$

Using this, we match the r.h.s. of (6.21) and (6.22) and obtain for the invariant amplitude:

$$T(q^2) = \frac{e^2(Q_u^2 - Q_d^2)N_c}{2\pi^2 q^2}\,. \tag{6.24}$$

Remarkably, the amplitude reveals a pole at $q^2 = 0$ which is exactly what is anticipated, namely the intermediate massless pion (remember that we still work in the chiral limit).

Finally, to link (6.24) with the $\pi^0 \to 2\gamma$ amplitude, we use a dispersion relation for (6.20) in the variable q^2 where we include only the π^0 intermediate state, keeping in mind that the decay constants of all other pseudoscalar states are suppressed in the chiral limit (in Chapter 8.6 this point will be substantiated). We also adopt the isospin symmetry limit, hence the η meson does not contribute either. The dispersion relation containing only the massless pion pole reads:

$$\langle\gamma(k_1)\gamma(k_2)|j_{\mu 5}^{I=1}|0\rangle = \frac{\langle\gamma(k_1)\gamma(k_2)|\pi^0\rangle\langle\pi^0|j_{\mu 5}^{I=1}|0\rangle}{-q^2}$$

$$= \frac{1}{-q^2}\left(4\pi\alpha_{em}\epsilon^{\mu\nu\alpha\beta}\varepsilon_{1\mu}\varepsilon_{2\nu}k_{1\alpha}k_{2\beta}\,g_{\pi\gamma\gamma}\right)\left(-i\sqrt{2}q_\mu f_\pi\right)\,. \tag{6.25}$$

where in the last step we used the representations (6.5) and (6.18) for the matrix elements in the residue of the pion pole.

Comparing this equation with the parameterization in (6.20), where the anomaly induced result (6.24) for the invariant amplitude is substituted, we finally reach our goal, obtaining a relation for the decay constant of $\pi^0 \to 2\gamma$:

$$g_{\pi\gamma\gamma} = \frac{3(Q_u^2 - Q_d^2)}{2\sqrt{2}\pi^2 f_\pi}\,, \tag{6.26}$$

which, after substituting in (6.6), determines the width of the neutral pion,

$$\Gamma(\pi^0 \to 2\gamma) = \frac{\alpha_{em}^2 m_\pi^3}{32\pi^3 f_\pi^2}\,. \tag{6.27}$$

Using the measured value of f_π given in (2.64) and comparing the result of (6.27) with the inverse lifetime of π^0 (according to [1], in almost 100%, the neutral pion decays into two photons), one finds an impressive agreement. Note especially the role of the factor $N_c = 3$ in (6.26). Before the advent of QCD, the quark model used in the anomaly equation failed to explain the neutral pion width by a factor of nine. The restored agreement was one of the striking manifestations of the quark color charge.

---

[1] A spin one particle cannot decay into two photons, and this is known as the Landau-Yang theorem.

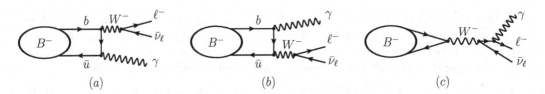

Figure 6.1 Diagrams of the photoleptonic $B^- \to \ell^- \bar{\nu}_\ell \gamma$ decay, corresponding to the photon emission from the (a) light u quark, (b) heavy b quark and (c) charged lepton.

6.2 PHOTOLEPTONIC WEAK DECAYS

In Chapter 2, the weak leptonic decays of pseudoscalar mesons, such as $\pi^- \to \ell^- \bar{\nu}_\ell$ or $B^- \to \ell^- \bar{\nu}_\ell$, were analyzed. Hadronic matrix elements of these decays are reduced to the decay constants, f_π or f_B, respectively. Suppose that the weak transition into a lepton-neutrino pair is accompanied by a photon emission. As an example of this *weak photoleptonic decay*, we consider

$$B^-(p+q) \to \ell^-(p_\ell)\bar{\nu}_\ell(p_\nu)\gamma(p), \tag{6.28}$$

where $q = p_\ell + p_\nu$ is the momentum of the lepton pair. We concentrate on the decay modes with $\ell = e, \mu$ and neglect all lepton masses, assuming that the photon is energetic enough to be resolved from the charged lepton. The light flavor photoleptonic decays $\pi^- \to \ell^- \bar{\nu}_\ell \gamma$ and $K^- \to \ell^- \bar{\nu}_\ell \gamma$, have been observed [1], whereas the branching fractions of (6.28) and of the analogous charmed meson decays $D_{(s)}^- \to \ell^- \bar{\nu}_\ell \gamma$ are not yet measured.

One could expect that a photon emission does not influence the hadronic structure of the $B \to \ell \bar{\nu}_\ell$ amplitude, so that the decay constant f_B remains the only essential parameter also for the $B \to \ell \bar{\nu}_\ell \gamma$ decay. In reality, the dynamics of the photoleptonic decay is largely determined by specific form factors.

The amplitude of (6.28) is derived, retaining in the S-matrix element the weak semileptonic $b \to u$ transition described by the effective Hamiltonian (1.29) and, simultaneously, the e.m. interaction with a photon emission; hence, we have to combine together the first orders of the S-matrices (B.3) and (B.4):

$$
\begin{aligned}
S_{fi}^{(B \to \ell \nu_\ell \gamma)} &= \langle \ell^-(p_\ell)\bar{\nu}_\ell(p_\nu)\gamma(p)|T\Big\{\Big(i\int d^4x \mathcal{L}_{QED}^{int}(x)\Big)\Big(-i\int d^4y \mathcal{H}_W(y)\Big)\Big\}|B^-(p+q)\rangle \\
&= e\frac{G_F}{\sqrt{2}}V_{ub}\int d^4x \int d^4y \langle \ell^-(p_\ell)\bar{\nu}_\ell(p_\nu)\gamma(p)|T\Big\{\big(j_\mu^{em}(x) - \bar{\ell}(x)\gamma_\mu \ell(x)\big)A_{em}^\mu(x) \\
&\quad \times \bar{\ell}(y)\Gamma_\rho \nu_\ell(y)\, \bar{u}(y)\Gamma^\rho b(y)\Big\}|B^-(p+q)\rangle,
\end{aligned}
\tag{6.29}
$$

where $\Gamma_\rho = \gamma_\rho(1 - \gamma_5)$ is our usual notation and in the quark e.m. current only the u- and b-quark components are essential:

$$j_\mu^{em}(x) = Q_u \bar{u}(x)\gamma_\mu u(x) + Q_b \bar{b}(x)\gamma_\mu b(x) + \cdots. \tag{6.30}$$

We then separate in (6.29) from each other the parts containing the quark and lepton e.m. current. The first part corresponds to a photon emission from the initial B meson (the diagrams in Figure 6.1(a),(b)), in which case the leptons are free and their weak current is factorized, employing the momentum decomposition of lepton fields (see (A.21)). In the second part, the photon is emitted from the final-state charged lepton (the diagram in Figure 6.1(c)). In addition, we use the expansion (4.8) and retain only the component of the e.m. field corresponding to the photon emission with the momentum p. We obtain:

$$S_{fi}^{(B \to \ell\nu_\ell\gamma)} = e\frac{G_F}{\sqrt{2}}V_{ub}\int d^4x\int d^4y\, \varepsilon^{\mu*}e^{ipx}$$

$$\times\left(\bar{u}(p_\ell)\Gamma_\rho v(p_\nu)e^{i(p_\ell+p_\nu)y}\langle 0|T\{j_\mu^{em}(x)\bar{u}(y)\Gamma^\rho b(y)\}|B^-(p+q)\rangle\right.$$

$$\left.-\langle\ell^-(p_\ell)\bar{\nu}_\ell(p_\nu)|T\{\bar{\ell}(x)\gamma_\mu\ell(x)\bar{\ell}(y)\Gamma_\rho\nu_\ell(y)\}|0\rangle\langle 0|\bar{u}(y)\Gamma^\rho b(y)|B^-(p+q)\rangle\right),\ (6.31)$$

where $\varepsilon\equiv\varepsilon(p,\lambda)$ is the photon polarization vector. After that, in both parts of the above expression, we apply a translation of the current operators, shifting $y\to 0$ and, correspondingly, $x\to x-y$. E.g., the hadronic matrix element in the second line is transformed as follows:

$$\langle 0|T\{j_\mu^{em}(x)\bar{u}(y)\Gamma^\rho b(y)\}|B^-(p+q)\rangle$$

$$= \langle 0|T\{e^{-i\hat{P}y}j_\mu^{em}(x)e^{i\hat{P}y}\bar{u}(0)\Gamma^\rho b(0)e^{-i\hat{P}y}\}|B^-(p+q)\rangle$$

$$= \langle 0|T\{j_\mu^{em}(x-y)\bar{u}(0)\Gamma^\rho b(0)\}|B^-(p+q)\rangle e^{-i(p+q)y}\,,\ (6.32)$$

where we use (A.78) and put an additional exponent of the momentum operator $\hat{P}$ in the second line, making use of $\langle 0|e^{-i\hat{P}y} = \langle 0|$. The integrand in the S-matrix element becomes a function of $x-y$. We interchange the order of integrations and transform the integral according to the following scheme:

$$\int d^4y\, e^{i(p_\ell+p_\nu)y-ip_By}\int d^4x\, e^{ipx}f(x-y) = \int d^4y e^{i(p_\ell+p_\nu)y+ipy-ip_By}\int d^4x' e^{ipx'}f(x')$$

where $p_B = p+q$ and $x' = x-y$ are introduced. The integral over d^4y yields a factor $(2\pi)^4\delta^{(4)}(p_B-p-p_\ell-p_\nu)$, reflecting the momentum conservation. Similar transformations are applied to the leptonic and hadronic matrix elements in the third line of (6.31). According to the convention (B.17), we separate the δ-functions and obtain from (6.31) the decay amplitude:

$$\mathcal{A}(B\to\ell\nu_\ell\gamma) = -ie\frac{G_F}{\sqrt{2}}V_{ub}\,\varepsilon^{\mu*}\Big(\bar{u}(p_\ell)\Gamma^\rho v(p_\nu)$$

$$\times\int d^4x e^{ipx}\langle 0|T\{j_\mu^{em}(x)\bar{u}(0)\Gamma_\rho b(0)\}|B^-(p+q)\rangle$$

$$-\int d^4x e^{ipx}\langle\ell^-(p_\ell)\bar{\nu}_\ell(p_\nu)|T\{\bar{\ell}(x)\gamma_\mu\ell(x)\bar{\ell}(0)\Gamma_\rho\nu_\ell(0)\}|0\rangle\langle 0|\bar{u}\Gamma^\rho b|B^-(p+q)\rangle\Big).\ (6.33)$$

The hadronic matrix element in the third line of the above equation reduces to the B meson decay constant:

$$\langle 0|\bar{u}\Gamma_\rho b|B^-(p+q)\rangle = -\langle 0|\bar{u}\gamma_\rho\gamma_5 b|B^-(p+q)\rangle = -if_B(p+q)_\rho\,,\ (6.34)$$

defined similar to (2.120). Furthermore, the purely leptonic matrix element is directly calculated, using the momentum expansion of the leptonic field operators $\bar{\ell}(x), \nu_\ell(0)$ and contracting the two remaining lepton fields in a propagator $\langle 0|T\{\ell(x)\bar{\ell}(0)\}|0\rangle = iS_\ell(x,0)$, for which we use the expression similar to (A.49), neglecting the lepton mass:

$$\int d^4x e^{ipx} \langle \ell^-(p_\ell)\bar{\nu}_\ell(p_\nu)| T\{\bar{\ell}(x)\gamma_\mu \ell(x)\bar{\ell}(0)\Gamma_\rho \nu_\ell(0)\}|0\rangle$$

$$= i \int d^4x\, e^{ipx+ip_\ell x} \bar{u}(p_\ell)\gamma_\mu S_\ell(x,0)\Gamma_\rho v(p_\nu)$$

$$= i \int d^4x\, e^{ipx+ip_\ell x} \bar{u}(p_\ell)\gamma_\mu \int \frac{d^4k}{(2\pi)^4} e^{-ikx} \frac{\slashed{k}}{k^2}\Gamma_\rho v(p_\nu)$$

$$= i \int d^4k\, \delta^{(4)}(p+p_\ell-k)\bar{u}(p_\ell)\gamma_\mu \frac{\slashed{k}}{k^2}\Gamma_\rho v(p_\nu) = i\bar{u}(p_\ell)\gamma_\mu \frac{\slashed{p}_\ell + \slashed{p}}{(p_\ell + p)^2}\Gamma_\rho v(p_\nu)\,. \quad (6.35)$$

Multiplied by the momentum $(p+q)_\rho$ originating from (6.34), this expression transforms into

$$i\bar{u}(p_\ell)\gamma_\mu \frac{\slashed{p}_\ell + \slashed{p}}{(p_\ell + p)^2}(\slashed{p}+\slashed{q})(1-\gamma_5)v(p_\nu)$$

$$= i\bar{u}(p_\ell)\gamma_\mu \frac{(\slashed{p}_\ell + \slashed{p})(\slashed{p}+\slashed{p}_\ell)}{(p_\ell + p)^2}(1-\gamma_5)v(p_\nu) = i\bar{u}(p_\ell)\Gamma_\mu v(p_\nu)\,, \quad (6.36)$$

where we substituted $q = p_\ell + p_\nu$ and used the Dirac equation for the massless neutrino bispinor.

The only unresolved and in fact the most nontrivial part in the amplitude of $B \to \ell \bar{\nu}_\ell \gamma$ is the hadronic matrix element in the second line of (6.33) for which we introduce the notation

$$T_{\mu\rho}(p,q) = i \int d^4x e^{ipx} \langle 0|T\{j_\mu^{em}(x)\bar{u}(0)\Gamma_\rho b(0)\}|B^-(p+q)\rangle\,. \quad (6.37)$$

Using (6.34) and (6.36), we write down the amplitude of the photoleptonic B decay in a compact form:

$$\mathcal{A}(B \to \ell \nu_\ell \gamma) = -e\frac{G_F}{\sqrt{2}}V_{ub}\,\varepsilon^{\mu*}\Big(\bar{u}(p_\ell)\Gamma^\rho v(p_\nu)T_{\mu\rho}(p,q) - if_B\bar{u}(p_\ell)\Gamma_\mu v(p_\nu)\Big). \quad (6.38)$$

The above amplitude has to be gauge invariant with respect to the photon field, that is, to vanish after the polarization vector ε_μ is replaced with the photon momentum p_μ. This seems not to be the case for the second part of this amplitude proportional to f_B and describing the photon emission from the final-state charged lepton. In fact, everything is correct, because a compensating, so-called contact term emerges from the first part of the amplitude that describes the photon emission off the B meson. In physical terms, the situation can be explained in terms of Feynman diagrams in QED. A set of diagrams is complete and gauge invariant if the photon is attached to all lines of electrically charged particles. In $B \to \ell \nu_\ell \gamma$, electric charge is transferred from the initial B-meson, via W boson, to a charged lepton; hence, it is the sum of all three photon emission diagrams that has to be gauge invariant.

Instead of calculating separate diagrams, here we use a more general procedure [57] to prove gauge invariance and to define the proper decomposition of the amplitude (6.38). The key element is the *Ward identity* for the e.m. current. To derive this identity, we multiply (6.37) with the photon momentum:

$$p^\mu T_{\mu\rho}(p,q) = i \int d^4x\, p^\mu e^{ipx} \langle 0|T\{j_\mu^{em}(x)J_\rho(0)\}|B^-(p+q)\rangle$$

$$= \int d^4x\, \partial^\mu\big(e^{ipx}\big) \langle 0|T\{j_\mu^{em}(x)J_\rho(0)\}|B^-(p+q)\rangle$$

$$= -\int d^4x\, e^{ipx} \langle 0|\partial^\mu\big(T\{j_\mu^{em}(x)J_\rho(0)\}\big)|B^-(p+q)\rangle\,, \quad (6.39)$$

where a short-hand notation $J_\rho = \bar{u}\Gamma_\rho b$ is introduced. In the above, the integration by parts was employed in which the surface terms at $x \to \pm\infty$ contain infinitely fast oscillating exponents and vanish. At first sight, the whole expression on the r.h.s. of (6.39) vanishes because the e.m. current is conserved and $\partial^\mu j_\mu^{em}(x) = 0$. In fact, there is an additional nonvanishing piece, stemming from the differentiation of the time-ordering operator. Representing the latter in a form of the step-function $\theta(x_0)$ of the time variable x_0, we transform the integrated T-product of current operators in (6.39):

$$- \int d^4x \, e^{ipx} \, \partial^\mu \big(T\{j_\mu^{em}(x)J_\rho(0)\} \big)$$

$$= - \int d^4x \, e^{ipx} \, \partial^\mu \big(\theta(x_0)j_\mu^{em}(x)J_\rho(0) + \theta(-x_0)J_\rho(0)j_\mu^{em}(x) \big)$$

$$= - \int dx_0 \int d\vec{x} \, e^{ipx} \, \delta(x_0) \big[j_0^{em}(x_0,\vec{x}), J_\rho(0,\vec{0}) \big]$$

$$= - \int d\vec{x} \, e^{-i\vec{p}\vec{x}} \big[j_0^{em}(0,\vec{x}), J_\rho(0,\vec{0}) \big], \tag{6.40}$$

where the derivative of the θ-function is used:

$$\partial^\mu \theta(x_0) = \delta^{0\mu}\delta(x_0),$$

and the sum of two terms with a different time ordering forms an equal-time commutator of two currents, so that:

$$p^\mu T_{\mu\rho}(p,q) = - \int d\vec{x} \, e^{-i\vec{p}\vec{x}} \, \langle 0| \big[j_0^{em}(0,\vec{x}), J_\rho(0,\vec{0}) \big] |B^-(p+q)\rangle. \tag{6.41}$$

Note that we only need the commutator with the axial-vector current:

$$\big[j_0^{em}(0,\vec{x}), \bar{u}(0,\vec{0})\gamma_\rho\gamma_5 b(0,\vec{0}) \big] = \delta^{(3)}(\vec{x})\bar{u}(0,\vec{x})\gamma_\rho\gamma_5 b(0,\vec{x}), \tag{6.42}$$

because the vector part of the weak current, being commuted with the e.m. current produces again a vector current which does not fit the B-meson spin-parity. The relation (6.42) is a part of the *current algebra*, a closed system of equal time commutators derived from the canonical anticommutation relations for the free quark-field operators (see (A.73)). Due to the presence of the delta-function in (6.42), the commutator vanishes outside a short distance interval; hence, we may argue that, due to asymptotic freedom of QCD, the current algebra is not influenced by the quark-gluon interactions.

Substituting (6.42) in (6.41) and integrating out the δ-function, we finally obtain the Ward identity:

$$p^\mu T_{\mu\rho}(p,q) = \langle 0|\bar{u}(0)\gamma_\rho\gamma_5 b(0)|B^-(p+q)\rangle = if_B(p+q)_\rho. \tag{6.43}$$

Returning to the amplitude (6.38) we can now confirm its gauge invariance. Replacing $\varepsilon^\mu \to p_\mu$ and using the Ward identity, we obtain:

$$p^\mu \big(\bar{u}(p_\ell)\Gamma^\rho v(p_\nu)T_{\mu\rho}(p,q) - if_B\bar{u}(p_\ell)\Gamma_\mu v(p_\nu) \big)$$

$$= \bar{u}(p_\ell)\Gamma^\rho v(p_\nu)if_B(p+q)_\rho - if_B p^\mu \bar{u}(p_\ell)\Gamma_\mu v(p_\nu) = 0, \tag{6.44}$$

where in the first term in the second line Dirac equation for massless leptons is taken into account.

To proceed, we write down a generic Lorentz-decomposition of the hadronic matrix element (6.37):

$$T_{\mu\rho}(p,q) = g_{\mu\rho}T^{(g)} + q_\mu p_\rho T^{(qp)} + p_\mu q_\rho T^{(pq)} + q_\mu q_\rho T^{(qq)} + p_\mu p_\rho T^{(pp)} + \epsilon_{\mu\rho\lambda\sigma}p^\lambda q^\sigma F_V, \tag{6.45}$$

consisting of all possible tensors built from the two independent four-vectors p and q, each one multiplied by an invariant amplitude which depends on q^2. All other kinematical invariants are fixed: $p^2 = 0$, $p \cdot q = (m_B^2 - q^2)/2$. Multiplying both parts of this decomposition by p_μ and applying the Ward identity (6.43), we obtain

$$p^\mu T_{\mu\rho}(p, q) = i f_B(p_\rho + q_\rho) = p_\rho T^{(g)} + (qp)p_\rho T^{(qp)} + (qp)q_\rho T^{(qq)}. \tag{6.46}$$

Comparing the coefficients at p_ρ and q_ρ on both sides of the above equality, we find that

$$T^{(g)} + (qp)T^{(qp)} = i f_B, \quad (qp)T^{(qq)} = i f_B, \tag{6.47}$$

fixing the amplitude $T^{(qq)}$. Concerning the two others, entering the first relation above, we can always redefine them:

$$T^{(g)} = (qp)A + i(qp)F_A, \quad T^{(qp)} = B - iF_A$$

with the condition that the additional amplitudes A and B fulfil the relation

$$A + B = i\frac{f_B}{(qp)}. \tag{6.48}$$

After that, we rewrite the decomposition (6.45) in terms of redefined amplitudes:

$$T_{\mu\rho}(p, q) = \left(g_{\mu\rho}(qp) - q_\mu p_\rho\right)iF_A + g_{\mu\rho}(qp)A + q_\mu p_\rho B + p_\mu q_\rho T^{(qp)}$$

$$+ i\frac{q_\mu q_\rho}{(qp)}f_B + p_\mu p_\rho T^{(pp)} + \epsilon_{\mu\rho\lambda\sigma}p^\lambda q^\sigma F_V. \tag{6.49}$$

The terms in this expression containing p_μ are inessential because they vanish in the decay amplitude (6.38), being multiplied by the photon polarization vector. The term proportional to f_B also vanishes in the product with the lepton current (in the adopted massless lepton limit). The terms in $T_{\mu\rho}$ proportional to F_A and F_B are retained in the decay amplitude. Both are gauge invariant, so that, being multiplied by p_μ, they vanish. The terms proportional to A and B are together forming the contact term which, as we have seen, restores gauge invariance with respect to the second term in (6.38) describing the photon emission from the final state. Altogether, there remains a certain freedom in choosing the value of B which would correspond to the freedom of redefining the amplitude F_A. One convenient choice is to choose $B = 0$, hence fixing $A = if_B/(qp)$. This leads to the final form adopted for the hadronic matrix element

$$T_{\mu\rho}(p, q) = \left(g_{\mu\rho}(qp) - q_\mu p_\rho\right)iF_A + \epsilon_{\mu\rho\lambda\sigma}p^\lambda q^\sigma F_V + g_{\mu\rho}if_B + \dots, \tag{6.50}$$

where only the relevant part is shown. Accordingly, the complete decay amplitude is

$$\mathcal{A}(B \to \ell\nu_\ell\gamma) = -e\frac{G_F}{\sqrt{2}}V_{ub}\left(\bar{u}(p_\ell)\Gamma^\rho v(p_\nu)\left[\left(\varepsilon_\rho(qp) - (q\varepsilon^*)p_\rho\right)iF_A(q^2)\right.\right.$$

$$\left.\left. + \epsilon_{\rho\mu\sigma\lambda}\varepsilon^{\mu*}q^\sigma p^\lambda F_V(q^2) + i\varepsilon_\rho^* f_B\right] - i\bar{u}(p_\ell)\Gamma^\rho v(p_\nu)\varepsilon_\rho^* f_B\right), \tag{6.51}$$

where the part in brackets corresponds to the photon emission from the valence constituents of the B meson and the remaining term to the photon emission off the charged lepton. The latter is explicitly canceled by the contact term in the former. The decay amplitude is gauge invariant and depends only on the two invariant amplitudes $F_{V,A}(q^2)$. We also restored the

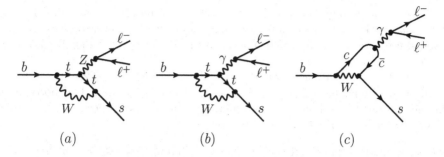

Figure 6.2 Diagrams of the semileptonic FCNC $b \to s\ell^+\ell^-$ transition: (a),(b)-with t-quark loops; (c) with intermediate $c\bar{c}$ pairs.

q^2 dependence of these amplitudes which can be interpreted as a sort of effective form factors of $B \to \gamma$ transition. However, in contrast to the "ordinary" form factors of the local quark currents, these are nonlocal objects because the photon emission and weak transition occur at a certain average distance.

Having the decomposition of the $B \to \ell\bar{\nu}_\ell\gamma$ decay amplitude in terms of form factors, it is straightforward to derive the decay probability. Basically, the kinematics in this process is the same as in a semileptonic decay $B \to h\ell\bar{\nu}_\ell$ where the lepton pair has an invariant mass q^2 and the mass of h vanishes. For the photoleptonic decay it is more convenient to use the photon energy in the B-meson rest frame instead of q^2:

$$E_\gamma = \frac{m_B^2 - q^2}{2m_B}, \quad 0 < E_\gamma < \frac{m_B}{2},$$

so that a small invariant mass of the lepton pair corresponds to a large photon energy (large recoil) and vice versa. We quote the expression for the differential width obtained squaring the amplitude (6.51):

$$\frac{d\Gamma}{dE_\gamma} = \frac{\alpha_{em}G_F^2|V_{ub}|^2}{6\pi^2}m_B\left(1 - \frac{2E_\gamma}{m_B}\right)E_\gamma^3\left(|F_A|^2 + |F_V|^2\right), \tag{6.52}$$

which tells us that the region of small photon recoil is kinematically suppressed. The general form of the amplitude (6.51) in terms of form factors does not depend on the flavor of the decaying hadron and can be used for other photoleptonic decays.

6.3   THE ANATOMY OF $B \to K^{(*)}\ell^+\ell^-$ AMPLITUDES

An overlap of weak and e.m. interactions generates nonlocal hadronic effects in the FCNC $b \to s\ell^+\ell^-$ decays, such as $B \to K\ell^+\ell^-$ and $B \to K^*\ell^+\ell^-$. In the decay amplitudes, these effects form an important background for the dominant contributions originating from the short-distance loops with virtual t-quarks and W, Z bosons. In Figure 6.2, three typical diagrams of the $b \to s\ell^+\ell^-$ transition are shown. The first two of them contain loops with the t-quark. Since the average distances of t-quark propagation are much smaller than any other distance in this process, e.g., $1/m_t \ll 1/m_b \ll 1/\Lambda_{QCD}$, it is possible to approximate the t-quark contributions as a pointlike $\bar{s}b\bar{\ell}\ell$ ($\bar{s}b\gamma$) vertex in the first (second) diagram. These vertices are described by the effective FCNC interaction (1.44) with the local operators $O_{9,10}$ ($O_{7\gamma}$) given in (1.43). The effective operators are multiplied by the numerical coefficients $C_{9,10}$ ($C_{7\gamma}$) obtained calculating the loop diagrams involving t, W, Z.

In the third diagram, Figure 6.2(c), the $b \to s\ell^+\ell^-$ transition is generated in a different way, as a combination of an "ordinary" weak and e.m. interactions. The weak interaction is in this case a nonleptonic transition $b \to c\bar{c}s$. The intermediate $c\bar{c}$ state in this diagram annihilates via quark e.m. current into a virtual photon which then produces a lepton pair in the final state. This mechanism known as the *charm-loop effect* will be discussed in more detail below.

To describe the weak interaction part of the diagram in Figure 6.2(c), we use an effective Hamiltonian similar to (1.29) for semileptonic decays. The virtual W-boson exchange between the two quark weak currents is replaced by four-quark operators. The dominant operators contributing to the $b \to c\bar{c}s$ transition are

$$O_1(x) = \bar{s}_i(x)\Gamma_\rho c^i(x)\bar{c}_k(x)\Gamma^\rho b^k(x), \quad O_2(x) = \bar{c}_i(x)\Gamma_\rho c^i(x)\bar{s}_k(x)\Gamma^\rho b^k(x),$$

with $\Gamma_\rho = \gamma_\rho(1-\gamma_5)$ and with color indices shown explicitly, so that the effective nonleptonic Hamiltonian is

$$\mathcal{H}_W^{b\to c\bar{c}s}(x) = \frac{G_F}{\sqrt{2}}V_{cb}V_{cs}^*\big(C_1 O_1(x) + C_2 O_2(x)\big). \tag{6.53}$$

In the operator O_1 we recognize the product of the quark weak currents from (1.31) and (1.30), whereas the operator O_2 emerges due to QCD effects. In a pure electroweak theory, the weak Hamiltonian is given by (6.53) with $C_1 = 1.0$ and $C_2 = 0$. Perturbative QCD corrections generated by the gluons with large momenta (typically, between m_W and m_b) exchanged between quarks in the weak vertex, renormalize this coefficient to a certain value $C_1 \neq 1.0$ and, simultaneously, generate a different color structure of four quarks that can be written in a form of the operator O_2 with the coefficient $C_2 \neq 0$. Further details including the resummation of gluon corrections and the renormalization scale dependence of the Wilson coefficients are beyond our scope and can be found e.g., in the review [14]; hence, here we do not show the scale-dependence of these coefficients and operators, assuming that they are normalized at a characteristic scale $\mu = m_b$.

To summarize, in order to include all contributions to the $b \to s\ell^+\ell^-$ transition, exemplified by the diagrams in Figure 6.2, one needs to extend the effective Hamiltonian (1.44) adding also the nonleptonic four-quark operators:

$$\mathcal{H}_{eff}(x) = -\frac{G_F}{\sqrt{2}}V_{tb}V_{ts}^*\bigg(\sum_{i=1,2} C_i O_i(x) + \sum_{i=7\gamma,9,10} C_i O_i(x) + \dots\bigg). \tag{6.54}$$

Indicated by ellipsis are additional operators, the so-called QCD penguins[2]. They include the four-quark operators $O_{3,4,5,6}$ which originate from the short-distance diagrams similar to the one in Figure 6.2(a), where the lepton, antilepton and Z lines are replaced, respectively, by the quark, antiquark and gluon lines. In addition, one has to add the operator O_{8g} presented in (1.45) which is the gluon analog of $O_{7\gamma}$. The effective operators O_{3-6} and O_{8g} have small Wilson coefficients and induce minor contributions. Moreover, their nonlocal contributions have the same structure as the ones of $O_{1,2}$, hence, for simplicity, we ignore the QCD penguin operators in the further discussion. We also neglect the $b \to u\bar{u}s$ weak transitions and the corresponding operators obtained from (6.53) replacing $c \to u$. Their contributions are heavily suppressed by the combination $V_{ub}V_{us}^* \sim \lambda^4$ of the CKM matrix elements (1.33); hence, we can safely use the approximation $V_{cb}V_{cs}^* = -V_{tb}V_{ts}^*$ following from the unitarity of the CKM matrix (1.32). This explains the CKM coefficients multiplying $O_{1,2}$ in (6.54). To compare various contributions, we quote the values of the Wilson coefficients at $\mu = m_b$:

$$C_1(m_b) = 1.117, \quad C_2(m_b) = -0.267,$$
$$C_{7\gamma}(m_b) = -0.319, \quad C_9(m_b) = 4.228, \quad C_{10}(m_b) = -4.410. \tag{6.55}$$

---

[2] A detailed description of all effective operators can be found e.g., in [14].

It is straightforward to write down the parts of the $B \to K^{(*)}\ell^+\ell^-$ decay amplitude generated by the local operators $O_{9,10}$. One has to insert the corresponding term of the effective Hamiltonian (6.54) between the initial B-meson and final $K^{(*)}\ell^+\ell^-$ states. Let us first consider the decay with a kaon in the final state, taking for definiteness the mode

$$B^-(p_1) \to K^-(p_2)\ell^+(k_1)\ell^-(k_2),$$

where $q = k_1 + k_2 = p_1 - p_2$ is the momentum transfer to the lepton pair. Note that, due to a sufficiently large phase space, the decays with all lepton flavors $\ell = e, \mu, \tau$ are possible. To derive the decay amplitude, we start from the S-matrix element in which only the part of the effective Hamiltonian (6.54) with the operator O_9 is included:

$$S_{fi}^{(B \to K\ell^+\ell^-;O_9)} = \langle K^-(p_2)\ell^+(k_1)\ell^-(k_2)| -i\int d^4x \mathcal{H}_{eff}^{(O_9)}(x)|B^-(p_1)\rangle,$$

$$= \frac{iG_F}{\sqrt{2}}\frac{\alpha_{em}}{2\pi}V_{tb}V_{ts}^*C_9\int d^4x \langle K^-(p_2)\ell^+(k_1)\ell^-(k_2)|\bar{s}(x)\gamma_\rho(1-\gamma_5)b(x)\bar{\ell}(x)\gamma^\rho\ell(x)|B^-(p_1)\rangle.$$

$$(6.56)$$

Factorizing the lepton fields, using their momentum decomposition, performing the quark current translation and separating the overall factor with $\delta^{(4)}(p_1 - p_2 - k_1 - k_2)$, as we have done several times before, we obtain the contribution of the operator O_9 to the $B \to K\ell^+\ell^-$ decay amplitude:

$$\begin{aligned} \mathcal{A}^{(O_9)}(B \to K\ell^+\ell^-) &= \frac{G_F}{\sqrt{2}}\frac{\alpha_{em}}{2\pi}V_{tb}V_{ts}^*C_9\bar{u}_\ell(k_2)\gamma^\rho v_\ell(k_1)\langle K^-(p_2)|\bar{s}\gamma_\rho b|B^-(p_1)\rangle \\ &= \frac{G_F}{\sqrt{2}}\frac{\alpha_{em}}{2\pi}V_{tb}V_{ts}^*C_9\bar{u}_\ell(k_2)\gamma_\rho v_\ell(k_1)\,p_2^\rho f_{BK}^+(q^2), \end{aligned} \quad (6.57)$$

where we use the definition (2.94) for the $B \to K$ form factor of the vector $b \to s$ current. The terms proportional to the momentum transfer q vanish due to the lepton current conservation. A similar answer is obtained for the O_{10} contribution, the only difference being an extra γ_5 in the lepton current:

$$\mathcal{A}^{(O_{10})}(B \to K\ell^+\ell^-) = \frac{G_F\alpha_{em}}{\sqrt{2}2\pi}V_{tb}V_{ts}^*C_{10}\bar{u}_\ell(k_2)\gamma_\rho\gamma_5 v_\ell(k_1)\,p_2^\rho f_{BK}^+(q^2), \quad (6.58)$$

if one neglects the lepton masses.

The operator $O_{7\gamma}$ triggers a semileptonic decay due to an additional e.m. interaction of the lepton pair with photon. The corresponding S-matrix element is:

$$\begin{aligned} S_{fi}^{(B \to K\ell^+\ell^-;O_{7\gamma})} &= \langle K^-(p_2)\ell^+(k_1)\ell^-(k_2)|T\Big\{\Big(i\int d^4y\,\mathcal{L}_{QED}^{int}(y)\Big) \\ &\times \Big(-i\int d^4x\,\mathcal{H}_{eff}^{(O_{7\gamma})}(x)\Big)\Big\}|B^-(p_1)\rangle = \frac{G_F}{\sqrt{2}}\frac{e}{8\pi^2}V_{tb}V_{ts}^*C_{7\gamma}\int d^4x\int d^4y \\ &\times \langle K^-(p_2)\ell^+(k_1)\ell^-(k_2)|T\Big\{(-e)A_{em}^\rho(y)\bar{\ell}(y)\gamma_\rho\ell(y) \\ &\times \bar{s}(x)\sigma_{\mu\nu}[m_b(1+\gamma_5) + m_s(1-\gamma_5)]b(x)F^{\mu\nu}(x)\Big\}|B^-(p_1)\rangle. \end{aligned} \quad (6.59)$$

We factorize out the lepton current together with the photon propagator between the effective $O_{7\gamma}$ vertex and the lepton-pair emission point:

$$\begin{aligned} S_{fi}^{(B \to K\ell^+\ell^-;O_{7\gamma})} &= -\frac{G_F}{\sqrt{2}}\frac{e^2}{8\pi^2}V_{tb}V_{ts}^*C_{7\gamma}\int d^4x\int d^4y\,\bar{u}_\ell(k_2)\gamma_\rho v_\ell(k_1)e^{i(k_1+k_2)y} \\ &\times \langle 0|T\{A_{em}^\rho(y)F^{\mu\nu}(x)\}|0\rangle(m_b + m_s)\langle K^-(p_2)|\bar{s}(x)\sigma_{\mu\nu}b(x)|B^-(p_1)\rangle. \end{aligned} \quad (6.60)$$

Here we take into account that only the tensor current without γ_5 contributes to the $B \to K$ matrix element. Differentiating the photon propagator in the momentum space, we obtain its modification involving the e.m. field strength tensor:

$$\langle 0|T\{A_{em}^\rho(y)F^{\mu\nu}(x)\}|0\rangle = \int \frac{d^4k}{(2\pi)^4} e^{-ik(y-x)} \frac{k^\mu g^{\nu\rho} - k^\nu g^{\mu\rho}}{k^2}. \qquad (6.61)$$

Substituting this expression in (6.60) and translating the current in the hadronic matrix element, we obtain, after contracting the indices,

$$S_{fi}^{(B \to K\ell^+\ell^-;O_{7\gamma})} = -\frac{G_F}{\sqrt{2}} \frac{e^2}{8\pi^2} V_{tb}V_{ts}^* C_{7\gamma} \int \frac{d^4k}{(2\pi)^4} \int d^4x\, e^{i(p_2-p_1+k)x} \int d^4y\, e^{i(k_1+k_2-k)y}$$

$$\times \frac{-2k^\nu}{k^2} \bar{u}_\ell(k_2)\gamma^\mu v_\ell(k_1)(m_b + m_s)\langle K^-(p_2)|\bar{s}\sigma_{\mu\nu}b|B^-(p_1)\rangle. \qquad (6.62)$$

One of the δ-functions appearing after the integration over coordinates, being integrated out with the momentum k, yields the replacement $k \to q$, whereas the second one is reduced to $\delta^{(4)}(p_1 - p_2 - k_1 - k_2)$. Separating this factor according to our usual convention, we obtain the decay amplitude

$$\mathcal{A}^{(O_{7\gamma})}(B \to K\ell^+\ell^-) = -i\frac{G_F}{\sqrt{2}} \frac{\alpha_{em}}{\pi} V_{tb}V_{ts}^* C_{7\gamma}$$

$$\times \frac{q^\nu}{q^2} \bar{u}_\ell(k_2)\gamma^\mu v_\ell(k_1)(m_b + m_s)\langle K^-(p_2)|\bar{s}\sigma_{\mu\nu}b|B^-(p_1)\rangle$$

$$= \frac{G_F}{\sqrt{2}} \frac{\alpha_{em}}{\pi} V_{tb}V_{ts}^* C_{7\gamma} \frac{2(m_b + m_s)}{m_B + m_K} \bar{u}_\ell(k_2)\gamma^\mu v_\ell(k_1)p_{2\mu}f_{BK}^T(q^2), \qquad (6.63)$$

where we use the $B \to K$ tensor form factor defined in (2.119) in which the part proportional to q_μ again vanishes due to the lepton current conservation. Note that the propagator $1/q^2$ (the photon pole) is compensated by the factor q^2 emerging in the decomposition of the hadronic matrix element. Adding together the contributions (6.57), (6.58) and (6.63) of all three operators, we obtain

$$\mathcal{A}^{(O_{9,10,7\gamma})}(B \to K\ell^+\ell^-) = \frac{G_F}{\sqrt{2}} \frac{\alpha_{em}}{\pi} V_{tb}V_{ts}^* \left[\bar{u}_\ell(k_2)\gamma_\mu v_\ell(k_1)p_2^\mu \left(C_9 f_{BK}^+(q^2) \right.\right.$$

$$\left.\left. + \frac{2(m_b + m_s)}{m_B + m_K} C_{7\gamma}f_{BK}^T(q^2) \right) + \bar{u}_\ell(k_2)\gamma_\mu\gamma_5 v_\ell(k_1)p_2^\mu C_{10}f_{BK}^+(q^2) \right], \qquad (6.64)$$

with a hadronic part fully reduced to the $B \to K$ form factors of the local quark currents.

We have to complete the expression (6.64) with nonlocal hadronic contributions of the operators $O_{1,2}$. To this end, we combine the weak nonleptonic Hamiltonian (6.53) with the second-order e.m. interaction, including the quark and lepton currents and the photon propagator. The initial S-matrix element is:

$$S_{fi}^{(B \to K\ell^+\ell^-;O_{1,2})} = \langle K^-(p_2)\ell^+(k_1)\ell^-(k_2)|T\left\{ \frac{1}{2!}\left(i\int d^4x' \mathcal{L}_{QED}^{int}(x') \right) \right.$$

$$\times \left(i\int d^4y \mathcal{L}_{QED}^{int}(y) \right)\left(-i\int d^4z \mathcal{H}_W^{b\to c\bar{c}s}(z) \right)\left.\right\}|B^-(p_1)\rangle$$

$$= -\frac{ie^2}{2!} \frac{G_F}{\sqrt{2}} V_{tb}V_{ts}^* \int d^4z \int d^4y \int d^4x' \langle K^-(p_2)|\ell^+(k_1)\ell^-(k_2)|T\left\{ \left(j_\mu^{em}(x') - \bar{\ell}(x')\gamma_\mu\ell(x') \right)A^\mu(x') \right.$$

$$\times \left(j_\nu^{em}(y) - \bar{\ell}(y)\gamma_\nu\ell(y) \right)A^\nu(y)\overline{O}_{12}(z)\left.\right\}|B^-(p_1)\rangle, \qquad (6.65)$$

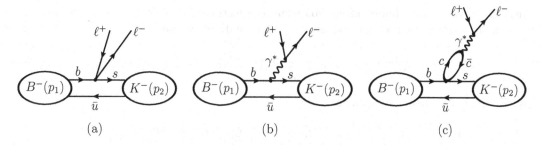

Figure 6.3 Diagrams of the $B \to K\ell\ell$ decays: (a) and (b) contributions of the effective operator $O_{9,10}$ and $O_{7\gamma}$; (c) nonlocal contribution with c-quark loop.

where the part $Q_c \bar{c} \gamma_\mu c$ of the quark e.m. contributes[3] and we introduce a compact notation for the linear combination of the effective operators:

$$\overline{O}_{12}(x) \equiv \sum_{i=1,2} C_i O_i(x) \,.$$

Combining the quark and lepton e.m. currents, we encounter two equal contributions to (6.65), hence, we retain one of them, removing the factor $1/2$. The next step is to factorize the lepton current and to contract the photon fields into a propagator:

$$S_{fi}^{(B \to K\ell^+\ell^-; O_{1,2})} = -ie^2 \frac{G_F}{\sqrt{2}} V_{tb} V_{ts}^* \, \bar{u}_\ell(k_2) \gamma_\mu v_\ell(k_1) \int d^4x' \int d^4y \, e^{i(k_1+k_2)y}$$

$$\times i \int d^4z \int \frac{d^4k}{(2\pi)^4} \frac{g^{\mu\nu}}{k^2} e^{-ik(x'-y)} \langle K^-(p_2) | T\{ j_\nu^{em}(x') \overline{O}_{12}(z) \} | B^-(p_1) \rangle$$

$$= -ie^2 \frac{G_F}{\sqrt{2}} V_{tb} V_{ts}^* \, \bar{u}_\ell(k_2) \gamma_\mu v_\ell(k_1) \int \frac{d^4k}{k^2} \delta^{(4)}(k_1 + k_2 + k) I^\mu(k, p_1, p_2) \,. \tag{6.66}$$

In the above, we converted the y-integration into a delta-function and introduced a compact notation for the hadronic matrix element integrated over the two remaining coordinates:

$$I_\mu(k, p_1, p_2) = i \int d^4z \int d^4x' e^{-ikx'} \langle K^-(p_2) | T\{ j_\nu^{em}(x') \overline{O}_{12}(z) \} | B^-(p_1) \rangle \,. \tag{6.67}$$

We transform this expression, inserting the exponents of momentum operators, and translating the coordinates with the help of (A.78):

$$I_\mu(k, p_1, p_2) = i \int d^4z \, e^{-ip_1 z + ip_2 z} \int d^4x' e^{-ikx'} \langle K^-(p_2) | T\{ e^{-i\hat{P}z} j_\nu^{em}(x') e^{i\hat{P}z}$$

$$\times e^{-i\hat{P}z} \overline{O}_{12}(z) e^{i\hat{P}z} \} | B^-(p_1) \rangle$$

$$= i \int d^4z \, e^{-ip_1 z + ip_2 z} \int d^4x' e^{-ikx'} \langle K^-(p_2) | T\{ j_\nu^{em}(x' - z) \overline{O}_{12}(0) \} | B^-(p_1) \rangle$$

$$= (2\pi)^4 \delta^{(4)}(-p_1 + p_2 - k) i \int d^4x \, e^{-ikx} \langle K^-(p_2) | T\{ j_\nu^{em}(x) \overline{O}_{12}(0) \} | B^-(p_1) \rangle, \tag{6.68}$$

where at the last step, the variable was transformed, $x' = x + z$, and the z integration produced a δ-function. Substituting (6.68) in (6.66), we integrate with this function over k,

---

[3]For the sake of generality, we retain the whole e.m. current, keeping in mind that the resulting expressions are also valid for the contributions of subleading operators $O_{3-6,8g}$ with different flavors.

resulting in $k = -q$. Finally, after separating the factor $i(2\pi)^4\delta^{(4)}(k_1 + k_2 + p_2 - p_1)$, we obtain the contribution of the operators $O_{1,2}$ to the decay amplitude:

$$\mathcal{A}^{(O_{1,2})}(B \to K\ell^+\ell^-) = -4\pi\alpha_{em}\frac{G_F}{\sqrt{2}}V_{tb}V_{ts}^*\frac{\bar{u}_\ell(k_2)\gamma_\mu v_\ell(k_1)}{q^2}\mathcal{H}_{BK}^\mu(q, p_2)\,, \tag{6.69}$$

where the hadronic matrix element

$$\mathcal{H}_{BK}^\mu(q, p_2) = i\int d^4x\, e^{iq\cdot x}\langle K(p_2)|T\{j_\mu^{em}(x), \overline{O}_{12}(0)\}|B(p_2 + q)\rangle$$

$$= [(p_2 \cdot q)q^\mu - q^2 p_2^\mu]\mathcal{H}_{BK}(q^2) \tag{6.70}$$

describes the nonlocal effect of the charmed quark-antiquark pair emitted from the weak vertex and annihilated via photon into a lepton pair. The above Lorentz decomposition takes into account the e.m. current conservation.

Adding (6.69) to (6.64), we complete the decay amplitude taking into account the non-local effects (see the diagrams in Figure 6.3):

$$\mathcal{A}(B \to K\ell^+\ell^-) = \frac{G_F}{\sqrt{2}}\frac{\alpha_{em}}{\pi}V_{tb}V_{ts}^*\left[\bar{u}_\ell(k_2)\gamma_\mu v_\ell(k_1)p_2^\mu\left(C_9 f_{BK}^+(q^2)\right.\right.$$

$$\left.\left. + \frac{2(m_b + m_s)}{m_B + m_K}C_{7\gamma}f_{BK}^T(q^2) + 4\pi^2\mathcal{H}_{BK}(q^2)\right) + \bar{u}_\ell(k_2)\gamma^\mu\gamma_5 v_\ell(k_1)p_2^\mu C_{10}f_{BK}^+(q^2)\right]. \tag{6.71}$$

Note that in the nonlocal contribution, similar to the $O_{7\gamma}$ one, the photon propagator $1/q^2$ is canceled by the factor q^2, reflecting the longitudinal helicity of the virtual photon in the $B \to K\ell\ell$ decay.

In contrast to f_{BK}^+ and f_{BK}^T, the invariant function $\mathcal{H}_{BK}(q^2)$ in (6.71) is not a form factor in a conventional sense, because it contains a product of two operators separated by a certain average interval $\langle x^2\rangle$. Depending on q^2, this interval is spacelike or timelike. E.g., when q^2 approaches the mass of the charmonium resonances, starting from the lowest one , J/ψ, the amplitude (6.69) describes a real physical process: the weak nonleptonic $B \to J/\psi K$ decay followed by an e.m. decay $J/\psi \to \ell^+\ell^-$. This process represents a prominent background for the FCNC interactions. Therefore, the interval around charmonium resonances is usually isolated in the experimental studies and the region $q^2 \ll 4m_c^2$, far from the c-quark pair production threshold, is preferred. In fact, the amplitude $\mathcal{H}_{BK}(q^2)$ is not vanishing in the low q^2 region, because a virtual c-quark loop still contributes to the decay amplitude. A useful tool to describe this effect is the dispersion relation for the amplitude $\mathcal{H}_{BK}(q^2)$ in the q^2 variable. More details can be found in [58, 59].

In conclusion, we present the analogous expressions for the amplitude of $B \to K^*\ell\ell$ decay, valid in the narrow K^* approximation. The procedure of identifying the contributions stemming from various operators in this decay is very similar to the one used above. The only essential difference is the proliferation of form factors and, correspondingly, of nonlocal invariant amplitudes, due to the presence of additional degrees of freedom, namely, the polarization of the vector meson K^*.

For $B \to K^* \ell^+ \ell^-$ decay amplitude the following expression is obtained:

$$
\begin{aligned}
A(B \to K^* l^+ l^-) &= \frac{G_F}{2\sqrt{2}} \frac{\alpha_{em}}{\pi} V_{tb} V_{ts}^* \Bigg[\bar{u}_\ell(k_2) \gamma^\mu v_\ell(k_1) \Big(\epsilon_{\mu\nu\rho\sigma} \varepsilon^{*\nu} q^\rho p^\sigma \mathcal{M}_1(q^2) \\
&\quad - i\varepsilon_\mu^* \mathcal{M}_2(q^2) + i(\varepsilon^* \cdot q) p_\mu \mathcal{M}_3(q^2) - \frac{8\pi^2}{q^2} \mathcal{H}_\mu(p,q) \Big) \\
&\quad + \bar{u}_\ell(k_2) \gamma^\mu \gamma_5 v_\ell(k_1) \Big(\epsilon_{\mu\nu\rho\sigma} \varepsilon^{*\nu} q^\rho p^\sigma \mathcal{N}_1(q^2) - i\varepsilon_\mu^* \mathcal{N}_2(q^2) \\
&\quad + i(\varepsilon^* \cdot q) p_\mu \mathcal{N}_3(q^2) \Big) \Bigg],
\end{aligned}
\tag{6.72}
$$

where the invariant amplitudes $\mathcal{M}_{1,2,3}(\mathcal{N}_{1,2,3})$ contain the contributions of the effective operators O_9 (O_{10}) reduced to combinations of the $B \to K^*$ form factors defined according to (2.136) and (2.142), and ε is the polarization vector of the K^*-meson. The expressions for these invariant amplitudes are:

$$
\begin{aligned}
\mathcal{M}_1(q^2) &= C_9 \frac{2 V^{BK^*}(q^2)}{m_B + m_{K^*}} + 4 C_7^{eff} \frac{m_b + m_s}{q^2} T_1^{BK^*}(q^2), \\
\mathcal{M}_2(q^2) &= C_9 (m_B + m_{K^*}) A_1^{BK^*}(q^2) \\
&\quad + 2 C_7^{eff} (m_B^2 - m_{K^*}^2) \frac{m_b + m_s}{q^2} T_2^{BK^*}(q^2), \\
\mathcal{M}_3(q^2) &= 2 C_9 \frac{A_2^{BK^*}(q^2)}{m_B + m_{K^*}} \\
&\quad + 4 C_7^{eff} \frac{m_b - m_s}{q^2} \left(T_2^{BK^*}(q^2) + \frac{q^2}{m_B^2 - m_{K^*}^2} T_3^{BK^*}(q^2) \right),
\end{aligned}
\tag{6.73}
$$

and

$$
\begin{aligned}
\mathcal{N}_1(q^2) &= 2 C_{10} \frac{V^{BK^*}(q^2)}{m_B + m_{K^*}}, \qquad \mathcal{N}_2(q^2) = C_{10} (m_B + m_{K^*}) A_1^{BK^*}(q^2), \\
\mathcal{N}_3(q^2) &= 2 C_{10} \frac{A_2^{BK^*}(q^2)}{m_B + m_{K^*}}.
\end{aligned}
\tag{6.74}
$$

Finally, the term $\mathcal{H}_\mu$ decomposed in three invariant amplitudes:

$$
\begin{aligned}
\mathcal{H}^\mu(p,q) &= \epsilon^{\mu\alpha\beta\gamma} \epsilon_\alpha^* q^\beta p^\gamma \mathcal{H}_1(q^2) + i[(m_B^2 - m_{K^*}^2) \epsilon^{*\mu} - (\epsilon^* \cdot q)(2p+q)_\mu] \mathcal{H}_2(q^2) \\
&\quad + i(\epsilon^* \cdot q) \left[q_\mu - \frac{q^2}{m_B^2 - m_{K^*}^2} (2p+q)_\mu \right] \mathcal{H}_3(q^2),
\end{aligned}
\tag{6.75}
$$

parameterizes the contribution of nonlocal effects of operators $O_{1,2}$ to this decay amplitude.

CHAPTER 7

FORM FACTOR ASYMPTOTICS

7.1 CAN WE CALCULATE HADRON FORM FACTORS IN QCD ?

The variety of hadron form factors considered in previous chapters can be defined in the following general form:

$$\langle h_f(p)|\bar{q}_1\Gamma_A q_2|h_i(p+q)\rangle = L_A(p,q)F(q^2)\,, \qquad (7.1)$$

where on l.h.s the matrix element of a local quark-current operator $\bar{q}_1\Gamma_A q_2$ is sandwiched between the initial and final hadronic states with four-momenta $p+q$ and p. Denoted with Γ_A is a certain combination of Dirac matrices. On the r.h.s. the form factor $F(q^2)$ is multiplied by the Lorentz structure L_A, determined by the spin-parities of the current and hadrons. In (7.1), depending on the amount of independent Lorentz structures allowed by symmetries, there could also be additional form factors. If the final state contains more than one hadron, then the form factors, apart from q^2, depend on other kinematical invariants formed by the hadron momenta.

Currently, there is a continuous progress in calculating many of the hadron form factors in the spacelike region, employing numerical simulation of QCD on the lattice. E.g., the results and perspectives on the light-meson form factors can be found in [60]. In essence, the question in the title of this section refers to an analytical calculation in QCD. In the definition (7.1), we deal with a matrix element, which consists of the operator and the physical states. Clearly, the quark current operator in (7.1) is well defined in QCD. The major challenge in calculating the form factor $F(q^2)$ is our very limited ability to describe the hadronic states h_i and h_f in analytical terms.

To understand the origin of these difficulties, let us return to the form factor of the electron-atom interaction discussed in the Introduction and presented there in a similar compact form of the matrix element (18). Importantly, this form factor has another, far more detailed expression (17), where the initial and final states are explicitly described by the atomic wave functions well defined in quantum mechanics. The underlying quantum field theory in this case is QED. In this theory, it is possible to consider an initial state of two isolated electrons and calculate their scattering amplitude in terms of perturbative diagrams, starting from the one-photon exchange. The result of this calculation can be tested, measuring the cross section of the Møller scattering $e^-e^- \to e^-e^-$. Importantly, to describe the electron-atom scattering in QED, we use the nonrelativistic limit for the electrons bound in the atom. In this limit, the photon exchange between the electrons and atomic nuclei is converted into the Coulomb potential. One solves the corresponding Schrödinger equation, and obtains the atomic energy spectrum and wave functions. Applying this well-defined calculational procedure based on QED, one benefits, on the one hand, from a smallness of the coupling α_{em}, and, on the other hand, from a large mass of the electron in comparison with its $O(\alpha_{em} m_e)$ binding energy.

From the above discussion, we conclude that, in order to calculate the form factor (7.1), the wave functions of hadrons h_i and h_f have to be defined in terms of their quark and antiquark constituents, and it should be possible to calculate these wave functions in QCD. It suffices to know their momentum representations, or, in other words, to calculate how the momentum of a hadron is distributed among its constituents.

Needless to say, the task outlined above has no straightforward solution for several reasons. As explained in Chapter 1.7, additional "nonvalence" quark-antiquark pairs and gluons are spontaneously created and annihilated inside hadrons. These quantum effects are nonperturbative and inseparable from the QCD vacuum fluctuations; hence, it is not possible to attribute a fixed number of constituents to a given hadron. The state with the valence quark content is only one of the components of a hadronic state (cf. (1.79)). A related difficulty is that light quarks and antiquarks confined in a hadron are purely relativistic since $m_{u,d} \ll \Lambda_{QCD}$ and $m_s \sim \Lambda_{QCD}$; hence, an interquark potential cannot be defined even for the valence component of a hadron[1]. Simple gluon-exchange diagrams are also not a valid replacement for an interquark potential because hadrons are formed by nonperturbative interactions. We conclude that a standard method of calculating form factors with the use of the bound state wave functions is generally not applicable to hadrons in QCD. Still, an important exception does exist and will be discussed in the next section.

7.2   ASYMPTOTICS OF THE PION FORM FACTOR IN QCD

A systematic description of hadron form factors in QCD is possible at asymptotically large spacelike momentum transfers,

$$Q^2 = -q^2 \to \infty,$$

where the obstacles mentioned in the previous section are, to a large extent, absent. As an important study case, we consider the pion e.m. form factor, repeating its definition in (2.26) in terms of the underlying hadronic matrix element:

$$\langle \pi^+(p_2)|j_\mu^{em}|\pi^+(p_1)\rangle = (p_1 + p_2)_\mu F_\pi(q^2). \tag{7.2}$$

It is convenient to choose a specific reference frame, which is known as the *Breit frame*, sometimes called also the "brick wall" frame. We choose two independent four-vectors

$$p_1^\mu = (p_1^0, \vec{p}_1) \quad \text{and} \quad p_2^\mu = (p_2^0, \vec{p}_2),$$

fixing the momentum transfer $q^\mu = p_2^\mu - p_1^\mu$. With the two on-shellness conditions, $p_1^2 = p_2^2 = m_\pi^2$, we are left with six independent momentum components, to be fixed by the choice of the Lorentz frame and its orientation in three dimensions. In the Breit frame, the following conditions are put on the components:

$$p_1^1 = p_1^2 = 0; \quad p_2^3 = -p_1^3; \quad q^0 = q^1 = q^2 = 0.$$

Hereafter, we also neglect the pion mass and the light u, d quark masses. The adopted conditions define all momenta as functions of the large scale $Q = \sqrt{Q^2}$:

$$p_1^\mu = \left(\frac{Q}{2}, 0, 0, -\frac{Q}{2}\right), \quad p_2^\mu = \left(\frac{Q}{2}, 0, 0, \frac{Q}{2}\right), \quad q = (0, 0, 0, Q). \tag{7.3}$$

---

[1]Heavy quarkonia represent an exception: to a certain degree of accuracy they can be treated as two-body bound states where a confining potential between heavy quark and antiquark is a viable concept. However, the form of this potential is not analytically calculable from QCD.

In the chosen frame, an energetic pion reverses its longitudinal direction after absorbing a virtual photon with a purely longitudinal momentum, so that the outgoing pion has no transverse momentum. In the adopted massless approximation, the pion momenta are lightlike and can be rewritten as:

$$p_1^\mu = \frac{Q}{2}n^\mu, \quad p_2^\mu = \frac{Q}{2}\bar{n}^\mu, \tag{7.4}$$

introducing the two lightlike unit vectors:

$$n^\mu = (1,0,0,-1), \quad \bar{n}^\mu = (1,0,0,1); \quad n^2 = \bar{n}^2 = 0, \quad n \cdot \bar{n} = 2. \tag{7.5}$$

After defining the kinematics, we consider the valence quark-antiquark state of the pion. Assigning the momenta k_1 and k_1' (k_2 and k_2'), respectively, to the quark and antiquark in the initial (final) pion, we specify their three-momentum components

$$\vec{k}_1 = \left(\vec{k}_{1\perp}, -\alpha_1\frac{Q}{2}\right), \quad \vec{k}_1' = \left(-\vec{k}_{1\perp}, -\bar{\alpha}_1\frac{Q}{2}\right),$$

$$\vec{k}_2 = \left(\vec{k}_{2\perp}, \alpha_2\frac{Q}{2}\right), \quad \vec{k}_2' = \left(-\vec{k}_{2\perp}, \bar{\alpha}_2\frac{Q}{2}\right), \tag{7.6}$$

where $\bar{\alpha}_{1(2)} \equiv 1 - \alpha_{1(2)}$. The parameters $\alpha_{1,2}$, varying within the interval

$$0 < \alpha_{1,2} < 1,$$

determine the share of the longitudinal momentum carried by the valence quark and antiquark inside a fast moving pion. The energies of these constituents are

$$E_i = \sqrt{\frac{\alpha_i^2 Q^2}{4} + k_{i\perp}^2}, \quad E_i' = \sqrt{\frac{\bar{\alpha}_i^2 Q^2}{4} + k_{i\perp}^2}, \quad i = 1,2. \tag{7.7}$$

Note that the average transverse momentum $k_\perp = |\vec{k}_\perp|$ of the quark and antiquark inside a pion is of the same $O(\Lambda_{QCD})$ as in the pion rest frame, because the boost to the Breit frame acts in the longitudinal direction.

We consider first a valence state of the initial pion in which both quark and antiquark carry a finite share of the longitudinal momentum:

$$\alpha_1 \sim \bar{\alpha}_1 = O(1), \tag{7.8}$$

and are almost collinear:

$$|\vec{k}_\perp| \ll Q. \tag{7.9}$$

The difference ΔE between the sum of the quark and antiquark energies $E_1 + E_1'$ and the initial pion energy $p_1^0 = Q/2$ determines the virtuality of the valence state considered as a quantum fluctuation. We calculate this difference at $Q \to \infty$, where, due to (7.9), only the first term of the expansion in $k_\perp/Q$ is retained. Employing the energy-time uncertainty relation, we estimate the average duration of the valence state:

$$\Delta\tau \sim \frac{1}{\Delta E} = \frac{1}{E_1 + E_1' - Q/2} = \frac{1}{\sqrt{\frac{\alpha_1^2 Q^2}{4} + k_\perp^2} + \sqrt{\frac{\bar{\alpha}_1^2 Q^2}{4} + k_\perp^2} - Q/2}$$

$$\simeq \frac{\alpha_1\bar{\alpha}_1 Q}{k_T^2} \sim \frac{Q}{k_T^2}, \tag{7.10}$$

revealing a considerable time dilation. At asymptotically large Q, the valence configuration of a fast moving pion has a long lifetime and is the most probable one. All nonvalence states

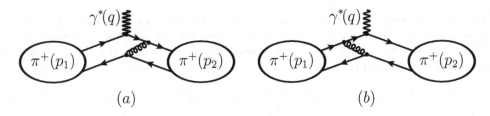

Figure 7.1 Diagrams of the hard scattering contribution to the pion form factor with $O(\alpha_s)$ gluon exchange diagrams.

involving extra soft gluons or quark-antiquark pairs, have a parametrically shorter duration. This can be seen by simply adding a generic soft component of $O(\Lambda_{QCD})$ to the energy of the quark-antiquark state, yielding at $Q \to \infty$:

$$\Delta\tau \sim \frac{1}{\Delta E} \sim \frac{1}{E_1 + E_1' + \Lambda_{QCD} - Q/2} \simeq \frac{1}{\Lambda_{QCD}}, \tag{7.11}$$

where we used the same expansion of E_1, E_1' as in (7.10).

We conclude that a virtual photon with large Q finds a fast moving initial pion predominantly in a state consisting of a quasifree collinear quark and antiquark, and nonperturbative effects have a minor influence during the process of the virtual photon absorption. Note that this process has an asymptotically short duration $\sim 1/Q$ acting as a "snapshot" with respect to the "frozen" quark-antiquark state. Here we should emphasize that, although the photon absorption is a short-distance process with quasifree quark and antiquark involved, the probability amplitude to form a valence configuration inside a pion cannot be simply obtained from a perturbative calculation and is determined by the long-distance structure of the pion.

Using a terminology of the parton model of hadrons[2], we can say that in the asymptotic regime of the photon-pion interaction the initial pion finds itself in a two-parton state. Remember that the parton model is mainly used to describe the processes with a multihadron final state, such as the deep inelastic electron scattering on a hadron or the decay of heavy quarkonium into multiple hadrons which was discussed in Chapter 1. Contrary to that, in the case of the pion form factor there is a single pion in the final state[3]. Moreover, we expect that in the Breit frame a fast moving final pion also finds itself predominantly in a collinear state of two partons, moving in the opposite direction with respect to the initial pion. The only possibility to produce such a state after the virtual photon transfers its whole momentum to a valence quark or antiquark, is to exchange a hard (i.e., perturbative) gluon with a momentum of $O(Q)$ between the struck quark and spectator, so that both partons change their direction. The corresponding diagrams of this so-called *hard-scattering mechanism* are shown in Figure 7.1(a),(b).

Guided by the above qualitative arguments, we calculate these diagrams at very large Q^2 and obtain the dominant contribution to the hadronic matrix element (7.2) of the pion-to-pion e.m. transition. Our main assumption is that the quark and antiquark are the only constituents (partons) of both initial and final pions. The further important simplification is that we neglect the relative transverse momenta of the initial partons with respect to the large longitudinal components. Since the momentum transfer is purely longitudinal, the

---

[2]For an introduction, see, e.g., [61].

[3]In fact, pion-to-pion transition is a rather rare process, more probable is that the struck pion transforms into a multihadron final state.

quark and antiquark in the final state pion also do not acquire a transverse momentum; hence, we can approximate the momenta of the quarks and antiquarks specified in (7.6) and (7.7) as:

$$k_1 = \alpha_1 p_1, \; k_1' = \bar{\alpha}_1 p_1, \quad k_2 = \alpha_2 p_2, \; k_2' = \bar{\alpha}_2 p_2, \tag{7.12}$$

and represent the initial and final pion states in a form of superpositions of free quark-antiquark states with various collinear momenta:

$$|\pi^+(p_1)\rangle = \int_0^1 d\alpha_1 \, \tilde{\phi}_\pi(\alpha_1) |\{u(\alpha_1 p_1)\bar{d}(\bar{\alpha}_1 p_1)\}_\pi\rangle,$$

$$\langle \pi^+(p_2)| = \int_0^1 d\alpha_2 \, \tilde{\phi}_\pi(\alpha_2) \, \langle\{u(\alpha_2 p_2)\bar{d}(\bar{\alpha}_2 p_2)\}_\pi|. \tag{7.13}$$

The amplitude $\tilde{\phi}_\pi$ determines the momentum distribution between two partons in the initial (α_1) or final (α_2) pion. In Chapter 9 we will give a more rigorous, QCD-based definition of the so-called *pion distribution amplitude* (DA), which is closely related to $\tilde{\phi}_\pi$. For the sake of brevity, we will also call $\tilde{\phi}_\pi$ a pion DA in our subsequent discussion. We normalize this amplitude as

$$\int_0^1 d\alpha \, \tilde{\phi}_\pi(\alpha) = 1, \tag{7.14}$$

and the states as

$$\langle\{u(\alpha_2 p_2)\bar{d}(\bar{\alpha}_2 p_2)\}_\pi|\{u(\alpha_1 p_1)\bar{d}(\bar{\alpha}_1 p_1)\}_\pi\rangle = 2p_1^0 (2\pi)^3 \delta^{(3)}(\vec{p}_1 - \vec{p}_2), \tag{7.15}$$

to fulfil the usual normalization condition (1.81) for the pion states. The index π inside the bra and ket vectors in (7.13) indicates that the quark-antiquark state has the same quantum numbers as the pion, in particular, the same spin-parity $J^P = 0^-$.

After the pion states (7.13) are specified, we proceed to the derivation of the hadronic matrix element. We will not repeat the whole procedure with the S-matrix (2.12) of the electron-pion scattering already presented in Chapter 2. The only difference is that now we have to expand the S matrix also in powers of $\mathcal{L}_{QCD}^{int}$. The result will be a series of perturbative QCD diagrams with gluon exchanges between the quarks. The value of the QCD coupling α_s in these diagrams depends on the characteristic energy-momentum in the quark-gluon vertices which is of order of the momentum transfer $Q = \sqrt{Q^2}$. Since we consider large $Q \gg \Lambda_{QCD}$, the corresponding small value of $\alpha_s(Q)$ suppresses higher-order corrections in perturbative QCD and the lowest, $O(\alpha_s)$ diagrams in Figure 7.1(a),(b) provide a reasonable approximation for the hard-scattering mechanism.

To calculate these diagrams, we first take the part of the hadronic matrix element (7.2) where only the u-quark component of the e.m. current contributes, denoting this part as $\mathcal{F}_\mu^{(u)}$. We then insert the quark-gluon interaction at the second order in the coupling g_s:

$$\mathcal{F}_\mu^{(u)} = \langle\pi^+(p_2)|T\left\{\left[ig_s \int d^4x \, \bar{u}(x)\slashed{A}^a(x)\frac{\lambda^a}{2}u(x)\right]\right.$$

$$\left.\times Q_u \bar{u}(0)\gamma_\mu u(0)\left[ig_s \int d^4y \, \bar{d}(y)\slashed{A}^b(y)\frac{\lambda^b}{2}d(y)\right]\right\}|\pi^+(p_1)\rangle, \tag{7.16}$$

where the Dirac and color indices of quark fields are not shown. The next step is to contract the gluon fields into a propagator using (A.54):

$$i\langle 0|T\{A^a_\nu(x)A^b_\rho(y)\}|0\rangle = \delta^{ab}\int\frac{d^4k}{(2\pi)^4}e^{-ik(x-y)}\frac{g_{\nu\rho}}{k^2}\,,\qquad(7.17)$$

As far as the u-quark fields are concerned, in (7.16) there are two ways to contract them in a propagator:

$$\langle 0|T\{u^j_\alpha(x)\bar{u}_{\beta\,i}(0)\}|0\rangle = i\delta^j_i\delta_{\alpha\beta}\int\frac{d^4f}{(2\pi)^4}e^{-ifx}\frac{\not{f}}{f^2}\,,$$

$$\langle 0|T\{u^j_\alpha(0)\bar{u}_{\beta\,i}(x)\}|0\rangle = i\delta^j_i\delta_{\alpha\beta}\int\frac{d^4f'}{(2\pi)^4}e^{if'x}\frac{\not{f}'}{f'^2}\,.\qquad(7.18)$$

taken in the momentum representation (A.49) and corresponding to the two diagrams in Figure 7.1(a) and (b), respectively.

We continue then with a detailed derivation of the diagram (a), while obtaining the diagram (b) is essentially a repetition, with a different order of quark fields and different virtual momentum f'. Applying the contractions (7.17) and (7.18) of the gluon and quark fields, we obtain from the initial expression (7.16):

$$\mathcal{F}^{(u)}_\mu = 4\pi\alpha_s Q_u\int d^4x\int d^4y\Big[\langle\pi^+(p_2)|\bar{u}_{i\delta}(x)d^m_\epsilon(y)\rangle\Big]\{\dots\}\Big[\bar{d}_{n\delta'}(y)u^l_{\epsilon'}(0)|\pi^+(p_1)\rangle\Big],\quad(7.19)$$

where the insertion $\{\dots\}$ is a combination of matrices and propagators:

$$\{\dots\} = \frac{(\lambda^a)^i_j\,(\lambda^a)^n_m}{4}\int\frac{d^4f}{(2\pi)^4}e^{-ifx}\frac{\delta^j_l}{f^2}[\gamma^\nu\not{f}\gamma_\mu]_{\delta\epsilon'}\int\frac{d^4k}{(2\pi)^4}e^{-ik(x-y)}\frac{g_{\nu\rho}}{k^2}[\gamma^\rho]_{\delta'\epsilon}\,.$$

The states in squared brackets in (7.19) contain quark and antiquark operators acting on the initial and final pions. Replacing the pions by the states (7.13), we obtain:

$$\bar{d}_{n\delta'}(y)u^l_{\epsilon'}(0)|\pi^+(p_1)\rangle = \int_0^1 d\alpha_1\,\tilde{\phi}_\pi(\alpha_1)\bar{d}_{n\delta'}(y)u^l_{\epsilon'}(0)|\{u(\alpha_1 p_1)\bar{d}(\bar{\alpha}_1 p_1)\}_\pi\rangle$$

$$= \left(-i\delta^l_n\frac{(\not{p}_1\gamma_5)_{\epsilon'\delta'}}{4N_c}f_\pi\int_0^1 d\alpha_1 e^{-i\bar{\alpha}_1 p_1 y}\tilde{\phi}_\pi(\alpha_1)\right)|0\rangle\,,\qquad(7.20)$$

$$\langle\pi^+(p_2)|\bar{u}_{i\delta}(x)d^m_\epsilon(y) = \int_0^1 d\alpha_2\,\tilde{\phi}_\pi(\alpha_2)\langle\{u(\alpha_2 p_2)\bar{d}(\bar{\alpha}_2 p_2)\}_\pi|\bar{u}_{i\delta}(x)d^m_\epsilon(y)$$

$$= \left(i\delta^m_i\frac{(\not{p}_2\gamma_5)_{\epsilon\delta}}{4N_c}f_\pi\int_0^1 d\alpha_2 e^{i\alpha_2 p_2 x + i\bar{\alpha}_2 p_2 y}\tilde{\phi}_\pi(\alpha_2)\right)\langle 0|\,.\qquad(7.21)$$

To explain the above expressions, let us concentrate on the first one, since the second one is just a conjugation of the first. To derive (7.20), the momentum expansion of the quark field operators $\bar{d}(y)$ and $u(0)$ is used, where only the components with the momenta $\bar{\alpha}_1 p_1$ and $\alpha_1 p_1$ are relevant. The action of these operators on a quark-antiquark state with the same momenta yields a vacuum state multiplied by the exponential factor originating from the momentum expansion. In addition, the coefficient contains a Dirac matrix structure $\not{p}_1\gamma_5$ with the pion spin-parity. Note that other possible structures containing the coordinate y are neglected because the average distance y between the quark operators, being of $O(1/Q)$, is small[4]. Furthermore, in the local limit $y \to 0$ we can use a certain normalization condition for this coefficient. To see that, we multiply both sides of (7.20) by $\gamma_\rho\gamma_5$ and take the trace over Dirac and color indices. After that, adding a vacuum bra-state, we reduce the emerging matrix element to the pion decay constant defined according to (2.48):

$$\delta_l^n \langle 0|\bar{d}_n(0)\gamma_\rho\gamma_5 u^l(0)|\pi^+(p_1)\rangle = -i\frac{\mathrm{Tr}[\gamma_\rho\gamma_5\not{p}_1\gamma_5]}{4}f_\pi \int_0^1 d\alpha_1 \tilde{\phi}_\pi(\alpha_1)\langle 0|0\rangle = ip_{1\rho}f_\pi, \qquad (7.22)$$

where the normalization condition (7.14) is used. It is also instructive to perform the same procedure as above but with nonzero coordinates, obtaining, e.g., from (7.21):

$$\langle \pi^+(p_2)|\bar{u}(x)\gamma_\rho\gamma_5 d(y)|0\rangle = -ip_\rho f_\pi \int_0^1 d\alpha_2 e^{i\alpha_2 p_2 x + i\bar{\alpha}_2 p_2 y}\tilde{\phi}_\pi(\alpha_2). \qquad (7.23)$$

This equation will be used later in Chapter 9.

Substituting (7.20), and (7.21) in (7.19), contracting the color, Dirac and Lorentz indices and using the normalization $\langle 0|0\rangle = 1$ of the vacuum state, we obtain:

$$\mathcal{F}_\mu^{(u)(a)} = \pi Q_u \alpha_s \frac{\mathrm{Tr}(\lambda^a\lambda^a)}{16N_c^2}f_\pi^2 \int_0^1 d\alpha_1\,\tilde{\phi}_\pi(\alpha_1)\int_0^1 d\alpha_2\tilde{\phi}_\pi(\alpha_2)$$

$$\times \int \frac{d^4 f}{(2\pi)^4 f^2}\int \frac{d^4 k}{(2\pi)^4 k^2}\mathrm{Tr}\left(\not{p}_2\gamma_5\gamma^\rho\not{f}\gamma_\mu\not{p}_1\gamma_5\gamma_\rho\right),$$

$$\times \int d^4 x e^{-i(f-\alpha_2 p_2 + k)x}\int d^4 y e^{i(k-\bar{\alpha}_1 p_1+\bar{\alpha}_2 p_2)y}. \qquad (7.24)$$

Replacing the x, y integrals by the δ-functions of momenta and integrating out these functions, we replace the momenta of the virtual quark and gluon with linear combinations of $p_{1,2}$:

$$k = \bar{\alpha}_1 p_1 - \bar{\alpha}_2 p_2, \quad f = p_2 - \bar{\alpha}_1 p_1. \qquad (7.25)$$

Noticing that, in the adopted massless pion approximation, $p_{1,2}^2 = 0, q^2 = -2(p_1 p_2)$, we obtain

$$f^2 = -2\bar{\alpha}_1(p_1 p_2) = \bar{\alpha}_1 q^2, \quad k^2 = -2\bar{\alpha}_1\bar{\alpha}_2(p_1 p_2) = \bar{\alpha}_1\bar{\alpha}_2 q^2.$$

Furthermore, we simplify the trace in (7.24) considerably, employing the properties of γ_5, the relation for the Lorentz-contracted γ-matrices and the equality (7.25) for f:

$$\mathrm{Tr}\left(\not{p}_2\gamma_5\gamma^\rho\not{f}\gamma_\mu\not{p}_1\gamma_5\gamma_\rho\right) = -2\mathrm{Tr}\left(\not{p}_2\not{f}\gamma_\mu\not{p}_1\right)$$

$$= -2\mathrm{Tr}\left(\not{p}_2(\not{p}_2 - \bar{\alpha}_1\not{p}_1)\gamma_\mu\not{p}_1\right) = 16\bar{\alpha}_1(p_1 p_2)p_{1\mu} = -8\bar{\alpha}_1 q^2 p_{1\mu}.$$

After the integrations in (7.24) are performed, we observe a cancellation of the factor q^2 emerging in the numerator with the same factor stemming from the propagators[5]. The final

---

[4]This is a rather qualitative assessment. A more systematic expansion near the light-cone $y^2 = 0$ will be discussed in Chapter 9.

[5]This cancellation reflects the presence of the quark and gluon spin, because in the (hypothetical) case of scalar particles the trace is absent, resulting in an extra power of q^2 in the denominator.

answer for the diagram in Figure 7.1(a) can be written as

$$\mathcal{F}_\mu^{(u),(a)} = \frac{8\pi Q_u \alpha_s}{9Q^2} f_\pi^2 \int\limits_0^1 d\alpha_1 \frac{\tilde{\phi}_\pi(\alpha_1)}{\bar{\alpha}_1} \int\limits_0^1 d\alpha_2 \frac{\tilde{\phi}_\pi(\alpha_2)}{\bar{\alpha}_2} p_{1\mu} \,. \tag{7.26}$$

The contribution of the diagram in Figure 7.1(b) is equal, apart from the replacement $p_{1\mu} \to p_{2\mu}$. Thus, the sum of the two diagrams is proportional to $(p_1 + p_2)_\mu$ restoring gauge invariance of the e.m. interaction and yielding the Lorentz structure multiplying the form factor in (7.2). It remains to take into account similar diagrams where the photon is attached to the d-quark. They (in the isospin limit) yield the same expressions with the coefficient $-Q_d$. Extra minus is due to the fact that a photon emission takes place from the antiquark line. The overall factor $Q_u - Q_d = 1$ corresponds to the unit electric charge of the π^+. We finally obtain

$$\mathcal{F}_\mu^{(u),(a+b)} = (p_1 + p_2)_\mu F_\pi^{asymp}(q^2) \,, \tag{7.27}$$

where the invariant function of momentum transfer:

$$F_\pi^{asymp}(q^2) = \frac{8\pi\alpha_s}{9Q^2} f_\pi^2 \left(\int\limits_0^1 d\alpha \frac{\tilde{\phi}_\pi(\alpha)}{\bar{\alpha}} \right)^2 \,, \tag{7.28}$$

is the asymptotic part of the pion e.m. form factor for which a $1/Q^2$ behavior is predicted.

The asymptotics of the pion form factor obtained above illustrates the concept of *collinear factorization*, valid for the exclusive hadronic processes with a large momentum transfer. In (7.28), the perturbative part, given at the leading order (LO) by a gluon exchange diagram, forms what is called a *hard-scattering amplitude* or hard-scattering kernel. This part is integrated (or, in other words, forms a convolution) with the DAs, and the latter describe collinear parton states of the initial and final pions.

A more general form of this factorization formula[6]

$$F_\pi(Q^2) = f_\pi^2 \int\limits_0^1 d\alpha_2 \int\limits_0^1 d\alpha_1 \tilde{\phi}_\pi(\alpha_1, \mu) T(Q^2, \alpha_1, \alpha_2, \mu) \tilde{\phi}_\pi(\alpha_2, \mu) \,, \tag{7.29}$$

implies next-to-leading gluon radiative corrections, described by the diagrams with extra gluon exchanges added to the ones in Figure 7.1. Starting from the one-loop level, these diagrams generate the so-called evolution of the functions $\tilde{\phi}_\pi(\alpha)$ and of the hard scattering kernel, that is, a logarithmic dependence on the renormalization scale μ. In the leading order considered here, the hard-scattering kernel has a simple expression:

$$T^{LO}(Q^2, \alpha_1, \alpha_2) = \frac{8\pi\alpha_s}{9Q^2 \bar{\alpha}_1 \bar{\alpha}_2} \,, \tag{7.30}$$

where the only scale-dependent quantity is the QCD coupling, and it is natural to take α_s at a scale of $O(Q)$.

One important question remains, concerning the convergence of the factorization formula already at the LO level (7.28). Integrating the DAs in the whole interval $0 < \alpha_{1,2} < 1$, we tacitly assume that at the endpoint $\alpha = 1$ the function $\tilde{\phi}_\pi(\alpha)$ vanishes at least linearly in $\bar{\alpha}$

---

[6]The history of this formula and of similar ones (e.g., for the nucleon form factors) dates back to the first studies of hard-scattering hadron processes on the basis of the parton model [62, 63]. In QCD the asymptotics of hadron form factors was derived in [64, 65, 66, 67, 68] and the scale dependence is known as the Efremov-Radyushkin-Brodsky-Lepage (ERBL) evolution.

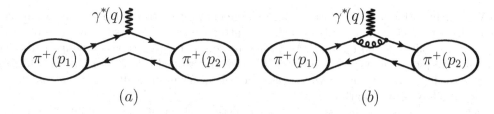

Figure 7.2 Diagrams of the (a) soft overlap contribution to the pion form factor and (b) the $O(\alpha_s)$ correction.

to render the integral finite. This expectation is confirmed if one uses a QCD-based definition of these functions, to be discussed in Chapter 9. However, including power-suppressed terms, e.g., the effects of $O(k_\perp/Q^2)$ that we have neglected, one encounters divergent integrals in the collinear factorization, as was found in [69]. Attempts to regularize these divergences lead to various modifications of the factorization framework, e.g., including also the transverse-momentum dependence in the pion distribution amplitudes:

$$\tilde{\phi}_\pi(\alpha) \to \tilde{\phi}_\pi(\alpha, k_\perp).$$

It is however difficult to realize this extension on the basis of QCD, without resorting to the models of quark wave functions in hadrons.

Intrinsically, the problem of end-point singularities and power suppressed terms is related to a more general question concerning the mechanism of momentum transfer at large Q^2. Do we necessarily need a perturbative gluon exchange which costs an extra $\alpha_s(Q)$ suppression of the form factor? The point is that, apart from the leading two-parton state with both partons (quark and antiquark) carrying a finite share of the pion momentum, a different configuration is conceivable. Suppose that the initial pion considered in the Breit frame at large Q^2, contains a single parton (quark or antiquark) carrying the whole momentum of the pion, that is, $\alpha_1 \simeq 1$ in (7.13). The second parton and, eventually, also the nonvalence components are then soft, with tiny momenta $\sim \Lambda_{QCD} \ll Q$. We assume that such an extremely asymmetric configuration has a nonvanishing probability. Then, after absorbing a virtual photon, the parton with a large momentum reverses its direction. Importantly, the final pion can now be formed with the same parton configuration as the one of the initial pion. This is due to the fact that, being very soft, the spectator partons have no definite direction. The resulting form factor is then an overlap of two configurations, both consisting of an energetic parton and soft spectators. Because of the absence of α_s, this *soft overlap mechanism*[7] apparently gains an enhancement with respect to the hard scattering. At the same time, at $Q^2 \to \infty$, the soft overlap part of the form factor is expected to decrease faster than $\sim 1/Q^2$; hence, a quantitative comparison of two mechanisms at moderately large Q^2 is far from being straightforward.

The diagram of the soft overlap form factor is shown in Figure 7.2(a), representing an essentially nonperturbative object. In Figure 7.2(b), one of the perturbative corrections to this mechanism is shown. This diagram contributes to an important effect [70] of $O(\alpha_s \log^2 Q^2)$, which, after resummation of all similar diagrams, decreases the form factor[8]. The presence of this so-called Sudakov suppression is frequently used as an argument to neglect the soft-overlap form factor with respect to the hard-spectator one. However, this argument is not convincing if one stays within the factorization approach where the leading-order diagram of the soft overlap is not accessible.

---

[7]known also as the Feynman mechanism [61],

[8]An introductory discussion of this effect can be found in the review [38].

We conclude that a hadron form factor at large momentum transfers is essentially incomplete without a soft-overlap contribution. The latter, being not accessible in terms of hard-scattering amplitudes and hadron DAs, should be considered as a separate nonperturbative quantity. A systematic way of doing that has been developed [71, 72] for the form factors of heavy-to-light transitions, in which case the role of asymptotically large scale is played by the infinitely heavy quark mass.

In Chapter 10, we will discuss a QCD-based method which offers a possibility to calculate both the soft-overlap and hard-scattering parts of the pion form factor.

CHAPTER 8

QCD SUM RULES

8.1 TWO-SIDED USE OF THE CORRELATION FUNCTION

The method of *QCD sum rules*[1] originally developed in [73, 74] is used to calculate various characteristics of hadrons, including their masses, decay constants and form factors. A sum rule is an analytic relation derived from a certain underlying object in QCD: the *correlation function*. In this section we present a general outline of this derivation, postponing many important details to the subsequent sections of this chapter.

The simplest, two-point correlation function in the momentum representation is defined as:

$$\Pi_{AB}(q) = i \int d^4x \, e^{iqx} \langle 0|T\{j_A(x)j_B(0)\}0\rangle = L_{AB}\,\Pi(q^2)\,, \tag{8.1}$$

where the time-ordered product of the two quark current operators

$$j_A(x) = \bar{q}(x)\,\Gamma_A\,q'(x) \ \text{ and } \ j_B(0) = \bar{q}'(0)\,\Gamma_B\,q(0)\,, \quad (q,q' = \{u,d,s,c,b\})\,, \tag{8.2}$$

is sandwiched between the QCD vacuum states. In (8.1) we isolate the kinematical Lorentz-structure $L_{AB}(q)$ from the invariant amplitude. The latter is a function of q^2, the only invariant variable in the two-point correlation function. In case there are multiple structures and, correspondingly, more than one invariant amplitude, a separate sum rule can in principle be derived for each of them.

The currents $j_{A,B}$ are color-neutral, with sums over the color indices not shown explicitly in (8.2). The spin and parity of these currents are determined by the choice of the Dirac matrices denoted as $\Gamma_{A,B}$ with A, B replacing the Lorentz indices. In the following, we assume that Γ_A and Γ_B are the same matrices with different indices. Note that $\Gamma_A \neq \Gamma_B$ is possible, provided both currents contain a component with the same spin-parity, as in the case of axial-vector and pseudoscalar currents, where $\Gamma_A = \gamma_\mu \gamma_5$ and $\Gamma_B = \gamma_5$. If $q' \neq q$, the currents in (8.2) are flavor nondiagonal but mutually conjugated, so that in (8.1) both flavors are conserved, as it should be for a vacuum-to-vacuum transition. Apart from the electromagnetic and weak currents discussed in Chapters 1.2 and 1.3, any effective current (see Chapter 1.4), that is, any local color-neutral operator built from quark, antiquark and gluon fields, can also serve as j_A and j_B.

It is possible to calculate the correlation function (8.1) at large spacelike momenta, such that

$$-q^2 \equiv Q^2 \gg \Lambda_{QCD}^2\,, \tag{8.3}$$

in terms of the diagrams describing propagation of quarks and gluons in the QCD vacuum. These diagrams are introduced in the following sections of this chapter, where their computation for a particular choice of currents is carried out to a certain accuracy.

---

[1]Known also as the Shifman-Vainshtein-Zakharov (SVZ) sum rules.

As a result, we obtain for the correlation function (8.1) an analytic expression, which includes purely perturbative contributions and terms with the QCD vacuum condensates, schematically:

$$\Pi^{(QCD)}(q^2) = \Pi\big(q^2; m_q, m_{q'}, \alpha_s, \langle 0|\hat{O}_{d_{min}}|0\rangle, ..., \langle 0|\hat{O}_{d_{max}}|0\rangle\big). \qquad (8.4)$$

Indicated above is the dependence of the correlation function on the universal QCD parameters: the quark-gluon coupling, quark masses and vacuum condensate densities. The latter are vacuum averages of color-neutral, scalar ($J^P = 0^+$) and local operators $\hat{O}_d$ with dimensions $d_{min} \leq d \leq d_{max}$. In particular, the vacuum averages with $d = 3$ and $d = 4$ are, respectively, the quark and gluon condensate densities (1.16) already discussed in Chapter 1. We emphasize that (8.4) is a reliable approximation for the correlation function only if the variable q^2 lies in the spacelike region (8.3). In particular, this expression is not applicable at $q^2 \sim 0$ because, as explicit calculations demonstrate, the terms with condensates are proportional to the inverse powers of q^2 and diverge[2].

The next step is to relate the correlation function (8.1) to hadronic degrees of freedom. Here the key observation is that in the timelike region of positive q^2 the amplitude $\Pi_{AB}(q)$ describes real physical processes: on-shell hadronic states are created by the quark current j_B from the vacuum and annihilated by the current j_A. Clearly, these states have the same quantum numbers (flavor content, J^{PC}, I^G) as the correlated currents. Suppose the lightest hadron h with these quantum numbers has a mass m_h. Then, at $q^2 = m_h^2$ or, equivalently, at $q_0 = m_h$ in the rest frame $\vec{q} = 0$, the amplitude $\Pi_{AB}(q)$ describes creation and annihilation of the single on-shell hadron h. Gradually increasing q^2, one subsequently crosses the thresholds for heavier hadronic states including excited hadrons and multihadron states with the quantum numbers of h.

All intermediate hadronic states in the correlation function are represented and summed up in the expression for its imaginary part which follows from the general unitarity relation (see Appendix B for derivation):

$$\begin{aligned} 2\,\mathrm{Im}\Pi_{AB}(q) &= 2\pi\delta(q^2 - m_h^2)\langle 0|j_A\overline{|h(q)\rangle\langle h(q)|}j_B|0\rangle)^* \\ &+ \sum_{h'}\!\int\! d\tau_{h'}\,\langle 0|j_A\overline{|h'(q)\rangle\langle h'(q)|}j_B|0\rangle)^*, \end{aligned} \qquad (8.5)$$

where we deliberately isolated the contribution of the lightest hadron h from the rest. In the above equation, the sum goes over all possible heavier hadronic states h' and the overlines indicate summation over the polarizations. The phase space integration is performed in each contribution, in particular, the δ-function in the first term accounts for the phase space of the single hadron h (see (5.73)).

The unitarity relation (8.5) can be rewritten for the invariant amplitude, factorizing out the Lorentz structure on both sides. On the l.h.s. we have:

$$\mathrm{Im}\,\Pi_{AB}(q) = L_{AB}(q)\,\mathrm{Im}\Pi(q^2), \qquad (8.6)$$

provided L_{AB} is chosen properly and does not contain additional, so-called kinematical singularities in the variable q^2, which may induce a spurious imaginary part. On the r.h.s., after extracting the same Lorentz structure, the product of hadronic matrix elements in the first term is reduced to the square of decay constants of the hadron h:

$$\langle 0|j_A\overline{|h(q)\rangle\langle h(q)|}j_B|0\rangle) = L_{AB}(q)f_h^2, \qquad (8.7)$$

---

[2]Here we tacitly assume that the quarks q, q' are light. In the case of heavy quarks Q, Q', the region of applicability can be shifted towards the timelike region up to $q^2 \ll (m_Q + m_{Q'})^2$, see Section 8.6.3.

and a compact parametrization for the sum over heavier hadronic states is introduced

$$\frac{1}{2\pi}\sum_{h'}\int d\tau_{h'}\langle 0|j_A\overline{|h'(q)\rangle\langle h'(q)|}j_B|0\rangle\rangle = L_{AB}(q)\rho_{h'}(q^2)\,,\tag{8.8}$$

with the invariant function $\rho_{h'}(q^2)$ which is the spectral density of heavier states. The resulting unitarity relation for the invariant amplitude is:

$$\frac{1}{\pi}\mathrm{Im}\Pi(q^2) = \delta(q^2 - m_h^2)f_h^2 + \theta(q^2 - s_{h'})\rho_{h'}(q^2)\,,\tag{8.9}$$

where the θ-function stems from the phase space and reflects the lowest threshold $q^2 = s_{h'}$ of the heavier than h hadronic states.

At this stage our knowledge of the correlation function $\Pi(q^2)$ covers two disconnected regions of the variable q^2: the interval (8.3) of negative q^2, where the QCD-based expression (8.4) is valid and the interval of positive $q^2 \geq m_h^2$ where the imaginary part of $\Pi(q^2)$ is given by the hadronic sum (8.9). To establish a link between spacelike and timelike regions, we treat $\Pi(q^2)$ as a function of the complex variable q^2. The situation is very similar to the connection between spacelike and timelike pion form factors discussed in Chapter 5. The correlation function is analytic everywhere in the complex q^2-plane, except on the part of the positive real axis $q^2 = s > 0$, where $\Pi(q^2)$ has singularities generated from the unitarity relation. There is a simple pole at $s = m_h^2$, and, starting from $s = s_{h'}$, quite a complicated interplay of poles and branch points corresponding, respectively, to excited hadrons and thresholds of multihadron states. To circumvent these singularities, we introduce the same contour on the complex plane as in Figure 5.1, where now $s_0 = m_h^2$ and $s_1 = s_{h'}$. Applying the Cauchy theorem and following the steps listed after (5.29), we arrive at the dispersion relation for the function $\Pi(q^2)$ similar to (5.34) for the pion form factor. In this way, the correlation function calculated in the q^2-region (8.3) is linked to the sum over hadronic contributions specified in (8.9):

$$\Pi^{(QCD)}(q^2) = \frac{1}{\pi}\int_{m_h^2}^{\infty} ds\frac{\mathrm{Im}\,\Pi(q^2)}{s - q^2} = \frac{f_h^2}{m_h^2 - q^2} + \int_{s_{h'}}^{\infty} ds\frac{\rho_{h'}(s)}{s - q^2}\,.\tag{8.10}$$

Thus, the name of the method can now be understood almost literally: a two-sided use of a correlation function in **QCD** yields for the **sum** over hadronic parameters a certain **rule**.

An important difference should be mentioned between the dispersion relation (8.10) and the one in (5.34). In the latter, the integral over imaginary part is convergent, due to the QCD asymptotics of the pion form factor, whereas the integral on the r.h.s. of (8.10) is not always convergent, depending on the choice of currents in the correlation function. The divergence can be removed by performing subtractions in the dispersion relations, as explained in Chapter 5.2. In fact, as we shall see below, subtraction terms are irrelevant after a special transformation is applied to both sides of (8.10).

Sum rules of the type (8.10) are applied in two different ways. First, if there are sufficient and accurate data on hadron masses and hadronic matrix elements saturating the r.h.s. of (8.10), it is possible to estimate one of the universal parameters entering $\Pi^{(QCD)}$ – a quark mass or a condensate density – provided all other parameters are known, e.g., from different QCD sum rules. The second possible way of using (8.10) is to extract parameters of the lightest (ground-state) hadron h, in this case the mass m_h and decay constant f_h. To realize this task, an estimate of the dispersion integral over the hadronic spectral density $\rho(s)$ is necessary. To this end, one has to employ an important property of the correlation function known as the *quark-hadron duality* or parton-hadron duality.

To understand the origin of duality, we consider the asymptotic region of the spacelike momentum transfer, $q^2 \to -\infty$. In this region the correlation function coincides with the asymptotic limit of the QCD result:

$$\lim_{q^2 \to -\infty} \Pi(q^2) = \lim_{q^2 \to -\infty} \Pi^{(QCD)}(q^2) \equiv \Pi^{(pert)}(q^2) . \tag{8.11}$$

The function $\Pi^{(pert)}(q^2)$ is entirely determined by the perturbative contributions. The vacuum condensate terms in $\Pi^{(QCD)}(q^2)$, being proportional to inverse powers of q^2, vanish at $q^2 \to -\infty$. The perturbative part of the correlation function, considered below in more detail, is given by quark loop diagrams and has a predominantly logarithmic dependence on q^2.

To proceed, we analytically continue the function $\Pi^{(pert)}(q^2)$ in the complex q^2 plane. On the positive real axis $q^2 = s$ this function develops an imaginary part caused by the on-shell quark propagators in the loop diagrams. The quark-antiquark threshold $s_{min} = (m_q + m_{q'})^2$ serves as a branch point. We emphasize that – in contrast to the hadronic spectral density (8.9) – the imaginary part of $\Pi^{(pert)}(q^2)$ is nothing more than a mathematical property of this function, because in reality quarks are confined in hadrons and cannot go on shell. A dispersion relation for the perturbative part can then be written as:

$$\Pi^{(pert)}(q^2) = \int\limits_{s_{min}}^{\infty} ds \, \frac{\operatorname{Im} \Pi^{(pert)}(q^2)}{s - q^2} . \tag{8.12}$$

Substituting the above relation and the hadronic dispersion relation (8.10) in the asymptotic equality (8.11), we obtain an equation between the two integrals[3]:

$$\int\limits_{m_h^2}^{\infty} ds \, \frac{\operatorname{Im} \Pi(s)}{s - q^2} = \int\limits_{s_{min}}^{\infty} ds \frac{\operatorname{Im} \Pi^{(pert)}(s)}{s - q^2} , \tag{8.13}$$

valid at large negative q^2. The above property of the correlation function is known as the *global* quark-hadron duality. From the same relation we can infer that the integrands on both sides should have equal asymptotic behavior at $s \to \infty$:

$$\lim_{s \to \infty} \operatorname{Im} \Pi(s) = \lim_{s \to \infty} \operatorname{Im} \Pi^{(pert)}(s) . \tag{8.14}$$

This relation, applied at sufficiently large s as an approximation

$$\operatorname{Im} \Pi(s) \simeq \operatorname{Im} \Pi^{(pert)}(s) , \tag{8.15}$$

is what is usually called the *local* quark-hadron duality. Evidently, this approximation involves a stronger assumption than the global duality (8.13). Indeed, (8.15) implies that the asymptotic behavior calculated from QCD diagrams at large s coincides point-by-point with the hadronic spectral density, whereas both the limiting condition (8.14) and the equality of integrals in (8.13) do not exclude oscillations of the latter density while approaching the asymptotic limit $\operatorname{Im} \Pi^{(pert)}(s)$.

What we actually use, is the condition of *semi-local* duality that is more general than (8.15): The part of the integral on the l.h.s. of (8.13) which includes the hadronic spectral density $\rho_{h'}(s)$ at $s \geq s_h'$ is equal to a part of the integral on the r.h.s.:

$$\int\limits_{s_{h'}}^{\infty} ds \frac{\rho_{h'}(s)}{s - q^2} = \frac{1}{\pi} \int\limits_{s_0}^{\infty} ds \frac{\operatorname{Im} \Pi^{(pert)}(s)}{s - q^2} , \tag{8.16}$$

---

[3]If subtractions are needed in the dispersion relations, we may simply differentiate both parts of this equation over q^2 to render the integrals on both sides convergent.

where s_0 is an effective threshold, not necessarily equal to $s_{h'}$. The new parameter s_0 has then to be determined or adjusted.

Substituting (8.16) in (8.10), we obtain a relation between the contribution of the ground-state hadron h and the correlation function calculated in QCD:

$$\frac{f_h^2}{m_h^2 - q^2} = \Pi^{(QCD)}(q^2) - \frac{1}{\pi} \int\limits_{s_0}^{\infty} ds \frac{\text{Im}\,\Pi^{(pert)}(s)}{s - q^2}, \tag{8.17}$$

to be used at sufficiently large $q^2 < 0$.

To get the final form of QCD sum rule, one usually applies to both parts of the above relation a transformation suppressing the second term on the r.h.s. with respect to the rest and making the whole procedure less sensitive to the semi-local duality approximation. One straightforward option is a multiple differentiation in q^2 of both parts of (8.17) at a certain large $q^2 = -Q_0^2$, yielding:

$$\frac{f_h^2}{(m_h^2 + Q_0^2)^{n+1}} = \frac{1}{n!} \left[\frac{d^n}{d(q^2)^n} \Pi^{(QCD)}(q^2) \right]_{q^2 = -Q_0^2} - \frac{1}{\pi} \int\limits_{s_0}^{\infty} ds \frac{\text{Im}\,\Pi^{(pert)}(s)}{(s + Q_0^2)^{n+1}}. \tag{8.18}$$

Far more effective and more frequently used is the *Borel transformation* defined in (A.103). Applying it to (8.17), we obtain:

$$f_h^2\, e^{-m_h^2/M^2} = \hat{\mathcal{B}}_{M^2} \Pi^{(QCD)}(q^2) - \frac{1}{\pi} \int\limits_{s_0}^{\infty} ds\, e^{-s/M^2} \text{Im}\,\Pi^{(pert)}(s). \tag{8.19}$$

This transformation exponentially suppresses the integral on the r.h.s., improving the accuracy of the duality approximation. Simultaneously, it makes superfluous the subtractions in the initial dispersion relation, removing the polynomial in q^2 subtraction terms and rendering all integrals convergent. Most importantly, instead of the spacelike momentum squared $q^2 = -Q^2$, a new effective scale – the Borel parameter squared M^2 – emerges. As follows from the definition (A.103), M^2 roughly corresponds to the ratio Q_0^2/n in the differentiated sum rule (8.18) where both Q_0^2 and n are taken very large.

Note that an auxiliary sum rule for the ground-state hadron mass can be easily obtained from (8.19), taking the derivative over $(-1/M^2)$ of both sides and dividing the result to (8.19), so that the decay constant f_h drops out in the ratio:

$$m_h^2 = \frac{\frac{d}{d(-1/M^2)} \hat{\mathcal{B}}_{M^2} \Pi^{(QCD)}(q^2) - \frac{1}{\pi} \int\limits_{s_0}^{\infty} ds\, s\, e^{-s/M^2} \text{Im}\,\Pi^{(pert)}(s)}{\hat{\mathcal{B}}_{M^2} \Pi^{(QCD)}(q^2) - \frac{1}{\pi} \int\limits_{s_0}^{\infty} ds\, e^{-s/M^2} \text{Im}\,\Pi^{(pert)}(s)}. \tag{8.20}$$

This relation is frequently used to fix the threshold parameter s_0 by fitting the r.h.s. to the experimentally measured value of m_h.

In the following sections of this chapter we discuss in more detail the basic elements of a sum rule method, considering various examples, mostly in the form of the Borel sum rule (8.19).

8.2 SHORT DISTANCE DOMINANCE

In order to closely investigate the correlation function (8.1), let us rewrite it in a slightly simplified form. It is always possible to multiply both sides of (8.1) by a certain four-dimensional tensor, e.g., by $g_{\mu\nu}$, in order to contract the Lorentz indices. The contracted correlation function becomes a Lorentz-invariant quantity, hence it depends only on q^2, schematically:

$$\Pi(q^2) = i \int d^4x\, e^{iqx}\, \langle 0|T\{j_A(x)j^A(0)\}0\rangle\,. \tag{8.21}$$

Correspondingly, the integrand depends only[4] on x^2:

$$i\langle 0|T\{j_A(x)j^A(0)\}0\rangle \equiv \Pi(x^2)\,. \tag{8.22}$$

It is instructive to determine the average size of the space-time integration region in (8.21) in which the integrand $\Pi(x^2)$ dominates.

The amplitude (8.22) describes propagation of the quark-antiquark pair $\bar{q}q'$ emitted at $x = 0$ by an external pointlike "source" and absorbed at point x by an external "sink." Note that the integration in (8.21) is largely regulated by the oscillating exponential factor depending on the external momentum q. Correspondingly, the actual size of the space-time region, where the dominant contribution of the function $\Pi(x^2)$ is accumulated, essentially depends on the value of q^2. Let us assume that the momentum transfer is spacelike and $Q^2 = -q^2$ is large, as in (8.3). In practice, $Q^2 \sim 1$ GeV^2 already satisfies this condition.

Performing the Fourier transformation of the function $\Pi(x^2)$:

$$\Pi(x^2) = \int d\tau\, e^{ix^2\tau}\Pi(\tau)\,, \tag{8.23}$$

we obtain:

$$\Pi(q^2) = \int d^4x\, e^{iqx} \int d\tau\, e^{ix^2\tau}\Pi(\tau) = \int d\tau\, \Pi(\tau) \int d^4x\, e^{i(qx+\tau x^2)}\,. \tag{8.24}$$

Completing the power of the exponent to a full square:

$$q\cdot x + x^2\tau = \tau\left(x + \frac{q}{2\tau}\right)^2 - \frac{q^2}{4\tau}\,,$$

and shifting the x-integration, $d^4x \to d^4(x + q/(2\tau)) \to d^4x'$, yields after renaming the coordinates $x' \to x$:

$$\Pi(q^2) = \int d^4x \int d\tau\, e^{ix^2\tau} e^{iQ^2/(4\tau)}\Pi(\tau)\,. \tag{8.25}$$

At least one of the exponential factors in the integral over τ strongly oscillates in the regions where $|\tau x^2|$ or $|Q^2/(4\tau)|$ is large. The contributions of these regions to the integral are therefore strongly suppressed[5]. Thus, the dominant contribution to (8.25) stems from the interval constrained simultaneously by the two conditions:

$$|\tau x^2| \lesssim 1,\quad \frac{Q^2}{4|\tau|} \lesssim 1\,, \tag{8.26}$$

limiting the oscillations of both exponents in (8.25). Combining the two above inequalities, we obtain:

$$|x^2| \lesssim 1/Q^2\,. \tag{8.27}$$

---

[4]Here we assume that the currents $j_{A,B}$ contain only light quarks $q, q' = \{u, d, s\}$, so that the dependence of $\Pi(x^2)$ on quark masses can safely be neglected, at least in the short-distance region where $1/x \gg \Lambda_{QCD}$.

[5]This property of the integrals with oscillating factors follows from the Riemann-Lebesgue theorem.

Hence, at asymptotically large $Q^2 \to \infty$, the interval near the light-cone, $|x^2| \sim 0$, dominates in (8.21). To narrow down this region further, we notice that at $q^2 < 0$ it is always possible to choose a reference frame with the time component of the momentum equal to zero, $q = (0, \vec{q})$, so that $Q^2 = \vec{q}^2$. In this frame the exponent in (8.21) is equal to $\exp(-i\vec{q} \cdot \vec{x})$, and, in order to avoid strong oscillations, the condition

$$|\vec{x}| \lesssim \frac{1}{|\vec{q}|} = 1/Q, \quad (Q \equiv \sqrt{Q^2}) \tag{8.28}$$

has to be valid. Combining the latter with (8.27) leads to the dominance of short distances and time intervals at large Q^2:

$$x_0 \sim |\vec{x}| \lesssim 1/Q. \tag{8.29}$$

We conclude that at $Q^2 \to \infty$ the product of currents in (8.21) can be approximately replaced by a local composite operator at $x \sim 0$. Note that if we proceed in the opposite direction and gradually decrease Q^2 up to the values of $O(\Lambda_{QCD}^2)$, both constraints (8.27) and (8.29) are then removed, expanding the dominant space-time region in the correlation function (8.21) to long distances and time intervals of $O(1/\Lambda_{QCD})$.

8.3 PERTURBATIVE CONTRIBUTIONS

At large momentum transfers, $Q^2 \gg \Lambda_{QCD}^2$, the correlation function (8.21) describes propagation of highly virtual quark and antiquark in the vacuum, from the emission to the absorption point. The effective quark-gluon coupling $\alpha_s(\mu)$, taken at a characteristic normalization scale $\mu \sim Q$ is sufficiently small starting already from $Q \gtrsim 1$ GeV. This enables us to calculate the correlation function in terms of Feynman diagrams containing quark and gluon propagators and quark-gluon vertices.

We return to the initial definition (8.1) of the correlation function with uncontracted Lorentz indices and substitute the current operators (8.2) in the explicit form:

$$\Pi_{AB}(q^2) = i \int d^4x\, e^{iqx} \langle 0|T\{\bar{q}(x)\,\Gamma_A q'(x)\bar{q}'(0)\,\Gamma_B q(0)\}|0\rangle. \tag{8.30}$$

Replacing the T-products of the quark and antiquark fields by the quark propagators (see (A.47)),

$$S_{(q)}(0, x) = -i\langle 0|T\{q(0)\bar{q}(x)\}|0\rangle, \quad S_{(q')}(x, 0) = -i\langle 0|T\{q'(x)\bar{q}'(0)\}|0\rangle, \tag{8.31}$$

yields:

$$\Pi_{AB}(q^2) = -i \int d^4x\, e^{iqx}\text{Tr}\{iS_{(q)}(0, x)\Gamma_A iS_{(q')}(x, 0)\,\Gamma_B\}, \tag{8.32}$$

where the trace over Dirac and color indices is taken. The extra minus sign accounts for anticommutation of the fermion field operators, in this case $q(0)$ and $\bar{q}(x)$.

The leading-order (LO) contribution to (8.32) is of $O(\alpha_s^0)$ and corresponds to the asymptotic freedom limit of QCD. No gluons are emitted and absorbed at this order, hence, the free-quark propagators for $S_{(q)}^{(0)}(-x)$ and $S_{(q')}^{(0)}(x)$ should be used in (8.32), forming a one-loop diagram with two vertices of external currents and two virtual quark lines. This diagram for a particular choice of quark flavors $q = s$, $q' = u$ is shown in Figure 8.1(a) and will be calculated in Section 8.5. Taking into account emission and absorption of a virtual gluon by the propagating quarks, yields radiative corrections of $O(\alpha_s)$ and provides the next-to-leading order (NLO) contribution to the correlation function. The corresponding two-loop diagrams

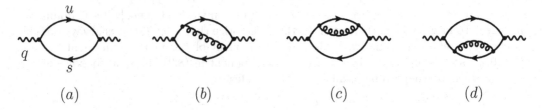

Figure 8.1 Diagrams corresponding to the perturbative part of the two-point correlation function (8.41): (a) the leading-order quark loop, (b)–(d) the gluon radiative corrections in $O(\alpha_s)$. Wavy lines denote external currents.

are shown in Figures 8.1(b)–(d). Note that the gluon exchange diagram 8.1(b) corresponds to the perturbative interaction between quark and antiquark, whereas the diagrams 8.1(c) and (d) describe perturbative corrections to the quark propagators and also contribute to the running of quark masses. The sum over the perturbative diagrams:

$$\Pi_{AB}^{(pert)}(q^2) = \Pi_{AB}^{(0)}(q^2; m_q, m_{q'}, \mu) + \frac{\alpha_s(\mu)}{\pi} \Pi_{AB}^{(1)}(q^2, m_q, m_{q'}, \mu) + ... , \tag{8.33}$$

contains the LO and NLO contributions denoted as $\Pi_{AB}^{(0)}$ and $\Pi_{AB}^{(1)}$, respectively, where the dependence on the quark masses is explicitly shown, and the ellipsis indicates the higher-order gluon radiative corrections. The presence of the normalization scale μ in the above expression indicates that the loop diagrams are generally divergent and need to undergo the regularization and renormalization procedures. Note that in the correlation functions used in QCD sum rules, the divergent parts, once properly regularized and isolated, are unimportant for the rest of the procedure.

8.4   VACUUM CONDENSATES AND OPERATOR-PRODUCT EXPANSION

As already mentioned in Chapter 1.1.3, in QCD the vacuum state contains fluctuations of quark-antiquark and gluon fields with nonvanishing average density – the vacuum condensates. In fact, these nonperturbative effects influence the correlation function already in the spacelike region. Virtual quark and antiquark, propagating from the emission to the absorption point, interact with the vacuum fields, yielding contributions to the correlation function (8.30) not accounted for in the perturbative part (8.33). At large Q^2, the short-distance dominance of the correlation function allows for a systematic account of these contributions. The key point is that at large virtualities the average time and distance of the quark propagation in (8.30) is much smaller than the characteristic $O(1/\Lambda_{QCD})$ size of vacuum fluctuations; hence, the correlation function serves as an effective local probe providing a "snapshot" of the QCD vacuum. This probe is only sensitive to the averaged local densities of the vacuum fields – the universal vacuum condensates.

The set of condensates in QCD includes the vacuum averages (vacuum expectation values) of all possible color-neutral, flavorless and scalar local operators built from quark and gluon fields. The simplest ones – quark and gluon condensates – were presented in (1.16). The operators

$$\hat{O}_3(x) = \bar{q}_i(x) q^i(x) , \quad \hat{O}_4(x) = G_{\mu\nu}^a(x) G^{a\mu\nu}(x) , \tag{8.34}$$

forming these condensates have the lowest possible dimensions in energy units: $d = 3$ and $d = 4$, respectively.

Note that in QCD there are no operators with vacuum quantum numbers and $d < 3$. Evidently, we cannot take a vacuum average of a gluon-field operator $A_\mu^a(x)$ with dimension

$d = 1$, because it has a color charge and spin. A single quark-field operator $q^i(x)$ with $d = 3/2$ has, in addition to color and spin, also flavor, electric charge and baryon number. Less trivial is the absence of the vacuum condensate with $d = 2$. The only suitable operator $A^a_\mu(x)A^{a\mu}(x)$, which possesses vacuum quantum numbers, violates the local $SU(3)_c$ gauge symmetry.

The list of essential operators with vacuum quantum numbers includes also:

$$\hat{O}_5(x) = \bar{q}_i(x)\sigma_{\mu\nu}\frac{(\lambda^a)^i_k}{2}q^k(x)G^{a\mu\nu}(x),$$

$$\hat{O}_6(x) = \bar{q}_i(x)(\Gamma_A)^i_k q^k(x)\bar{q}'_j(x)(\Gamma^A)^j_m q'^m(x) \tag{8.35}$$

with $d = 5$ and $d = 6$, respectively, where Γ_A is a generic combination of Dirac and Gell-Mann matrices, and in general $q' \neq q$. The vacuum averages of $\hat{O}_5$ and $\hat{O}_6$ form, respectively, the *quark-gluon* and *four-quark condensates*. The operators of higher than $d = 6$ dimension constructed by adding extra gluon field-strengths or quark-antiquark pairs will play a minor role in our discussion.

Taking into account the contributions of vacuum condensates, the correlation function (8.30) calculated in QCD at large $Q^2 = -q^2$ becomes:

$$\Pi^{(QCD)}_{AB}(q^2) = \Pi^{(pert)}_{AB}(q^2) + \sum_{d=3,4,5,6,\ldots} C^{AB}_d(q^2)\langle 0|\hat{O}_d|0\rangle, \tag{8.36}$$

where $\hat{O}_{3,4,5,6}$ are defined in (8.34) and (8.35). This expression is more explicit than the schematic form (8.4). The coefficients $C^{AB}_d(q^2)$ are calculated in terms of specific diagrams describing interactions of propagating quarks with condensates, representing the latter in a form of external static fields. Examples of these diagrams are shown in Figures 8.2–8.5. A detailed calculation of some of them will be presented in the next section. We emphasize that the perturbative part $\Pi^{(pert)}_{AB}$ and the coefficients $C^{AB}_d(q^2)$ in (8.36) depend on the choice of currents in the correlation function, whereas the vacuum condensates $\langle 0|\hat{O}_d|0\rangle$ are universal parameters of nonperturbative QCD.

The sum in (8.36) formally contains an infinite series of vacuum condensate terms. Note however that the correlation function has a certain fixed dimension which we denote as d_0. Being of nonperturbative origin, the vacuum condensate of an operator O_d, has a characteristic value in the ballpark of $(\Lambda_{QCD})^d$; hence, the coefficient $C^{AB}_d(q^2)$ of this condensate is proportional to $1/Q^{(d-d_0)}$, compensating the condensate dimension. Here we assume that the currents $j_{A,B}$ consist only of light quarks in which case $m_{q,q'} \ll Q$ and the dependence of $C^{AB}_d(q^2)$ on the quark masses is inessential. We conclude that the part of the correlation function originating from QCD vacuum effects represents a series of power corrections $(\Lambda_{QCD}/Q)^d$ to the perturbative part. Therefore, the sum over condensates in (8.36) can be truncated at a certain dimension d_{max}, similar to retaining the terms up to a certain order in α_s in the perturbative part.

The expression (8.36) can be interpreted in a more general way, in terms of the Wilson operator-product expansion (OPE). The time-ordered product of two local operators built from quantum fields and taken at short distances can be expressed via sum of local operators ordered according to their dimension:

$$T\{\hat{O}_a(x)\hat{O}_b(0)\} = \sum_{d\geq 0} C^{ab}_d(x^2)\hat{O}_d(0), \tag{8.37}$$

where the term with $d = 0$ corresponds to the dimensionless unit operator $\hat{O}_0 = \mathbb{1}$ with no fields, normalized as $\langle 0|\mathbb{1}|0\rangle = 1$. Note that all operators on the r.h.s. of this expansion should have the same quantum numbers as the combination of the operators on the l.h.s.

In QCD, the expansion (8.37) is, strictly speaking, proved only within perturbation theory, where the coefficients $C_d^{ab}(x^2)$ are calculable in terms of Feynman diagrams. Importantly, (8.37) is an operator equation, and remains valid after taking matrix elements between any $\langle f|$ and $|i\rangle$ states on both sides.

Applying the operator expansion to the product of the quark-current operators in the correlation function (8.1), we have:

$$T\{j_A(x)j_B(0)\} = \sum_{d\geq 0} C_d^{AB}(x^2)O_d(0).\tag{8.38}$$

A subsequent vacuum average and Fourier-transformation yields an expansion for the vacuum-averaged correlation function, which formally coincides with (8.36) with

$$C_d^{AB}(q^2) = i\int d^4 x e^{iqx} C_d^{AB}(x^2)\tag{8.39}$$

and

$$C_0^{AB}(q^2) = \Pi_{AB}^{(pert)}(q^2).\tag{8.40}$$

The above equation is consistent with the fact that only the loop diagrams with no external fields contribute to the perturbative part of the correlation function.

The systematic account of nonperturbative vacuum condensates, developed in [73], has qualitatively extended the framework of the OPE beyond QCD perturbation theory. This generalized OPE involves certain important details which remain out of our scope. One is the role of short-distance vacuum fluctuations appearing at higher than $d = 8$ dimensions in OPE ("direct instantons"). Another issue is the interplay of the regions of small virtualities in the perturbative loop diagrams with respect to the low momenta quarks and gluons belonging to the vacuum fluctuations. A comprehensive discussion of these nontrivial aspects of the generalized OPE can be found in [75] (or in a more pedagogical way, in the review [76]) where a "pragmatic version" of OPE was formulated which we also follow here.

8.5 CALCULATING A TWO-POINT CORRELATION FUNCTION

8.5.1 Perturbative loop

In this section we consider, as a study case, a correlation function with certain light quark currents and calculate the OPE terms, starting from the perturbative LO term. More specifically, we introduce the vacuum-to-vacuum transition amplitude:

$$\Pi_5(q^2) = i\int d^4 x e^{iqx}\langle 0|T\{j_5^{(s)}(x)j_5^{(s)\dagger}(0)\}|0\rangle,\tag{8.41}$$

correlating the strangeness-changing pseudoscalar current and its conjugate, taken at two different points:

$$j_5^{(s)}(x) = (m_s + m_u)\bar{s}(x)i\gamma_5 u(x),$$
$$j_5^{(s)\dagger}(0) = (m_s + m_u)\bar{u}(0)i\gamma_5 s(0).\tag{8.42}$$

This particular choice allows us to maintain a certain degree of generality with two different quark flavors and nonvanishing masses m_s and m_u. On the other hand, since the currents have spin zero, we do not need to extract Lorentz structures. The correlation function $\Pi_5(q^2)$ itself is an invariant function of q^2, the momentum-transfer squared. Note that the current in (8.42) can be represented in a form of four-divergence of the axial-vector current:

$$j_5^{(s)}(x) = \partial_\mu\big(\bar{s}(x)\gamma^\mu\gamma_5 u(x)\big).\tag{8.43}$$

That is in fact the same relation as (2.52) derived in Chapter 2 for a different axial-vector current. At spacelike q^2, the perturbative LO contribution to (8.41) is a quark loop of the form (8.32):

$$\Pi_5^{(0)}(q^2) = -i(m_s + m_u)^2 \int d^4x \, e^{iqx} \text{Tr}\{iS_{(s)}^{(0)}(-x)i\gamma_5 iS_{(u)}^{(0)}(x)\, i\gamma_5\}, \qquad (8.44)$$

where the free s- and u-quark propagators are substituted between the two γ_5 vertices. This diagram is shown in Figure 8.1(a). We use the propagators (A.49) in the momentum representation:

$$
\begin{aligned}
\Pi_5^{(0)}(q^2) &= iN_c(m_s + m_u)^2 \int d^4x \, e^{iqx} \text{Tr}\Big\{ \int \frac{d^4p}{(2\pi)^4} e^{ipx} \frac{(\not p + m_s)}{p^2 - m_s^2} i\gamma_5 \\
&\quad \times \int \frac{d^4p'}{(2\pi)^4} e^{-ip'x} \frac{(\not p' + m_u)}{p'^2 - m_u^2} \, i\gamma_5 \Big\},
\end{aligned}
\qquad (8.45)
$$

where the factor $N_c = 3$ originates from the summation over color indices and the remaining trace is taken over Dirac matrices. Collecting together all exponential factors, we perform the x- and p'-integrations, making use of the relations:

$$\int \frac{d^4p'}{(2\pi)^4} f(p') \int d^4x \, e^{i(q+p-p')x} = \int d^4p' f(p')\delta^{(4)}(p' - q - p) = f(p + q),$$

so that

$$\Pi_5^{(0)}(q^2) = -iN_c(m_s + m_u)^2 \int \frac{d^4p}{(2\pi)^4} \frac{\text{Tr}\{(\not p + m_s)\gamma_5((\not p + \not q) + m_u)\gamma_5\}}{(p^2 - m_s^2)((p+q)^2 - m_u^2)}. \qquad (8.46)$$

Note that the integral over momentum p flowing in the loop is divergent in the ultraviolet domain $p \to \infty$. The most elaborated way to handle this divergence is *dimensional regularization*, that is, to shift the dimension of integration:

$$4 \to D \equiv 4 - \epsilon, \qquad (8.47)$$

rendering the integral finite.

After dimensional regularization the trace is computed[6]. Here we use $\{\gamma_5, \gamma_\mu\} = 0$ and $\gamma_5^2 = 1$, so that:

$$\text{Tr}\{(\not p + m_s)\gamma_5((\not p + \not q) + m_u)\gamma_5\} = 4(-p(p+q) + m_s m_u). \qquad (8.48)$$

Substituting this simple expression in the numerator of the integrand in (8.46), we then use the parametrization (A.87) for the denominator:

$$\Pi_5^{(0)}(q^2) = 4N_c i(m_s + m_u)^2 \mu^{4-D} \int \frac{d^D p}{(2\pi)^D} \int_0^1 dx \frac{p(p+q) - m_s m_u}{[x(p^2 - m_s^2) + \bar x((p+q)^2 - m_u^2)]^2}, \qquad (8.49)$$

where $\bar x \equiv 1 - x$. The factor μ^{4-D} with an auxiliary energy-momentum scale μ restores the physical dimension of the correlation function. To proceed further, we transform the denominator in the above expression:

$$x(p^2 - m_s^2) + \bar x((p+q)^2 - m_u^2) = (p + \bar x q)^2 + x\bar x q^2 - xm_s^2 - \bar x m_u^2,$$

---

[6]Nominally, the number of 4×4 γ_μ-matrices in D dimensions is equal to D, but this circumstance does not influence the trace considered here, which has no summation over Lorentz indices. There are also subtleties related with using the γ_5 matrix in D-dimensions. Here we tacitly adopt the so-called naive dimensional regularization scheme.

and shift the integration variable:

$$p + \bar{x}q \equiv \tilde{p}, \quad d^D p = d^D \tilde{p},$$

obtaining:

$$\Pi_5^{(0)}(q^2) = 12i(m_s + m_u)^2 \int\limits_0^1 dx \, \mu^{4-D} \int \frac{d^D \tilde{p}}{(2\pi)^D} \frac{\tilde{p}^2 - (\bar{x} - x)\tilde{p}q - (x\bar{x}q^2 + m_s m_u)}{[\tilde{p}^2 - (xm_s^2 + \bar{x}m_u^2 - x\bar{x}q^2)]^2}. \tag{8.50}$$

To perform the integration over D-dimensional momentum space, we use the standard Wick rotation to the Euclidean space:

$$\tilde{p}_0 \to i\tilde{p}_4, \quad \tilde{p}^2 \to -\sum_{\alpha=1,..4} \tilde{p}_\alpha^2, \tag{8.51}$$

followed by the integration over spherical coordinates in that space. The first and third terms in (8.50) are reduced to the standard integrals:

$$\int \frac{d^D \tilde{p}}{(2\pi)^D} \frac{\tilde{p}^2}{[\tilde{p}^2 - R^2]^2} = \frac{-iD\Gamma\left(2 - \frac{D}{2}\right)}{(16\pi^2)^{D/4}(2-D)}(R^2)^{D/2-1},$$

$$\int \frac{d^D \tilde{p}}{(2\pi)^D} \frac{1}{[\tilde{p}^2 - R^2]^2} = \frac{i\Gamma\left(2 - \frac{D}{2}\right)}{(16\pi^2)^{D/4}}(R^2)^{D/2-2}, \tag{8.52}$$

where in our case

$$R^2 = xm_s^2 + \bar{x}m_u^2 - x\bar{x}q^2,$$

so that $R^2 > 0$ at $q^2 < 0$. The remaining second term, which is linear in $\tilde{p}$, vanishes after integration, due to antisymmetry. Substituting the integrals (8.52) in (8.50), we replace D by $4 - \epsilon$ according to (8.47). The result:

$$\Pi_5^{(0)}(q^2) = -\frac{12(m_s + m_u)^2}{(16\pi^2)^{1-\epsilon/4}} \mu^\epsilon \Gamma(\epsilon/2) \int\limits_0^1 dx \left[\frac{4-\epsilon}{2-\epsilon}(R^2)^{1-\epsilon/2} \right.$$

$$\left. - (x\bar{x}q^2 + m_s m_u)(R^2)^{-\epsilon/2} \right], \tag{8.53}$$

is then expanded near $\epsilon = 0$, employing the formula:

$$\Gamma(\epsilon/2) = \frac{2}{\epsilon} - \gamma_E + O(\epsilon), \quad a^\epsilon = 1 + \epsilon \ln a + O(\epsilon^2), \tag{8.54}$$

where $\gamma_E = 0.5772...$ is the Euler-Mascheroni constant, and yielding:

$$\Pi_5^{(0)}(q^2) = \frac{3}{4\pi^2}(m_s + m_u)^2 \int\limits_0^1 dx \left\{ \left[\ln\left(\frac{xm_s^2 + \bar{x}m_u^2 - x\bar{x}q^2}{\mu^2} \right) - \left(\frac{2}{\epsilon} - \gamma_E + \ln(4\pi) \right) \right] \right.$$

$$\left. \times (2xm_s^2 + 2\bar{x}m_u^2 - m_s m_u - 3x\bar{x}q^2) - (xm_s^2 + \bar{x}m_u^2 - x\bar{x}q^2) \right\}. \tag{8.55}$$

The term proportional to $\sim 1/\epsilon$ and stemming from the pole of the Γ function represents the divergent part of the loop diagram at $\epsilon \to 0$. Being polynomial in q^2, this term together with other polynomial terms in (8.55) will vanish after Borel transformation or multiple differentiation over q^2. Thus, in the QCD sum rule both the ultraviolet divergence of the loop diagram and the μ- dependence introduced by regularization procedure do not play a role. In what follows, we only consider the logarithmic part of (8.55) taken at large $Q^2 \equiv -q^2 \gg \Lambda_{QCD}^2$, where a perturbative calculation of the correlation function is possible. Importantly, $\Pi_5^{pert}(q^2)$, consisting of $\Pi_5^{(0)}(q^2)$ and NLO corrections, fully determines the asymptotic behavior of $\Pi_5(q^2)$ at $q^2 \to -\infty$.

Retaining the global quark-mass factors in the currents (8.42) we may neglect the masses in the light-quark propagators, keeping in mind that both s and u quark masses are small with respect to characteristic moments $\sim Q$ flowing through the loop. In this case, an alternative and much simpler way of calculation is to employ the quark propagator (A.50) in the coordinate representation. Instead of (8.45) we obtain

$$\Pi_5^{(0)}(q^2) = -\frac{i^3 N_c (m_s + m_u)^2}{4\pi^4} \int d^4x\, e^{iqx} \frac{x_\alpha x_\beta}{(x^2)^4} Tr\{\gamma^\alpha \gamma_5 \gamma^\beta \gamma_5\}. \tag{8.56}$$

Here we can use one of the integrals in (A.95). They are valid up to a (divergent) polynomial in q^2, which is however irrelevant for a sum rule. The result, after taking trace and contracting the Lorentz indices, reads:

$$\Pi_5^{(0)}(q^2) = -\frac{3}{8\pi^2}(m_s + m_u)^2 q^2 \ln(-q^2/\mu^2). \tag{8.57}$$

It coincides (up to a polynomial in q^2) with the limit $m_{s,u} \to 0$ of the expression in curly brackets in (8.55) integrated over x.

For the QCD sum rule of the type (8.17), based on the correlation function $\Pi_5(q^2)$, we also need the imaginary part of $\Pi_5^{(pert)}(q^2)$. It will be used for the quark-hadron duality approximation (see (8.19)). In the LO, neglecting the light quark masses in the propagators, we easily obtain from (8.57):

$$Im\Pi_5^{(0)}(s) = \frac{3}{8\pi}(m_s + m_u)^2 s\, \theta(s), \tag{8.58}$$

analytically continuing the logarithmic function towards positive $q^2 = s$ and using the formula:

$$\ln(-q^2) = \ln|q^2| - i\pi\theta(q^2). \tag{8.59}$$

Higher order corrections in m_s^2/s to (8.58) are numerically small in the region $s > s_0$, and the mass of the u quark can be safely neglected.

From (8.58) we see that the imaginary part of the LO perturbative contribution grows linearly with s; hence, also for the complete correlation function we can state that

$$\lim_{s\to\infty} Im\,\Pi_5(s) \sim s,$$

and at least two subtractions are needed to render the dispersion integral finite. We already discussed that there are basically two procedures available – the multiple differentiation (8.18) and Borel transformation (8.19) – that remove subtraction terms and improve the convergence of the dispersion integrals. Here we use, following the analyses of this particular sum rule in the literature (see, e.g., [77] and references therein), a combination of both procedures. First, we simply differentiate the expression twice over q^2. For the LO part we obtain:

$$\Pi_5^{(0)''}(q^2) = \frac{d^2}{d(q^2)^2}\Pi_5^{(0)}(q^2) = \frac{2}{\pi}\int_0^\infty ds\, \frac{Im\Pi_5^{(0)}(s)}{(s-q^2)^3} = -\frac{3(m_s + m_u)^2}{8\pi^2 q^2}, \tag{8.60}$$

where the strange-quark mass corrections are neglected on the r.h.s.

Perturbative NLO contributions to the correlation function (8.41) are described by the diagrams with gluon radiative corrections shown in Figures 8.1 b,c,d. Their calculation is technically more involved, as compared to a simple loop, especially in the case of massive quarks. The details are far beyond our scope. Just to mention that the two-loop diagrams are computed step by step, regularizing divergences, isolating $O(1/\epsilon)$ parts first at the one-loop level and performing then a renormalization procedure, which also introduces scale-dependence of the quark masses in the LO contribution[7]. The color factor in these diagrams is reduced to a trace of two λ-matrices located in quark-gluon vertices. Importantly, this trace (see (A.30)) can be factorized out and separated from the trace over Dirac matrices, yielding a numerical coefficient.

The resulting $O(\alpha_s)$ correction is relatively simple in the massless case. Note that, assuming massless quarks in the diagrams of radiative corrections, we in fact introduce an uncertainty of $O(\alpha_s(Q)m_s/Q)$ which is numerically small. For completeness, we quote perturbative contribution to the correlation function (8.41) including the $O(\alpha_s)$ and $O(m_s^2)$ corrections:

$$\Pi_5^{(pert)''}(q^2, \mu) = \Pi_5^{(0)''}(q^2)\left(1 + \frac{\alpha_s(\mu)}{\pi}\left[\frac{11}{3} - 2\ln\left(\frac{-q^2}{\mu^2}\right)\right] + 2\frac{m_s^2}{q^2} + \dots\right), \quad (8.61)$$

where the scale $\mu \sim Q$ is an optimal choice and all higher-order corrections are indicated by ellipsis. In fact, corrections to $\Pi_5^{(pert)}(q^2)$ up to $O(\alpha_s^4)$ have been evaluated [78], see also [77] where these results are used in a QCD sum rule.

8.5.2 Condensate terms

The perturbative part (8.61) represents only a part of the correlation function (8.41). There are also contributions of vacuum condensates suppressed by inverse powers of Q^2. They form a nonperturbative part of the OPE (8.37). Calculating these contributions, we adopt the fixed-point (Fock-Schwinger) gauge for the vacuum gluon field as defined in (A.55), with $x_0 = 0$[8].

We first compute the $d = 3$ contribution of the quark condensate. It is possible to perform this calculation in a rather general way, deriving the coefficient C_3 at the quark-antiquark operator $\hat{O}_3 = \bar{q}q$ in the OPE of the type (8.37) and after that taking the vacuum average. In practice, it is more convenient to treat the quark condensate as an external static field distributed in the vacuum with a constant density equal to $\langle 0|\bar{q}q|0\rangle$, where in our case $q = u, s$. The virtual quark-antiquark pair interacts with that field after being emitted and before being absorbed by an external current. The two diagrams depicted in Figure 8.2 represent this interaction in the leading $O(\alpha_s^0)$. In these diagrams, one quark propagates perturbatively, whereas the other one is absorbed and emitted by the u- or s-quark condensate field with vanishing momentum. Accordingly, in the initial expression for the correlation function (8.41) we substitute the currents in terms of quark fields and free propagators in the two possible ways:

$$\Pi_5^{\langle\bar{q}q\rangle}(q^2) = i^3(m_s + m_u)^2 \int d^4x e^{iqx}\left(\langle 0|\bar{u}_{\alpha i}(0)u_\beta^k(x)|0\rangle[\gamma_5 iS_{(s)k}^{(0)i}(-x)\gamma_5]_{\alpha\beta}\right.$$

$$\left. + \langle 0|\bar{s}_{\alpha i}(x)s_\beta^k(0)|0\rangle[\gamma_5 iS_{(u)k}^{(0)i}(x)\gamma_5]_{\alpha\beta}\right). \quad (8.62)$$

---

[7]The renormalization and scale dependence is also relevant if the currents entering the correlation function have anomalous dimensions, which is however not the case for the current (8.42).

[8]Since each term of OPE is separately gauge-invariant, we can use a different, e.g., Feynman gauge in the perturbative contributions considered in the previous section.

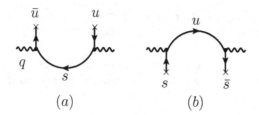

Figure 8.2 Diagrams corresponding to the (a) $\bar{u}u$-quark condensate and (b) $\bar{s}s$-quark condensate contribution to the correlation function (8.41). The crossed lines denote vacuum fields.

Here all indices are shown explicitly, and quark field operators sandwiched between vacuum states are separated from the coefficient functions containing combinations of Dirac matrices and color factors δ_k^i. Since only local operators have to be taken into account, we have to expand the quark fields entering the vacuum averages in (8.62):

$$
\begin{aligned}
u_{\alpha i}(x) &= u_{\alpha i}(0) + x^\mu \overrightarrow{D}_\mu u_{\alpha i}(0) + \dots , \\
\bar{s}_{\alpha i}(x) &= \bar{s}_{\alpha i}(0) + \bar{s}_{\alpha i}(0) \overleftarrow{D}_\mu x^\mu + \dots .
\end{aligned}
\tag{8.63}
$$

Replacing the ordinary derivatives in the above expansion by the covariant ones, we maintain the local gauge invariance of QCD. Since we employ the fixed point gauge, this replacement does not make a difference. Subsequent terms in (8.63) denoted by ellipsis are relevant for the contributions of vacuum condensates with higher dimensions, hence we omit them. Inserting (8.63) we obtain for the vacuum matrix elements in (8.62) in the adopted accuracy:

$$
\begin{aligned}
\langle 0|\bar{u}_{\alpha i}(0) u_\beta^k(x)|0\rangle &= \langle 0|\bar{u}_{\alpha i}(0) u_\beta^k(0)|0\rangle + \langle 0|\bar{u}_{\alpha i}(0) \overrightarrow{D}_\mu u_\beta^k(0)|0\rangle x^\mu , \\
\langle 0|\bar{s}_{\alpha i}(x) s_\beta^k(0)|0\rangle &= \langle 0|\bar{s}_{\alpha i}(0) s_\beta^k(0)|0\rangle + \langle 0|\bar{s}_{\alpha i}(0) \overleftarrow{D}_\mu s_\beta^k(0)|0\rangle x^\mu .
\end{aligned}
\tag{8.64}
$$

The only possible way to parameterize the vacuum averages is:

$$
\langle 0|\bar{u}_{\alpha i} u_\beta^k|0\rangle = A\delta_i^k \delta_{\beta\alpha} , \quad \langle 0|\bar{u}_{\alpha i} \overrightarrow{D}_\mu u_\beta^k|0\rangle = B\delta_i^k (\gamma_\mu)_{\beta\alpha} ,
\tag{8.65}
$$

for the $\bar{u}u$-operators, where A and B are scalar quantities, and δ_i^k on the r.h.s. guarantees the color neutrality of the vacuum state. To obtain A and B, both parts of these equations are multiplied by $\delta_k^i \delta_{\alpha\beta}$ and $\delta_k^i (\gamma_\mu)_{\alpha\beta}$, respectively, and the traces are taken, yielding

$$
A = \frac{1}{4N_c} \langle 0|\bar{u}u|0\rangle , \quad B = \frac{1}{16N_c} \langle 0|\bar{u}\overrightarrow{D}u|0\rangle = -\frac{im_u}{16N_c} \langle 0|\bar{u}u|0\rangle ,
\tag{8.66}
$$

where the Dirac equation in the last relation was used. The analogous expressions are obtained for the vacuum averages of the operators with s-quark fields:

$$
\langle 0|\bar{s}_{\alpha i} s_\beta^k|0\rangle = \frac{1}{4N_c} \langle 0|\bar{s}s|0\rangle \delta_i^k \delta_{\beta\alpha} , \quad \langle 0|\bar{s}_{\alpha i} \overleftarrow{D}_\mu s_\beta^k|0\rangle = \frac{im_s}{16N_c} \langle 0|\bar{s}s|0\rangle \delta_i^k (\gamma_\mu)_{\beta\alpha} .
\tag{8.67}
$$

In the following we use a shorthand notation:

$$
\langle \bar{q}q\rangle \equiv \langle 0|\bar{u}u|0\rangle \simeq \langle 0|\bar{d}d|0\rangle , \quad \langle \bar{s}s\rangle \equiv \langle 0|\bar{s}s|0\rangle .
\tag{8.68}
$$

The u- and d-quark condensates are approximately equal due to the isospin symmetry of QCD, whereas the strange quark condensate $\langle \bar{s}s\rangle$ deviates from $\langle \bar{q}q\rangle$. The $SU(3)_{fl}$-symmetry violating ratio $\langle \bar{s}s\rangle/\langle \bar{q}q\rangle$, being intrinsically related to the quark mass difference $m_s - m_{u,d}$, is a nontrivial nonperturbative effect by itself. We may argue that, the heavier a quark-antiquark pair is, the more it decouples from the QCD vacuum, hence, one expects that $\langle \bar{s}s \rangle \lesssim \langle \bar{q}q \rangle$.

Continuing the calculation, we substitute in (8.62) the vacuum matrix elements (8.64), where the relations (8.65)–(8.67) are used, and replace the propagators by their momentum representations:

$$\Pi_5^{\langle \bar{q}q \rangle}(q^2) = (m_s + m_u)^2 N_c \int d^4 x \, e^{iqx}$$
$$\times \left\{ \frac{\langle \bar{q}q \rangle}{4N_c} \left(\delta_{\beta\alpha} - i \frac{m_u}{4} x^\mu (\gamma_\mu)_{\beta\alpha} \right) \left[\gamma_5 \int \frac{d^4 p}{(2\pi)^4} e^{ipx} \frac{\not{p} + m_s}{p^2 - m_s^2} \gamma_5 \right]_{\alpha\beta} \right.$$
$$\left. + \frac{\langle \bar{s}s \rangle}{4N_c} \left(\delta_{\beta\alpha} + i \frac{m_s}{4} x^\mu (\gamma_\mu)_{\beta\alpha} \right) \left[\gamma_5 \int \frac{d^4 p}{(2\pi)^4} e^{-ipx} \frac{\not{p} + m_u}{p^2 - m_u^2} \gamma_5 \right]_{\alpha\beta} \right\}, \qquad (8.69)$$

where the color trace is taken. Taking also the Dirac trace and integrating over coordinates and four-momentum with the help of the following relation:

$$\int \frac{d^4 p}{(2\pi)^4} \int d^4 x \, e^{iqx \pm ipx} x^\mu f(p) = -i \frac{\partial}{\partial q_\mu} f(\mp q), \qquad (8.70)$$

we finally obtain the quark condensate contribution:

$$\Pi_5^{\langle \bar{q}q \rangle}(q^2) = (m_s + m_u)^2 \left\{ \langle \bar{q}q \rangle \left(\frac{m_s}{q^2 - m_s^2} - \frac{m_u(q^2 - 2m_s^2)}{2(q^2 - m_s^2)^2} \right) \right.$$
$$\left. + \langle \bar{s}s \rangle \left(\frac{m_u}{q^2 - m_u^2} - \frac{m_s(q^2 - 2m_u^2)}{2(q^2 - m_u^2)^2} \right) \right\}. \qquad (8.71)$$

We notice that the quark condensate is always multiplied by a quark mass, forming effectively a $d = 4$ term in OPE which is also a renormalization-scale independent quantity. To apply (8.71) in a QCD sum rule, we can safely neglect in the expression in curly brackets the terms of $O(m_u) \ll m_s$ and the terms proportional to $m_s^2 \ll Q^2$ in the denominator. Taking the second derivative, we obtain:

$$\Pi_5^{\langle \bar{q}q \rangle \prime\prime}(q^2) = \frac{(m_s + m_u)^2}{(q^2)^3} m_s \left(2\langle \bar{q}q \rangle - \langle \bar{s}s \rangle \right). \qquad (8.72)$$

The next, $d = 4$ contribution to the correlation function (8.41) is generated by gluon condensate and is described by the diagrams in Figure 8.3. As in the previous case of quark condensate, these diagrams can be interpreted as a propagation of virtual quarks in an external gluon field. Detailed structure of the external gluon field is inessential, because at the end of the calculation the field is averaged over the QCD vacuum state and parameterized by the gluon condensate $\langle 0|G_{\mu\nu}^a G^{a\mu\nu}|0\rangle$.

The simplest way is to use the quark propagator in which the external field insertions are expressed via gluon field-strength. Use of the fixed point gauge for the gluon field (see Appendix A) enables this task. Fixing $x_0 = 0$, we have from (A.64) for an arbitrary point z:

$$A_\nu(z) = \frac{1}{2} z^\mu G_{\mu\nu}(0), \qquad (8.73)$$

where we also use the compact notation (A.45) for the gluon field and field-strength. Note that we skip other terms of the expansion (A.64), because, containing derivatives of $G_{\mu\nu}$, they do not contribute to the gluon-condensate vacuum density. The latter is obtained

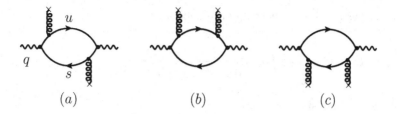

(a) (b) (c)

Figure 8.3 Diagrams corresponding to the gluon condensate contribution to the correlation function (8.41).

from combining two gluon field-strengths in three possible ways, as shown in Figure 8.3. Therefore, the quark propagator to a necessary accuracy is obtained, attaching one or two gluon lines with the external field (8.73) to the virtual quark line, resulting in the expansion

$$S(x,y) = S^{(0)}(x-y) + S^{(1)}(x,y) + S^{(2)}(x,y) \tag{8.74}$$

with the terms

$$S^{(1)}(x,y) = -\int d^4z\, S^{(0)}(x-z)\rlap{A}(z)S^{(0)}(z-y), \tag{8.75}$$

$$S^{(2)}(x,y) = \int d^4z' \int d^4z\, S^{(0)}(x-z')\rlap{A}(z')S^{(0)}(z'-z)\rlap{A}(z)S^{(0)}(z-y) \tag{8.76}$$

with one and two gluon insertions, respectively. We omit the flavor index at the propagator for brevity, so that the above expressions are valid for both $q = s, u$, and adopt the approximation of massless virtual quarks, in which case the derivation is simpler than for massive quarks.

Let us concentrate on the one-gluon term (8.75). We use the expression (A.50) for a massless free-quark propagator and the gluon field (8.73). Furthermore, to avoid possible divergencies, the integration dimension is shifted to $D = 4 - \epsilon$. We obtain

$$S^{(1)}(x,y) = -\frac{1}{2(2\pi^2)^2}\int d^D z \frac{\rlap{z}-\rlap{x}}{[(x-z)^2]^2}z^\mu\gamma^\nu G_{\mu\nu}(0)\frac{\rlap{z}-\rlap{y}}{[(z-y)^2]^2}$$

$$= -\frac{1}{8\pi^4}I^\mu_{\alpha\beta}(x,y)\gamma^\alpha\gamma^\nu\gamma^\beta G_{\mu\nu}(0), \tag{8.77}$$

where the integral over z

$$I^\mu_{\alpha\beta}(x,y) = \int d^D z \frac{(x-z)_\alpha z^\mu (z-y)_\beta}{\left[(x-z)^2\right]^2\left[(z-y)^2\right]^2} \tag{8.78}$$

can be represented as a derivative:

$$I^\mu_{\alpha\beta}(x,y) = -\frac{1}{4}\frac{\partial}{\partial x_\alpha}\frac{\partial}{\partial y_\beta}\tilde{I}^\mu(x,y) \tag{8.79}$$

of a simpler integral

$$\tilde{I}^\mu(x,y) \equiv \int d^D z \frac{z^\mu}{(x-z)^2(z-y)^2} \tag{8.80}$$

which is computed using a standard technique. First, we parameterize it, using (A.88):

$$\tilde{I}^\mu(x,y) = \int_0^1 dv \int d^D z \frac{z^\mu}{\left[z^2 - 2(vx + \bar{v}y)z + vx^2 + \bar{v}y^2\right]^2}, \tag{8.81}$$

and then, denoting $\bar{v} = 1 - v$, rearrange the denominator:

$$\tilde{I}^{\mu}(x, y) = \int_0^1 dv \int d^D z \frac{z^{\mu}}{\left[(z - (vx + \bar{v}y))^2 + v\bar{v}(x - y)^2 \right]^2} \tag{8.82}$$

and shift the four-dimensional variable:

$$z_{\mu} = z'_{\mu} + (vx + \bar{v}y)_{\mu}.$$

In the resulting integral, the term proportional to z'_{μ} vanishes due to antisymmetry of the integrand. The rest is a standard integral (see (A.92)) calculated employing a transformation to the Euclidean coordinate space, similar to the calculation in the D-dimensional momentum space presented in detail in Section 8.5:

$$\tilde{I}^{\mu}(x, y) = \int_0^1 dv \left(vx^{\mu} + \bar{v}y^{\mu} \right) \int \frac{d^D z'}{\left[z'^2 + v\bar{v}(x - y)^2 \right]^2}$$

$$= \int_0^1 dv \left(vx^{\mu} + \bar{v}y^{\mu} \right) \frac{-i\pi^{D/2} \Gamma(2 - \frac{D}{2})}{[-v\bar{v}(x - y)^2]^{2 - D/2}}. \tag{8.83}$$

Using this expression in (8.79), we notice that after differentiation all terms acquire a factor $2 - D/2 = \epsilon/2$, compensating the pole $\sim 1/\epsilon$ in the expansion (8.54) of $\Gamma(\epsilon/2)$; hence, we can safely take the $\epsilon \to 0$ limit. The differentiated coordinate-dependent factor $1/[(x - y)^2]^{\epsilon/2}$ yields $1/(x - y)^2$ or $1/[(x - y)^2]^2$, whereas in the absence of the $1/\epsilon$ pole,

$$\frac{1}{(v\bar{v})^{\epsilon/2}} = \frac{1}{1 + \frac{\epsilon}{2} \ln[\bar{v}v] + ...} \to 1,$$

and integration over v becomes trivial. The result is:

$$I_{\alpha\beta}^{\mu}(x, y) = \frac{i\pi^2}{4} \left(\frac{g_{\alpha\beta}(x + y)^{\mu} + g_{\alpha}^{\mu}(x - y)_{\beta} - g_{\beta}^{\mu}(x - y)_{\alpha}}{(x - y)^2} \right.$$

$$\left. - 2 \frac{(x + y)^{\mu}(x - y)_{\alpha}(x - y)_{\beta}}{\left[(x - y)^2 \right]^2} \right). \tag{8.84}$$

Using it in (8.77) and employing the decomposition of three γ-matrices (see (A.15)), we finally obtain the $O(G_{\mu\nu})$ term of the massless quark propagator:

$$S^{(1)}(x, y) = g_s \frac{\lambda^a}{2} \left[-\frac{1}{8\pi^2} \frac{(x - y)^{\mu}}{(x - y)^2} \gamma^{\nu} \gamma_5 \tilde{G}_{\mu\nu}^a(0) + \frac{i}{4\pi^2} y^{\mu} x^{\nu} \frac{(\not{x} - \not{y})}{[(x - y)^2]^2} G_{\mu\nu}^a(0) \right], \tag{8.85}$$

where we return to the explicit form of the field-strength tensor and use its dual as defined in (A.43).

Derivation of the second-order term $S^{(2)}(x, y)$ from (8.76) is performed along the same lines, but with somewhat lengthy algebra. Note that only the averaged product of the two gluon field-strenghts contributes to $S^{(2)}(x, y)$, hence, the vacuum averaging relation (A.66) can be applied beforehand, simplifying the derivation. Without going into more detail (see, e.g., the review [79]), the following expression is established for the propagator of massless quark q in external gluon field:

$$
\begin{aligned}
S_{(q)}(x,y) &= \frac{1}{2\pi^2}\frac{(\not{x}-\not{y})}{[(x-y)^2]^2} - \frac{1}{8\pi^2}\frac{(x-y)^\mu}{(x-y)^2}\gamma^\nu\gamma_5 g_s\frac{\lambda^a}{2}\tilde{G}^a_{\mu\nu}(0) \\
&+ \frac{i}{4\pi^2}y^\mu x^\nu\frac{(\not{x}-\not{y})}{[(x-y)^2]^2}g_s\frac{\lambda^a}{2}G^a_{\mu\nu}(0) \\
&- \frac{1}{192\pi^2}\frac{(\not{x}-\not{y})}{[(x-y)^2]^2}\left(x^2 y^2-(xy)^2\right)g_s^2 G^a_{\mu\nu}(0)G^{a\mu\nu}(0)\,,
\end{aligned}\tag{8.86}
$$

to the accuracy specified in (8.74).

Note that, as a consequence of the fixed point gauge, this propagator is not invariant with respect to the translation of coordinates. Needless to say, for a physical amplitude translational symmetry can only be violated in separate diagrams and is restored in their sum. Furthermore, if the correlation function is defined as in (8.41), where one of the points coincides with the fixed point (e.g., $y=0$), the parts of the propagator in the second and third line of (8.86) vanish, in particular, the two-gluon term does not contribute. Consequently, to obtain the gluon condensate contribution we only need to calculate the diagram in Figure 8.3(a).

To this end, we start from the expression (8.44), in which we replace both free-quark propagators by the one-gluon part of the propagator (8.86) and take the vacuum average of the gluon field-strengths:

$$
\begin{aligned}
\Pi_5^{\langle GG\rangle}(q^2) &= -i(m_s+m_u)^2\int d^4x\, e^{iqx}\langle 0|\mathrm{Tr}\{S_{(s)}^{(1)}(0,x)\gamma_5 S_{(u)}^{(1)}(x,0)\,\gamma_5\}|0\rangle \\
&= \frac{ig_s^2}{64\pi^4}(m_s+m_u)^2\mathrm{Tr}\Big(\frac{\lambda^a\lambda^b}{4}\Big)\int d^4x\, e^{iqx}\frac{x_{\mu'}x_{\alpha'}}{(x^2)^2} \\
&\times \frac{1}{4}\mathrm{Tr}\{\gamma_{\nu'}\gamma_{\beta'}\}\epsilon^{\mu'\nu'\mu\nu}\epsilon^{\alpha'\beta'\alpha\beta}\langle 0|G^a_{\mu\nu}G^b_{\alpha\beta}|0\rangle\,.
\end{aligned}\tag{8.87}
$$

With the help of the averaging relation (A.66) for the gluon condensate, we simplify this expression. Contracting the indices and employing one of the integration formulas from (A.95), yields

$$
\Pi_5^{\langle GG\rangle}(q^2) = -\frac{(m_s+m_u)^2}{8q^2}\langle GG\rangle\,,\tag{8.88}
$$

and, after double differentiation in q^2, we obtain:

$$
\Pi_5^{\langle GG\rangle''}(q^2) = -\frac{(m_s+m_u)^2}{4(q^2)^3}\langle GG\rangle\,,\tag{8.89}
$$

where a compact notation

$$
\langle GG\rangle \equiv \langle 0|\frac{\alpha_s}{\pi}G^a_{\mu\nu}G^{a\mu\nu}|0\rangle\tag{8.90}
$$

is introduced for the gluon condensate. Note that including α_s into this definition yields a scale independent quantity.

To achieve the level of accuracy adopted in most of the QCD sum rule applications, we include in the OPE of the correlation function (8.41) the contributions of condensates with dimensions $d=5$ and $d=6$. They take into account, respectively, interactions of the propagating quarks with the quark-gluon and four-quark condensates, expressed via

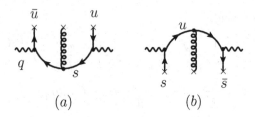

Figure 8.4 Diagrams contributing to the quark-gluon condensate term in the correlation function (8.41).

the vacuum averages of the operators in (8.35). We introduce a compact notation for the quark-gluon condensate:

$$\langle \bar{q}Gq \rangle \equiv \langle 0 | g_s \, \bar{q}\sigma_{\mu\nu} \frac{\lambda^a}{2} G^{a\mu\nu} q \| 0 \rangle. \tag{8.91}$$

For the vacuum averages of four-quark operators we assume a factorization into a product of the two quark condensate densities, schematically:

$$\langle 0 | \bar{q}\Gamma_A q \bar{q}'\Gamma^A q' | 0 \rangle \sim \langle \bar{q}q \rangle \langle \bar{q}'q' \rangle \,, \tag{8.92}$$

where $q, q' = u, d, s$ and the coefficient depends on the matrices Γ_A. A general averaging formula based on this conjecture is presented in (A.69).

The diagrams contributing to $d = 5$ and $d = 6$ terms of OPE are shown, respectively in Figures 8.4 and 8.5. Their calculation is technically more involved, although the same basic elements are employed as for the quark and gluon condensate contributions; hence, below we only present the resulting expressions from the literature. Only a few comments are in order. The complete quark-gluon condensate contribution includes not only the diagrams shown in Figure 8.4, where quark, antiquark and gluon lines directly form the vacuum condensate. Also contributions stemming from the diagrams in Figure 8.2 have to be taken into account. In these diagrams the term with the second covariant derivative in the expansion (8.63) of the quark field around $x = 0$ has to be used. A product of two covariant derivatives, via QCD equation of motion, is then reduced to a gluon field strength, and, combined with the quark and antiquark fields, forms the condensate. We may think of this contribution as if a gluon is emitted from the vacuum quark or antiquark line on the diagram in Figure 8.2. This is just a qualitative interpretation because the whole effect lies beyond QCD perturbation theory. Similarly, the complete four-quark condensate term in OPE, apart from the diagrams in Figure 8.5, receives the following additional contributions: from the diagrams in Figure 8.2 with three covariant derivatives of the quark field and from the diagrams in Figure 8.4 with one derivative of the gluon field strength. The terms with derivatives, via QCD equation of motion, are reduced to a color-charged quark-antiquark density which is combined with the initial quark-antiquark pair to form the four-quark condensate. Some useful averaging relations are presented in Appendix A.

Collecting all terms of the OPE up to $d = 6$, we obtain an expression for the second derivative of the correlation function (8.41) valid at sufficiently large $Q^2 = -q^2$:

$$\Pi_5^{(OPE)''}(q^2) = \frac{d^2}{d(q^2)^2}\Pi_5^{(OPE)}(q^2) = \frac{(m_s + m_u)^2}{Q^2} \left[\frac{3}{8\pi^2} - \frac{m_s \left(2\langle \bar{q}q \rangle - \langle \bar{s}s \rangle \right)}{(Q^2)^2} \right.$$

$$\left. + \frac{\langle GG \rangle}{4(Q^2)^2} - \frac{3m_s \langle \bar{q}Gq \rangle}{(Q^2)^3} - \frac{32\pi\alpha_s \left(\langle \bar{q}q \rangle^2 + \langle \bar{s}s \rangle^2 - 9\langle \bar{q}q \rangle \langle \bar{s}s \rangle \right)}{(Q^2)^3} \right] + \dots, \tag{8.93}$$

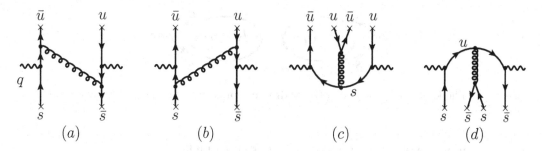

Figure 8.5 Diagrams contributing to the four-quark condensate term in the correlation function (8.41). The diagrams obtained from (c) and (d) with, respectively, the $s\bar{s}$ and $u\bar{u}$ pair emitted from the gluon line, are not shown.

where $q = u, d$. The $d = 5, 6$ terms are taken from [77]. There one can also find important perturbative corrections, up to $O(\alpha_s^4)$ in the perturbative part, as well as the corrections of $O(m_s^2/Q^2)$, all indicated by ellipsis in the above equation.

In (8.93) we confirm the general structure of OPE given in (8.36) in which the universal condensate densities are multiplied by the coefficients suppressed by inverse powers of Q^2 with respect to the perturbative part. The latter is identified as the coefficient of the $d = 0$ term. Note that the OPE coefficients in (8.93) are essentially nonuniversal and depend on quantum numbers of the currents correlated in (8.41). An important feature of the condensates with odd dimensions is that their contributions to OPE are always multiplied by a quark mass factor, to render the overall dimension of the correlation function; hence, $d = 3$ and $d = 5$ terms are usually merged with the $d = 4$ and $d = 6$ terms, respectively, since their coefficients have the same inverse power of Q^2. The perturbative part represented in (8.93) by the LO loop diagram, according to (8.60), can also be rewritten as a dispersion integral over the OPE spectral density, to be used for the quark-hadron duality approximation. Starting from $O(\alpha_s)$, this spectral density receives contributions not only from the perturbative loops but also from radiative corrections to the condensate contributions. A general rule is that the terms included in the spectral density are those with nonvanishing imaginary part above the duality threshold.

To summarize the calculations presented in this section, we may interpret them in several different ways:

- Vacuum condensates are static external fields and we account for the interactions between virtual quarks in the correlation function and these fields;

- There is a separation of distances in the correlation function characterized by a large scale $\mu \gg \Lambda_{QCD}$. All QCD interactions with energy-momenta larger (smaller) than μ are included in the coefficient functions (in the condensate densities) of OPE;

- The condensates take into account modifications of quark and gluon propagators at small virtualities caused by the QCD vacuum fluctuations.

These interpretations are interrelated and they all have certain limitations, reflecting the approximate character of the OPE.

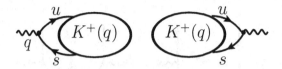

Figure 8.6 The kaon contribution to the hadronic representation of the correlation function (8.41).

8.6 VARIOUS APPLICATIONS OF QCD SUM RULES

8.6.1 The s-quark mass determination

Having at hand the OPE expression (8.93), we proceed with derivation of the QCD sum rule. Since the correlation function depends quadratically on the s-quark mass (the u-quark mass can safely be neglected), this sum rule is well suited to determine m_s. The next step is to obtain the hadronic representation in a form (8.9), applying unitarity relation to the correlation function (8.41). The lightest hadron (the ground state) with quantum numbers of the strange pseudoscalar current is the K^+ meson. We use the definition of the kaon decay constant,

$$\langle 0|(m_s + m_u)\bar{s}i\gamma_5 u|K^+(q)\rangle = f_K m_K^2 \,, \tag{8.94}$$

similar to the one in (2.57) for f_π. The ground-state contribution to the correlation function shown in Figure 8.6 enters the unitarity relation:

$$\frac{1}{\pi}\mathrm{Im}\Pi_5(s) = \delta(s - m_K^2)f_K^2 m_K^4 + \theta(s - s_{K'})\rho_{K'}(s) \,, \tag{8.95}$$

where heavier hadronic states are parameterized by a generic spectral density $\rho_{K'}(s)$. Kaon is a stable hadron with respect to strong decays, hence we neglect its tiny total width given by the inverse lifetime.

Since we aim at using a sum rule to determine one of the QCD parameters in the OPE (8.93), we need as much available information as possible on the hadronic states contributing to $\rho_{K'}(s)$. A state with two pseudoscalar mesons cannot be produced by the pseudoscalar current, due to the P-parity conservation. Therefore, the next state heavier than kaon in the hadronic spectral density is $K\pi\pi$, so that the threshold in (8.95) is

$$s_{K'} = (m_K + 2m_\pi)^2.$$

However, there is no straightforward way to parameterize the $K\pi\pi$ term in $\rho_{K'}(s)$, also because it involves resonance contributions such as $K\rho$ or $K^*\pi$. On the other hand, it is conceivable that the hadronic spectral density $\rho_{K'}(s)$ is saturated by a series of resonances generated by the radial excitations of the kaon. This is similar to multiresonance models of the pion and kaon form factors considered in Chapter 5.5. Moreover, $K\pi\pi$ and other multihadron states, being strongly coupled to radially excited kaons, are implicitly taken into account if one includes total widths of the resonances.

The two first radial excitations of the kaon are $K_1 \equiv K(1460)$ and $K_2 \equiv K(1830)$ [1]. Importantly, their decay constants are suppressed with respect to f_K as can be shown using chiral perturbation theory (see more details e.g., in [80]). The decay constant f_{K_1} can also be measured in $\tau \to K_1\nu_\tau$ decay, whereas the second excitation is too heavy for a τ decay. Finally, for the spectral density of excited states the following ansatz is used:

$$\rho_{K'}(s) = \frac{1}{\pi}\sum_{i=1,2} f_{K_i}^2 m_{K_i}^4 \left(\frac{\Gamma_{K_i} m_{K_i}}{(s - m_{K_i}^2)^2 + (\Gamma_{K_i} m_{K_i})^2}\right) + \theta(s - s_0)\frac{1}{\pi}\mathrm{Im}\Pi_5^{(OPE)}(s) \,, \tag{8.96}$$

where in the first term the factors in parentheses reproduce Breit-Wigner formulas after inserting in the dispersion integral. These factors approach $\delta(s - m_{K_i}^2)$ in the narrow-width limit $\Gamma_{K_i} \to 0$, as follows from the representation

$$\delta(x - a) = \frac{1}{\pi} \lim_{\epsilon \to 0} \frac{\epsilon}{(x - a)^2 + \epsilon^2}. \tag{8.97}$$

Furthermore, for the hadronic states heavier than K_2, a quark-hadron duality approximation is adopted in (8.96) with a threshold parameter s_0. Matching the double differentiated dispersion relation to the QCD result, we obtain an initial form of a sum rule:

$$\Pi_5^{(OPE)''}(q^2) = 2 \int\limits_0^\infty \frac{ds}{(s - q^2)^3} \left[f_K^2 m_K^4 \delta(s - m_K^2) + \rho_{K'}(s) \theta(s - s_{k'}) \right]. \tag{8.98}$$

Substituting in l.h.s. the OPE from (8.93), and in r.h.s. the spectral density $\rho_{K'}(s)$ defined in (8.96), we apply to both parts the Borel transformation defined in (A.103), and obtain the final form of QCD sum rule:

$$(m_s + m_u)^2 \left\{ \frac{3}{8\pi^2} \left[1 - \left(1 + \frac{s_0}{M^2}\right) e^{-s_0/M^2} \right] - \frac{1}{2M^4} \left[m_s \left(2\langle \bar{q}q \rangle - \langle \bar{s}s \rangle \right) - \frac{\langle GG \rangle}{4} \right] \right.$$
$$\left. + \frac{1}{6M^6} \left[-3m_s \langle \bar{q}Gq \rangle - 32\pi\alpha_s \left(\langle \bar{q}q \rangle^2 + \langle \bar{s}s \rangle^2 - 9\langle \bar{q}q \rangle \langle \bar{s}s \rangle \right) \right] + \ldots \right\}$$
$$= \frac{f_K^2 m_K^4}{M^4} e^{-m_K^2/M^2} + \sum_{i=1,2} \frac{f_{K_i}^2 m_{K_i}^4}{M^4} \int_0^\infty ds\, e^{-s/M^2} \frac{\Gamma_{K_i} m_{K_i}}{\pi \left[(s - m_{K_i}^2)^2 + (\Gamma_{K_i} m_{K_i})^2 \right]}, \tag{8.99}$$

where dots indicate the gluon radiative corrections as well as suppressed terms proportional to the powers of m_s^2/M^2. The complete expressions can be found in [77]. Note that, according to the quark-hadron duality approximation adopted in (8.96), in the first term on the l.h.s. of (8.99), we subtracted from the perturbative loop contribution the integral over the loop spectral density (8.58) taken from s_0 to infinity.

The sum rule (8.99) is used to extract the value of m_s, fitting the l.h.s. (OPE) to the r.h.s. (hadronic representation). For this application, all other input parameters in (8.99) should be fixed or constrained. Some of them are specific for this particular sum rule, such as the Borel parameter M, the effective threshold s_0 and the renormalization scale μ, whereas the others (α_s, condensates) are universal.

The variable M characterizes a virtuality scale at which OPE for the correlation function is matched to the hadronic dispersion relation. The values of M have to be large enough to render the power corrections in OPE under numerical control, so that the neglected corrections of higher dimensions are small. On the other hand, M has to be not too large, in order to minimize the uncertainty introduced by the quark-hadron duality approximation. Indeed, at large M the contributions of heavier resonances in r.h.s. of (8.99) are not sufficiently suppressed with respect to the ground-state kaon contribution, making the result more sensitive to the choice of duality threshold. Both above criteria determine a certain region of M (*Borel window*) within which the sum rule can be trusted. For the sum rule (8.99), the optimal values of M are in the ballpark of 1 GeV. An additional useful procedure is to differentiate (8.99) over $1/M^2$ and divide the result by the original sum rule. The factor $(m_s + m_q)^2$ will cancel in this ratio, which then can be used to constrain s_0. Varying M within the Borel window, and adjusting s_0, yields an interval of resulting s-quark masses. This interval is usually counted as a part of the parametric uncertainty, to be combined with the errors caused by other input parameters.

The perturbative part of OPE in (8.99) was calculated adopting the $\overline{MS}$ scheme for quark masses. An optimal scale at which m_s and α_s are normalized is $\mu \sim M$. To take into account radiative corrections to the perturbative part, one has to evaluate $\alpha_s(\mu)$ running it from the value quoted in (1.17). The other universal inputs entering (8.99) are condensate densities. The most important one is the quark condensate which is usually determined from the relation (1.95), employing the average intervals of u, d quark masses from [1] quoted in (1.20). Using also the value of f_π in (2.64), we obtain

$$\langle \bar{q}q \rangle = -\left(276^{+12}_{-10}\,\text{MeV}\right)^3 , \tag{8.100}$$

at the normalization scale $\mu = 2$ GeV. Employing the factorization (8.92), one uses this value also for the four-quark condensate contribution in (8.99). To determine the strange quark condensate $\langle \bar{s}s \rangle$, one cannot simply use the relation (1.96), since, as opposed to (1.95), it receives relatively large unaccounted corrections of $O(m_s^2)$. Instead, it is possible to combine QCD sum rules for strange and nonstrange baryons and extract the $SU(3)_{fl}$-violating ratio $\langle \bar{s}s \rangle / \langle \bar{q}q \rangle$. These sum rules are dominated by quark condensate contributions. Referring to the review [81], where one can find more details, we quote the value of this scale-independent ratio

$$\frac{\langle \bar{s}s \rangle}{\langle \bar{q}q \rangle} = 0.8 \pm 0.1 . \tag{8.101}$$

Furthermore, the QCD sum rule for charmonium states considered in one of the following subsections, determines the interval for the gluon condensate

$$\langle GG \rangle = 0.012^{+0.006}_{-0.012}\,\text{GeV}^4 . \tag{8.102}$$

which is also scale-independent. It remains to specify the quark-gluon condensate which is usually parameterized as $\langle \bar{q}Gq \rangle = m_0^2 \langle \bar{q}q \rangle$. The numerical value of the rescaling coefficient is again determined from comparing the QCD sum rules for light-quark baryons with experiment (see, e.g., [81]) yielding

$$m_0^2 = 0.8 \pm 0.2\,\text{GeV}^2 . \tag{8.103}$$

We notice that the variety of QCD sum rules provides a self-sufficient set of relations that, being combined with experimental input, provides, to a large extent, the whole set of vacuum condensate parameters.

Further details on practical applications of QCD sum rules can be found in the original paper [74] and dedicated reviews such as [82, 83, 84].

8.6.2 Pion decay constant

As another instructive example, we consider the QCD sum rule for the pion decay constant. It is obtained from the correlation function of two axial-vector currents,

$$\Pi_{5\mu\nu}(q) = i\int d^4x\, e^{iqx} \langle 0|T\{j_{\mu5}(x)j_{\nu5}^\dagger(0)\}|0\rangle = -g_{\mu\nu}\Pi_{(g)}(q^2) + q_\mu q_\nu \Pi_{(q)}(q^2), \tag{8.104}$$

where $j_{\mu5} = \bar{u}\gamma_\mu\gamma_5 d$ is also a part of the weak current. The Lorentz decomposition of this correlation function consists of two independent tensors and contains two invariant amplitudes $\Pi_{(g)}$ and $\Pi_{(q)}$. This is consistent with nonconservation of the axial-vector current. Before turning to the actual derivation of a sum rule, we make a digression. Following [74] and combining the correlation function (8.104) with the chiral limit of vanishing u, d quark masses, we demonstrate the existence of a light pion state in QCD and derive the fundamental Gell-Mann-Oakes-Renner relation (1.95) between the pion mass and light quark masses.

The first step is a multiplication of (8.104) by q_μ, leaving one vector index uncontracted:

$$q^\mu \Pi_{5\mu\nu}(q) = i\int d^4x q^\mu e^{iqx} \langle 0|T\{j_{\mu 5}(x)j_{\nu 5}^\dagger(0)\}|0\rangle = \int d^4x \frac{\partial}{\partial x_\mu}\left(e^{iqx}\right)\langle 0|T\{j_{\mu 5}(x)j_{\nu 5}^\dagger(0)\}|0\rangle$$

$$= -\int d^4x\, e^{iqx} \frac{\partial}{\partial x_\mu}\left(\langle 0|T\{j_{\mu 5}(x)j_{\nu 5}^\dagger(0)\}|0\rangle\right)$$

$$= -\int d^4x\, e^{iqx}\left(\langle 0|T\{\partial^\mu j_{\mu 5}(x)j_{\nu 5}^\dagger(0)\}|0\rangle + \delta(x_0)\langle 0|[j_{05}(x), j_{\nu 5}^\dagger(0)]|0\rangle\right), \quad (8.105)$$

where we perform similar steps as in the derivation of the Ward identity (6.41) in Chapter 6.2. In particular, the second term in the last line in (8.105) is a *contact term*, resulting from differentiation over the factor $\theta(x_0)$ which emerges from time-ordering. Note that this particular term in (8.105) can simply be discarded because, according to (A.73), the commutator is reduced to the local vector currents $\bar{u}(0)\gamma_\nu u(0)$, $\bar{d}(0)\gamma_\nu d(0)$ and vanishes after averaging over vacuum. At this stage, it is convenient to make a translation of the coordinates in the vacuum matrix element using the operator relation (A.78):

$$\langle 0|\partial^\mu j_{\mu 5}(x)j_{\nu 5}^\dagger(0)|0\rangle = \langle 0|e^{i\hat{P}\cdot x}\partial^\mu j_{\mu 5}(0)e^{-i\hat{P}\cdot x}j_{\nu 5}^\dagger(0)e^{i\hat{P}\cdot x}|0\rangle = \langle 0|\partial^\mu j_{\mu 5}(0)j_{\nu 5}^\dagger(-x)|0\rangle,$$

where the vanishing momentum of the vacuum allows to insert or remove exponents of the four-momentum operator acting onto the vacuum state. As a result, we have

$$q^\mu \Pi_{5\mu\nu}(q) = -\int d^4x\, e^{iqx}\langle 0|T\{\partial^\mu j_{\mu 5}(0)j_{\nu 5}^\dagger(-x)\}|0\rangle,$$

$$= -\int d^4x\, e^{-iqx}\langle 0|T\{\partial^\mu j_{\mu 5}(0)j_{\nu 5}^\dagger(x)\}|0\rangle,$$

$$= -(m_u + m_d)\int d^4x\, e^{-iqx}\langle 0|T\{j_5(0)j_{\nu 5}^\dagger(x)\}|0\rangle, \quad (8.106)$$

where the integration variable was transformed, $x \to -x$. In the above, we also used (2.52) and replaced the divergence of the axial-vector current by the pseudoscalar current $j_5 \equiv \bar{u}i\gamma_5 d$ multiplied by the sum of the quark masses.

The next stage is to multiply (8.106) by q^ν. Repeating essentially the same steps as above, we obtain:

$$q^\mu q^\nu \Pi_{5\mu\nu}(q) = +i\int d^4x\, e^{-iqx}\left((m_u + m_d)^2\langle 0|T\{j_5(0)j_5^\dagger(x)\}|0\rangle,\right.$$

$$\left. -(m_u + m_d)\delta(x_0)\langle 0|[j_5(0), j_{05}^\dagger(x)]|0\rangle\right). \quad (8.107)$$

Using (A.73), we obtain for the commutator of quark currents

$$[j_5(0), j_{05}^\dagger(x)]\big|_{x_0=0} = i\delta(\vec{x})\left(\bar{u}(0)u(x) + \bar{d}(x)d(0)\right), \quad (8.108)$$

which after integration with δ-functions reduces the contact term in (8.107) to the quark condensate density (in the isospin symmetry limit). The final answer for the contracted correlation function is:

$$q^\mu q^\nu \Pi_{5\mu\nu}(q) = -q^2 \Pi_{(g)}(q^2) + (q^2)^2 \Pi_{(q)}(q^2)$$

$$= i(m_u + m_d)^2\int d^4x\, e^{iqx}\langle 0|T\{j_5(x)j_5^\dagger(0)\}|0\rangle + 2(m_u + m_d)\langle \bar{q}q\rangle. \quad (8.109)$$

The contact term dominates in this equation because it is proportional to the first power of quark masses. Neglecting the higher powers of quark masses, we derive an asymptotic relation for the linear combination of invariant amplitudes:

$$\Pi_{(g)}(q^2) - q^2\Pi_{(q)}(q^2) = -\frac{2(m_u + m_d)\langle \bar{q}q \rangle}{q^2} + O((m_u + m_d)^2). \tag{8.110}$$

Importantly, this relation is exact at $O(m_u, m_d)$ and does not depend on any assumption about OPE, in particular all other condensates do not contribute. We only used the quark current commutators which are based on the free-field commutation relations for quark fields, hence, strictly speaking, this relation is valid at sufficiently large $Q^2 = -q^2$. The asymptotic behavior of the amplitude (8.110) allows us to use an unsubtracted dispersion relation. To obtain that, we need the imaginary part of the correlation function (8.104). For simplicity, in the unitarity relation we only retain the pseudoscalar mesons π and $\pi' \equiv \pi(1300)$ and the axial meson $a_1(1260)$. All other radial excitations and continuum states with the same quantum numbers are omitted for brevity. The unitarity relation has the following form:

$$2\,\mathrm{Im}\Pi_{5\mu\nu}(q) = \langle 0|j_{\mu5}|\pi\rangle\langle\pi|j_{\nu5}^\dagger|0\rangle d\tau_\pi + \langle 0|j_{\mu5}|\pi'\rangle\langle\pi'|j_{\nu5}^\dagger|0\rangle d\tau_{\pi'}$$
$$+\langle 0|j_{\mu5}|a_1\rangle\langle a_1|j_{\nu5}^\dagger|0\rangle d\tau_{a_1} + \dots, \tag{8.111}$$

where

$$d\tau_h = 2\pi\delta(s - m_h^2)$$

is the phase space for a single hadron h (see, e.g., (5.73)) and $q^2 = s$ in the timelike region. For the pion states we use the hadronic matrix element (2.48), whereas for the axial meson the analogous definition of the decay constant is

$$\langle 0|j_{\mu5}|a_1\rangle = \epsilon_\mu^{(a_1)} m_{a_1} f_{a_1}, \tag{8.112}$$

where $\epsilon^{(a_1)}$ is the polarization vector which satisfies the same orthogonality and summation conditions (4.12) as for a vector meson. Substituting these definitions in (8.111) and summing over the a_1 polarizations, we obtain:

$$\frac{1}{\pi}\mathrm{Im}\Pi_{5\mu\nu}(q) = q_\mu q_\nu\left[f_\pi^2\delta(s - m_\pi^2) + f_{\pi'}^2\delta(s - m_{\pi'}^2) + \dots\right]$$
$$+\left(-g_{\mu\nu}m_{a_1}^2 + q_\mu q_\nu\right)f_{a_1}^2\delta(s - m_{a_1}^2) + \dots, \tag{8.113}$$

so that only the pseudoscalar states contribute to the combination of invariant amplitudes in (8.110):

$$\frac{1}{\pi}\mathrm{Im}\left(\Pi_{(g)}(s) - s\Pi_{(q)}(s)\right) = -\left(f_\pi^2 m_\pi^2\delta(s - m_\pi^2) + f_{\pi'}^2 m_{\pi'}^2\delta(s - m_{\pi'}^2) + \dots\right), \tag{8.114}$$

where dots indicate heavier than π' radial excitations of the pion and continuum states. The corresponding dispersion relation is:

$$\Pi_{(g)}(q^2) - q^2\Pi_{(q)}(q^2) = \frac{1}{\pi}\int\frac{ds}{s - q^2}\mathrm{Im}\left(\Pi_{(g)}(s) - s\Pi_{(q)}(s)\right) = \frac{-f_\pi^2 m_\pi^2}{m_\pi^2 - q^2} + \frac{-f_{\pi'}^2 m_{\pi'}^2}{m_{\pi'}^2 - q^2} + \dots. \tag{8.115}$$

Substituting in l.h.s. of this relation the exact asymptotic expression (8.110) at $O(m_u + m_d)$, we conclude that a consistent matching of both sides demands that the pion mass squared is proportional to the quark masses and the pion decay constant does not vanish in the chiral limit:

$$m_\pi^2 \sim (m_u + m_d), \quad f_\pi \sim (m_u + m_d)^0.$$

In addition, to maintain the balance between (8.110) and (8.115), all heavier states with pion quantum numbers in the dispersion relation should have residues proportional to the quark masses, e.g., the decay constant of π',

$$f_{\pi'} \sim (m_u + m_d),$$

vanishes in the chiral limit. Finally, after equating (8.110) and (8.115), and comparing the coefficients at $O((m_u + m_d)/q^2)$ on both sides, we reproduce the Gell-Mann-Oakes-Renner relation (1.95).

Repeating the same derivation for the correlation function of axial-vector currents with strangeness, it is possible to obtain the analogous relation for the kaon mass squared given in (1.96). The fact that the decay constants of radially excited kaons are suppressed and vanish in the chiral limit was already used in the sum rule for the correlation function (8.115) considered in the previous subsection.

Let us now proceed to the QCD sum rule for the pion decay constant. As we realized, this hadronic parameter does not vanish in the limit of massless quarks; hence, we can safely put m_u, m_d and m_π^2 to zero in the correlation function (8.104), to simplify the calculation at the expense of neglecting very small corrections. In this (chiral) limit, the axial-vector current is conserved and, as follows from (8.109),

$$\Pi_{(g)}(q^2) - q^2 \Pi_{(q)}(q^2) = 0, \tag{8.116}$$

hence there is only one independent invariant amplitude. We choose $\Pi_{(q)}$, keeping in mind that the pion state contributes only to this amplitude. From (8.113) we can easily read off the imaginary part and obtain the hadronic dispersion relation:

$$\Pi_{(q)}(q^2) = \frac{1}{\pi} \int \frac{ds}{s - q^2} \mathrm{Im}\Pi_{(q)}(s) = \frac{f_\pi^2}{-q^2} + \frac{f_{a_1}^2}{m_{a_1}^2 - q^2} + \dots, \tag{8.117}$$

where the excited states (only the axial ones in the chiral limit) are denoted by ellipsis. At large $Q^2 = -q^2$, this hadronic representation is matched to the OPE for $\Pi_{(q)}(q^2)$ calculated from the two-point diagrams. The latter are obtained from the ones in Figures 8.1–8.5, replacing $s \to u$, $u \to d$, the vertices $(m_s + m_u)i\gamma_5 \to \gamma_\mu \gamma_5, \gamma_\nu \gamma_5$ and neglecting the quark masses.

Calculation of the LO loop diagram closely resembles the one explained in detail in Section 8.5. In fact, we can start directly from the expression (8.46), replacing the trace and putting quark masses to zero:

$$\Pi_{5\mu\nu}^{(0)}(q^2) = -iN_c \int \frac{d^4p}{(2\pi)^4} \frac{Tr\{\slashed{p}\gamma_\mu\gamma_5(\slashed{p} + \slashed{q})\gamma_\nu\gamma_5\}}{p^2(p+q)^2}. \tag{8.118}$$

Note that in the chiral limit there is no difference between perturbative parts of correlation functions with axal-vector and vector currents. At the LO loop level this is seen explicitly in the trace, where the γ_5 matrices merge into a unit matrix. After dimensional regularization and integration, the divergent and q^2-independent part of the loop diagram is separated in the form of a $1/\epsilon$ term from the q^2-dependent part:

$$\Pi_{5\mu\nu}^{(0)} = -\frac{3}{2\pi^2}\left(q_\mu q_\nu - g_{\mu\nu}q^2\right) \int_0^1 du\, u(1-u) \ln\left(-q^2 u(1-u)\right) + O(1/\epsilon), \tag{8.119}$$

which obeys the relation (8.116). From this expression, using (8.59), we obtain a dispersion relation for the loop contribution to the invariant amplitude:

$$\Pi_{(q)}^{(0)}(q^2) = \frac{1}{4\pi^2} \int\limits_0^\infty \frac{ds}{s - q^2}, \tag{8.120}$$

where the divergence is ignored since it is removed by Borel transformation. Note that the quark and quark-gluon condensate contributions, being multiplied by the u, d -quark masses, both vanish in the chiral limit. The nonvanishing $d = 4$ term in the OPE is the gluon condensate contribution. Here we can use the expression (8.87), replacing the vertices of interpolating currents:

$$
\begin{aligned}
\Pi_{5\mu\nu}^{\langle GG \rangle}(q^2) &= \frac{-ig_s^2}{64\pi^4} \mathrm{Tr}\left(\frac{\lambda^a \lambda^b}{4}\right) \int d^4 x\, e^{iqx} \frac{x_{\mu'} x_{\alpha'}}{(x^2)^2} \\
&\quad \times \frac{1}{4} \mathrm{Tr}\{\gamma_{\nu'} \gamma_\mu \gamma_5 \gamma_{\beta'} \gamma_\nu \gamma_5\} \epsilon^{\mu' \nu' \lambda \rho} \epsilon^{\alpha' \beta' \alpha \beta} \langle 0 | G_{\lambda\rho}^a G_{\alpha\beta}^b | 0 \rangle \\
&= (q_\mu q_\nu - g_{\mu\nu} q^2) \frac{\langle GG \rangle}{12(q^2)^2},
\end{aligned}
\tag{8.121}
$$

where we repeated the vacuum averaging, integration and contraction of indices in the same way that it was done in Section 8.5 and used the notation (8.90). This expression provides the gluon condensate term in the invariant amplitude:

$$\Pi_{(q)}^{\langle GG \rangle}(q^2) = \frac{\langle GG \rangle}{12(q^2)^2}. \tag{8.122}$$

Returning to the hadronic representation (8.117), we now have to follow a different strategy as compared to the previous example of the s-quark mass determination. Here we aim at isolating the ground-state pion contribution containing f_π^2. Therefore, we essentially rely on the quark-hadron duality approximation, replacing the contribution of all heavier than pion states by the integral over perturbative spectral density:

$$\Pi_{(q)}(q^2) = \frac{f_\pi^2}{-q^2} + \frac{1}{4\pi^2} \int\limits_{s_0^\pi}^\infty \frac{ds}{s - q^2} + \dots, \tag{8.123}$$

introducing a specific threshold parameter s_0^π.

At this point all elements of the QCD sum rule are obtained. We substitute in r.h.s. the OPE which in our case is the sum of the loop and gluon condensate contributions given in (8.120) and (8.122). To improve the accuracy, we also include in the OPE the NLO correction to the perturbative part and the $d = 6$ four-quark condensate contribution taking it from the original calculation in [74]. After Borel transformation we finally obtain the sum rule:

$$f_\pi^2 = \frac{1}{4\pi^2} \int\limits_0^{s_0^\pi} ds\, e^{-s/M^2} \left(1 + \frac{\alpha_s}{\pi}\right) + \frac{\langle GG \rangle}{12 M^2} + \frac{176}{81 M^4} \pi \alpha_s \langle \bar{q} q \rangle^2 + \dots. \tag{8.124}$$

relating the pion decay constant to universal QCD parameters (α_s and condensate densities) and depending on the two auxiliary parameters: the variable Borel scale M and the process-dependent duality threshold s_0^π. The measured value of f_π is indeed reproduced in the region $M^2 \sim 1$ GeV^2 and at $s_0^\pi \simeq 0.7$ GeV^2. The alternative methods to use this sum rule, e.g., combining it with the sum rule for the axial meson decay constant, can be found in [74].

8.6.3 Heavy quarkonium

To present further important examples of QCD sum rules and their applications, we switch to the heavy quark sector and consider a correlation function of two heavy quark-antiquark currents with the same flavor. For definiteness, we concentrate on a well-studied case of the vector currents of charmed quarks:

$$\Pi_{\mu\nu}^{(c)}(q) = i \int d^4x e^{iqx} \langle 0|T\{\bar{c}(x)\gamma_\mu c(x)\bar{c}(0)\gamma_\nu c(0)\}|0\rangle \ . \tag{8.125}$$

The new element in this correlation function is the intrinsic large scale

$$m_c \gg \Lambda_{QCD},$$

in addition to the external momentum transfer q. This guarantees a short-distance dominance of the correlation function not only in the spacelike region, but also at $q^2 \geq 0$. To determine the region where the OPE is valid for (8.125), we use the formalism of heavy quark expansion considered in Chapter 1.8. We then choose the frame where $\vec{q} = 0$ and $q^2 = q_0^2$, defining a unit velocity vector $v = (1,0,0,0)$. At $m_c \to \infty$, we decompose the momenta of the virtual c-quark field and its conjugate propagating in the correlation function:

$$p_c = m_c v + \tilde{k}, \ \ p_{\bar{c}} = -m_c v - \tilde{k} + q \ . \tag{8.126}$$

splitting the parts that scale with m_c from the residual momenta, so that the total momentum is conserved:

$$p_c + p_{\bar{c}} = q \ .$$

We then transform the c quark fields into the HQET effective fields similar to (1.130), factorizing out the dependence on the heavy scale according to the decomposition in (8.126),

$$c(x) = h_v(x)e^{-im_c vx} \ , \ \ \bar{c}(x) = \bar{h}_v(x)e^{-im_c vx} \ ,$$

and obtain

$$\Pi_{\mu\nu}^{(c)}(q) = i \int d^4x e^{i(q-2m_c v)x} \langle 0 \mid T\{\bar{h}_v(x)\gamma_\mu h_v(x), \bar{h}_v(0)\gamma_\nu h_v(0)\} \mid 0\rangle \ . \tag{8.127}$$

The effective fields inside the vacuum average are now decoupled from the heavy scale m_c; hence, this correlation function is similar to (8.1), we only have to replace the external momentum

$$q \to q - 2m_c v.$$

Following the derivation in Section 8.2, we can prove that the average distance in the correlation function is

$$x_0 \sim |\vec{x}| \lesssim \frac{1}{\sqrt{(2m_c v - q)^2}} = \frac{1}{2m_c - \sqrt{q^2}};$$

hence, for the correlation function (8.125) the OPE in local operators is valid if

$$q^2 \ll 4m_c^2 \ , \tag{8.128}$$

including also the whole spacelike region $q^2 < 0$.

The loop diagrams including the LO loop and NLO perturbative corrections are the same as in Figure 8.1, where both quark flavors are now replaced by c. The expression for the LO loop is calculable using the technique described in Section 8.5. It is also instructive to perform this calculation in a different way, employing the unitarity relation for the quark loop diagram, and obtaining its imaginary part. We stress that the singularities generated by on-shell quarks and forming this imaginary part have no direct physical sense because in reality the quarks are confined. Only the integrals over the imaginary part of $\Pi_{\mu\nu}$ are needed, in order to maintain the quark-hadron duality approximation for the resulting QCD sum rule.

We start from decomposing the correlation function (8.125) into Lorentz structures,

$$\Pi_{\mu\nu}^{(c)}(q) = \left(- q^2 g_{\mu\nu} + q_\mu q_\nu\right)\Pi^{(c)}(q^2)\,, \tag{8.129}$$

which yields a single invariant amplitude $\Pi^{(c)}(q^2)$, to obey the vector-current conservation condition:

$$q^\mu \Pi_{\mu\nu}^{(c)}(q) = q^\nu \Pi_{\mu\nu}^{(c)}(q) = 0\,, \tag{8.130}$$

Note that the invariant amplitude is obtained multiplying both sides of (8.129) by the metric tensor:

$$g^{\mu\nu}\Pi_{\mu\nu}^{(c)}(q) = -3q^2\Pi^{(c)}(q^2)\,. \tag{8.131}$$

Repeating the same steps as in the case of the correlation function of light-quark currents in Section 8.5, we arrive at the loop integral

$$\Pi_{\mu\nu}^{(c,0)}(q^2) = iN_c \int \frac{d^4p}{(2\pi)^4} \frac{Tr\{(\not{p} + m_c)\gamma_\mu((\not{p} + \not{q}) + m_c)\gamma_\nu\}}{(p^2 - m_c^2)((p+q)^2 - m_c^2)}\,. \tag{8.132}$$

A direct transition to the imaginary part is possible with the help of the *Cutkosky rule*. According to this prescription, in order to obtain the imaginary part of the two-point diagram we have to replace each propagator by its imaginary part as given in (A.102) and add a general factor i. We obtain

$$
\begin{aligned}
2\mathrm{Im}\,\Pi_{\mu\nu}^{(c,0)}(q) &= -N_c \int \frac{d^4p}{(2\pi)^2} Tr\{(\not{p} + m_c)\gamma_\mu((\not{p} + \not{q}) + m_c)\gamma_\nu\} \\
&\times \delta(p^2 - m_c^2)\theta(-p_0)\delta((p+q)^2 - m_c^2)\theta(p_0 + q_0)\,,
\end{aligned} \tag{8.133}
$$

or, using (8.131),

$$
\begin{aligned}
2\mathrm{Im}\left(3q^2\Pi^{(c,0)}(q^2)\right) &= N_c \int \frac{d^4p}{(2\pi)^2} 4(q^2 + 2m_c^2) \\
&\times \delta(p^2 - m_c^2)\theta(-p_0)\delta((p+q)^2 - m_c^2)\theta(p_0 + q_0)\,,
\end{aligned} \tag{8.134}
$$

where we substitute in r.h.s. the expression for the Lorentz-contracted trace calculated taking into account the mass-shell conditions $(p + q)^2 = m_c^2$ and $p^2 = m_c^2$:

$$Tr\{(\not{p} + m_c)\gamma_\mu((\not{p} + \not{q}) + m_c)\gamma^\mu\} = 4(q^2 + 2m_c^2)\,. \tag{8.135}$$

Since the imaginary part is nonvanishing at positive q^2, we can choose the rest frame of the currents (or the c.m. frame of the $c\bar{c}$ pair) where $\vec{q} = 0$ and $q^2 = q_0^2$. The remaining phase-space integration is then gradually reduced to a kinematical factor:

$$\int d^4p\, \delta(p^2 - m_c^2)\theta(-p_0)\delta((p+q)^2 - m_c^2)\theta(p_0 + q_0)$$

$$= \int_{-q_0}^{0} dp_0 \int d\vec{p}\, \delta(p_0^2 - \vec{p}^2 + 2p_0q_0 + q^2 - m_c^2)\delta(p_0^2 - \vec{p}^2 - m_c^2)$$

$$= 4\pi \int_{-q_0}^{0} dp_0\, \delta(2p_0q_0 + q^2) \int_0^\infty |\vec{p}|^2 d|\vec{p}|\delta(p_0^2 - \vec{p}^2 - m_c^2)$$

$$= 4\pi \int_{-q_0}^{0} dp_0\, \delta(2p_0q_0 + q^2)\frac{p_0^2 - m_c^2}{2\sqrt{p_0^2 - m_c^2}}\theta(p_0^2 - m_c^2)$$

$$= \frac{\pi}{\sqrt{q^2}} \int_0^{\sqrt{q^2}} dp_0'\, \delta(p_0' - \frac{\sqrt{q^2}}{2})\sqrt{p_0'^2 - m_c^2}\,\theta(p_0'^2 - m_c^2),$$

$$= \frac{\pi}{2}\sqrt{1 - \frac{4m_c^2}{q^2}}\,\theta(q^2 - 4m_c^2)\,. \tag{8.136}$$

Substituting this factor in (8.134), we finally obtain the imaginary part (spectral density) of the massive loop diagram at $q^2 = s$:

$$\operatorname{Im}\Pi^{(c,0)}(s) = \frac{N_c}{24\pi}(3 - v^2(s))v(s)\theta(s - 4m_c^2)\,, \tag{8.137}$$

where we use the notation

$$v(s) = \sqrt{1 - \frac{4m_c^2}{s}}$$

for the velocity of the c-quark in the c.m. frame. This expression allows us to obtain the correlation function in a form of the dispersion relation. One has however to take into account that the asymptotics of the imaginary part at $q^2 \to \infty$ is a constant, hence there is a need to perform at least one subtraction to render the dispersion integral convergent. The most convenient way is to use the fact[9] that $\Pi(0) = 0$ and perform the subtraction at $q^2 = 0$:

$$\Pi^{(c,0)}(q^2) = \frac{q^2}{\pi} \int_{4m_c^2}^\infty ds\frac{\operatorname{Im}\Pi^{(c,0)}(s)}{s(s - q^2)}\,. \tag{8.138}$$

The gluon radiative corrections to the loop diagram are currently calculated up to $O(\alpha_s^3)$.
We only quote the NLO correction to the perturbative part:

$$\operatorname{Im}\Pi^{(c,pert)}(s) = \operatorname{Im}\Pi^{(c,0)}(s)\left(1 + \alpha_s C_F\left[\frac{\pi}{2v(s)} - \frac{v(s)+3}{4}\left(\frac{\pi}{2v(s)} - \frac{3}{4\pi}\right)\right]\right)\,. \tag{8.139}$$

The most important contribution to the $O(\alpha_s)$ correction is the term inversely proportional to the quark velocity which becomes large at $s \to 4m_c^2$ (at $v(s) \to 0$), near the $\bar{c}c$ threshold. In fact, this term originates from the Coulomb part of the one-gluon exchange between

---

[9]The same loop diagram, up to a factor of $4\pi\alpha_{em}Q_c^2$, describes the contribution of the c-quark to the polarization operator of the photon in QED, which has to vanish at $q^2 = 0$ to keep the photon massless.

quark and antiquark. At each $O(\alpha_s^n)$, a corresponding term of $O(\alpha_s^n/v^n)$ emerges. Methods to resum these corrections in the nonrelativistic region $v \to 0$ use a dedicated QCD-based effective theory of N(on)R(elativistic)QCD[10].

Another peculiarity of the correlation function (8.125) concerns its condensate expansion. Since the c-quark mass is much larger than Λ_{QCD}, the average lifetimes and propagation distances of $\bar{c}c$ fluctuations in the QCD vacuum are much shorter than typical scales of the light quark-antiquark and gluonic fluctuations forming the vacuum condensates. Due to asymptotic freedom, a spontaneous $c\bar{c}$ fluctuation produced by gluon fields in the vacuum is also less probable, being suppressed by $\alpha_s(m_c)$ and/or by a power of the inverse c-quark mass. Therefore, separate terms describing nonperturbative vacuum averages of the $\bar{c}c$ operators in OPE (i.e., the c-quark condensates) are absent. The light-quark condensate contributions to the correlation of $c\bar{c}$ currents are also suppressed, containing extra orders of α_s caused by additional gluon exchanges between the propagating heavy quarks and the light quark vacuum fields. We conclude that at the usual accuracy level, that is, retaining $d \leq 6$ terms in OPE, only the gluon condensate has to be taken into account.

One has to calculate the same diagrams as the ones depicted in Figure 8.3. The main ingredient of this calculation is the propagator of a massive quark in the external gluon field. To obtain it, one uses the fixed-point gauge, similar to the massless case considered in Section 8.5. The derivation, more conveniently done in the momentum representation, does not contain essentially new elements but is more cumbersome and we skip the details.

The c-quark propagator up to the first order in the (constant) gluon field-strength has the following expression:

$$
S_{(c)}(x,y) = \int \frac{d^4k}{(2\pi)^4} e^{-ik(x-y)} \left\{ \frac{\slashed{k} + m_c}{k^2 - m_c^2} \right.
$$

$$
\left. + \frac{g_s}{2} G_{\mu\nu}^a \frac{\lambda^a}{2} \left[\frac{\epsilon^{\mu\nu\alpha\beta}\gamma_\alpha\gamma_5 k_\beta - m_c\sigma^{\mu\nu}}{(k^2 - m_c^2)^2} + y^\mu \frac{(k^2 - m_c^2)\gamma^\nu - 2k^\nu(\slashed{k} + m_c)}{(k^2 - m_c^2)^2} \right] \right\}. \quad (8.140)
$$

The resulting gluon condensate term

$$
\Pi^{(c,GG)}(q^2) = \frac{f_G(v^2(q^2))}{48(q^2)^2} \langle GG \rangle, \quad (8.141)
$$

where [73]

$$
f_G(x) = \frac{3(x+1)(x-1)^2}{2x^{5/2}} \log \frac{\sqrt{x}+1}{\sqrt{x}-1} - \frac{3x^2 - 2x + 3}{x^2},
$$

is then added to the perturbative part of the OPE for the invariant amplitude:

$$
\Pi^{(c,OPE)}(q^2) = \frac{q^2}{\pi} \int\limits_{4m_c^2}^\infty ds \frac{\text{Im}\,\Pi^{(c,pert)}(s)}{s(s-q^2)} + \Pi^{(c,GG)}(q^2). \quad (8.142)
$$

According to the unitarity condition, the dispersion relation for (8.125) receives contributions from all hadronic states with $c\bar{c}$ flavor content and spin-parity $J^P = 1^-$, starting

---

[10]For a comprehensive review see, e.g., [45].

from the lightest one[11], the J/ψ meson (hereafter we omit $(1S)$ for brevity):

$$\Pi_{\mu\nu}^{(c,hadr)}(q) = q^2 \sum_{\psi=J/\psi,\psi(2S)} \frac{\langle 0|\bar{c}\gamma_\mu c|\psi(q)\rangle\langle\psi(q)|\bar{c}\gamma_\nu c|0\rangle}{m_\psi^2(m_\psi^2 - q^2)} + q^2 \int_{4m_D^2}^{\infty} \frac{ds}{s(s - q^2)} \rho_{\mu\nu}^h(s). \quad (8.143)$$

There are two isolated poles corresponding to J/ψ and $\psi(2S)$ and the dispersion integral over $\rho_{\mu\nu}^h$ accumulates the contributions of hadronic states above the open charm threshold $4m_D^2$ corresponding to the lightest $D\bar{D}$-meson state. Using on the r.h.s. the definition of the ψ-meson decay constant:

$$\langle 0|\bar{c}\gamma_\mu c|\psi(q)\rangle = m_\psi \varepsilon_\mu^{(\psi)} f_\psi, \quad (8.144)$$

where $\varepsilon^{(\psi)}$ is the polarization vector, and assuming quark-hadron duality for the open charm states with an effective threshold s_0^ψ:

$$\rho_{\mu\nu}^h(s) = (-s g_{\mu\nu} + q_\mu q_\nu)\frac{1}{\pi}\text{Im}\,\Pi^{(c,pert)}(s)\theta(s - s_0^\psi), \quad (8.145)$$

we replace l.h.s. by the OPE given in (8.142) and obtain:

$$\frac{q^2}{\pi} \int_{4m_c^2}^{s_0^\psi} ds \frac{\text{Im}\,\Pi^{(c,pert)}(s)}{s(s - q^2)} + \Pi^{(c,GG)}(q^2) = q^2 \sum_{\psi=J/\psi,\psi(2S)} \frac{f_\psi^2}{m_\psi^2 - q^2}, \quad (8.146)$$

where the integrals over the perturbative spectral density taken from s_0^ψ to infinity were subtracted from both sides. Note that the region above the open charm threshold is dominated by several broad ψ resonances, listed in the charmonium section of [1]; hence, one could modify the duality ansatz in (8.146), adding more resonances to the hadronic part and, correspondingly, increasing the duality threshold.

The conventional form of the QCD (SVZ) sum rule for vector charmonium is obtained [73, 74] by taking power moments of the invariant amplitude at $q^2 = 0$, defined as

$$M_n^{(c)} = \frac{1}{n!}\frac{d^n}{d(q^2)^n}\Pi^{(c)}(q^2)\Big|_{q^2=0}, \quad (8.147)$$

and equating the moments derived from the OPE expression with the ones obtained from the hadronic dispersion relation:

$$M_n^{(c,OPE)} = M_n^{(c,hadr)} = \frac{1}{\pi} \int_{m_{J/\psi}^2}^{\infty} \frac{ds}{s^{n+1}}\text{Im}\Pi^{(c,hadr)}(s). \quad (8.148)$$

With growing n, the subtracted part of the dispersion integral gets suppressed with respect to the rest, making the sum rule less sensitive to the quark-hadron duality ansatz (8.145). But at the same time, the condensate terms increase with respect to the perturbative part; hence, one should compromise between these two effects while choosing an optimal set of moments.

---

[11]Strictly speaking, quantum numbers allow also the light hadronic states with $I = 0$ and $J^P = 1^-$ located below the charmonium levels. Invoking quark-hadron duality, these states should be attributed to the OPE diagrams with at least three intermediate gluons exchanged between two $\bar{c}c$ loops, hence, suppressed at least by $O(\alpha_s^3)$.

The sum rule (8.148) can be used to estimate the J/ψ decay constant, attributing an appropriate duality interval for this resonance in the hadronic part. In fact, this sum rule is mainly used in the opposite direction, to determine the c-quark mass. The reason is that we do actually know the hadronic moments on the r.h.s. of (8.148). The latter are directly related to the integrals over the cross section of charm production in e^+e^- annihilation. We use the fact that the vector current $\bar{c}\gamma_\mu c$ coincides with the c-quark part of the quark e.m. current, up to the quark charge factor Q_c. Thus, the amplitudes to produce $c\bar{c}$ states via the virtual photon in e^+e^- annihilation contain the same hadronic matrix elements as those in the unitarity sum of the correlation function (8.125). As a result, the total cross section of charm production which contains a sum over all these amplitudes squared is expressed in terms of the imaginary part of the correlation function:

$$\sigma^{(e^+e^- \to charm)}(s) = \alpha_{em}^2 Q_c^2 k(s) \sum_{h_c} \int d\tau_{h_c} \overline{|(\bar{e}\gamma^\mu e)\langle 0|\bar{c}\gamma_\mu c|h_c\rangle|^2}$$

$$= \alpha_{em}^2 Q_c^2 k(s)\ell^{\mu\nu} \sum_{h_c} \int d\tau_{h_c} \langle 0|\bar{c}\gamma_\mu c|h_c\rangle\langle h_c|\bar{c}\gamma_\nu c|0\rangle = 2\alpha_{em}^2 Q_c^2 k(s)\ell^{\mu\nu} \mathrm{Im}\Pi_{\mu\nu}^{(c,hadr)}(q)$$

$$= 2\alpha_{em}^2 Q_c^2 k(s)\ell^{\mu\nu}\left(-g_{\mu\nu}s + q_\mu q_\nu\right)\mathrm{Im}\Pi^{(c,hadr)}(s)\,, \tag{8.149}$$

where we use hermiticity of the current and apply the unitarity relation (B.39) for the correlation function (8.125). In the above the following notation is used: the sum goes over all possible states with $c\bar{c}$ content, $h_c = \{J/\psi, \psi(2S), D\bar{D}, \dots\}$, integrated over the respective phase space; the squared electron-positron e.m. current transforms into a tensor $\ell_{\mu\nu}$; $k(s)$ denotes the intermediate photon propagator multiplied by a kinematical factor stemming from the cross section definition. We actually do not need to specify $k(s)$. Instead, we calculate the cross section of a purely leptonic process $e^+e^- \to \mu^+\mu^-$, neglecting a small (with respect to m_c) muon mass. At $O(\alpha_{em}^2)$, the muonic cross section will, as in (8.149), be reduced to the imaginary part of the correlation function of two muon e.m. currents, formally obtained replacing c-quark fields by muon fields in (8.125)):

$$\sigma^{(e^+e^- \to \mu^+\mu^-)}(s) = 2\alpha_{em}^2 k(s)\ell^{\mu\nu}\mathrm{Im}\Pi_{\mu\nu}^{(\mu)}(q) = 2\alpha_{em}^2 k(s)\ell^{\mu\nu}(-g_{\mu\nu}s + q_\mu q_\nu)\mathrm{Im}\Pi^{(\mu)}(s), \tag{8.150}$$

where all factors, apart from the quark charge factor, are the same as in (8.149). But in this case we can calculate the exact imaginary part of $\Pi^{(\mu)}$ because it is given by a two-point loop consisting of two muon propagators. In fact, the result is easily read off the expression (8.137) for the c-quark loop, removing the factor N_c and putting $m_c \to 0$ ($v \to 1$):

$$\mathrm{Im}\Pi^{(\mu)}(s) = \frac{1}{12\pi}\,. \tag{8.151}$$

The muon-pair production cross section is also well known from QED:

$$\sigma^{(e^+e^- \to \mu^+\mu^-)}(s) = \frac{4\pi\alpha_{em}^2}{3s}\,.$$

Forming a ratio of the two cross sections, we divide (8.149) by (8.150):

$$R^{(c)}(s) = \frac{\sigma^{(e^+e^- \to charm)}}{\sigma^{(e^+e^- \to \mu^+\mu^-)}} = \frac{Q_c^2 \mathrm{Im}\Pi^{(c,hadr)}(s)}{\mathrm{Im}\Pi^{(\mu)}(s)}\,, \tag{8.152}$$

and, using (8.151), obtain

$$\mathrm{Im}\Pi^{(c,hadr)}(s) = \frac{1}{12\pi Q_c^2}R_c(s)\,. \tag{8.153}$$

With this relation, the sum rule (8.148) is converted into

$$M_n^{(c,OPE)} = \frac{1}{12\pi^2 Q_c^2} \int\limits_{m_{J/\psi}^2}^{\infty} \frac{ds}{s^{n+1}} R^{(c)}(s).$$

(8.154)

This sum rule offers a possibility to extract the c-quark mass with an unprecedented accuracy, by matching the moments calculated with OPE with the ones saturated by the measured cross section. In parallel, a combination of moments is used to extract the value of the gluon condensate density quoted in (8.102).

We note, parenthetically, that another QCD prediction well confirmed by experiment concerns the asymptotics of the ratio $R^{(c)}(s)$. It follows from quark-hadron duality (8.145) at large energies:

$$\lim_{s\to\infty} \mathrm{Im}\Pi^{(c,hadr)}(s) = \mathrm{Im}\Pi^{(c,pert)}(s) = \frac{N_c}{12\pi},$$

where we again use (8.137) at $s \gg 4m_c^2$; hence, from (8.152) it follows that

$$\lim_{s\to\infty} R_c(s) = N_c Q_c^2 = \frac{4}{3}.$$

(8.155)

In conclusion, one has to mention that the sum rules for the vector $b\bar{b}$ currents represent a direct replica of (8.154) and are equally well suited to determine the b quark mass. The difference between the spectra of bottomonium and charmonium states is only in the amount of narrow levels below the open flavor threshold. Also the contribution of the gluon condensate (of the Coulomb part of the $O(\alpha_s)$ corrections) is smaller (larger) because $m_b > m_c$. A determination of m_c and m_b involving the most accurate to date OPE calculation of the moments can be found e.g., in [85]. QCD sum rules with the flavor-nondiagonal $\bar{b}c$ currents also have a considerable application potential for the B_c meson studies but are, so far, less explored. Further relevant details on QCD sum rules for heavy quarkonia can be found in the reviews [24, 82, 86, 45] .

8.6.4 Decay constants of heavy mesons

To apply QCD sum rules to heavy-flavored $\bar{B}, \bar{B}_s$ or D, D_s mesons, we need correlation functions of the quark currents with a heavy-light $Q\bar{q}$ flavor content, where Q is either b or c and $q = u, d, s$. There is an important aspect on which we focus here. Once the calculation of the correlation function is performed for a finite heavy-quark mass m_Q, the latter can be put to infinity, converting a sum rule to its analog in HQET. Doing that, one can also retain and quantitatively assess the inverse heavy-mass corrections which are in many cases not accessible in a pure HQET framework.

Two important examples are provided by QCD sum rules for the decay constants of the pseudoscalar and vector heavy-light mesons[12]. For these hadrons we use a generic notation

$$H = \{D_{(s)}, B_{(s)}\} \quad \text{and} \quad H^* = \{D_{(s)}^*, B_{(s)}^*\},$$

respectively. We start from defining the correlation functions:

$$\Pi_5(q) = i \int d^4x \, e^{iqx} \langle 0|T\{j_5(x)j_5^\dagger(0)\}|0\rangle$$

(8.156)

and

$$\Pi_{\mu\nu}(q) = i \int d^4x \, e^{iqx} \langle 0|T\{j_\mu(x)j_\nu^\dagger(0)\}|0\rangle = \left(-g_{\mu\nu} + \frac{q_\mu q_\nu}{q^2}\right)\Pi_T(q^2) + q_\mu q_\nu \Pi_L(q^2), \quad (8.157)$$

---

[12]One of the first applications is in [87].

where

$$j_5 = (m_Q + m_q)\bar{q}\,i\gamma_5 Q \quad \text{and} \quad j_\mu = \bar{q}\gamma_\mu Q$$

are, respectively, the pseudoscalar and vector heavy-light quark currents. Note that in (8.157) we use only the invariant amplitude $\Pi_T(q^2)$ at the transverse Lorentz structure, and ignore the second (longitudinal) structure whose appearance reflects nonconservation of the heavy-light quark current.

The invariant amplitudes obey double-subtracted dispersion relations in the variable q^2, so that

$$\Pi_5(q^2) - \Pi_5(0) - q^2 \left(\frac{d\,\Pi_5(q^2)}{dq^2}\right)\bigg|_{q^2=0} = \frac{(q^2)^2}{\pi} \int ds \frac{\mathrm{Im}\Pi_5(s)}{s^2(s-q^2)}, \tag{8.158}$$

and the same is valid for $\Pi_T(q^2)$. The imaginary part of $\Pi_5(q)$, in accordance with the unitarity condition, receives contributions of all hadronic states with the quantum numbers of heavy-light pseudoscalar mesons:

$$\frac{1}{\pi}\mathrm{Im}\Pi_5(s) = m_H^4 f_H^2 \delta(s - m_H^2) + \rho_5^h(s)\theta(s - (m_{H^*} + m_P)^2). \tag{8.159}$$

Quite a similar unitarity relation is valid for $\Pi_T(q)$:

$$\frac{1}{\pi}\mathrm{Im}\Pi_T(s) = m_{H^*}^2 f_{H^*}^2 \delta(s - m_{H^*}^2) + \rho_T^h(s)\theta(s - (m_H + m_P)^2). \tag{8.160}$$

In the above relations, the H and H^* ground states are separated from the contributions of excited and continuum states represented by the hadronic spectral densities $\rho_5^h(s)$ and $\rho_T^h(s)$, respectively. The θ-function multiplying $\rho_{5(T)}^h(s)$ takes into account the threshold of the lightest possible continuum state $|H^*P\rangle$ ($|HP\rangle$) in the pseudoscalar (vector) channel, where P is the lightest pseudoscalar meson with the valence $\bar{q}$ antiquark, e.g., pion (kaon) for $q = u, d\,(s)$. In (8.159) and (8.160) we use standard definitions of the heavy meson decay constants:

$$\langle 0|j_5|H(q)\rangle = m_H^2 f_H, \quad \langle 0|j_\mu|H^*(q)\rangle = m_{H^*}\varepsilon_\mu^{(H^*)} f_{H^*}, \tag{8.161}$$

where $\varepsilon^{(H^*)}$ is the polarization vector of H^*.

From what we already know about OPE, it follows that at $q^2 \ll m_Q^2$ the correlation functions (8.156) and (8.157) are well approximated by a truncated expansion consisting of the perturbative part and condensate terms. We write this expansion in a generic form, valid for both invariant amplitudes:

$$\Pi_{T(5)}^{(OPE)}(q^2) = \Pi_{5(T)}^{(pert)}(q^2) + \Pi_{5(T)}^{\langle\bar{q}q\rangle}(q^2) + \Pi_{5(T)}^{\langle GG\rangle}(q^2) + \Pi_{5(T)}^{\langle\bar{q}Gq\rangle}(q^2) + \Pi_{5(T)}^{\langle\bar{q}q\bar{q}q\rangle}(q^2), \tag{8.162}$$

including quark, gluon, quark-gluon and four-quark condensate terms and truncated at the dimension $d = 6$ of the latter term. The perturbative contributions are conveniently represented in terms of a (double subtracted) dispersion integral:

$$\Pi_{5(T)}^{(pert)}(q^2) = \frac{(q^2)^2}{\pi} \int\limits_{(m_Q+m_q)^2}^{\infty} ds \frac{\mathrm{Im}\Pi_{5(T)}^{(pert)}(s)}{s^2(s-q^2)}. \tag{8.163}$$

The imaginary parts at LO have the following expressions:

$$\mathrm{Im}\Pi_5^{(pert)}(s) = \frac{3m_Q^2}{8\pi} s \left(1 - \frac{m_Q^2}{s}\right)^2,$$

$$\mathrm{Im}\Pi_T^{(pert)}(s) = \frac{s}{8\pi} \left(1 - \frac{m_Q^2}{s}\right)^2 \left(2 + \frac{m_Q^2}{s}\right), \tag{8.164}$$

obtained from the one-loop diagrams calculated similar to the light-quark loop considered in Section 8.5. Note that hereafter we neglect the light quark mass. Gluon radiative corrections to the perturbative part have been calculated for both correlation functions up to $O(\alpha_s^2)$. A useful reference is [88] where the complete expressions, including the omitted light-quark mass terms and radiative corrections, are presented[13].

The quark and gluon condensate terms in (8.162) are obtained as described in Section 8.5, considering quark propagation in the external vacuum fields, with the gluon field taken in the fixed point gauge. The diagrams in Figures 8.2(a) and 8.3, with the corresponding replacements of quark flavors, $s \to Q$ and $u \to q$, describe these contributions whereas Figure 8.2(b) does not contribute due to the absence of a heavy quark condensate. Presence of the massive quark slightly alters the calculational procedure. The results are:

$$\Pi_5^{\langle \bar{q}q \rangle}(q^2) = -\frac{m_Q^3 \langle \bar{q}q \rangle}{m_Q^2 - q^2}, \quad \Pi_T^{\langle \bar{q}q \rangle}(q^2) = -\frac{m_Q \langle \bar{q}q \rangle}{m_Q^2 - q^2} \tag{8.165}$$

for the quark condensate terms, and

$$\Pi_5^{\langle GG \rangle}(q^2) = \frac{m_Q^2 \langle GG \rangle}{12(m_Q^2 - q^2)}, \quad \Pi_T^{\langle GG \rangle}(q^2) = -\frac{\langle GG \rangle}{12(m_Q^2 - q^2)} \tag{8.166}$$

for the gluon condensate ones. Differently from the case of light-quark currents, the quark condensate term in the OPE is now strongly enhanced, being multiplied by the heavy quark mass. The same is valid for the $d = 5$ quark-gluon condensate contributions:

$$\Pi_5^{\langle \bar{q}Gq \rangle}(q^2) = -\frac{m_Q^2}{2(m_Q^2 - q^2)^2}\left(1 - \frac{m_Q^2}{m_Q^2 - q^2}\right)m_Q \langle \bar{q}Gq \rangle,$$

$$\Pi_T^{\langle \bar{q}Gq \rangle}(q^2) = \frac{m_Q^3}{2(m_Q^2 - q^2)^3}\langle \bar{q}Gq \rangle, \tag{8.167}$$

which are described by the diagram in Figure 8.4(a) with an additional contribution stemming from the diagram in Figure 8.2(a). To complete the OPE, we present the expressions for the $d = 6$ four-quark condensate contribution obtained in the usual factorized approximation:

$$\Pi_5^{\langle \bar{q}q\bar{q}q \rangle}(q^2) = -\frac{8\pi m_Q^2}{27(m_Q^2 - q^2)^4}q^2\left(2q^2 - 3m_Q^2\right)\alpha_s \langle \bar{q}q \rangle^2,$$

$$\Pi_T^{\langle \bar{q}q\bar{q}q \rangle}(q^2) = -\frac{8\pi}{81(m_Q^2 - q^2)^4}\left(9m_Q^4 - 16m_Q^2 q^2 + 4q^4\right)\alpha_s \langle \bar{q}q \rangle^2. \tag{8.168}$$

The light-quark mass and $O(\alpha_s)$ corrections to the condensate terms are omitted for brevity, they can be found in [88]. Note that the OPE terms with $d = 5, 6$ in (8.167) and (8.168) are parametrically suppressed by extra inverse powers of $(m_Q^2 - q^2)$ with respect to the $d = 3, 4$ terms in (8.165) and (8.166), justifying the use of truncated OPE at $q^2 \ll m_Q^2$.

Considering first the pseudoscalar-current correlation function, we collect perturbative and condensate terms together, form the complete expression for the OPE in the adopted approximation (8.162) and substitute it in l.h.s. of dispersion relation (8.158), whereas in r.h.s. we use the hadronic representation (8.159). After that, we perform the Borel

---

[13]Note that in these expressions the quark masses are defined in the $\overline{MS}$ scheme.

transformation (A.103), removing subtraction terms and obtaining

$$\Pi_5^{OPE}(M^2) = \frac{1}{\pi} \int\limits_{m_Q^2}^{\infty} ds\, e^{-s/M^2} \mathrm{Im}\Pi_5^{(pert)}(s) + \Pi_5^{(cond)}(M^2)$$

$$= m_H^4 f_H^2 e^{-m_H^2/M^2} + \int\limits_{(m_H^*+m_P)^2}^{\infty} ds\, e^{-s/M^2} \rho_5^h(s)\,, \tag{8.169}$$

where the sum of the condensate terms in (8.165)–(8.168) after Borel transformation is equal to

$$\Pi_5^{(cond)}(M^2) = m_Q^2 \left[-m_Q\langle\bar{q}q\rangle + \frac{\langle GG\rangle}{12} - \frac{m_Q\langle\bar{q}Gq\rangle}{2M^2}\left(1 - \frac{m_Q^2}{2M^2}\right) \right.$$
$$\left. - \frac{16\pi\alpha_s\langle\bar{q}q\rangle^2}{27M^2}\left(1 - \frac{m_Q^2}{4M^2} - \frac{m_Q^4}{12M^4}\right) \right] e^{-\frac{m_Q^2}{M^2}}. \tag{8.170}$$

The next step is to apply the quark-hadron duality approximation to the hadronic spectral density:

$$\rho_5^h(s)\theta(s - (m_{H^*} + m_P)^2) = \frac{1}{\pi}\mathrm{Im}\Pi_5^{(pert)}(s)\theta(s - s_0^H)\,, \tag{8.171}$$

introducing the effective threshold s_0^H for the pseudoscalar channel. Using the above relation in r.h.s. of (8.169), we subtract from both parts of it the integral over perturbative imaginary part taken from s_0^H to infinity and reduce the resulting relation to an equation for the H-meson decay constant:

$$f_H^2 = \frac{e^{m_H^2/M^2}}{m_H^4}\left(\frac{1}{\pi}\int\limits_{m_Q^2}^{s_0^H} ds\, e^{-s/M^2}\mathrm{Im}\Pi_5^{(pert)}(s) + \Pi_5^{(cond)}(M^2)\right). \tag{8.172}$$

A similar derivation for the vector-current correlation function yields a sum rule for the H^*-meson decay constant:

$$f_{H^*}^2 = \frac{e^{m_{H^*}^2/M^2}}{m_{H^*}^2}\left(\frac{1}{\pi}\int\limits_{m_Q^2}^{s_0^{H^*}} ds\, e^{-s/M^2}\mathrm{Im}\Pi_T^{(pert)}(s) + \Pi_T^{(cond)}(M^2)\right), \tag{8.173}$$

where the sum of Borel-transformed condensate contributions is

$$\Pi_T^{(cond)}(M^2) = \left[-m_Q\langle\bar{q}q\rangle - \frac{\langle GG\rangle}{12} + \frac{m_Q^3\langle\bar{q}Gq\rangle}{4M^4} \right.$$
$$\left. - \frac{32\pi\alpha_s\langle\bar{q}q\rangle^2}{81M^2}\left(1 + \frac{m_Q^2}{M^2} - \frac{m_Q^4}{8M^4}\right) \right] e^{-\frac{m_Q^2}{M^2}}. \tag{8.174}$$

Both QCD sum rules (8.172) and (8.173) can now be used to calculate the decay constants of D, B and D^*, B^* mesons. To find an optimal interval of the Borel parameter, we compare the sum rule (8.172) or (8.173) with a typical QCD sum rule for a light unflavored meson with a mass m_h. In the latter sum rule we will have a similar expression for the corresponding decay constant squared with an exponential factor $\exp(m_h^2/M^2)$. For the light-meson sum rules the numerical analysis tells us that an optimal choice for the Borel parameter is in the ballpark of $M^2 \sim 1 \text{ GeV}^2 \sim m_h^2$. Returning to the heavy meson sum rules, we have to take into account also the exponential factor containing the heavy quark mass, hence in this case we anticipate that

$$M^2 \sim m_H^2 - m_Q^2 \sim (m_Q + \bar{\Lambda})^2 - m_Q^2 \sim 2\bar{\Lambda}m_Q \,, \tag{8.175}$$

where we used the scaling relation (1.119) for the heavy meson mass and neglected small corrections; hence, we can introduce a variable effective parameter τ parameterizing the Borel parameter:

$$M^2 = 2m_Q\tau \,. \tag{8.176}$$

Importantly, τ does not scale with m_Q but has to be kept parametrically larger than Λ_{QCD}, so that $\tau \sim 1 \text{ GeV}$ is comparable with a typical Borel scale for the light-meson sum rules.

Furthermore, as we already mentioned before, to constrain the duality threshold in the sum rule (8.172) or (8.173), it is useful to differentiate it over $1/M^2$ and divide the resulting relation to the initial sum rule. The decay constant cancels in the ratio and we obtain a relation which determines the heavy meson mass in terms of QCD parameters and duality threshold. Fixing the meson mass by its measured value, we can use this relation to adjust the threshold parameter and then employ its value in the sum rule for the decay constant.

All other parameters entering (8.172) and (8.173) are already known: the values of $\overline{MS}$ heavy quark masses are given in (1.20) and the condensates in (8.100), (8.102), (8.103). A more detailed analysis of QCD sum rules for the decay constants of heavy pseudoscalar and vector mesons can be found, e.g., in [88].

Our main interest is the $m_Q \to \infty$ limit of these sum rules. To obtain this limit, we use the HQET scaling (1.119) of the meson mass, assuming that, due to the heavy-quark spin symmetry, the m_Q-independent parameter $\bar{\Lambda}$ is the same for both H and H^* mesons:

$$m_H \simeq m_{H*} = m_Q + \bar{\Lambda} \,. \tag{8.177}$$

Furthermore, one can argue that the duality threshold s_0^H is close to the mass of the first radially excited state of H. The excitation energy denoted as ω_0 – and again assumed equal for pseudoscalar- and vector-meson channels – is determined by quark-gluon dynamics of the light degrees of freedom in a heavy meson and does not depend on m_Q. Therefore,

$$s_0^H \simeq s_0^{H^*} = (m_Q + \omega_0)^2 \simeq m_Q^2 + 2m_Q\omega_0 \,. \tag{8.178}$$

We substitute the above relation together with (8.176) and (8.177) in the sum rules (8.172) and (8.173) and rewrite them in a form where the dependence on the heavy mass scale[14] is explicit:

$$f_H^2 = \frac{1}{m_H}\left(\frac{m_Q}{m_H}\right)^3 e^{\frac{\bar{\Lambda}}{\tau}+\frac{\bar{\Lambda}^2}{2m_Q\tau}}\left[\frac{3\tau^3}{\pi^2}\int_0^{\frac{\omega_0}{\tau}} dz\, e^{-z}\left(\frac{z^2}{1+\frac{2z\tau}{m_Q}}\right) - \langle\bar{q}q\rangle + \frac{\langle GG\rangle}{12m_Q}\right.$$
$$\left. + \frac{\langle\bar{q}Gq\rangle}{16\tau^2}\left(1-\frac{4\tau}{m_Q}\right) + \frac{\pi\alpha_s\langle\bar{q}q\rangle^2}{162\tau^3}\left(1+\frac{6\tau}{m_Q}-\frac{48\tau^2}{m_Q^2}\right)\right] \,, \tag{8.179}$$

$$f_{H^*}^2 = \frac{1}{m_{H^*}}\left(\frac{m_Q}{m_{H^*}}\right)e^{\frac{\bar{\Lambda}}{\tau}+\frac{\bar{\Lambda}^2}{2m_Q\tau}}\left[\frac{\tau^3}{\pi^2}\int_0^{\frac{\omega_0}{\tau}} dz\, e^{-z}\left(\frac{z^2}{1+\frac{2z\tau}{m_Q}}\right)\left(2+\frac{1}{1+\frac{2z\tau}{m_Q}}\right)\right.$$
$$\left. - \langle\bar{q}q\rangle - \frac{\langle GG\rangle}{12m_Q} + \frac{\langle\bar{q}Gq\rangle}{16\tau^2} + \frac{\pi\alpha_s\langle\bar{q}q\rangle^2}{162\tau^3}\left(1-\frac{16\tau}{m_Q}-\frac{32\tau^2}{m_Q^2}\right)\right] \,. \tag{8.180}$$

---

[14]The radiative gluon corrections are not shown, their expressions can be found in [88].

For the infinitely heavy quark mass, we obtain

$$\lim_{m_Q \to \infty} (f_H) = \lim_{m_Q \to \infty} (f_{H^*}) = \frac{\hat{f}}{\sqrt{m_H}}, \tag{8.181}$$

reproducing the known HQET relation (2.146), where the static decay constant is given by the sum rule (without radiative corrections):

$$\hat{f} = e^{\frac{\bar{\Lambda}}{2\tau}} \left(\frac{3\tau^3}{\pi^2} \int_0^{\frac{\omega_0}{\tau}} dz\, z^2 e^{-z} - \langle \bar{q}q \rangle + \frac{\langle \bar{q}Gq \rangle}{16\tau^2} + \frac{\pi\alpha_s \langle \bar{q}q \rangle^2}{162\tau^3} \right)^{1/2}. \tag{8.182}$$

Note that also the inverse heavy-mass corrections to this limit are easily obtained from (8.179) and (8.180). On the other hand, the sum rule (8.182) can be independently obtained in the framework of HQET, starting from the correlation function of the currents containing an effective field h_v instead of Q (see, e.g., the review [25]).

8.7 THREE-POINT SUM RULES FOR HADRON FORM FACTORS

It is possible to extend the method of QCD sum rules to hadron form factors. To get access to a given hadron-to-hadron transition matrix element, one has to consider a vacuum correlation function containing a T-product of three quark-current operators. Two of them interpolate the initial and final hadronic states in the matrix element and the third one is the (e.m., weak or effective) transition current. If all external momenta squared flowing in the currents are far below hadronic thresholds, then the quarks and gluons propagating between the current vertices are highly virtual. As a result, an OPE similar to (8.38), consisting of perturbative and vacuum-condensate terms, is valid also for the three-point correlation function. Since condensates are universal, the differences between the three- and two-point correlation functions lie only in the perturbative coefficients. In particular, in three-point correlation functions these coefficients depend on several invariant variables. A more substantial modification takes place in the hadronic representation. As we shall see, for a three-point correlation function a double dispersion relation emerges.

Here we consider one example of a three-point QCD sum rule[15] allowing one to obtain the amplitude of the radiative transition

$$J/\psi \to \eta_c(1S) + \gamma, \tag{8.183}$$

between the vector and pseudoscalar charmonium levels. The advantage of this example is that it contains all essential elements of the method, whereas the calculation of the LO diagram is technically simple.

The correlation function that will be related to the amplitude of (8.183) is defined with the following expression:

$$\begin{aligned}
\Delta_{\mu\nu}^{(c)}(p,q) &= \int d^4y\, e^{ipy} \int d^4x\, e^{iqx} \langle 0 | T\{ j_5^{(c)}(y) j_\mu^{em(c)}(x) j_\nu^{(c)}(0) \} | 0 \rangle \\
&= \epsilon_{\mu\nu\alpha\beta} p^\alpha q^\beta \Delta^{(c)}((p+q)^2, p^2),
\end{aligned} \tag{8.184}$$

where $j_\nu^{(c)} = \bar{c}\gamma_\nu c$ and $j_5^{(c)} = \bar{c}i\gamma_5 c$ are the vector and pseudoscalar currents[16] with the quantum numbers of J/ψ and η_c, respectively, whereas $j_\mu^{em(c)} = Q_c \bar{c}\gamma_\mu c$ is the c-quark part

---

[15]It was also the first application [89] of this method.
[16]Since we neglect gluon radiative corrections, the pseudoscalar current is used without an additional factor m_c needed to compensate the renormalization scale dependence, cf. the definition (2.57).

of the e.m. current. The above correlation function depends on two independent momenta p and q and we put $q^2 = 0$, assuming a real photon emission from the e.m. current. Since one of the three currents is a pseudoscalar, in order to obey the P-parity conservation, the above correlation function has to transform as a pseudotensor and is therefore parameterized in terms of a single Lorentz structure with a Levi-Civita tensor multiplied by the invariant amplitude $\Delta^{(c)}$. The latter is a function of the two independent invariant variables: the momenta squared $(p+q)^2$ and p^2 flowing in the currents. Both variables are chosen to be far below the $c\bar{c}$ threshold:

$$(p+q)^2 \ll 4m_c^2, \quad p^2 \ll 4m_c^2, \tag{8.185}$$

to ensure a short-distance dominance in the correlation function (8.184); hence, $\Delta^{(c)}_{\mu\nu}$ is calculable in terms of the OPE, containing perturbative and condensate terms, similar to the correlation function (8.125) of two $c\bar{c}$ currents. Before turning to the OPE, let us convince ourselves that the correlation function (8.184) indeed leads to a sum rule for the $J/\psi \to \eta_c\gamma$ hadronic matrix element.

To this end, we need a hadronic representation of (8.184) which is evidently more complicated than the one used for two-point correlation functions in the previous section. We start from treating the invariant amplitude $\Delta^{(c)}((p+q)^2, p^2)$ as an analytic function of the variable p^2, continuing this variable towards the timelike region and keeping the second variable $(p+q)^2$ fixed at a certain value within the region defined in (8.185). Employing analyticity, we write down a usual dispersion relation[17] for this function:

$$\Delta^{(c)}((p+q)^2, p^2) = \frac{1}{\pi} \int\limits_{m_{\eta_c(1S)}^2}^{\infty} \frac{ds}{s - p^2} \operatorname{Im}_{[p^2]} \Delta^{(c)}((p+q)^2, s), \tag{8.186}$$

where the imaginary part, according to the unitarity relation (B.39), is

$$2\operatorname{Im}_{[p^2]}\left[\Delta^{(c)}_{\mu\nu}(p, q)\right] = \epsilon_{\mu\nu\alpha\beta}p^\alpha q^\beta \, 2\operatorname{Im}_{[p^2]}\Delta^{(c)}((p+q)^2, s)$$

$$= \sum_{\eta_c} d\tau_{\eta_c}\langle 0|j_5^{(c)}|\eta_c(p)\rangle\langle\eta_c(p)|\hat{\Pi}^{(c)}_{\mu\nu}(q)|0\rangle + \sum_h \int d\tau_h \langle 0|j_5^{(c)}|\overline{h(p)}\rangle\langle h(p)|\hat{\Pi}^{(c)}_{\mu\nu}(q)|0\rangle. \tag{8.187}$$

In (8.186), the lower limit of integration indicates that the lightest hadronic state is $\eta_c(1S)$. In the above, we also introduce a short-hand notation for the Fourier-integrated operator product

$$\hat{\Pi}^{(c)}_{\mu\nu}(q) = \int d^4x \, e^{iqx} T\{j_\mu^{em(c)}(x), j_\nu^{(c)}(0)\}, \tag{8.188}$$

which plays the role of the second current operator in (B.39).

The sum over η_c in (8.187) goes over the two charmonium levels [1],

$$\eta_c = \{\eta_c(1S), \eta_c(2S)\},$$

located below the open charm threshold. Each term in this sum contains the one-particle phase space

$$d\tau_{\eta_c} = 2\pi\delta(s - m_{\eta_c}^2), \tag{8.189}$$

the decay constant of η_c defined as

$$\langle 0|j_5^{(c)}|\eta_c(p)\rangle = m_{\eta_c}^2 f_{\eta_c}, \tag{8.190}$$

---

[17]We neglect subtractions in an anticipation of taking power moments or performing Borel transformation.

and a vacuum-to-η_c matrix element of the nonlocal operator (8.188) parameterized as:

$$\langle \eta_c(p)|\hat{\Pi}_{\mu\nu}^{(c)}(q)|0\rangle = \epsilon_{\mu\nu\alpha\beta}p^\alpha q^\beta \tilde{\Delta}^{(\eta_c)}((p+q)^2)\,, \qquad (8.191)$$

where $p^2 = s = m_{\eta_c}^2$. Choosing the same Lorentz-structure as in (8.184), we again follow the P-parity argument, having, instead of the pseudoscalar current, a pseudoscalar state. Using (8.189)–(8.191), we transform the first part of the unitarity sum in (8.187):

$$\sum_{\eta_c} d\tau_{\eta_c}\langle 0|j_5^{(c)}|\eta_c(p)\rangle\langle \eta_c(p)|\hat{\Pi}_{\mu\nu}^{(c)}(q)|0\rangle$$

$$= 2\pi\epsilon_{\mu\nu\alpha\beta}p^\alpha q^\beta \sum_{\eta_c} \delta(s - m_{\eta_c}^2)m_{\eta_c}^2 f_{\eta_c}\tilde{\Delta}^{(\eta_c)}((p+q)^2)\,. \qquad (8.192)$$

The second sum in (8.187) goes over hadronic states

$$h = \{D\bar{D}^*, D^*\bar{D}, \dots\}$$

with the quantum numbers of η_c, all located above the open charm threshold

$$s_{th} = (m_D + m_{D^*})^2.$$

Note that the $D\bar{D}$ state does not contribute because it cannot have the same spin-parity as η_c. Each term in the unitarity sum over h-states in (8.187) involves an integral over the phase space of h at fixed $p^2 = s = m_h^2$ and a sum over polarizations. The vacuum-to-h matrix elements of pseudoscalar current in this sum are, in general, not constants but timelike form factors depending on s. Therefore, a detailed description of the sum over h-states is not feasible. Moreover, the set of states above the open charm threshold is incomplete without radially excited broad η_c-resonances, strongly coupled to the continuum of the open charm states. These are pseudoscalar partners of the broad ψ resonances observed in $e^+e^- \to charm$ cross section. From our experience with the pion form factor in the timelike region, it is conceivable to model the sum over h-states with a set of broad η_c resonances, absorbing the interactions with charmed mesons into their widths. Actually, such a model is not needed here since we intend to use the quark-hadron duality approximation for the sum over the above-threshold states. In this respect, it is sufficient to represent each term of this sum in terms of an the invariant amplitude depending on the two variables $p^2 = s$ and $(p+q)^2$. The most general parameterization satisfying these conditions and consistent with the Lorentz decomposition is

$$\sum_h \int d\tau_h \langle 0|j_5^{(c)}|\overline{h(p)}\rangle\langle h(p)|\hat{\Pi}_{\mu\nu}^{(c)}(q)|0\rangle$$

$$= 2\pi\epsilon_{\mu\nu\alpha\beta}p^\alpha q^\beta \sum_h f_h(s)\tilde{\Delta}^{(h)}((p+q)^2, s)\theta(s - s_{th})\,, \qquad (8.193)$$

where the invariant amplitudes emerge from the hadronic matrix elements:

$$\langle 0|j_5^{(c)}|h(p)\rangle \to f_h(s)\,, \quad \langle h(p)|\hat{\Pi}_{\mu\nu}^{(c)}(q)|0\rangle \to \tilde{\Delta}^{(h)}((p+q)^2, s)\,,$$

respectively, after integrating over $d\tau_h$, summing over the polarizations and isolating the Lorentz structure. We substitute (8.192) and (8.193) in the imaginary part (8.187), and use the latter in the dispersion relation (8.186):

$$\Delta^{(c)}((p+q)^2, p^2) = \sum_{\eta_c} \frac{m_{\eta_c}^2 f_{\eta_c}\tilde{\Delta}^{(\eta_c)}((p+q)^2)}{m_{\eta_c}^2 - p^2} + \int_{s_{th}}^{\infty} \frac{ds}{s - p^2} \sum_h f_h(s)\tilde{\Delta}^{(h)}((p+q)^2, s)\,,$$

$$(8.194)$$

where the poles corresponding to η_c mesons are made explicit.

The next step concerns the invariant amplitudes $\widetilde{\Delta}^{(\eta_c)}((p+q)^2)$ and $\widetilde{\Delta}^{(h)}((p+q)^2, s)$ of the vacuum-to-η_c and vacuum-to-h transitions, respectively. We notice that these amplitudes are very similar to the two-point correlation functions of two $c\bar{c}$ currents considered in the previous section, but contain η_c or h instead of the vacuum state. Being analytic functions of the variable $(p+q)^2$, each of these amplitudes obeys a dispersion relation that we again write down in a simplified form without subtractions:

$$\widetilde{\Delta}^{(\eta_c)}((p+q)^2) = \frac{1}{\pi} \int_{m_{J/\psi}^2}^{\infty} \frac{ds'}{s' - (p+q)^2} \, \text{Im}_{[(p+q)^2]} \widetilde{\Delta}^{(\eta_c)}(s') \,, \tag{8.195}$$

and at fixed $p^2 = s$:

$$\widetilde{\Delta}^{(h)}((p+q)^2, s) = \frac{1}{\pi} \int_{m_{J/\psi}^2}^{\infty} \frac{ds'}{s' - (p+q)^2} \, \text{Im}_{[(p+q)^2]} \widetilde{\Delta}^{(h)}(s', s) \,. \tag{8.196}$$

In this channel, the imaginary part receives contributions of intermediate hadronic states with the quantum numbers of J/ψ:

$$\epsilon_{\mu\nu\alpha\beta} p^\alpha q^\beta \, 2\,\text{Im}_{[(p+q)^2]} \widetilde{\Delta}^{(\eta_c)}(s') = \sum_\psi d\tau_\psi \langle \eta_c(p)|j_\mu^{em(c)} \overline{|\psi(p+q)\rangle \langle \psi(p+q)|} j_\nu^{(c)} |0\rangle$$

$$+ \sum_{h'} d\tau_{h'} \langle \eta_c(p)|j_\mu^{em(c)} \overline{|h'(p+q)\rangle \langle h'(p+q)|} j_\nu^{(c)} |0\rangle \,. \tag{8.197}$$

$$f_h(s)\epsilon_{\mu\nu\alpha\beta} p^\alpha q^\beta \, 2\,\text{Im}_{[(p+q)^2]} \widetilde{\Delta}^{(h)}(s', s)$$

$$= \int d\tau_h \langle 0|j_5^{(c)} \overline{|h(p)\rangle} \left(\sum_\psi d\tau_\psi \overline{\langle h(p)|} j_\mu^{em(c)} \overline{|\psi(p+q)\rangle \langle \psi(p+q)|} j_\nu^{(c)} |0\rangle \right.$$

$$\left. + \sum_{h'} d\tau_{h'} \overline{\langle h(p)|} j_\mu^{em(c)} \overline{|h'(p+q)\rangle \langle h'(p+q)|} j_\nu^{(c)} |0\rangle \right) \,. \tag{8.198}$$

The sum over ψ in the above equations includes the two states,

$$\psi = \{J/\psi, \psi(2S)\},$$

located below the open charm threshold $s_{th}' = 4m_D^2$. The states entering the sum over h' lie above that threshold:

$$h' = \{D\bar{D}, D^*\bar{D}, D\bar{D}^*, \dots\}$$

and are strongly coupled to the broad ψ resonances.

In the first term in (8.197), we recognize the hadronic matrix elements of radiative transitions between the ψ and η_c levels and parameterize them as

$$\langle \eta_c(p)|j_\mu^{em(c)}|\psi(p+q)\rangle = \epsilon_{\alpha\beta\rho\mu} p^\alpha q^\beta \varepsilon^{(\psi)\rho} \frac{\mathcal{F}_{\psi\eta_c\gamma}}{m_\psi} \,, \tag{8.199}$$

introducing invariant constants $\mathcal{F}_{\psi\eta_c\gamma}$ of these transitions. They are dimensionless analogs of the $\rho \to \pi\gamma$ decay constant introduced in Chapter 4. In particular, expressing the amplitude of the $J/\psi \to \eta_c(1S)\gamma$) decay via the hadronic matrix element (8.199):

$$
\mathcal{A}(J/\psi \to \eta_c(1S)\gamma) = \sqrt{4\pi\alpha_{em}}\varepsilon^{*(\gamma)\mu}\langle \eta_c(1S)(p)|j_\mu^{em(c)}|J/\psi(p+q)\rangle
$$

$$
= \sqrt{4\pi\alpha_{em}}\epsilon_{\alpha\beta\rho\mu}p^\alpha q^\beta \varepsilon^{(J/\psi)\rho}\varepsilon^{*(\gamma)\mu}\frac{\mathcal{F}_{J/\psi\eta_c(1S)\gamma}}{m_{J/\psi}}, \qquad (8.200)
$$

and comparing with (4.13) and (4.16), we obtain the transition width

$$
\Gamma(J/\psi \to \eta_c\gamma) = \frac{\alpha_{em}}{24}|\mathcal{F}_{J/\psi\eta_c(1S)\gamma}|^2\left(1 - \frac{m_{\eta_c(1S)}^2}{m_{J/\psi}^2}\right)^3 m_{J/\psi}. \qquad (8.201)
$$

Returning to the unitarity relation (8.197), we use (8.199), introduce the decay constants of the vector ψ-mesons:

$$
\langle\psi|j_\nu^{(c)}|0\rangle = \varepsilon_\nu^{(\psi)*}m_\psi f_\psi, \qquad (8.202)
$$

insert the phase-space factors $d\tau_\psi$ and sum over the polarization vectors of intermediate ψ-states. For the first sum in (8.197) we obtain:

$$
\sum_\psi d\tau_\psi\langle \eta_c(p)|j_\mu^{em(c)}|\overline{\psi(p+q)}\rangle\langle\psi(p+q)|j_\nu^{(c)}|0\rangle
$$

$$
= 2\pi\epsilon_{\mu\nu\alpha\beta}p^\alpha q^\beta\sum_\psi \delta(s' - m_\psi^2)\mathcal{F}_{J/\psi\eta_c\gamma}f_\psi, \qquad (8.203)
$$

whereas the sum over h' is parameterized in a similar general form:

$$
\sum_{h'}\!\!\!\int d\tau_{h'}\langle \eta_c(p)|j_\mu^{em(c)}|\overline{h'(p+q)}\rangle\langle h'(p+q)|j_\nu^{(c)}|0\rangle
$$

$$
= 2\pi\epsilon_{\mu\nu\alpha\beta}p^\alpha q^\beta\sum_{h'}\mathcal{F}_{h'\eta_c\gamma}(s')f_{h'}(s')\theta(s' - s'_{th}), \qquad (8.204)
$$

where, after integration over $d\tau_{h'}$ and sum over polarizations, in each term we have, instead of decay constants, generic form factors depending on $(p+q)^2 = s'$. With (8.203) and (8.204) the imaginary part of the invariant amplitude in (8.197) can be written as

$$
\frac{1}{\pi}\text{Im}_{[(p+q)^2]}\widetilde{\Delta}^{(\eta_c)}(s') = \sum_\psi \delta(s' - m_\psi^2)\mathcal{F}_{J/\psi\eta_c\gamma}f_\psi + \sum_{h'}\mathcal{F}_{h'\eta_c\gamma}(s')f_{h'}(s')\theta(s' - s'_{th}), \quad (8.205)
$$

and used in the dispersion relation (8.195):

$$
\widetilde{\Delta}^{(\eta_c)}((p+q)^2) = \sum_\psi \frac{f_\psi\mathcal{F}_{\psi\eta_c\gamma}}{m_\psi^2 - (p+q)^2} + \int_{s'_{th}}^\infty \frac{ds'}{s' - (p+q)^2}\sum_{h'}\mathcal{F}_{h'\eta_c\gamma}(s')f_{h'}(s'). \quad (8.206)
$$

Along the same lines, we parameterize the first and then the second term on the r.h.s. of (8.198):

$$
\int d\tau_h\langle 0|j_5^{(c)}|\overline{h(p)}\rangle\sum_\psi d\tau_\psi\overline{\langle h(p)|}j_\mu^{em(c)}|\overline{\psi(p+q)}\rangle\langle\psi(p+q)|j_\nu^{(c)}|0\rangle
$$

$$
= 2\pi\epsilon_{\mu\nu\alpha\beta}p^\alpha q^\beta f_h(s)\sum_\psi \delta(s' - m_\psi^2)\mathcal{F}_{\psi h\gamma}(s)f_\psi,
$$

$$
\int d\tau_h\langle 0|j_5^{(c)}|\overline{h(p)}\rangle\sum_{h'}\!\!\!\int d\tau_{h'}\overline{\langle h(p)|}j_\mu^{em(c)}|\overline{h'(p+q)}\rangle\langle h'(p+q)|j_\nu^{(c)}|0\rangle
$$

$$
= 2\pi\epsilon_{\mu\nu\alpha\beta}p^\alpha q^\beta\sum_{h'}f_h(s)\mathcal{F}_{h'h\gamma}(s',s)f_{h'}(s')\theta(s' - s'_{th}), \qquad (8.207)
$$

and obtain a corresponding dispersion relation

$$\widetilde{\Delta}^{(h)}((p+q)^2,s) = \sum_{\psi} \frac{f_\psi \mathcal{F}_{\psi h\gamma}(s)}{m_\psi^2 - (p+q)^2} + \int_{s'_{th}}^{\infty} \frac{ds'}{s' - (p+q)^2} \sum_{h'} \mathcal{F}_{h'h\gamma}(s',s) f_{h'}(s'). \quad (8.208)$$

In the initial dispersion relation (8.194), we replace the functions $\widetilde{\Delta}^{(\eta_c)}$ and $\widetilde{\Delta}^{(h)}$ by their dispersion forms (8.206) and (8.208) and, finally, obtain the hadronic representation of the three-point correlation function in a form of a double dispersion relation:

$$\Delta^{(c)}((p+q)^2,p^2) = \sum_{\eta_c,\psi} \frac{m_{\eta_c}^2 f_{\eta_c} f_\psi \mathcal{F}_{\psi \eta_c \gamma}}{(m_{\eta_c}^2 - p^2)(m_\psi^2 - (p+q)^2)}$$

$$+ \sum_{\eta_c,h'} \frac{m_{\eta_c}^2 f_{\eta_c}}{m_{\eta_c}^2 - p^2} \int_{s'_{th}}^{\infty} ds' \frac{\mathcal{F}_{h'\eta_c\gamma}(s') f_{h'}(s')}{s' - (p+q)^2} + \sum_{h,\psi} \int_{s_{th}}^{\infty} ds \frac{f_h(s)}{s - p^2} \frac{f_\psi \mathcal{F}_{\psi h\gamma}(s)}{m_\psi^2 - (p+q)^2}$$

$$+ \sum_{h,h'} \int_{s_{th}}^{\infty} ds \frac{f_h(s)}{s - p^2} \int_{s'_{th}}^{\infty} ds' \frac{\mathcal{F}_{h'h\gamma}(s',s) f_{h'}(s')}{s' - (p+q)^2}. \quad (8.209)$$

The first term on r.h.s contains hadronic matrix elements of the four radiative transitions:

$$J/\psi \to \eta_c(1S)\gamma, \quad \psi(2S) \to \eta_c(1S)\gamma, \quad \eta_c(2S) \to J/\psi\gamma, \quad \psi(2S) \to \eta_c(2S)\gamma.$$

The first(second) term in the second line of (8.209) contains a sum over the transitions between the η_c-mesons and h'-states (between the ψ-mesons and h-states). Finally, the double integral in the third line accumulates transitions between the above-threshold states. The location of all these transitions on the plane of two variables s, s' is shown in Figure 8.7.

Importantly, the dispersion relation (8.209) is valid at any value of p^2 and $(p+q)^2$, provided an infinitesimal $+i\epsilon$ is added to both variables in the timelike regions at real valued $p^2 = s \geq m_{\eta_c}^2$ and $(p+q)^2 = s' \geq m_{J/\psi}^2$. Still, a practical use of this relation in a form of a sum rule is only possible in the region (8.185), where l.h.s. is calculated employing OPE and matched to r.h.s. and where the unresolved parts of the hadronic representation (8.209) are replaced by their quark-hadron duality counterparts. Furthermore, to suppress possible subtraction terms in both dispersion relations, we need to apply to (8.209) a double Borel transformation in both variables $(p+q)^2$ and p^2 or, alternatively, differentiate (8.209) over these variables a sufficient number of times. One frequently used method is to take the double moments of the correlation function near zero momenta transfers defined as

$$\mathcal{M}_{kn}^{(c)} = \frac{1}{k!n!} \frac{\partial^k}{\partial((p+q)^2)^k} \frac{\partial^n}{\partial(p^2)^n} \Delta^{(c)}((p+q)^2,p^2)\Big|_{p^2=(p+q)^2=0}. \quad (8.210)$$

In the resulting relation, all subtraction terms vanish and the relative weight of the transitions between heavier states is suppressed by powers of the ground-state to excited meson

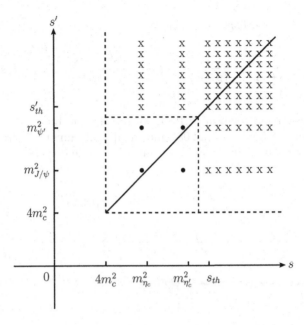

Figure 8.7 A schematic view of the plane of the two variables s and s' in the double dispersion relation. The points (crosses) indicate the four transitions between lower levels (transitions involving the above-threshold levels). The solid straight line along the diagonal $s = s'$ is the region where the LO double spectral density is nonvanishing. The square formed by the dashed lines indicates one of the possible quark-hadron duality regions.

mass ratios. Applying the moments to the hadronic representation (8.209) yields

$$
\mathcal{M}_{kn}^{(c,hadr)} = \frac{1}{m_{\eta_c(1S)}^{2n} m_{J/\psi}^{2k+2}} \left[f_{\eta_c(1S)} f_{J/\psi} \mathcal{F}_{J/\psi \eta_c(1S)\gamma} \right.
$$

$$
+ \left(\frac{m_{J/\psi}}{m_{\psi(2S)}} \right)^{2k+2} f_{\eta_c(1S)} f_{\psi(2S)} \mathcal{F}_{\psi(2S)\eta_c(1S)\gamma}
$$

$$
+ \left(\frac{m_{\eta_c(1S)}}{m_{\eta_c(2S)}} \right)^{2n} f_{\eta_c(2S)} f_{\psi(1S)} \mathcal{F}_{\eta_c(2S)J/\psi(1S)\gamma}
$$

$$
\left. + \left(\frac{m_{\eta_c(1S)}}{m_{\eta_c(2S)}} \right)^{2n} \left(\frac{m_{J/\psi}}{m_{\psi(2S)}} \right)^{2k+2} f_{\eta_c(2S)} f_{\psi(2S)} \mathcal{F}_{\psi(2S)\to\eta_c(2S)\gamma} \right]
$$

$$
+ \sum_{\eta_c, h'} \frac{f_{\eta_c}}{m_{\eta_c}^{2n}} \int_{s'_{th}}^{\infty} ds' \frac{\mathcal{F}_{h'\eta_c\gamma}(s') f_{h'}(s')}{s'^{k+1}} + \sum_{h,\psi} \int_{s_{th}}^{\infty} ds \frac{f_h(s)}{s^{n+1}} \frac{f_\psi \mathcal{F}_{\psi h\gamma}(s)}{m_\psi^{2k+2}}
$$

$$
+ \sum_{h,h'} \int_{s_{th}}^{\infty} ds \frac{f_h(s)}{s^{n+1}} \int_{s'_{th}}^{\infty} ds' \frac{\mathcal{F}_{h'h\gamma}(s',s) f_{h'}(s')}{s'^{k+1}}. \tag{8.211}
$$

After establishing the form of the hadronic representation, we turn to the OPE for the three-point correlation function. The perturbative part is described by triangle c-quark loop diagrams. The LO diagram is a simple loop shown in Figure 8.8(a). One should also take into account the equal diagram with an opposite direction of the internal quark line[18],

---

[18]A way to prove their equality is to insert the C-conjugation operators between the currents (as e.g.. in (2.27)), inverting the direction of the internal quark line. The total C-parity of the currents and of the

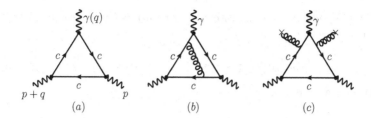

Figure 8.8 (a) The diagram corresponding to the LO contribution; (b) one of the NLO diagrams; (c) one of the gluon condensate contributions to the OPE for the three-point correlation function.

corresponding to the second possible way to contract quark and antiquark fields located at three different points.

In Figure 8.8(b), an example of NLO gluon correction is shown. Actually, only a partial calculation of these diagrams exists (see [90],[82]). In the condensate part of the OPE, we encounter only the gluon condensate contribution, in a full analogy with the correlation function of the two $c\bar{c}$ currents. The $d = 4$ gluon condensate term is obtained adding together the diagrams where two external gluon-field lines are emitted. One of those diagrams is shown in Figure 8.8(c). In what follows, we calculate the LO contribution to OPE and derive its double dispersion relation to be used for the quark-hadron duality ansatz.

The triangle diagram contribution to OPE is obtained from (8.184), contracting all c and $\bar{c}$ fields in propagators (cf. (8.32)):

$$\Delta_{\mu\nu}^{(c,0)}(p,q) = -2Q_c \int d^4y\, e^{ipy} \int d^4x\, e^{iqx} \text{Tr}\big[iS_{(c)}(0,y)i\gamma_5 iS_{(c)}(y,x)\gamma_\mu iS_{(c)}(x,0)\gamma_\nu\big]\,, (8.212)$$

where the trace is taken over Dirac and color indices and the factor two takes into account the second diagram. Using free propagators (A.49) and taking the color trace, leads to

$$\Delta_{\mu\nu}^{(c,0)}(p,q) = -2Q_c N_c \int d^4y\, e^{ipy} \int d^4x\, e^{iqx} \text{Tr}\bigg[\int \frac{d^4k}{(2\pi)^4} e^{iky} \frac{(\slashed{k}+m_c)}{k^2-m_c^2}\gamma_5$$

$$\times \int \frac{d^4k'}{(2\pi)^4} e^{-ik'(y-x)} \frac{(\slashed{k}'+m_c)}{k'^2-m_c^2}\gamma_\mu \int \frac{d^4k''}{(2\pi)^4} e^{-ik''x} \frac{(\slashed{k}''+m_c)}{k''^2-m_c^2}\gamma_\nu\bigg]. \quad (8.213)$$

In the above equation, the integrals over coordinates x and y form $\delta^{(4)}(q+k'-k'')$ and $\delta^{(4)}(p+k-k')$, respectively, and, after integrating out these δ-functions over k' and k'', we are left with a four-dimensional integral over the loop momentum k:

$$\Delta_{\mu\nu}^{(c,0)}(p,q) = -2Q_c N_c \int \frac{d^4k}{(2\pi)^4} \frac{\text{Tr}\big[(\slashed{k}+m_c)\gamma_5(\slashed{k}+\slashed{p}+m_c)\gamma_\mu(\slashed{k}+\slashed{p}+\slashed{q}+m_c)\gamma_\nu\big]}{(k^2-m_c^2)((k+p)^2-m_c^2)((k+p+q)^2-m_c^2)}. (8.214)$$

Here we do not need to shift the integration to $D \neq 4$ because the integral is convergent in the ultraviolet region. It is also convergent in the infrared region, due to the c-quark mass. Furthermore, we take into account that a nonvanishing trace with a γ_5 matrix should contain at least four additional four-dimensional γ-matrices. As a result, applying some Dirac algebra, the trace in (8.214) is greatly simplified:

$$\text{Tr}\big[(\slashed{k}+m_c)\gamma_5(\slashed{k}+\slashed{p}+m_c)\gamma_\mu(\slashed{k}+\slashed{p}+\slashed{q}+m_c)\gamma_\nu\big] = -4im_c\epsilon_{\mu\nu\alpha\beta}p^\alpha q^\beta\,.$$

vacuum state is positive, leading to the equality of the initial and inverted diagrams. Note that in case there is γ_α instead of γ_5 in the triangle loop, the C-insertion flips the sign and the two diagrams cancel each other – a fact known as the Furry theorem in QED, stating that the amplitudes with an odd number of photons vanish.

Since there is no dependence on the internal momentum k, the trace can be taken off the loop integral and we obtain:

$$\Delta_{\mu\nu}^{(c,0)}(p,q) = \epsilon_{\mu\nu\alpha\beta}p^\alpha q^\beta \Delta^{(c,0)}((p+q)^2, p^2)\,, \tag{8.215}$$

where the invariant amplitude is

$$\Delta^{(c,0)}((p+q)^2, p^2) = 16im_c \int \frac{d^4k}{(2\pi)^4(k^2 - m_c^2)((k+p)^2 - m_c^2)((k+p+q)^2 - m_c^2)}. \tag{8.216}$$

It is straightforward to calculate the above Feynman integral as a function of two variables $(p+q)^2$ and p^2 at $q^2 = 0$, applying the standard four-dimensional integration with a Wick rotation (see (8.51)). But that is not what we aim at. We need a double dispersion relation for the amplitude $\Delta^{(c,0)}$ in a generic form:

$$\Delta^{(c,0)}((p+q)^2, p^2) = \frac{1}{\pi^2} \int_{4m_c^2}^{\infty} \frac{ds}{s - p^2} \int_{f_{min}(s)}^{f_{max}(s)} \frac{ds'}{s' - (p+q)^2} \mathrm{Im}_{[(p+q)^2]}\mathrm{Im}_{[p^2]}\Delta^{(c,0)}(s', s), \tag{8.217}$$

obtained following the same steps as in the derivation of the hadronic representation (8.209). First, we have to write down a dispersion relation in one invariant variable at fixed second variable and, after that, replace the imaginary part by a dispersion relation in the second variable. Note that, in general, the limits in the second dispersion integral in (8.217) depend on s, being determined by the borders of a two-dimensional integration region in the (s, s')-plane. The integral (8.217) is convergent, hence, we can rearrange the order of integrations, provided the limits of the integration region are modified correspondingly. The double spectral density can be written in the two equivalent ways:

$$\rho^{(c,0)}(s', s) = \frac{1}{\pi^2}\mathrm{Im}_{[(p+q)^2]}\mathrm{Im}_{[p^2]}\Delta^{(c,0)}(s', s) = \frac{1}{\pi^2}\mathrm{Im}_{[p^2]}\mathrm{Im}_{[(p+q)^2]}\Delta^{(c,0)}(s', s). \tag{8.218}$$

Returning to (8.216), we start from the variable $(p+q)^2$ and calculate the imaginary part in this variable keeping p^2 fixed and far below the threshold $4m_c^2$. Applying the Cutkosky rule (A.102), we put on shell the quark and antiquark lines adjacent to the vertex of the vector current $j_\nu^{(c)}$ and obtain:

$$2\mathrm{Im}_{[(p+q)^2]}\Delta^{(c,0)}((p+q)^2, p^2) = -16m_c \int \frac{d^4k}{(2\pi)^2}\theta(-k_0)\theta(k_0 + p_0 + q_0)$$
$$\times \frac{\delta(k^2 - m_c^2)\delta((k+p+q)^2 - m_c^2)}{(k+p)^2 - m_c^2}\,. \tag{8.219}$$

The subsequent calculation largely repeats the one used in the previous section to obtain the imaginary part (8.137) of the two-point c-quark loop. But there is an essential difference: in (8.219) we now have a denominator which depends on the angular variable in the integral over k. To proceed, we consider the rest frame of the vector current: $\vec{p} + \vec{q} = 0$. From $(p+q)^2 = s'$ taken at $q^2 = 0$ it follows that in this frame

$$p_0 + q_0 = \sqrt{p^2 + |\vec{p}|^2} + |\vec{p}| = \sqrt{s'}$$

and

$$|\vec{p}| = \frac{s' - p^2}{2\sqrt{s'}}, \quad p_0 = \frac{s' + p^2}{2\sqrt{s'}}.$$

In the chosen frame, the denominator in (8.219) becomes

$$(k+p)^2 - m_c^2 = k^2 + 2k_0 p_0 - 2|\vec{k}||\vec{p}|\cos\theta + p^2 - m_c^2 \tag{8.220}$$

where θ is the angle between $\vec{p}$ and $\vec{k}$. Integrating out the δ-functions, we fix the components of the momentum k:

$$k_0 = -\frac{\sqrt{s'}}{2}, \quad |\vec{k}| = \frac{1}{2}\sqrt{s' - 4m_c^2},$$

and reduce the four-dimensional integral to an angular one:

$$\mathrm{Im}_{[(p+q)^2]}\Delta^{(c,0)}(s',p^2) = -\frac{4m_c}{\pi}\int_{-\infty}^{0}dk_0\int_{0}^{\infty}|\vec{k}|^2 d|\vec{k}|\delta(k_0^2 - \vec{k}^2 + s' + 2k_0\sqrt{s'} - m_c^2)$$

$$\times\delta(k_0^2 - \vec{k}^2 - m_c^2)\int_{-1}^{1}\frac{d\cos\theta}{k^2 + 2k_0 p_0 - 2|\vec{k}||\vec{p}|\cos\theta + p^2 - m_c^2}\theta(k_0 + p_0 + q_0)$$

$$= \frac{m_c v(s')}{\pi}\int_{-1}^{1}\frac{d\cos\theta}{(s' - p^2)(1 - v(s')\cos\theta)}\theta(\sqrt{s'} - 2m_c), \tag{8.221}$$

where

$$v(s') = \sqrt{1 - \frac{4m_c^2}{s'}}$$

is the velocity of the c-and $\bar{c}$ quark and negative k_0 corresponds to an on-shell antiquark. It remains to perform the angular integration to obtain the imaginary part:

$$\pi\mathrm{Im}_{[(p+q)^2]}\Delta^{(c,0)}(s',p^2) = \frac{m_c}{s' - p^2}\ln\frac{1 + v(s')}{1 - v(s')}\theta(\sqrt{s'} - 2m_c), \tag{8.222}$$

so that the first dispersion relation reads

$$\Delta^{(c,0)}((p+q)^2, p^2) = \frac{1}{\pi^2}\int_{4m_c^2}^{\infty}\frac{ds'}{s' - (p+q)^2}\left(\frac{m_c \ln\frac{1+v(s')}{1-v(s')}}{s' - p^2}\right). \tag{8.223}$$

The next step is to consider the imaginary part $\mathrm{Im}_{[(p+q)^2]}\Delta^{(c,0)}(s',p^2)$ as an analytic function of the variable p^2 at fixed s' and to take its imaginary part in $p^2 = s$ determining the double spectral density of $\Delta^{(c,0)}$. In our case this task is straightforward because in (8.222) the dependence on p^2 is limited to a simple pole producing a delta-function, so that the double spectral density is

$$\rho^{(c,0)}(s',s) = \frac{m_c}{\pi^2}\delta(s' - s)\ln\frac{1 + v(s')}{1 - v(s')}\theta(\sqrt{s'} - 2m_c), \tag{8.224}$$

allowing us to formally rewrite (8.223) as a the double dispersion relation:

$$\Delta^{(c,0)}((p+q)^2, p^2) = \int_{4m_c^2}^{\infty}\frac{ds'}{s' - (p+q)^2}\int_{4m_c^2}^{\infty}\frac{ds}{s - p^2}\rho^{(c,0)}(s',s). \tag{8.225}$$

In this case the region in the s, s' plane occupied by the double spectral density is reduced to a straight line on the diagonal $s = s'$ (see Figure 8.7). From this we calculate double moments of the LO contribution to the correlation function as defined in (8.210):

$$
\mathcal{M}_{kn}^{(c,0)} = \frac{1}{k!n!} \frac{\partial^k}{\partial((p+q)^2)^k} \frac{\partial^n}{\partial(p^2)^n} \Delta^{(c,0)}((p+q)^2, p^2) \Big|_{p^2 = (p+q)^2 = 0}
$$

$$
= \frac{m_c}{\pi^2} \int_{4m_c^2}^{\infty} \frac{ds}{s^{k+n+2}} \ln \frac{1 + v(s)}{1 - v(s)} = \frac{2m_c}{\pi^2 (4m_c^2)^{n+k+1}} \int_0^1 dv \, v(1 - v^2)^{n+k} \ln \frac{1 + v}{1 - v}. \qquad (8.226)
$$

The gluon condensate contribution to the OPE is calculated from the diagrams with vacuum gluons, one of them shown in Figure 8.8(c). The calculational procedure is again similar to the case of the two-point correlation function of c-quark currents considered in the previous section. The main element is the propagator of the c-quark in the external vacuum gluon field. However the calculation is technically more difficult. Adding the gluon condensate term to the LO result (8.226), we obtain, to the achieved accuracy, the expression for the OPE of the correlation function in a form of double moments:

$$
\mathcal{M}_{kn}^{(c,OPE)} = \mathcal{M}_{kn}^{(c,0)} \left(1 + C_{kn}^G \frac{\langle GG \rangle}{(4m_c^2)^2} \right), \qquad (8.227)
$$

where the dimensionless coefficient C_{kn}^G can be found e.g., in [91]. We see that the gluon condensate term is power suppressed with respect to the LO term by a fourth power of the ratio of $O(\Lambda_{QCD}/2m_c)$. What is also important is that the coefficient C_{kn}^G grows fast with n, k, signalling a breakdown of OPE for higher moments, which effectively corresponds to shifting s, s' towards the threshold $4m_c^2$.

Equating the calculated moments (8.227) to their hadronic representation (8.211),

$$
\mathcal{M}_{kn}^{(c,hadr)} = \mathcal{M}_{kn}^{(c,OPE)}
$$

at not too large k, n, we then apply quark-hadron duality. In the adopted approximation, we replace the part of the hadronic sum in (8.211) by a double dispersion integral taken from the perturbative LO spectral density over a certain region above the effective threshold parameters chosen for the s and s' variables. Assuming that all transitions in (8.211), involving the states above the open charm threshold, are covered by the duality approximation, we have the following generic expression for the duality ansatz:

$$
\sum_{\eta_c, h'} \frac{f_{\eta_c}}{m_{\eta_c}^{2n}} \int_{s'_{th}}^{\infty} ds' \frac{\mathcal{F}_{h'\eta_c\gamma}(s') f_{h'}(s')}{s'^{k+1}} + \sum_{h,\psi} \int_{s_{th}}^{\infty} ds \frac{f_h(s)}{s^{n+1}} \frac{f_\psi \mathcal{F}_{\psi h\gamma}(s)}{m_\psi^{2k+2}}
$$

$$
+ \sum_{h,h'} \int_{s_{th}}^{\infty} ds \frac{f_h(s)}{s^{n+1}} \int_{s'_{th}}^{\infty} ds' \frac{\mathcal{F}_{h'h\gamma}(s', s) f_{h'}(s')}{s'^{k+1}} = \int_{s_0}^{\infty} \frac{ds}{s^{k+1}} \int_{s'_0}^{f_{max}(s)} \rho^{(c,0)}(s', s) \frac{ds'}{s'^{n+1}}. \qquad (8.228)
$$

In our case the situation is simplified by the fact that a two-dimensional duality region in the s, s' plane is reduced to a one-dimensional diagonal $s = s'$, hence, $f_{max}(s) = \infty$ and $s_0 = s'_0$. Using (8.228) and the explicit form (8.224) of the double spectral density and subtracting on both sides the parts of the dispersion integral above duality thresholds, we obtain the final form of the double QCD sum rule involving the four lowest charmonium radiative transitions:

$$
\frac{1}{m_{\eta_c(1S)}^{2n} m_{J/\psi}^{2k+2}} \left[f_{\eta_c(1S)} f_{J/\psi} \mathcal{F}_{J/\psi \eta_c(1S)\gamma} \right.
$$

$$
+ \left(\frac{m_{J/\psi}}{m_{\psi(2S)}} \right)^{2k+2} f_{\eta_c(1S)} f_{\psi(2S)} \mathcal{F}_{\psi(2S)\eta_c(1S)\gamma}
$$

$$
+ \left(\frac{m_{\eta_c(1S)}}{m_{\eta_c(2S)}} \right)^{2n} f_{\eta_c(2S)} f_{\psi(1S)} \mathcal{F}_{\eta_c(2S)J/\psi(1S)\gamma}
$$

$$
+ \left(\frac{m_{\eta_c(1S)}}{m_{\eta_c(2S)}} \right)^{2n} \left(\frac{m_{J/\psi}}{m_{\psi(2S)}} \right)^{2k+2} \left. f_{\eta_c(2S)} f_{\psi(2S)} \mathcal{F}_{\psi(2S)\to\eta_c(2S)\gamma} \right]
$$

$$
= \frac{2m_c}{\pi^2(4m_c^2)^{n+k+1}} \int_0^{v_0} dv\, v(1-v^2)^{n+k} \ln\frac{1+v}{1-v} + \mathcal{M}_{kn}^{(c,0)} C_{kn}^G \frac{\langle GG \rangle}{(4m_c)^2}, \qquad (8.229)
$$

where $v_0 = \sqrt{1 - 4m_c^2/s_0}$. Note that our choice of duality region is not unique. We could have chosen another region, e.g., decreasing s_0 and including also the transitions involving $\eta_c(2S)$ and $\psi(2S)$ and obtaining a sum rule for the lowest $J/\psi\gamma$ transition amplitude. In more general cases, with a two-dimensional domain of nonvanishing double spectral density, also the form of the duality region can be chosen in various ways (square or triangle). This multiple choice generates additional "systematic" uncertainty of the double sum rules, hence, generally, they are less accurate than the two-point QCD sum rules.

Three-point correlation functions combined with double hadronic dispersion relations are also applied to obtain QCD sum rules for various e.m. form factors of light hadrons or for heavy-to-light weak transition form factors. One important application of this method is the pion form factor in the spacelike region calculated in [92, 93]. In this case, the LO and gluon condensate diagrams are obtained from the ones shown in Figure 8.8, if we replace the c quarks by light quarks and the real photon by a virtual one. Due to the presence of light quarks, also the four-quark condensate contributions have to be taken into account. At $q^2 = -Q^2 \neq 0$, the LO triangle loop contribution to the double spectral density, evaluated using Cutkosky rules, has a more complicated expression than the one obtained above, but this is just a technical difficulty. A real stumbling block in these sum rules is the Q^2 dependence of the condensate contributions, which, in a clear contradiction with the QCD asymptotics, remains constant or even grows with Q^2, so that the method is not applicable already at $Q^2 > 1$ GeV^2. A detailed description of three-point sum rules for the heavy-to-light form factors can also be found in [94].

In Chapter 10 we will describe another QCD based method of sum rules which overcomes this problem.

CHAPTER 9

LIGHT-CONE EXPANSION AND DISTRIBUTION AMPLITUDES

9.1 THE $\gamma^*\gamma^* \to \pi^0$ AMPLITUDE

In Chapter 7, while discussing the asymptotics of the pion form factor, we introduced a specific pion distribution amplitude (DA), defined as a probability amplitude to find quark and antiquark with certain collinear momenta in an energetic pion. In a more systematic way, these DAs emerge in hadronic matrix elements in which the products of quark-current operators are expanded near the light-cone. This operator-product expansion near the light-cone or, briefly, *light-cone OPE* generalizes the OPE in local operators employed in Chapter 8.

To explain the idea of the light-cone OPE and to define the related distribution amplitudes of the pion, we consider the following process of e.m. scattering:

$$e^+ e^- \to \pi^0 e^+ e^- , \tag{9.1}$$

observable in electron-positron collisions. The diagram of this process in the lowest order in e.m. coupling is shown in Figure 9.1. It includes a fusion of two virtual photons into a neutral pion in the final state:

$$\gamma^*(q)\gamma^*(p-q) \to \pi^0(p) . \tag{9.2}$$

The four-momenta q and $p-q$ of the photons are spacelike, $q^2 \leq 0$, $(p-q)^2 \leq 0$, and are determined by the momentum transfers from the initial electron and positron, respectively,

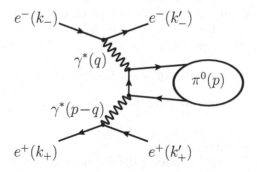

Figure 9.1   Diagram of the π^0 production via virtual photons.

so that $q = k_- - k'_-$ and $p - q = k_+ - k'_+$, whereas p is the four-momentum of the final pion with $p^2 = m_\pi^2$.

Factorizing the electron and positron currents and the photon propagators in a similar way to how it was done for the electron-pion scattering in Chapter 2, we obtain the amplitude of (9.1):

$$\mathcal{A}(e^+e^- \to \pi^0 e^+e^-) = 16\pi^2 \alpha_{em}^2 \frac{[\bar{u}(k'_-)\gamma^\mu u(k_-)]\,[\bar{v}(k_+)\gamma^\nu v(k'_+)]}{q^2(p-q)^2}$$

$$\times i \int d^4 x e^{-iqx} \langle \pi^0(p)|T\{j_\mu^{em}(x)j_\nu^{em}(0)\}|0\rangle . \tag{9.3}$$

The cross section of this process obtained by squaring the above amplitude is usually measured as a function of the photon virtuality $Q^2 = -q^2$, detecting only the final-state electron. Due to the inverse virtuality of the second photon in (9.3) this cross section is accumulated at $(p - q)^2 \simeq 0$, i.e., the second photon is by default almost real.

In what follows we concentrate on the hadronic matrix element

$$\mathcal{F}_{\mu\nu}(p,q) = i \int d^4 x\, e^{-iqx} \langle \pi^0(p)|T\{j_\mu^{em}(x)j_\nu^{em}(0)\}|0\rangle , \tag{9.4}$$

emerging as a part of the amplitude (9.3). It contains a product of two e.m. current operators taken at different points x and 0; hence, $\mathcal{F}_{\mu\nu}$ is not a simple form factor but rather belongs to the category of nonlocal hadronic amplitudes considered in Chapter 6. It is still considerably simpler than other nonlocal amplitudes, because the final state of (9.4) consists of a single on-shell hadron. On the other hand, the amplitude (9.4) containing a time-ordered product of two quark currents, represents a new type of correlation function. The only difference with respect to the vacuum-to-vacuum correlation functions considered in Chapter 8, is that there is now an on-shell pion state instead of the vacuum state.

The hadronic matrix element (9.4) is determined by two independent four-momenta p and q. At fixed $p^2 = m_\pi^2$, there are two independent kinematical invariants q^2 and $(p - q)^2$. The scalar product $p \cdot q$ is expressed via their combination:

$$p \cdot q = \frac{1}{2}[q^2 - (p - q)^2 + m_\pi^2].$$

Expanding (9.4) in invariant amplitudes, we encounter only one possible Lorentz structure:

$$\mathcal{F}_{\mu\nu}(p,q) = \epsilon_{\mu\nu\alpha\beta}p^\alpha q^\beta F_{\gamma^*\pi}(q^2, (p-q)^2) . \tag{9.5}$$

The Lorentz structure multiplying the invariant amplitude $F_{\gamma^*\pi}$ in the above transforms as a pseudotensor which is consistent with the combination of two four vectors and the pion state with spin-parity $J^P = 0^-$ in the amplitude (9.4).

As already mentioned, the process (9.1) is experimentally accessible when one of the intermediate photons is almost real. In this limit the invariant amplitude in (9.5) reduces to the photon-pion transition form factor

$$F_{\gamma^*\pi}(q^2, 0) \equiv F_{\gamma\pi}(q^2) . \tag{9.6}$$

Furthermore, when both photons are real, the amplitude

$$F_{\gamma^*\pi}(0,0) = F_{\gamma\pi}(0) = g_{\pi\gamma\gamma} \tag{9.7}$$

is a time-reversal of the $\pi^0 \to 2\gamma$ amplitude defined in (6.4), where $k_1 = p - q$, $k_2 = q$.

9.2 LIGHT-CONE DOMINANCE

In what follows, we consider the kinematical region, where the virtualities of both photons in the subprocess (9.2) are large and at the same time their difference is also large, so that

$$|2p \cdot q| \simeq ||(p - q)^2| - Q^2| \sim Q^2 \sim |(p - q)^2| \gg \Lambda_{QCD}^2 \,. \tag{9.8}$$

In terms of the dimensionless ratio

$$\xi = \frac{2p \cdot q}{Q^2} \simeq \frac{|(p - q)^2| - Q^2}{Q^2} \,, \tag{9.9}$$

the condition (9.8) implies that $\xi \sim 1$ but is not small. Note that hereafter we neglect in the amplitude $F_{\gamma^* \pi}(q^2, (p-q)^2)$ the pion mass squared $p^2 = m_\pi^2$ and the light-quark masses $m_{u,d}$ in comparison with the two large scales.

With the external momenta in the region (9.8), the integral over four-coordinate in (9.4) has a spectacular property: the oscillating exponential factor multiplying the hadronic matrix element effectively cuts away the regions in the four-dimensional space which are far from the light-cone $x^2 = 0$, so that the integration region near the light-cone dominates.

To prove that, we first notice that the dominant contribution to the integral in (9.4) stems from the region where the power of the exponent is not large,

$$q \cdot x \lesssim O(1) \,. \tag{9.10}$$

Outside that region fast oscillations of the exponential factor strongly suppress the integrand, as follows from the Riemann-Lebesgue theorem.

Furthermore, it is convenient to choose a reference frame in which the four-momenta q and p have only one spatial component each:

$$q = (q_0, 0, 0, q_3) \,, \quad p = (\mu, 0, 0, -\mu) \,,$$

where μ is a scale of $O(\Lambda_{QCD})$. In this frame

$$p \cdot q = p_0 q_0 - p_3 q_3 = \mu(q_0 + q_3) = \frac{Q^2 \xi}{2} \,. \tag{9.11}$$

The last equation is solved using the relation $q_3 = \sqrt{q_0^2 + Q^2}$ and we obtain:

$$q_0 = \frac{Q^2 \xi}{4\mu} - \frac{\mu}{\xi}, \quad q_3 = \frac{Q^2 \xi}{4\mu} + \frac{\mu}{\xi} \,. \tag{9.12}$$

As a result, the oscillating exponent in (9.4) becomes:

$$e^{-iq \cdot x} = e^{-iq_0 x_0 + iq_3 x_3} = e^{-i\frac{Q^2 \xi}{4\mu}(x_0 - x_3)} e^{i\frac{\mu}{\xi}(x_0 + x_3)} \,. \tag{9.13}$$

Hence, the region where the condition (9.10) is valid and where strong oscillations are absent, is simultaneously bounded by the two inequalities:

$$(x_0 - x_3) \lesssim \frac{4\mu}{Q^2 \xi}, \quad (x_0 + x_3) \lesssim \frac{\xi}{\mu} \,. \tag{9.14}$$

Multiplying these inequalities, leads to:

$$x_0^2 - x_3^2 \lesssim \frac{4}{Q^2} < \frac{4}{Q^2} + x_1^2 + x_2^2 \,, \tag{9.15}$$

yielding

$$x^2 \lesssim \frac{4}{Q^2}. \tag{9.16}$$

Thus, at asymptotically large Q^2, the region where the product of operators in (9.4) dominates, indeed shrinks towards the light-cone $x^2 \sim 0$. Note that (9.16) is a Lorentz-invariant condition and does not depend on the actual choice of the reference frame. This condition holds also if the components of the four-vector x are large; hence, the short-distance dominance $x \sim 0$ is generally not valid in the hadronic matrix element (9.4).

9.3   EXPANSION NEAR THE LIGHT-CONE

Having chosen the region of large virtualities (9.8) in the vacuum-to-pion matrix element (9.4), we effectively restrict the T-product of currents to the intervals near the light-cone:

$$\epsilon_{\mu\nu\alpha\beta} p^\alpha q^\beta \underbrace{F_{\gamma^* \pi}(q^2, (p-q)^2)}_{Q^2, |(p-q)^2| \gg \Lambda_{QCD}^2} = i \int d^4 x\, e^{-iqx} \langle \pi^0(p)| \underbrace{T\{j_\mu^{em}(x) j_\nu^{em}(0)\}}_{x^2 \to 0} |0\rangle. \tag{9.17}$$

At sufficiently small x^2, it is possible to use a general expansion of the product of current operators:

$$T\{j_\mu^{em}(x) j_\nu^{em}(0)\} = \sum_t \left[C_t(x^2) \mathcal{O}_t(x, 0) \right]_{\mu\nu}, \tag{9.18}$$

in a series of bilocal operators $\mathcal{O}_t(x, 0)$ composed from the quark, antiquark and gluon fields and multiplied[1] by the coefficient functions $C_t(x^2)$. This expansion contains an infinite amount of terms, so that all operators allowed by quantum numbers contribute.

There are two main reasons why the expansion (9.18) is functioning at $x^2 \to 0$. First of all, due to large virtualities, the coefficients $C_t(x^2)$ multiplying the bilocal operators $\mathcal{O}_t(x, 0)$ are calculable in terms of perturbative QCD diagrams. In particular, at the leading zeroth order in α_s, the coefficient $C_{t_{min}}(x^2)$ which we shall calculate explicitly, corresponds to a free-quark propagator. Secondly, near the light-cone, one can retain a finite number of terms in the expansion (9.18) and still achieve a reasonable accuracy, because, after substituting (9.18) in (9.4), the higher orders in x^2 convert into inverse powers of the large momentum scale Q^2; hence, the practical version of the OPE near the light-cone contains a finite amount of terms:

$$T\{j_\mu^{em}(x) j_\nu^{em}(0)\}\big|_{x^2 \to 0} = \sum_{t=t_{min}}^{t_{max}} \left[C_t(x^2) \mathcal{O}_t(x, 0) \right]_{\mu\nu}. \tag{9.19}$$

Note that this OPE is qualitatively different from the expansion in local operators discussed in Chapter 8, since it contains bilocal operators. Also the index t does not simply count the dimension of the operator, as will be explained below.

From all possible operators contributing to the expansion (9.18), we only have to take into account those which do not vanish, being sandwiched between the vacuum and π^0 state. The simplest operator consists of the combination of quark and antiquark fields with the valence content and spin-parity of the neutral pion:

$$\mathcal{O}_t(x, 0) = \left(\bar{u}(x) \Gamma_t u(0) - \bar{d}(x) \Gamma_t d(0) \right) \oplus \{x \leftrightarrow 0\}, \tag{9.20}$$

---

[1]The way we put the Lorentz indices on the r.h.s. of (9.18) indicates that in each term they are distributed between the operators and coefficients.

where the Dirac matrices Γ_t correspond either to an axial-vector ($J^P = 1^+$) or to a pseudoscalar ($J^P = 0^-$) spin-parity.

To demonstrate how the expansion (9.18) works at the leading order, we include only the u, d quark components of e.m. currents. Adding other flavor components in one of the currents corresponds to taking into account nonvalence components of the pion which are beyond our approximation. Employing the isospin decomposition (1.54) of the e.m. current we retain only the relevant combinations of isovector and isoscalar currents in the T-product:

$$
\begin{aligned}
T\{j_\mu^{em}(x) j_\nu^{em}(0)\} &= T\{[j_\mu^{I=1}(x) + j_\mu^{I=0}(x)][j_\nu^{I=1}(0) + j_\nu^{I=0}(0)]\} \\
&= T\{j_\mu^{I=1}(x) j_\nu^{I=0}(0)\} + T\{j_\mu^{I=0}(x) j_\nu^{I=1}(0)\} + \dots .
\end{aligned}
\tag{9.21}
$$

The remaining terms denoted by ellipsis do not contribute to the vacuum-to-pion matrix element. To prove that, we invoke the G-parity conservation and notice that the $I = 1$ ($I = 0$) component of the e.m. current has a positive (negative) G-parity; hence, only the product of isovector and isoscalar components has a negative G-parity, the same as the pion state.

The next step involves applying the usual Wick theorem, that is, contracting the u, d quark fields in all possible ways and forming the free quark propagators. It is sufficient to perform this procedure for the first T-product in the second line of (9.21):

$$
\begin{aligned}
&T\{j_\mu^{I=1}(x) j_\nu^{I=0}(0)\}\big|_{x^2 \to 0} = \\
&\frac{1}{12} T\{[\bar{u}(x)\gamma_\mu u(x) - \bar{d}(x)\gamma_\mu d(x)][\bar{u}(0)\gamma_\nu u(0) + \bar{d}(0)\gamma_\nu d(0)]\}\big|_{x^2 \to 0} \\
&= \frac{i}{12}\Big(\bar{u}(x)\gamma_\mu S_{(u)}^{(0)}(x,0)\gamma_\nu u(0) - \bar{d}(x)\gamma_\mu S_{(d)}^{(0)}(x,0)\gamma_\nu d(0) \\
&\quad + \bar{u}(0)\gamma_\nu S_{(u)}^{(0)}(0,x)\gamma_\mu u(x) - \bar{d}(0)\gamma_\nu S_{(d)}^{(0)}(0,x)\gamma_\nu d(x)\Big)\Big|_{x^2 \to 0},
\end{aligned}
\tag{9.22}
$$

while the second T-product yields an equal result. Note that the terms, in which all quark fields are contracted, generate a unit operator and do not contribute to the pion-to-vacuum matrix element. Inserting the massless free-quark propagator (A.50) in the above expression and multiplying it by factor 2, we obtain:

$$
\begin{aligned}
T\{j_\mu^{em}(x) j_\nu^{em}(0)\}_{x^2 \to 0} &= \frac{i x^\alpha}{12\pi^2 (x^2)^2} \\
&\times \Big[\Big(\bar{u}(x)\gamma_\mu \gamma_\alpha \gamma_\nu u(0) - \bar{d}(x)\gamma_\mu \gamma_\alpha \gamma_\nu d(0)\Big) \\
&\quad - \Big(\bar{u}(0)\gamma_\nu \gamma_\alpha \gamma_\mu u(x) - \bar{d}(0)\gamma_\nu \gamma_\alpha \gamma_\mu d(x)\Big)\Big] \\
&= -\frac{\epsilon_{\mu\nu\alpha\rho} x^\alpha}{12\pi^2 (x^2)^2}\Big[\Big(\bar{u}(x)\gamma^\rho \gamma_5 u(0) - \bar{d}(x)\gamma^\rho \gamma_5 d(0)\Big) \\
&\quad + \Big(\bar{u}(0)\gamma^\rho \gamma_5 u(x) - \bar{d}(0)\gamma^\rho \gamma_5 d(x)\Big)\Big] + \dots ,
\end{aligned}
\tag{9.23}
$$

where in the last equation only the relevant axial-vector part is retained in the product of three γ-matrices (see (A.15)). Note that the above expression already has the structure (9.18) of OPE, that is, a quark-antiquark bilocal operator of the type (9.20) is multiplied by a coefficient depending on the light-cone separation. Finally, replacing the product of currents in (9.17) by (9.23), we obtain the leading-order result for the correlation function

$$\epsilon_{\mu\nu\alpha\beta}p^\alpha q^\beta F^{(0)}_{\gamma^*\pi}(q^2,(p-q)^2) = \frac{-i}{12\pi^2}\epsilon_{\mu\nu\alpha\rho}\int d^4x \frac{x_\alpha}{(x^2)^2}e^{-iqx}$$

$$\times\left[\langle\pi^0(p)|(\bar{u}(x)\gamma^\rho\gamma_5 u(0) - \bar{d}(x)\gamma^\rho\gamma_5 d(0))|0\rangle\right.$$

$$\left.+\langle\pi^0(p)|\bar{u}(0)\gamma^\rho\gamma_5 u(x) - \bar{d}(0)\gamma^\rho\gamma_5 d(x)|0\rangle\right]_{x^2\to 0}. \tag{9.24}$$

This expression contains a new type of hadronic amplitude: a matrix element of a bilocal quark-antiquark operator sandwiched between the vacuum and π^0 state. This amplitude corresponds to the part of the diagram in Figure 9.1, in which the quark and antiquark lines emitted from the currents at an interval x^2 form the valence content of the pion. The virtual quark line on this diagram is included in the coefficient of the OPE.

It is instructive to further expand the bilocal operators entering (9.24) in local quark-antiquark operators. As an example, we consider the first operator in (9.24) and expand the u-quark field in a power series at the origin, $x=0$,

$$\bar{u}(x)\gamma^\rho\gamma_5 u(0) = \sum_{n=0}^\infty \frac{1}{n!}x_{\mu_1}x_{\mu_2}\ldots x_{\mu_n}\bar{u}(0)\overleftarrow{\partial}^{\mu_1}\overleftarrow{\partial}^{\mu_2}\ldots\overleftarrow{\partial}^{\mu_n}\gamma^\rho\gamma_5 u(0), \tag{9.25}$$

introducing an infinite series

$$\bar{u}(0)\overleftarrow{D}^{\mu_1}\overleftarrow{D}^{\mu_2}\ldots\overleftarrow{D}^{\mu_n}\gamma^\rho\gamma_5 u(0), \quad n = 0,1,2\ldots, \tag{9.26}$$

of local quark-antiquark operators with growing dimension. Replacing in the above the ordinary derivatives by the covariant ones, we restore the QCD gauge invariance for these operators[2]. Their vacuum-to-pion matrix elements depend on the single four-vector p and have the following generic decomposition:

$$\langle\pi^0(p)|\bar{u}(0)\overleftarrow{D}^{\mu_1}\overleftarrow{D}^{\mu_2}\ldots\overleftarrow{D}^{\mu_n}\gamma^\rho\gamma_5 u(0)|0\rangle = (-i)^{n+1}p^\rho p^{\mu_1}p^{\mu_2}\ldots p^{\mu_n}\mathcal{M}^d_n$$

$$-(-i)^{n+1}g^{\mu_1\mu_2}p^\rho p^{\mu_3}\ldots p^{\mu_n}\mathcal{M}^{d+2}_n + \{\ldots\}, \tag{9.27}$$

where, by construction of the series, $\mathcal{M}^{d+2}_0 = \mathcal{M}^{d+2}_1 = 0$ and $\{\ldots\}$ indicates the remaining combinations of p with the metric tensors $g^{\mu_j\mu_k}$ $(j,k=1,\ldots n)$, rendering the decomposition totally symmetric in $\mu_1,\mu_2,\ldots\mu_n$ as follows from the symmetry of the operators (9.26). The invariant coefficients $\mathcal{M}^d_n$ and $\mathcal{M}^{d+2}_n$ differ by two units of dimension. Note that these parameters, as well as the similar ones emerging in the rest of this expansion, are of a nonperturbative origin, being determined by the energy-momentum scales of $O(\Lambda_{QCD})$. E.g., the coefficient at $n=0$ in the above expansion:

$$\langle\pi^0(p)|\bar{u}(0)\gamma^\rho\gamma_5 u(0)|0\rangle = -ip^\rho\mathcal{M}^d_0, \tag{9.28}$$

is directly related to the pion decay constant, $\mathcal{M}^d_0 = f_\pi/\sqrt{2}$ in the isospin symmetry limit.

Inserting both parts of (9.25) between the vacuum and pion states and using (9.27), we obtain:

$$\langle\pi^0(p)|\bar{u}(x)\gamma^\rho\gamma_5 u(0)|0\rangle = \sum_{n=0}^\infty \frac{1}{n!}x_{\mu_1}x_{\mu_2}\ldots x_{\mu_n}$$

$$\times\left((-i)^{n+1}p^\rho p^{\mu_1}p^{\mu_2}\ldots p^{\mu_n}\mathcal{M}^d_n - (-i)^{n+1}g^{\mu_1\mu_2}p^\rho p^{\mu_3}\ldots p^{\mu_n}\mathcal{M}^{d+2}_n + \{\ldots\}\right), \tag{9.29}$$

---

[2]For a bilocal operator on the l.h.s. of (9.25) this replacement is equivalent to an insertion of the gauge link $[x,0]$ (defined in (A.46)) between the quark and antiquark fields,. An alternative is to use the fixed point gauge $x_\mu A^\mu = 0$ for gluon fields in which case $[x,0]=1$ and ordinary derivatives on the r.h.s. of (9.25) are equivalent to the covariant ones.

where the second term with $g^{\mu_1\mu_2}$ generates one extra power of x^2 with respect to the first term.

Returning to (9.24) we consider, as a sample, the contribution of the u-quark bilocal current:

$$\epsilon_{\mu\nu\alpha\beta}p^\alpha q^\beta F^{(0,u)}_{\gamma^*\pi}(q^2,(p-q)^2)$$
$$= -\frac{i}{12\pi^2}\epsilon_{\mu\nu\alpha\rho}\int d^4x\frac{x^\alpha}{(x^2)^2}e^{-iqx}\langle\pi^0(p)|\bar{u}(x)\gamma^\rho\gamma_5 u(0)|0\rangle, \qquad (9.30)$$

where we can use the expansion (9.29) for the matrix element. The integral over x is calculated separately:

$$\int d^4x\frac{x_\alpha}{(x^2)^2}x_{\mu_1}x_{\mu_2}\dots x_{\mu_n}e^{-iqx} = i^{n+1}\frac{\partial}{\partial q_\alpha}\frac{\partial}{\partial q_{\mu_1}}\frac{\partial}{\partial q_{\mu_2}}\cdots\frac{\partial}{\partial q_{\mu_n}}\int d^4x\frac{e^{-iqx}}{(x^2)^2}$$
$$= 2\pi^2(-2i)^n n!\frac{q_\alpha q_{\mu_1}q_{\mu_2}\dots q_{\mu_n}}{(q^2)^{n+1}} + \dots, \qquad (9.31)$$

employing the integration formulas given in (A.95). In the above, the terms containing at least one metric tensor $g_{\mu_j\mu_k}$ or $g_{\alpha\mu_k}$, are indicated by the ellipsis. Putting everything together, we obtain:

$$\epsilon_{\mu\nu\alpha\beta}p^\alpha q^\beta F^{(0,u)}_{\gamma^*\pi}(q^2,(p-q)^2) = -\frac{i}{6}\epsilon_{\mu\nu\alpha\rho}\sum_{n=0}^\infty\left[(-2i)^n\frac{q_\alpha q_{\mu_1}q_{\mu_2}\dots q_{\mu_n}}{(q^2)^{n+1}} + \dots\right]$$
$$\times\left((-i)^{n+1}p^\rho p^{\mu_1}p^{\mu_2}\dots p^{\mu_n}\mathcal{M}^d_n - (-i)^{n+1}g^{\mu_1\mu_2}p^\rho p^{\mu_3}\dots p^{\mu_n}\mathcal{M}^{d+2}_n + \{\dots\}\right). \qquad (9.32)$$

The terms with metric tensors, indicated by ellipsis in the square bracket, vanish, being multiplied by the structure proportional to $\mathcal{M}^d_n$, either because this multiplication yields $p^2 = m^2_\pi \simeq 0$, or because of the Levi-Civita tensor. Contracting the indices in (9.32) and using the notation (9.9) for the ratio of large scales, we finally obtain the invariant amplitude:

$$F^{(0,u)}_{\gamma^*\pi}(q^2,(p-q)^2) = \frac{1}{Q^2}\left(\frac{1}{6}\sum_{n=0}^\infty\xi^n\mathcal{M}^d_n\right) + \frac{1}{(Q^2)^2}\left(\frac{2}{3}\sum_{n=0}^\infty\xi^n\mathcal{M}^{d+2}_{n+2}\right) + \dots, \qquad (9.33)$$

where the term with the leading power and one of the subleading terms are shown. Since the parameter ξ is not small, in each of the infinite series, the terms, stemming from the local operators with growing dimensions, have comparable contributions with the same power of $1/Q^2$. We conclude that in the vacuum-to-pion matrix elements in (9.24) one has to retain the bilocal quark-antiquark operators, without re-expanding them in local operators. In what follows, these matrix elements will be treated as universal characteristics of the pion. Another observation is that the term in (9.33), originating from the part of the expansion (9.29) with an extra power of x^2, is suppressed by $1/Q^2$, making possible the expansion of bilocal operators near the light-cone.

9.4   THE PION DISTRIBUTION AMPLITUDE

Returning to the hadronic matrix elements in (9.24), it is more convenient to consider a similar one with a charged pion. It is related to its neutral pion counterpart via isospin symmetry:

$$\langle\pi^+(p)|\bar{u}(x)\gamma^\rho\gamma_5 d(0)|0\rangle = \frac{1}{\sqrt{2}}\langle\pi^0(p)|(\bar{u}(x)\gamma^\rho\gamma_5 u(0) - \bar{d}(x)\gamma^\rho\gamma_5 d(0))|0\rangle. \qquad (9.34)$$

It is also convenient to replace $x \to x_1$, $0 \to x_2$, with $x_{1,2} = \xi_{1,2} x$, where $\xi_{1,2}$ are arbitrary numbers, and $x^2 \simeq 0$.

The Lorentz decomposition of the matrix element (9.34) is a linear combination of two independent four-vectors:

$$\langle \pi^+(p)|\bar{u}(x_1)\gamma^\rho\gamma_5 d(x_2)|0\rangle_{x^2\to 0} = ip^\rho f(p \cdot x_1, p \cdot x_2) + (x_1 - x_2)^\rho g(p \cdot x_1, p \cdot x_2), \quad (9.35)$$

where in the second term the translational invariance is taken into account. The functions f and g can only depend on the two invariant variables $p \cdot x_1$, and $p \cdot x_2$, since all other invariants are fixed: $x_{1,2}^2 = 0$ and $p^2 = m_\pi^2 \simeq 0$. To derive a generic representation of these functions, we use the translation $x_{1,2} \to x_{1,2} + a$:

$$\langle \pi^+(p)|\bar{u}(x_1+a)\gamma^\rho\gamma_5 d(x_2+a)|0\rangle = \langle \pi^+(p)|e^{i\hat{P}\cdot a}\bar{u}(x_1)\gamma^\rho\gamma_5 d(x_2)e^{-i\hat{P}a}|0\rangle$$
$$= e^{ip\cdot a}\langle \pi^+(p)|\bar{u}(x_1)\gamma^\rho\gamma_5 d(x_2)|0\rangle. \quad (9.36)$$

In terms of the invariant functions defined in (9.35), the above relation yields

$$f(p \cdot (x_1+a), p \cdot (x_2+a)) = e^{ip\cdot a} f(p \cdot x_1, p \cdot x_2), \quad (9.37)$$

and the same equation for the function g. Differentiating (9.37) over the scalar product $p \cdot a$ and putting $a_\mu \to 0$, we obtain a differential equation:

$$\frac{\partial f(p \cdot x_1, p \cdot x_2)}{\partial(p \cdot x_1)} + \frac{\partial f(p \cdot x_1, p \cdot x_2)}{\partial(p \cdot x_2)} = if(p \cdot x_1, p \cdot x_2). \quad (9.38)$$

Applying the double Fourier transformation,

$$f(p \cdot x_1, p \cdot x_2) = \int\limits_{-\infty}^{\infty} du\, e^{iu(p\cdot x_1)} \int\limits_{-\infty}^{\infty} dv\, e^{iv(p\cdot x_2)} \bar{\phi}(u,v), \quad (9.39)$$

inserting it in (9.38) and comparing both sides of the resulting equation, we conclude that it is satisfied only at

$$u + v = 1,$$

that is, along the direct line $u = 1 - v$ in the $\{u, v\}$-plane. This condition fixes the general form of the Fourier-transform:

$$\bar{\phi}(u,v) = \delta(1 - u - v)\phi(u,v), \quad (9.40)$$

which, after substituting it back in the transformation (9.39) and integrating over the variable v, yields:

$$f(p \cdot x_1, p \cdot x_2) = \int\limits_{-\infty}^{\infty} du\, e^{iu(p\cdot x_1)+i(1-u)(p\cdot x_2)}\phi(u, 1-u). \quad (9.41)$$

Using this, we can write down a general representation for the hadronic matrix element (9.35):

$$\langle \pi^+(p)|\bar{u}(x_1)\gamma_\rho\gamma_5 d(x_2)|0\rangle_{x^2\to 0} = ip_\rho \int\limits_{-\infty}^{\infty} du\, e^{iu(p\cdot x_1)+i\bar{u}(p\cdot x_2)}\phi(u,\bar{u}) + \dots, \quad (9.42)$$

where $\bar{u} \equiv 1 - u$ and dots indicate the contribution of the invariant function g which has a similar form.

The ansatz (9.42) follows from symmetry considerations and contains a function of the parameter u. To further constrain this function, we notice that (9.42) closely resembles the representation (7.23) of the same hadronic matrix element in terms of the function $\tilde{\phi}$ describing the distribution of parton momenta in the fast pion. The only difference is that the integration region in (7.23) is limited by the unit interval. This motivates us to consider in (9.42) only those functions which are nonvanishing within the same interval:

$$\phi(u, \bar{u}) = N_\pi \varphi_\pi(u)\, \theta(u)\, \theta(1 - u)\,, \tag{9.43}$$

where N_π is a normalization factor. The function $\varphi_\pi(u)$ is then identified with the probability amplitude to form a two-parton state of the pion, consisting of the quark u and antiquark $\bar{d}$ which are emitted at a lightlike separation with the momenta up and $\bar{u}p$, respectively. Since we neglect the pion and light-quark masses, these momenta are also collinear and lightlike. Adopting (9.43), we obtain for the vacuum-to-pion matrix element:

$$\langle \pi^+(p)|\bar{u}(x_1)[x_1, x_2]\gamma_\rho\gamma_5 d(x_2)|0\rangle_{x^2 \to 0} = -ip_\rho f_\pi \int_0^1 du\, e^{iu(p \cdot x_1) + i\bar{u}(p \cdot x_2)} \varphi_\pi(u)\,, \tag{9.44}$$

where $N_\pi = -f_\pi$ follows from the local limit, $x_1 = x_2 = 0$, in which the above matrix element is reduced to the pion decay constant. The function $\varphi_\pi(u)$ is then normalized to a unit:

$$\int_0^1 du\, \varphi_\pi(u) = 1\,. \tag{9.45}$$

To make the definition (9.44) fully consistent with QCD, we added the gauge link defined as in (A.46). Hereafter it is omitted, adopting the fixed-point gauge for the gluon field. In (9.44), we also omit the part originating from the second term in (9.35) proportional to the difference of the coordinates. This term is actually subleading in the light-cone expansion and will be discussed later. We concentrate first on the leading term given by (9.44).

Note that, in contrast to a somewhat qualitative ansatz (7.23), the definition (9.44) stems from a systematic light-cone expansion of the bilocal quark-antiquark operator. The function $\varphi_\pi(u)$ is named the *twist-2 light-cone distribution amplitude (DA)* of the pion. Putting aside some details of the full definition, the twist t is a characteristic of a local operator, equal to the difference between its dimension and spin. Take, as an example, the local operator

$$\bar{u}(0)\gamma_\rho\gamma_5 d(0)\,.$$

It represents the lowest term in the expansion of the bilocal operator in (9.44) near $x_1 = x_2 = 0$. Counting the twist in this case is straightforward: $t = 2$ follows from subtracting the spin $J = 1$ from the dimension $d = 3$. Adding n derivatives:

$$\bar{u}(0)\overleftarrow{D}_{\mu_1}\overleftarrow{D}_{\mu_2}\dots\overleftarrow{D}_{\mu_n}\gamma_\rho\gamma_5 d(0)\,, \tag{9.46}$$

where, for simplicity, we assume that $x_2 = 0$ and the derivatives act on $\bar{u}(x_1)$ at $x_1 = 0$, we increase the dimension of the operator by n units: $d = 3 + n$. At each n, it is possible to arrange from (9.46) an operator with the maximal spin $J = 1 + n$, corresponding to the totally symmetric and traceless combination of derivatives. Therefore, at each n, there is an operator with the same $t = 2$ as the lowest dimension operator. Attributing twist 2 to the DA in (9.44) means that it is formed from an infinite series of matrix elements of $t = 2$ operators. Other combinations of derivatives obtained from (9.46), e.g., by contracting two of them:

$$\bar{u}(0)\overleftarrow{D}_\alpha\overleftarrow{D}^\alpha\overleftarrow{D}_{\mu_3}\overleftarrow{D}_{\mu_4}\dots\overleftarrow{D}_{\mu_n}\gamma_\rho\gamma_5 d(0)\,,$$

have a lower spin, hence, a higher twist, $t > 2$. In the light-cone OPE, the twist[3] plays practically the same role as the operator dimension plays in the local OPE considered in Chapter 8.

For the further discussion it is important to determine the relation of the pion DA to the matrix elements of the local operators (9.46). We start from the definition (9.44) and, putting for simplicity $x_1 \to x$ and $x_2 \to 0$, expand both sides in the power series around $x = 0$:

$$\langle \pi^+(p)|\bar{u}(x)\gamma_\rho\gamma_5 d(0)|0\rangle_{x^2 \to 0}$$

$$= \sum_{n=0}^{\infty} \frac{1}{n!} x_{\mu_1} x_{\mu_2} \ldots x_{\mu_n} \langle \pi^+(p)|\bar{u}(0)\overleftarrow{D}^{\mu_1}\overleftarrow{D}^{\mu_2}\ldots\overleftarrow{D}^{\mu_n}\gamma^\rho\gamma_5 d(0)|0\rangle_{x^2 \to 0}$$

$$= \sum_{n=0}^{\infty} \frac{1}{n!} x_{\mu_1} x_{\mu_2} \ldots x_{\mu_n} (-i)^{n+1} p^\rho p^{\mu_1} p^{\mu_2} \ldots p^{\mu_n} \sqrt{2}\mathcal{M}_n^d$$

$$= -ip^\rho f_\pi \sum_{n=0}^{\infty} \frac{i^n}{n!} x_{\mu_1} x_{\mu_2} \ldots x_{\mu_n} p^{\mu_1} p^{\mu_2} \ldots p^{\mu_n} \int_0^1 du\, u^n \varphi_\pi(u) \,,$$

where the same (isospin symmetric) definition of the matrix elements as in (9.29) is used. Matching the n-th terms of the above equation, we obtain:

$$f_\pi \int_0^1 du\, u^n \varphi_\pi(u) = (-1)^n \sqrt{2}\mathcal{M}_n^d \,. \tag{9.47}$$

Furthermore, in the isospin symmetry limit, the equality

$$\varphi_\pi(u) = \varphi_\pi(\bar{u}) \tag{9.48}$$

holds. To prove that, we apply the invariance of the matrix element on the l.h.s. of (9.44) with respect to the G-parity transformation:

$$\langle \pi^+(p)|\bar{u}(x_1)\gamma^\rho\gamma_5 d(x_2)|0\rangle \xrightarrow{\hat{I}_3} \langle \pi^-(p)|\bar{d}(x_1)\gamma^\rho\gamma_5 u(x_2)|0\rangle \xrightarrow{\hat{C}} \langle \pi^+(p)|\bar{u}(x_2)\gamma^\rho\gamma_5 d(x_1)|0\rangle$$

$$= -ip^\rho f_\pi \int_0^1 du\, e^{iu(p\cdot x_2)+i\bar{u}(p\cdot x_1)}\varphi_\pi(u) \,, \tag{9.49}$$

transform the integration variable $u \to \bar{u}$ and compare the result with the r.h.s. of (9.44).

Since $\varphi_\pi(u)$ is defined in the unit interval $0 < u < 1$, we can try to approximate this function using a simple Taylor expansion around $u = 1/2$:

$$\varphi_\pi(u) = \sum_{k=0,2,4,\ldots} c_k (u - 1/2)^k = \sum_{k=0,2,4,\ldots} \frac{c_k}{2^k}(u - \bar{u})^k \,, \tag{9.50}$$

where we have taken into account the symmetry relation (9.48) and included only the even powers of the difference $u - \bar{u}$. There is no compelling reason for the coefficients c_k to grow

---

[3]The twist determination for an arbitrary local operator is not always straightforward. One has to expand the quark and gluon fields in the longitudinal and transverse projections with respect to the light-cone. A systematic description including the details of the twist expansion can be found e.g., in [95, 96].

with the number k; hence, a polynomial in the even powers of $u - \bar{u}$, obtained truncating the above series at a certain (even) k_{max}, with the only constraint

$$\sum_{k=0,2,4,\ldots}^{k_{max}} \frac{c_k}{k+1} = 1 \, ,$$

following from the normalization condition (9.45), should presumably serve as a reasonable parameterization of the pion DA. However, the endpoint behavior of the DA,

$$\lim_{u \to 0} \varphi_\pi(u) = \lim_{u \to 1} \varphi_\pi(u) = \sum_{k=0,2,4,\ldots}^{k_{max}} \frac{c_k}{2^k}$$

is then left uncertain, being important, e.g., for the convergence of the factorization formula (7.29) discussed in Chapter 7. Moreover, the gluonic radiative corrections to the hard scattering diagrams induce a residual dependence on the renormalization scale μ in the DAs, as indicated in (7.29); hence, there should be a way to determine the corresponding μ dependence of the coefficients c_k in the polynomial parameterization.

The scale dependence of the DA $\varphi_\pi(u, \mu)$ can be inferred, via (9.47), from the renormalization properties of the local operators (9.46). For the latter, the one-loop diagrams of the gluonic radiative corrections are similar to the ones in Figure 1.4, with the vertex containing one of the operators (9.46) instead of the quark current. The calculation of these diagrams reveals that, in general, each operator gets an $O(\alpha_s)$ admixture of certain other quark-antiquark operators with a different structure of derivatives. As a result, after resumming the corrections, the operator (9.46) itself does not have a simple multiplicative renormalization with a certain anomalous dimension, as does the current operator in (1.46); hence, also the scale dependence of the pion DA – or of its expansion coefficients in (9.50) – cannot be inferred in a simple form.

Both problems – determining the end-point behavior and maintaining the polynomial structure of the function φ_π after renormalization – are solved, applying the conformal symmetry of massless QCD[4]. Leaving the details beyond our scope, let us only mention that certain polynomial combinations of local quark-antiquark operators transform as irreducible representations under specific conformal transformations, hence they do not mix with each other. Remarkably, this property is valid also after the one-loop renormalization in QCD. The conformal symmetry prescribes a special form of the polynomial expansion for the DAs, so that the coefficients of the polynomials have a mutiplicative scale dependence. The ansatz for the pion twist-2 DA inspired by the conformal symmetry is:

$$\varphi_\pi(u, \mu) = 6u\bar{u} \left[1 + \sum_{n=2,4,\ldots} a_n(\mu) C_n^{3/2}(u - \bar{u}) \right] , \qquad (9.51)$$

where $C_n^{3/2}(u - \bar{u})$ are the *Gegenbauer polynomials* (see (D.15) for their explicit form and (D.16) for the orthogonality relation). The coefficients have the following scale dependence

$$a_n(\mu) = a_n(\mu_0) \left(\frac{\alpha_s(\mu)}{\alpha_s(\mu_0)} \right)^{\gamma_n/\beta_0} , \qquad (9.52)$$

where

$$\gamma_n = \left[\frac{4}{3} - \frac{8}{3(n+1)(n+2)} + \frac{16}{3} \left(\sum_{k=2}^{n+1} \frac{1}{k} \right) \right] \qquad (9.53)$$

---

[4]Applications of the conformal symmetry in QCD are presented in the review [97]. A proof of the DA endpoint behavior, based on this symmetry for a two-point correlation function can be found in the review [98].

are the anomalous dimensions that are positive and grow with n. The coefficient β_0 was defined in (1.9). Multiplying both sides of (9.51) with the polynomial $C_m^{3/2}(u - \bar{u})$ and integrating over u, we use the orthogonality relation (D.16) and obtain:

$$a_n(\mu) = \frac{2(2n+3)}{3(n+1)(n+2)} \int_0^1 du \, C_n^{3/2}(u - \bar{u}) \varphi_\pi(u) \,. \tag{9.54}$$

This relation explains why a_n is called the n-th Gegenbauer moment of the pion twist-2 DA.

Due to asymptotic freedom of QCD, from (9.52) it follows that

$$\lim_{\mu \to \infty} a_n(\mu) = 0 \,.$$

Moreover, the larger the number of the coefficient is, the faster it vanishes at large scales. As a result, at $\mu \to \infty$ the DA asymptotically approaches to the simple function

$$\varphi_\pi^{asymp}(u) = 6u\bar{u} \,, \tag{9.55}$$

which is called the asymptotic pion twist-2 DA.

The fact that (9.51) vanishes linearly at the endpoints $u = 0$ and $\bar{u} = 0$, supports the use of the asymptotic formula (7.29) for the pion form factor. The only parameters we need as inputs for the pion twist-2 DA, apart from the overall normalization given by the pion decay constant, are the coefficients $a_n(\mu_0)$ at a certain low scale. Usually they are taken at $\mu_0 = 1.0$ GeV, where we can still trust the light-cone OPE. Truncating the series in (9.51) and adopting an ansatz for $\varphi_\pi(u)$ with a few coefficients, e.g., only with $a_{2,4}(\mu_0) \neq 0$, is a reasonable approximation, provided we apply this ansatz at sufficiently large scale.

Having defined the twist-2 pion DA in (9.44), we are in a position to calculate the correlation function (9.24) with two virtual photons. Using the isospin relation (9.34), we obtain the DA for the neutral pion:

$$\langle \pi^0(p) | (\bar{u}(x)\gamma^\rho\gamma_5 u(0) - \bar{d}(x)\gamma^\rho\gamma_5 d(0)) | 0 \rangle = -i\sqrt{2} p_\rho f_\pi \int_0^1 du \, e^{iu(p \cdot x)} \varphi_\pi(u) \,. \tag{9.56}$$

Substituting it in (9.24) and taking into account that $\varphi_\pi(\bar{u}) = \varphi_\pi(u)$, we obtain

$$\epsilon_{\mu\nu\alpha\beta} p^\alpha q^\beta F_{\gamma^*\pi}^{(0)}(q^2, (p-q)^2) = \frac{-\sqrt{2} f_\pi}{12\pi^2} \epsilon_{\mu\nu\alpha\rho} p^\rho \int d^4x \, \frac{x^\alpha}{(x^2)^2} e^{-iqx}$$

$$\times \int_0^1 du \left(e^{iu(p \cdot x)} + e^{i\bar{u}(p \cdot x)} \right) \varphi_\pi(u)$$

$$= \frac{-2\sqrt{2} f_\pi}{12\pi^2} \epsilon_{\mu\nu\alpha\rho} p^\rho \int_0^1 du \, \varphi_\pi(u) \int d^4x \, \frac{x^\alpha}{(x^2)^2} e^{-iq \cdot x + iup \cdot x}$$

$$= \frac{-2\sqrt{2} f_\pi}{12\pi^2} \epsilon_{\mu\nu\alpha\rho} p^\rho \int_0^1 du \, \varphi_\pi(u) \frac{2\pi^2 (up-q)^\alpha}{(up-q)^2} \,, \tag{9.57}$$

where we used one of the integrals from (A.95). Isolating the invariant amplitude, and expressing the denominator in terms of the invariant variables $Q^2 = -q^2$ and $(p-q)^2$, we finally obtain the LO twist-2 term of the correlation function (9.4):

$$F_{\gamma^*\pi}^{(0)}(Q^2, (p-q)^2) = \frac{\sqrt{2}}{3} f_\pi \int_0^1 du \, \frac{\varphi_\pi(u)}{\bar{u}Q^2 - u(p-q)^2} \,. \tag{9.58}$$

Here, the hard-scattering amplitude in the momentum space is convoluted with the pion twist-2 DA which, together with f_π, accumulates nonperturbative effects. In the next chapter this expression will be used to obtain a specific QCD sum rule for the form factor $F_{\gamma\pi}^{(0)}$ with one real photon, that is, at $(p-q)^2 = 0$.

A further expansion in powers of the light-cone separation x^2, e.g., inclusion of the $O(x^2)$ term in the expansion (9.29) of the vacuum-to-pion matrix element, brings us to the pion DAs with higher twists. The $t = 3, 4, ..$ contributions to the light-cone OPE correspond to taking into account two different effects: the transverse motion in the collinear two-parton state and the states with more than two partons, e.g., the quark-antiquark-gluon states. The latter correspond to adding a low virtuality gluon emitted from the virtual quark line and absorbed by the pion on the diagram in Figure 9.1. In the applications of light-cone OPE, such as the sum rules considered in the next chapter, the pion DAs up to $t = 4$ for both two-parton (quark-antiquark) and three-parton (quark-antiquark-gluon) states are usually included.

For a practical use, a convenient form of the twist expansion of the vacuum-to-pion matrix element is given in Appendix D, in (D.1) and (D.2). It includes all two-parton and three-parton DAs up to twist-4, respectively. Their Lorentz-structures and nomenclature are chosen according to [99], where one can find more details. To obtain a definition of a certain DA, one has to multiply both parts of (D.1) and (D.2) by the corresponding combination of Dirac matrices and take a trace.

Concentrating on the two-parton amplitudes first, we multiply the decomposition (D.1) by γ_5, take the trace and obtain the definition of the twist-3 DA:

$$\langle \pi^+(p) \mid \bar{u}(x_1) i\gamma_5 d(x_2) \mid 0 \rangle = f_\pi \mu_\pi \int_0^1 du \, e^{iu(p \cdot x_1) + i\bar{u}(p \cdot x_2)} \phi_{3\pi}^p(u) \tag{9.59}$$

with the normalization parameter

$$\mu_\pi = \frac{m_\pi^2}{m_u + m_d}. \tag{9.60}$$

In the local limit of this matrix element, we recognize the alternative definition (2.57) of the pion decay constant. The twist $t = 3$ of the DA $\phi_{3\pi}^p(u)$ is evident because e.g., the local operator $\bar{u}(0) i\gamma_5 d(0)$ contained in the expansion of (9.59), has $d = 3$ and $J = 0$. The definition of the second twist-3 DA of the pion is obtained contracting both parts of (D.1) with $\sigma_{\mu\nu}\gamma_5$:

$$\langle \pi^+(p) \mid \bar{u}(x_1) \sigma_{\mu\nu}\gamma_5 d(x_2) \mid 0 \rangle = i \left[p_\mu (x_1 - x_2)_\nu - p_\nu (x_1 - x_2)_\mu \right]$$
$$\times \frac{f_\pi \mu_\pi}{6} \int_0^1 du \, e^{iup \cdot x_1 + i\bar{u}(p \cdot x_2)} \phi_{3\pi}^\sigma(u). \tag{9.61}$$

It is less straightforward to explain why this DA has twist 3 and we refer to [95, 96]. To reproduce the normalization in (9.61), a differentiation over the coordinates on both parts is needed, before taking the local limit $x \to 0$ and relating the resulting operators, via equations of motion, to a pseudoscalar current.

The normalization parameter (9.60) deserves a special comment. Being essentially nonperturbative in nature, it is exceptionally large with respect to typical QCD scales of $O(\Lambda_{QCD})$. Using (1.95), we can relate μ_π to the quark condensate density and pion decay constant:

$$\mu_\pi = -\frac{2\langle \bar{q}q \rangle}{f_\pi^2}. $$

Note that the twist-3 contributions to the vacuum-to-pion hadronic matrix element (9.4) vanish in the chiral limit $m_{u,d} \sim m_\pi^2 \to 0$. The easiest way to see that is to return to the

T-product of the e.m. currents written in (9.22) in the form of bilocal operators. Taking the vacuum-to-pion matrix element of this product and retaining the twist-3 components from (D.1), leads to vanishing traces of an odd number of Dirac matrices. If we restore the light-quark masses in the quark propagators, the twist-3 contributions appear, but with $O(m_{u,d})$ coefficients. Multiplying them with the parameter μ_π, yields contributions proportional to m_π^2. The role of the twist-3 terms in the light-cone OPE is therefore greatly enhanced if one has a heavy quark propagator, in which case there is no chiral suppression. We will encounter this enhancement in Chapter 10, considering a sum rule for the heavy-light form factors.

Continuing the twist expansion in the sector of two-parton DAs, according to (D.1), there are two twist-4 DAs $\phi_{4\pi}(u)$ and $\psi_{4\pi}(u)$. Their normalization brings one more dimensionful parameter δ_π^2 which is defined via local quark-antiquark-gluon operator given in (D.12).

Listed in Appendix D are the specific polynomial expansions for each twist-3 and twist-4 DA, consisting of the asymptotic and nonasymptotic parts, similar to the Gegenbauer expansion (9.51) for the pion twist-2 DA. One should mention that, in terms of the conformal symmetry approach to the light-cone OPE, a specific conformal spin is attributed to each component of the polynomial expansion of DAs, in a full analogy with the conventional partial-wave expansion based on the angular momentum; hence, retaining the first few polynomials is quite similar to truncating a partial wave expansion, and the asymptotic part of DA is, in this sense analogous to the lowest partial wave.

Turning to multiparton DAs, it is straightforward to write down a general definition of the pion DA with $n = 3, 4, \ldots$ partons:

$$\langle \pi(p) | \psi_1(\xi_1 x) \psi_1(\xi_2 x) \ldots \psi_n(\xi_1 x) | 0 \rangle_{x^2 \to 0}$$
$$= \int \mathcal{D}\alpha \, e^{ip \cdot x (\xi_1 \alpha_1 + \xi_2 \alpha_2 + \ldots \xi_n \alpha_n)} \Phi(\alpha_1, \alpha_1, \ldots \alpha_n), \qquad (9.62)$$

where the parameters $\xi_1, \ldots, \xi_n$ correspond to the light-cone positions of the fields $\psi_1, \ldots \psi_n$ (quarks, antiquarks or gluons) and the n-dimensional integration element is:

$$\mathcal{D}\alpha = d\alpha_1 d\alpha_2 \ldots d\alpha_n \delta \left(\sum_{i=1}^{n} \alpha_i - 1 \right).$$

In practical applications, the quark-antiquark-gluon DAs, presented in Appendix D, are the most important ones, not only because they are inseparable parts of the light-cone OPE, but also because the twist-3 and twist-4 quark-antiquark-gluon DAs largely determine the parameters of their quark-antiquark counterparts. The QCD equations of motion impose exact relations between the nonlocal quark-antiquark and quark-antiquark-gluon operators. These relations are obtained (see, e.g., [95]), differentiating the quark-antiquark bilocal operators and replacing step by step the simple derivatives with a linear combination of covariant derivative and the gluon field, according to the basic definition (1.4). In this way, e.g., the coefficients in the nonasymptotic part of the twist-3 two-parton DAs in (D.5) are given by the parameters of the three-parton DA $\Phi_{3\pi}(\alpha_i)$ (see (D.2) and (D.6)). Relations of this type are also valid for the twist-4 DAs, where, e.g., the normalization of all twist-4 DAs is given by one and the same parameter δ_π^2, originally, a normalization factor of the quark-antiquark-gluon DAs given in (D.11); hence, the number of independent input parameters to describe the full set of ten pion DAs with $t \leq 4$ is substantially reduced.

Similar to the pion, other hadrons are also described with the sets of light-cone DAs. First of all, the kaon has a very similar structure of DAs that largely repeats the pattern we described above for the pion. The same concerns the η meson, whereas the η' is a more complex object from the point of view of the light-cone OPE. Here one has to take into account the mixing with the two-gluon operators which is absent in the pion and kaon cases.

The most important feature of the kaon DAs is a systematic account for the effects of the $SU(3)_{fl}$-symmetry violation at $O(m_s)$ and, accordingly, at $O(m_K^2)$. The corresponding DAs can be found in [99, 100]. In particular, the Gegenbauer expansion for the kaon twist-2 DA $\varphi_K(u)$ is the same as (9.51), but contains also the odd Gegenbauer moments:

$$\varphi_K(u,\mu) = 6u\bar{u}\left[1 + \sum_{n=1,2,3,4,..} a_n^K(\mu)C_n^{3/2}(u - \bar{u})\right], \qquad (9.63)$$

including the lowest moment a_1^K. The symmetry with respect to the middle point $u = 1/2$ is then violated,

$$\varphi_K(u) \neq \varphi_K(\bar{u}).$$

The same effect takes place in the kaon higher-twist DAs, parameterized including additional (odd) polynomials. If we define e.g., the charged kaon twist-2 DA with

$$\langle K^-(p)|\bar{s}(x_1)[x_1,x_2]\gamma_\rho\gamma_5 u(x_2)|0\rangle_{x^2\to 0} = -ip_\rho f_K \int_0^1 du\, e^{iu(p\cdot x_1)+i\bar{u}(p\cdot x_2)}\varphi_K(u), \qquad (9.64)$$

then a_1^K can be related to the local ($x_{1,2} \to 0$) operator

$$\langle K^-(p)|\bar{s}\gamma_\mu\gamma_5 i\overleftrightarrow{D}_\lambda u|0\rangle = -ip_\mu p_\lambda f_K \frac{3}{5}a_1^K, \qquad (9.65)$$

where the derivative is defined as $\overleftrightarrow{D}_\lambda = \overrightarrow{D}_\lambda - \overleftarrow{D}_\lambda$. Physically, we expect that a heavier s-quark in the two-parton configuration has a larger average momentum than the u-quark, since both are the parts of a bound state moving with the same velocity. That will correspond to a positive value of a_1^K, if the s-quark momentum is assigned as in (9.64). Estimates of the lowest moments of the pion and kaon twist-2 DAs are available from the QCD on the lattice (see, e.g., [101]). Applying QCD sum rules for the two-point correlation functions involving the local operators defining the Gegenbauer moments, it is also possible to estimate these parameters[5] of the pion and kaon. One example [102] is the calculation of the kaon moment a_1^K at a scale of $O(1\text{GeV})$, applying the correlation function of the operator (9.65) and the strange axial current. The resulting estimate is a positive value, consistent with the lattice QCD calculations.

It might sound unusual, but also the real photon has its own light-cone DAs. The reason one introduces them, follows from the long-distance (or hadronic) structure inherent to the photon and revealed, e.g., in the radiative decays of light mesons such as $\rho \to \pi\gamma$ discussed in Chapter 4. It is conceivable to have a process where a light quark-antiquark pair, emitted from the interval near the light-cone, annihilates into a photon, but, before the annihilation takes place, this quark-antiquark pair propagates at an average long distance of $O(1/\Lambda_{QCD})$. The amplitude of this process can then be parameterized by the vacuum-to-photon matrix element of the bilocal quark-antiquark current. The lowest, twist-2 photon DA for the quark flavor q has the following definition:

$$\langle \gamma(p)|\bar{q}(x_1)[x_1,x_2]\sigma_{\alpha\beta}q(x_2)|0\rangle = iQ_q\chi\langle\bar{q}q\rangle(\varepsilon_\alpha p_\beta - \varepsilon_\beta p_\alpha)\int_0^1 du\, e^{iu(p\cdot x_1)+i\bar{u}(p\cdot x_2)}\varphi_\gamma(u),$$

$$(9.66)$$

where ε is the polarization vector of the photon with momentum p, so that $\varepsilon\cdot p = 0$, $p^2 = 0$. The Lorentz structure is chosen to obey the e.m. gauge invariance, also the gauge link in

---

[5]The earlier calculations are described in the review [98].

this case contains an e.m. part. The normalization is, as usual, related to the matrix element of a certain local quark-antiquark operator. Here it is a specific one, describing the quark condensate in the presence of the real photon field. It is parameterized in the units of the vacuum quark condensate with a coefficient χ, which is called the *e.m. susceptibility of the QCD vacuum*. There is however an essential difference as compared to the case of the pion, kaon or other hadrons. The photon emission, apart from taking place at long distances, can also occur at short distances, when the quark and antiquark emitted at a lightlike interval are highly virtual. This is possible simply because the e.m. quark-photon interaction is pointlike; hence, if one describes an amplitude to produce a photon from a quark and antiquark emitted at a lightlike separation, then there are two separate contributions: one of them is a perturbative diagram of the photon emission involving the quark propagators and the other one is described by the photon DAs. A comprehensive description of the latter can be found in [103].

Here we skip a discussion of the vector meson $(\rho, \omega, K^*, \phi)$ DAs, for which the main elements[6], including the twist and polynomial expansion of the two-parton and three-parton states, are largely the same as for the pions and kaons. We only notice that introducing the DAs one assumes the narrow widths of vector mesons.

A more general approach defines *dimeson DAs* which describe the states of two stable mesons that are formed from the quark-antiquark partons emitted at lightlike distances. The dipion DAs were introduced in [104] for the $\gamma\gamma^* \to 2\pi$ process[7] and worked out in detail in [106]. Their definitions, for the charged dipion state, which only has the isospin $I = 1$, are:

$$\langle \pi^+(k_1)\pi^0(k_2)|\bar{u}(x)\gamma_\mu d(0)|0\rangle = -\sqrt{2}k_\mu \int_0^1 du\, e^{iu(k\cdot x)} \Phi_\parallel^{I=1}(u, \zeta, k^2)\,,$$

$$\langle \pi^+(k_1)\pi^0(k_2)|\bar{u}(x)\sigma_{\mu\nu} d(0)|0\rangle = 2\sqrt{2}i\frac{k_{1\mu}k_{2\nu} - k_{2\mu}k_{1\nu}}{2\zeta - 1} \int_0^1 du\, e^{iu(k\cdot x)} \Phi_\perp^{I=1}(u, \zeta, k^2)\,, \quad (9.67)$$

The twist-2 DAs $\Phi_{\parallel,\perp}^{I=1}(u, k^2, \zeta)$ depend on the dipion invariant mass squared $k^2 = (k_1+k_2)^2$, which has to be not too far above the threshold $k_{min}^2 = 4m_\pi^2$, and on the longitudinal momentum fraction u ($\bar{u}$) carried by the u-quark ($\bar{d}$-quark). The parameter ζ describes an additional variable, related to the angle θ_π of the pions in their c.m. frame:

$$(2\zeta - 1) = \sqrt{1 - \frac{4m_\pi^2}{k^2}}cos\theta_\pi\,.$$

Importantly, the first of these DAs in the local limit is related, via the isospin symmetry, to the pion e.m. form factor in the timelike region:

$$\int_0^1 du\, \Phi_\parallel^{I=1}(u, \zeta, k^2) = (2\zeta - 1)F_\pi^{em}(k^2)\,, \quad (9.68)$$

whereas the second DA is normalized to the form factor of the $\bar{u}(x)\sigma_{\mu\nu} d(0)$ current which cannot be directly measured. Furthermore, these DAs can be doubly expanded in Legendre and Gegenbauer polynomials with multiplicatively renormalizable coefficient functions of k^2. Due to the unitarity, these functions determining the dipion DAs, develop imaginary part at $k^2 > 4m_\pi^2$ and contain resonance (in this case ρ^+) contributions.

---

[6]see, e.g., [95, 96].

[7]Dimeson wave functions were also considered earlier, in a different context, in [105].

The list of existing applications of the light-cone OPE and DAs is incomplete without the nucleon DAs analyzed in detail in [107]. Their generic form is similar to (9.62), with $n = 3$ and three quark operators distributed over the light cone. Counting independent Dirac structures and twist components including $t = 3, 4, 5, 6$ yields altogether 27 nucleon DAs, much larger than in the case of mesons. The normalization coefficients are equal to the vacuum-nucleon matrix elements of the three-quark local currents similar to the matrix element of the Ioffe current in (3.33).

CHAPTER **10**

QCD LIGHT-CONE SUM RULES

10.1 QUARK PROPAGATOR NEAR THE LIGHT-CONE

We start our discussion of QCD light-cone sum rules (LCSRs) from expansion of the quark propagator

$$S_{(q)}(x_1, x_2) = -i\langle 0|T\{q(x_1)\bar{q}(x_2)|0\rangle \tag{10.1}$$

near the light-cone $(x_1 - x_2)^2 = 0$. Note that we use the same notation as for the propagator in an external gluon field considered in Chapter 8. As shown below, this propagator is restored in the local limit of the light-cone expansion. For simplicity, we consider a massless quark, which is a reasonable approximation for the light quarks $(q = u, d, s)$ at large virtualities. Translating the coordinates, we also choose $x_1 = x$, $x_2 = 0$.

The leading-order term of the expansion near $x^2=0$ is given by the free-quark propagator

$$S_{(q)}^{(0)}(x) = \frac{1}{2\pi^2} \frac{\slashed{x}}{(x^2)^2}, \tag{10.2}$$

which has the strongest singularity on the light cone and coincides with the leading-order propagator used in the local OPE near $x = 0$. Differences arise at the next-to-leading order, when interaction of the virtual quark with a gluon field is switched on.

Let us consider a gluon emission from a virtual quark at the first order in g_s, described by the diagram in Figure 10.1 and obtained inserting a quark-gluon vertex between two free-quark propagators:

$$S_{(q)}^{(1)}(x, 0) = -\int d^4 y S_{(q)}^{(0)}(x - y)\gamma_\rho A^\rho(y) S_{(q)}^{(0)}(y). \tag{10.3}$$

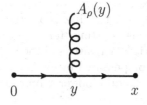

Figure 10.1 Diagram of the gluon emission from the quark propagator.

Color indices are hereafter omitted for brevity, and a compact matrix form (A.45) of the gluon field A^ρ is used. The gluon in (10.3) could in general have a nonvanishing momentum, e.g., being a component of a hadron distribution amplitude (DA). Our task is to expand (10.3) near $x^2 = 0$. We assume that the propagator $S_{(q)}(x, 0)$ enters a certain correlation function characterized by an external momentum $Q^2 \gg \Lambda_{QCD}^2$. As shown in the previous Chapter 9, in this case the intervals near the light-cone, $x^2 \sim 1/Q^2$, dominate in the quark propagation.

We closely follow the method developed in [108, 109] where one can find more details. In particular, for the gluon field it is convenient to adopt the fixed-point gauge (A.55) taken at $x_0 = 0$:

$$x^\rho A_\rho(x) = 0 \,. \tag{10.4}$$

Substituting (10.2) in (10.3) and applying the integral parametrizaton (A.88) to the product of numerators, we obtain

$$S_{(q)}^{(1)}(x, 0) = -\frac{3!}{(2\pi^2)^2} \int_0^1 du\, u\bar{u} \int d^4 y \frac{(\not{x} - \not{y})\gamma_\rho \not{y} A^\rho(y)}{\left[u(x-y)^2 + \bar{u}y^2\right]^4} \,, \tag{10.5}$$

where $\bar{u} \equiv 1 - u$. Rewriting the denominator:

$$u(x-y)^2 + \bar{u}y^2 = (y - ux)^2 + u\bar{u}x^2 \,,$$

and transforming the integration variable to $z = y - ux$ yields

$$S_{(q)}^{(1)}(x, 0) = -\frac{3\, g_s}{2\pi^4} \int_0^1 du\, u\bar{u} \int d^4 z \frac{(\bar{u}x_\alpha - z_\alpha)(ux_\beta + z_\beta)}{\left[z^2 + u\bar{u}x^2\right]^4} A_\rho(ux + z)\gamma^\alpha \gamma^\rho \gamma^\beta \,. \tag{10.6}$$

The next step involves expanding gluon field in powers of the deviation z from the (near light-cone) coordinate ux:

$$A_\rho(ux + z) = A_\rho(ux) + \partial_\mu A_\rho(ux)z^\mu + \frac{1}{2!}\partial_\mu \partial_\nu A_\rho(ux)z^\mu z^\nu + \dots \,, \tag{10.7}$$

where

$$\partial_\mu A_\rho(ux) \equiv \left.\frac{\partial A_\rho(w)}{\partial w^\mu}\right|_{w=ux} .$$

The advantage of the fixed-point gauge (10.4) is that, according to (A.63), a gluon field is directly expressed in terms of the field-strength tensor:

$$A_\rho(ux) = \int_0^1 d\alpha\, \alpha\, ux^\nu G_{\nu\rho}(\alpha ux) \,. \tag{10.8}$$

From this relation it follows that only the first two terms in the expansion (10.7) contain $G_{\mu\nu}$, whereas the subsequent terms, starting from the second derivative of A_ρ, can only generate $D_\alpha G_{\mu\nu}$ and higher covariant derivatives of the field-strength tensor. Having higher dimension, derivative terms are compensated by extra powers of x^2 in the expansion of the propagator and, correspondingly, produce power suppressed contributions to the light-cone OPE of the correlation functions.

In the rest of this section, we will reproduce the leading $O(G_{\mu\nu})$ term in the quark propagator. To this end, we substitute the expansion (10.7) in (10.6), retaining the first two terms. After that it is possible to integrate over z. The integrals with an odd number of four-coordinates z_α in the numerator vanish on symmetry grounds. For the remaining two integrals the formulas (A.92) and (A.93) are used[1], yielding:

$$\int \frac{d^4z}{(z^2 + u\bar{u}x^2)^4} = \frac{-i\pi^2}{6u^2\bar{u}^2(-x^2)^2}, \quad \int \frac{d^4z\, z_\alpha z_\beta}{(z^2 + u\bar{u}x^2)^4} = \frac{i\pi^2 g_{\alpha\beta}}{12u\bar{u}(-x^2)}. \tag{10.9}$$

After integration we obtain:

$$\begin{aligned}
S_{(q)}^{(1)}(x,0) &= \frac{i}{8\pi^2} \int_0^1 du \left[\left(\frac{2x_\alpha x_\beta}{(-x^2)^2} + \frac{g_{\alpha\beta}}{(-x^2)} \right) A_\rho(ux) \right. \\
&\quad + \left. \frac{ux_\beta \partial_\alpha A_\rho(ux) - \bar{u}x_\alpha \partial_\beta A_\rho(ux)}{(-x^2)} \right] \gamma^\alpha \gamma^\rho \gamma^\beta + \dots,
\end{aligned} \tag{10.10}$$

where dots replace terms beyond the $O(G_{\mu\nu})$ approximation. In the first line of the above expression we then use the anticommutation relation for γ-matrices, leading to:

$$(2x_\alpha x_\beta - g_{\alpha\beta}x^2)\gamma^\alpha \gamma^\rho \gamma^\beta = 4\slashed{x}x^\rho, \tag{10.11}$$

whereas in the first and second term of the second line it is more convenient to transform the product of three matrices, respectively, as:

$$\begin{aligned}
\gamma^\alpha \gamma^\rho \gamma^\beta &= (g^{\alpha\rho} - i\sigma^{\alpha\rho})\gamma^\beta, \\
\gamma^\alpha \gamma^\rho \gamma^\beta &= \gamma^\alpha (g^{\rho\beta} - i\sigma^{\rho\beta})
\end{aligned} \tag{10.12}$$

yielding:

$$\begin{aligned}
S_{(q)}^{(1)}(x,0) = \frac{i}{8\pi^2} \int_0^1 du \Bigg\{ & \frac{4\slashed{x}}{(-x^2)^2} x_\rho A^\rho(ux) \\
& + \frac{1}{(-x^2)} \left[u\left(\partial_\alpha A^\alpha(ux) - i\sigma^{\alpha\rho}\partial_\alpha A_\rho(ux) \right)\slashed{x} \right. \\
& \left. - \bar{u}\slashed{x}\left(\partial_\alpha A^\alpha(ux) + i\sigma^{\alpha\rho}\partial_\alpha A_\rho(ux) \right) \right] \Bigg\} + \dots.
\end{aligned} \tag{10.13}$$

Due to the gauge condition (10.4), the first term in the above expression vanishes[2]. In the second and third line of (10.13) we encounter two types of terms. The ones containing divergence of the gluon field $\partial_\alpha A^\alpha(ux)$ are reduced, with the help of (10.8), to covariant derivatives of the field-strength tensor:

$$\begin{aligned}
\partial_\beta A^\beta(ux) &= \frac{\partial}{\partial(ux^\beta)} \left(\int_0^1 d\alpha\, \alpha\, ux_\nu G^{\nu\beta}(\alpha ux) \right) \\
&= \int_0^1 d\alpha\, \alpha \left(g_{\nu\beta} G^{\nu\beta}(\alpha ux) + \alpha ux_\nu \partial_\beta G^{\nu\beta}(\alpha ux) \right) \\
&= \int_0^1 d\alpha\, \alpha^2 ux_\nu D_\beta G^{\nu\beta}(\alpha ux) + \dots,
\end{aligned} \tag{10.14}$$

---

[1] Being convergent, these integrals are taken at the physical dimension $D = 4$.
[2] In other gauges it contributes to the gauge link (A.46).

hence, should be omitted from the adopted approximation for quark propagator. Note that in the last equation above we replaced the ordinary derivative by the covariant one. The difference between these two derivatives in the fixed-point gauge produces a field-strength tensor and yields terms of $O(GG)$ in the propagator, which are also beyond our approximation. By the same token, we can replace the ordinary derivative in the remaining two terms emerging in the second line of the expansion (10.13), e.g.,

$$\sigma^{\alpha\rho}\partial_\alpha A_\rho(ux) = \frac{1}{2}\sigma^{\alpha\rho}\big(\partial_\alpha A_\rho(ux) - \partial_\rho A_\alpha(ux)\big) = \frac{1}{2}\sigma^{\alpha\rho}G_{\alpha\rho}(ux). \tag{10.15}$$

Finally, we obtain the following expression for the $O(G_{\mu\nu})$ term of the propagator:

$$S_{(q)}^{(1)}(x,0) = \frac{1}{16\pi^2}\int_0^1 du\frac{u\sigma^{\rho\beta}G_{\rho\beta}(ux)\slashed{x} + \bar{u}\slashed{x}\sigma^{\rho\beta}G_{\rho\beta}(ux)}{(-x^2)} + \dots. \tag{10.16}$$

It enters the following expression [109] for the massless quark propagator expanded near the light-cone, starting from the free-quark part and including also the terms with derivatives of the gluon field-strength:

$$S_{(q)}(x,0) = \frac{\slashed{x}}{2\pi^2(-x^2)^2} + \frac{g_s\lambda^a}{32\pi^2(-x^2)}\int_0^1 du\Big\{u\sigma_{\mu\nu}G^{a\mu\nu}(ux)\slashed{x} + \bar{u}\slashed{x}\sigma_{\mu\nu}G^{a\mu\nu}(ux)$$

$$+2iu\bar{u}\slashed{x}x_\rho D_\lambda G^{a\rho\lambda}(ux)\Big\} + \frac{g_s\lambda^a}{32\pi^2}\ln(-x^2)\int_0^1 du\Big\{i\Big(u\bar{u} - \frac{1}{2}\Big)D_\mu G^{a\mu\nu}(ux)\gamma_\nu$$

$$+\frac{i}{2}u\bar{u}(1-2u)x_\mu\slashed{D}D_\nu G^{a\mu\nu}(ux) + \frac{1}{2}u\bar{u}\epsilon_{\mu\nu\alpha\beta}x_\mu D^\alpha D_\lambda G^{a\lambda\beta}\gamma^\nu\gamma_5\Big\}. \tag{10.17}$$

Considering the OPE in local operators in Chapter 8, we calculated the quark propagator in the external field of the gluon condensate, which corresponds to a constant gluon field-strength, $G_{\mu\nu}(x) \simeq G_{\mu\nu}(0)$. Putting this condition on the gluon field in (10.17) and leaving only the terms without covariant derivative, we obtain, after integrating over the parameter u:

$$S_{(q)}^{(1)}(x,0) = \frac{g_s\lambda^a}{64\pi^2(-x^2)}\big(\sigma_{\mu\nu}\slashed{x} + \slashed{x}\sigma_{\mu\nu}\big)G^{a\mu\nu}(0).$$

Combining the relations (A.16) for σ matrices, we find that the above expression is equal to the one-gluon part of the local OPE propagator in (8.85).

10.2 CALCULATION OF THE PION FORM FACTOR

Asymptotics of the pion e.m. form factor at large spacelike momentum transfer was already discussed in Chapter 7. The factorization formula (7.28) reproduced there is an unambiguous prediction of QCD. In the $Q^2 \to \infty$ limit, the leading, $\sim 1/Q^2$ term of the spacelike form factor is calculable, convoluting perturbative quark-gluon scattering diagrams with the DAs of the initial and final pion states consisting of collinear quark and antiquark partons. These amplitudes are identified with the leading twist-2 light-cone DAs of the pion defined in (9.44).

Still, several important questions concerning the pion form factor remain unanswered:

- How large are the power suppressed soft-overlap contributions?

- At which Q^2 does the asymptotic $\sim 1/Q^2$ regime become dominant?

- How important are the higher-twist DAs of the pion?

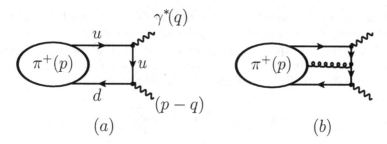

Figure 10.2 Contributions of (a) two-particle and (b) three-particle DAs to the correlation function (10.18). Blobs (vertices with wavy lines) represent the pion DAs (the currents).

Here we address these questions, applying the method of light-cone sum rules (LCSRs) in QCD[3]. The LCSR for the pion e.m. form factor [113, 114, 115], valid in the spacelike region $Q^2 \gg \Lambda_{QCD}^2$, includes both soft-overlap and hard-scattering contributions. In this section we present a detailed derivation of this sum rule.

The underlying object of the LCSR method is a correlation function containing the T-product of two quark currents sandwiched between the QCD vacuum and a hadron state. A correlation function of this type is the transition amplitude (9.4) of the virtual photons to pion, already explored in Chapter 9. With a proper choice of external momenta, the vacuum-to-hadron correlation function is calculated to a certain accuracy, employing OPE near the light-cone in terms of hadron DAs. To relate the form factor with the result of this calculation, the basic tools of the QCD (SVZ) sum rule method presented in Chapter 8 are used, including the dispersion relation, quark-hadron duality and Borel transformation.

To obtain the LCSR for the pion e.m. form factor, we define the correlation function

$$\mathcal{F}_{5\mu\nu}(p,q) = i \int d^4x e^{iqx} \langle 0|T\{j_{\mu5}^\dagger(0)j_\nu^{em}(x)\}|\pi^+(p)\rangle, \qquad (10.18)$$

of the quark e.m. current and the axial-vector current

$$j_{\mu5}^\dagger = \bar{d}\gamma_\mu\gamma_5 u.$$

The latter, being Hermitian conjugated to the current in (2.48) and acting on the vacuum state, interpolates the state π^+. What are the basic differences between (10.18) and the correlation functions used to obtain QCD sum rules in Chapter 8? First of all, the initial vacuum state is replaced by an on-shell pion with the 4-momentum p and $p^2 = m_\pi^2$; hence, there are two independent momenta, q and p, carried, respectively, by the e.m. current and pion, so that the external momentum $p - q$ is flowing off the axial-vector current. At fixed $q^2 = -Q^2$, the correlation function depends on the invariant variable $(p-q)^2$. Note also that the quark currents in (10.18) have different quantum numbers, chosen to match different initial and final states in the correlation function.

The correlation function (10.18) is decomposed in several independent Lorentz structures, and we single out one of them:

$$\mathcal{F}_{5\mu\nu}(p,q) = ip_\mu p_\nu \mathcal{F}(Q^2, (p-q)^2) + ..., \qquad (10.19)$$

where $\mathcal{F}$ is the corresponding invariant amplitude depending on the two independent variables. The remaining kinematical structures proportional to $p_\mu q_\nu$, $q_\mu p_\nu$, $q_\mu q_\nu$, $g_{\mu\nu}$ are denoted by ellipsis and will not be used.

---

[3]This method was originally suggested in [110, 111, 112].

Before starting the calculation based on light-cone OPE, we establish a relation between the correlation function $\mathcal{F}$ and the pion form factor. This part of LCSR derivation largely follows the SVZ sum rules explained in Chapter 8. At fixed Q^2, we use the analyticity of the correlation function (10.18) and, in particular, of the amplitude $\mathcal{F}(Q^2, (p-q)^2)$ in the variable $(p-q)^2$ and write down the corresponding dispersion relation:

$$\mathcal{F}(Q^2, (p-q)^2) = \frac{1}{\pi} \int ds \frac{\operatorname{Im}\mathcal{F}(Q^2, s)}{s - (p-q)^2}, \tag{10.20}$$

where possible subtractions are ignored, being, in fact, irrelevant due to the subsequent Borel transformation.

The imaginary part in (10.20) is given by the unitarity relation (B.39) adapted for correlation functions, where we only have to permute the initial and final states. A complete set of hadronic states between the currents in (10.18) includes all possible isospin $I = 1$ states with spin-parities $J^P = 0^-$ and $J^P = 1^+$ that are generated from the vacuum by the axial-vector current $j_{\mu 5}$. The lightest hadron in this set is the pion, then comes the continuum of multimeson states with $J^P = 0^-$ and $J^P = 1^+$, overlapping, respectively, with the radial excitations of the pion, such as $\pi(1300)$, and with the axial mesons, starting from $a_1(1260)$ [1]. The resulting unitarity relation has the form:

$$2\operatorname{Im}_{(p-q)^2}\mathcal{F}_{5\mu\nu}(p,q) = \int d\tau_\pi \langle 0|j_{\mu 5}^\dagger|\pi^+(p-q)\rangle \langle \pi^+(p-q)|j_\nu^{em}|\pi^+(p)\rangle$$

$$+ \underbrace{\sum_h \int d\tau_h \langle 0|j_{\mu 5}^\dagger|h(p-q)\rangle \langle h(p-q)|j_\nu^{em}|\pi^+(p)\rangle}_{2\pi\rho_{5\mu\nu}(p,q)}, \tag{10.21}$$

where we isolate the ground-state contribution of the pion with the one-particle phase space

$$\int d\tau_\pi = (2\pi)\delta((p-q)^2 - m_\pi^2),$$

(see (5.73)), and denote by

$$\rho_{5\mu\nu}(p,q) = ip_\mu p_\nu \rho_5^h(Q^2, (p-q)^2)\theta((p-q)^2 - s_{th}) + \dots, \tag{10.22}$$

the spectral density of all heavier than pion hadronic states, where s_{th} is the lowest threshold. Since the emission of two pions by the axial-vector current is forbidden by the P-parity conservation, the lightest contribution to the sum in (10.21) is the three-pion state, so that $s_{th} = (3m_\pi)^2$.

In the one-pion term in (10.21), we recognize two familiar hadronic matrix elements: (2.48) and (2.26), parameterized, respectively, with the pion decay constant and e.m. form factor. Substituting their definitions in (10.21), we obtain:

$$2\operatorname{Im}\mathcal{F}_{5\mu\nu}^{(\pi)}(p,q) = \left[i(p-q)_\mu f_\pi\right]\left[(2p-q)_\nu F_\pi(Q^2)\right]2\pi\delta(m_\pi^2 - (p-q)^2)$$

$$= 2ip_\mu p_\nu \left[2\pi f_\pi F_\pi(Q^2)\delta(m_\pi^2 - (p-q)^2)\right] + \dots, \tag{10.23}$$

Using the above equation together with (10.22), yields the imaginary part of the invariant amplitude:

$$\frac{1}{\pi}\operatorname{Im}\mathcal{F}(s) = 2f_\pi F_\pi(Q^2)\delta(m_\pi^2 - s) + \rho_5^h(Q^2, s)\theta(s - s_{th}), \tag{10.24}$$

transforming the dispersion relation (10.20) into

$$\mathcal{F}(Q^2, (p-q)^2) = \frac{2 f_\pi F_\pi(Q^2)}{m_\pi^2 - (p-q)^2} + \int\limits_{s_{th}}^{\infty} ds \frac{\rho_5^h(Q^2, s)}{s - (p-q)^2} \,. \tag{10.25}$$

Apart from an isolated pole at m_π^2, we encounter a superposition of branch points and poles contributing to the integral over the spectral density $\rho_5^h(Q^2, s)$. Note also that $\text{Im}\,\mathcal{F}(Q^2, s)$ is in general not a positive definite function, because the two currents in the correlation function are different.

Importantly, the dispersion relation (10.25) is valid at any $(p-q)^2$. From now on, we consider the deep spacelike[4] region:

$$Q^2, \; |(p-q)^2| \gg \Lambda_{QCD}^2,$$

and, simultaneously, keep a large but unequal momentum transfer,

$$Q^2 \neq |(p-q)|^2.$$

Note that the correlation function (10.18), after replacing $j_{\mu 5} \to j_\mu^{em}$, coincides with the amplitude of the $\gamma^* \gamma^* \to \pi^0$ transition (9.4), for which the light-cone dominance is valid in the same region (9.8) as the one defined above; hence, in this region also in the correlation function (10.18) the intervals near the light-cone, $x^2 \sim 1/Q^2$, dominate. Consequently, the products of quark and antiquark fields entering the currents in (10.18) can be expanded at $x^2 \simeq 0$ in nonlocal operators, leading to the pion DAs.

Turning to the actual calculation of (10.18), we retain only the u and d quarks in the e.m. current. The other flavor components of j_μ^{em} are isoscalars and do not contribute. Writing the currents $j_{\mu 5}$ and j_μ^{em} in explicit form, we encounter the products of two quark-antiquark operators taken at the points 0 and x:

$$\mathcal{F}_{5\mu\nu}(p, q) = i \int d^4 x e^{iqx} \langle 0 | T\{\bar{d}(0)\gamma_\mu\gamma_5 u(0)$$
$$\times [Q_u \, \bar{u}(x)\gamma_\nu u(x) + Q_d \, \bar{d}(x)\gamma_\nu d(x)]\} | \pi^+(p) \rangle \,. \tag{10.26}$$

We then form all possible combinations of u- and $\bar{d}$-fields matching the valence content of π^+, whereas the remaining quark fields are contracted into propagators. The result is:

$$\mathcal{F}_{5\mu\nu}(p, q) = -\int d^4 x \, e^{iqx} \langle 0 | Q_u \, \bar{d}(0)\gamma_\mu\gamma_5 S_u(0, x)\gamma_\nu u(x)$$
$$+ Q_d \, \bar{d}(x)\gamma_\nu S_d(x, 0)\gamma_\mu\gamma_5 u(0) | \pi^+(p) \rangle \,. \tag{10.27}$$

In the leading order (LO) of light-cone OPE, the propagators $S_{u,d}(x, 0)$ coincide with the massless free-quark propagator, provided we neglect the u, d-quark masses. The LO expression for the correlation function is then obtained:

$$\mathcal{F}_{5\mu\nu}^{(LO)}(p, q) = \frac{1}{2\pi^2} \int d^4 x \frac{x^\alpha e^{iq\cdot x}}{(x^2)^2} \Big(Q_u \langle 0 | \bar{d}_\omega(0) u_\xi(x) | \pi^+(p) \rangle \big[\gamma_\mu\gamma_\alpha\gamma_\nu\gamma_5\big]_{\omega\xi}$$
$$- Q_d \langle 0 | \bar{d}_\omega(x) u_\xi(0) | \pi^+(p) \rangle \big[\gamma_\nu\gamma_\alpha\gamma_\mu\gamma_5\big]_{\omega\xi} \Big) \,. \tag{10.28}$$

Each of the two similar terms in the above expression represents an overlap of quark propagator with a matrix element, in which the quark and antiquark operators separated by an

---

[4]Therefore, we ignore infinitesimal $i\epsilon$ in the denominator of the dispersion integral which has to be added at timelike $(p-q)^2 > 0$.

interval $x^2 \sim 1/Q^2$ are sandwiched between the pion and vacuum states. More specifically, the term with Q_u in (10.28) corresponds to the diagram in Figure 10.2(a) and the term with Q_d to its counterpart with an opposite direction of the quark lines and $u \leftrightarrow d$ replaced. Furthermore, in the pion-to-vacuum matrix elements in (10.28) we recognize the light-cone DAs of the pion, discussed in the previous Chapter 9 and presented in Appendix D. To adjust the definition of the DA to a matrix element with an initial-state pion, the expression (D.1) has to be Hermitian conjugated and summed over colors, yielding

$$\langle 0|\bar{d}_\omega(x_1)u_\xi(x_2)|\pi^+(p)\rangle_{(x_1-x_2)^2\to 0} = \frac{-if_\pi}{4}\int_0^1 du\, e^{-iup\cdot x_1 - i\bar{u}p\cdot x_2}[\slashed{p}\gamma_5]_{\xi\omega}\varphi_\pi(u) + \dots, \quad (10.29)$$

where we retain only the leading at $(x_1 - x_2)^2 \to 0$ twist-2 part. Higher-twist pion DAs, generating power-suppressed terms in the correlation function, will be added later. Inserting (10.29) with $x_1 = 0$, $x_2 = x$ ($x_1 = x$, $x_2 = 0$) in the first (second) term in (10.28), we obtain:

$$\mathcal{F}_{5\mu\nu}^{(LO,tw2)}(p,q) = \frac{if_\pi}{8\pi^2}\int_0^1 du \int d^4x\, \frac{x^\alpha e^{iq\cdot x}}{(x^2)^2}$$

$$\times\Big(Q_u\text{Tr}[\slashed{p}\gamma_\mu\gamma_\alpha\gamma_\nu]e^{-i\bar{u}p\cdot x}\varphi_\pi(u) - Q_d\text{Tr}[\slashed{p}\gamma_\nu\gamma_\alpha\gamma_\mu]e^{-iup\cdot x}\varphi_\pi(u)\Big), \quad (10.30)$$

where the γ_5 matrices were contracted using $\gamma_5^2 = 1$. Furthermore, we employ the symmetry relation (9.48) valid for the pion DAs in the adopted isospin-symmetry limit. We replace $\varphi_\pi(u) \to \varphi_\pi(\bar{u})$ in the first integral over u and, after that, transform the integration variable $u \to \bar{u}$, so that the two integrals over x in (10.30) coincide. Use of the integration formula from (A.95) brings us to:

$$\begin{aligned}
\mathcal{F}_{5\mu\nu}^{(LO,tw2)}(p,q) &= \frac{if_\pi}{8\pi^2}\int_0^1 du\, \varphi_\pi(u)\frac{2\pi^2(q-up)^\alpha}{(q-up)^2}\\
&\quad \times \Big(Q_u\text{Tr}[\slashed{p}\gamma_\mu\gamma_\alpha\gamma_\nu] - Q_d\text{Tr}[\slashed{p}\gamma_\nu\gamma_\alpha\gamma_\mu]\Big)\\
&= -2ip_\mu p_\nu f_\pi(Q_u - Q_d)\int_0^1 du\, \frac{u\varphi_\pi(u)}{(q-up)^2} + \dots, \quad (10.31)
\end{aligned}$$

where, after taking traces, we singled out the same Lorentz-structure as in (10.19), denoting the rest by the ellipsis. As we see, the two contributions to the correlation function merge and the quark charges add up to a unit, $Q_u - Q_d = 1$. Transforming the denominator in the above integral into a linear function of two invariant variables:

$$(q - up)^2 = -Q^2 - 2u(qp) = -Q^2\bar{u} + u(p-q)^2, \quad (10.32)$$

we hereafter neglect the terms containing $p^2 = m_\pi^2$, as compared to the large scales Q^2 and $|(p-q)^2|$. This is consistent with neglecting the u,d quark masses, i.e., adopting the chiral limit of QCD, since $m_\pi^2 \sim O(m_{u,d})$. The resulting expression for the invariant amplitude at the LO ($O(\alpha_s^0)$) and to the twist-2 accuracy is

$$\mathcal{F}^{(LO,tw2)}(Q^2,(p-q)^2) = 2f_\pi\int_0^1 du\, \frac{u\varphi_\pi(u)}{\bar{u}Q^2 - u(p-q)^2}. \quad (10.33)$$

Following the same path of derivation as for the QCD (SVZ) sum rules, explained in Chapter 8, we equate the hadronic dispersion relation (10.25) to the above expression:

$$\frac{2f_\pi F_\pi(Q^2)}{-(p-q)^2} + \int_{s_{th}}^\infty ds \frac{\rho_5^h(Q^2,s)}{s-(p-q)^2} = 2f_\pi \int_0^1 du \frac{u\varphi_\pi(u)}{\bar{u}Q^2 - u(p-q)^2} . \tag{10.34}$$

Importantly, the above equation is only valid at large Q^2 and spacelike $(p-q)^2 < 0$, far from the poles and cuts located on the real axis at $(p-q)^2 = s \geq m_\pi^2$; hence, we safely neglected the pion mass squared also in the pole term.

The next step is to apply the quark-hadron duality approximation. We first convert the integral on the r.h.s. of (10.34) into a dispersion integral in the variable $(p-q)^2$, transforming the integration variable:

$$u \to s = \frac{\bar{u}Q^2}{u} , \quad u = \frac{Q^2}{Q^2+s} , \tag{10.35}$$

so that

$$2f_\pi \int_0^1 du \frac{u\varphi_\pi(u)}{\bar{u}Q^2 - u(p-q)^2} = \int_0^\infty ds \frac{\rho^{(LO,2)}(Q^2,s)}{s-(p-q)^2} , \tag{10.36}$$

where

$$\rho^{(LO,2)}(Q^2,s) = 2f_\pi \frac{Q^2}{(Q^2+s)^2} \varphi_\pi(u(s)) \tag{10.37}$$

is the LO, twist-2 part of the spectral density determined from OPE. Then, we assume that the integral on the l.h.s. of (10.34), accumulating the contributions of the hadronic states heavier than pion, is approximated by the part of the dispersion integral (10.36) above a certain threshold:

$$\int_{s_{th}}^\infty ds \frac{\rho_5^h(Q^2,s)}{s-(p-q)^2} = \int_{s_0^\pi}^\infty ds \frac{\rho^{(LO,2)}(Q^2,s)}{s-(p-q)^2} . \tag{10.38}$$

Being in the same ballpark, s_0^π and s_{th} are not necessarily equal. Assuming the above equality between the integrals over the hadronic and OPE spectral densities, we adopt the semi-local duality, a weaker assumption than the local duality, which would demand

$$\rho_5^h(Q^2,s) \simeq \rho^{(LO,2)}(Q^2,s)$$

in the region $s > s_0^\pi$.

Substituting (10.38) in (10.34), we subtract the integral over $\rho^{(LO,2)}(Q^2,s)$ on both sides, leaving on the r.h.s. only the part of that integral at $s < s_0^\pi$. The dispersion relation is transformed into an approximate equation for the pion pole term:

$$\frac{2f_\pi F_\pi(Q^2)}{-(p-q)^2} = \int_0^{s_0^\pi} ds \frac{\rho^{(LO,2)}(Q^2,s)}{s-(p-q)^2} . \tag{10.39}$$

After that, we apply the Borel transformation (A.103) in the variable $(p-q)^2$, which in this case leads to an exponentiation of denominators, e.g.,

$$\frac{1}{s-(p-q)^2} \to \exp\left(-\frac{s}{M^2}\right) , \tag{10.40}$$

effectively replacing the variable $|(p-q)^2|$ by the Borel parameter squared M^2. The exponential weight appearing in the integrals suppresses (at sufficiently small M^2) the large s

region, making a sum rule less dependent on the accuracy of duality approximation. Simultaneously, the Borel transformation, as in other sum rules, removes possible subtraction terms from the dispersion relation. Finally, we obtain the LCSR for the pion form factor in the LO, twist-2 approximation:

$$F_\pi(Q^2) = \int_0^{s_0^\pi} ds \exp\left(-\frac{s}{M^2}\right) \frac{Q^2}{(Q^2+s)^2}\, \varphi_\pi(u(s))\,. \tag{10.41}$$

Note that the pion decay constant, appearing as a factor on both sides of this equation, cancels out. After transforming back to the variable u, the sum rule has an alternative and simpler form:

$$F_\pi(Q^2) = \int_{u_0^\pi}^\infty du\, \varphi_\pi(u) \exp\left(-\frac{\bar{u}Q^2}{uM^2}\right)\,, \tag{10.42}$$

where

$$u_0^\pi = \frac{Q^2}{Q^2+s_0^\pi}\,.$$

The duality threshold s_0^π for the pion channel can be taken from the QCD sum rule (8.124) for the pion decay constant.

The form factor obtained from the LCSR (10.42) with the LO and twist-2 accuracy does not yet contain any hard-scattering contribution, simply because it is of zeroth order in α_s. To access these contributions within LCSR approach, we have to include the $O(\alpha_s)$ radiative corrections to the correlation function providing the next-to-leading order (NLO) terms in the correlation function (10.18). The $O(\alpha_s)$ effects are given by the sum of the one-loop diagrams in Figure 10.3 generated by inserting perturbative gluon exchanges into the LO diagram. Note that in these diagrams the on-shell pion state is a long-distance object with respect to the light-cone separation between the two currents; hence, quark and antiquark lines with small virtualities[5] of $O(\Lambda_{QCD}^2)$, forming the pion DA, should be separated from the quarks with large virtuality propagating between the currents. Similarly, we should distinguish between virtual gluons in the loops and gluons flowing from the pion and absorbed by quarks. The latter are by default soft and have to be counted as a part of the pion DA. One such diagram with a three-particle pion DA is shown in Figure 10.2(b).

The $O(\alpha_s)$ contributions to the dominant twist-2 term of the correlation function (10.18) were calculated in [114]. After taking them into account, the invariant amplitude $\mathcal{F}$ maintains its factorized form

$$\mathcal{F}^{(LO\oplus NLO, tw2)}(Q^2, (p-q)^2) = 2f_\pi \int_0^1 du\, \varphi_\pi(u, \mu) \Big[T_0(Q^2, (p-q)^2, u)$$

$$+ \frac{\alpha_s C_F}{4\pi} T_1(Q^2, (p-q)^2, u, \mu) \Big]\,, \tag{10.43}$$

containing a product of two functions integrated over their common variable u. The function in squared brackets is the hard-scattering amplitude or hard-scattering kernel. It is calculable in terms of perturbative diagrams describing quarks and gluons propagating near the light-cone with characteristic virtualities of the order of $Q^2 \sim |(p-q)^2|$. The LO part of the kernel was already presented in (10.33):

$$T_0(Q^2, (p-q)^2, u) = \frac{u}{\bar{u}Q^2 - u(p-q)^2}\,, \tag{10.44}$$

---

[5]Small virtuality quarks and gluons are often called *soft*. Note that in the context of light-cone OPE they do not necessarily have small momentum components.

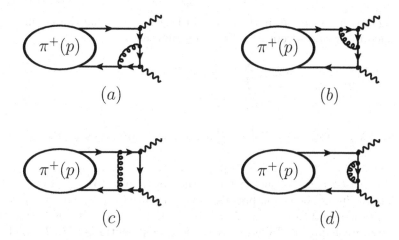

Figure 10.3 Diagrams of the $O(\alpha_s)$ radiative corrections to the correlation function (10.18).

and the analytic expression for the NLO part $T_1(Q^2, (p-q)^2, u, \mu)$ can be found in [114]. The hard-scattering kernel in (10.43) is convoluted with the twist-2 pion DA accumulating the soft, nonperturbative QCD effects.

The scale μ appearing in (10.43) in the NLO hard-scattering kernel emerges after the dimensional regularization of one-loop diagrams. This procedure is applied to avoid infrared (collinear) divergences in these diagrams which are compensated by a redefinition of the pion DA, more precisely, by renormalization of its Gegenbauer moments in the expansion (9.51). As a result, the dependence on the same scale μ appears in the pion DA entering the convolution in (10.43). Cancellation of divergences makes the LCSR approach meaningful, substantiating the use of the factorization formula (10.43). Simultaneously, μ can be interpreted as a scale that separates small and large virtuality contributions in the correlation function. Ideally, since it is an arbitrary scale, the μ-dependences in the two factors in (10.43) should mutually cancel each other. But in a truncated perturbation theory, in our case, at NLO, this cancellation is not complete; hence, a residual μ-dependence inevitably remains and manifests itself in the resulting LCSR. Usually, the same scale μ is adopted for the effective coupling α_s in (10.43) and also for quark masses if the latter are taken into account in the hard-scattering part of the correlation function. The mass renormalization is most conveniently done in the $\overline{MS}$ scheme. One usually finds an optimal interval of μ in the ballpark of the two large scales characterizing the correlation function, the momentum transfer Q and Borel parameter M.

In order to include the $O(\alpha_s)$ contribution in the sum rule (10.42), we need the imaginary part of the NLO hard-scattering kernel $T_1(Q^2, (p-q)^2, u, \mu)$ in the variable $(p-q)^2 = s > 0$ at fixed Q^2. This function has an involved expression presented in [114]. In fact, the resulting sum rule significantly simplifies for the asymptotic pion DA

$$\varphi_\pi(u(s)) = 6u(s)(1 - u(s)) = \frac{6Q^2 s}{(Q^2 + s)^2}, \qquad (10.45)$$

and can be written as

$$F_\pi(Q^2) = 6 \int_0^{s_0^\pi} ds\, e^{-s/M^2} \frac{Q^4 s}{(s + Q^2)^4} \left\{ 1 + \frac{\alpha_s}{3\pi} \left[\frac{\pi^2}{3} - 6 - \ln^2 \frac{Q^2}{s} + \frac{s}{Q^2} + \frac{Q^2}{s} \right] \right\}. \qquad (10.46)$$

Note that μ-dependence indeed cancels out in this case as expected, because the asymptotic DA (10.45) is scale-independent.

Expanding the r.h.s of (10.46) in the powers of $1/Q^2$, we find that the leading asymptotics of the form factor originates from the $O(\alpha_s)$ contribution:

$$F_\pi^{\mathrm{asymp}}(Q^2) = \frac{2\alpha_s}{\pi Q^2} \int\limits_0^{s_0^\pi} ds\, e^{-s/M^2}$$

$$+\frac{6}{Q^4} \int\limits_0^{s_0} ds\, s\, e^{-s/M^2} \left\{ 1 - \frac{\alpha_s}{3\pi} \left[10 - \frac{\pi^2}{3} + \ln^2 \frac{Q^2}{s} \right] \right\} + O(1/Q^6)\,, \qquad (10.47)$$

whereas the LO twist-2 part vanishes faster, being proportional to $1/Q^4$. Remarkably, LCSR predicts not only the Q^2-behavior but also the normalization of the pion form factor in agreement with the asymptotic limit (7.28). To see that, we use the QCD sum rule for f_π^2 in LO presented in Chapter 8. Taking only the loop contribution in (8.124) and neglecting the condensate terms – which is apparently consistent with the LCSR without higher twists – we replace in (10.47)

$$\int\limits_0^{s_0} ds\, e^{-s/M^2} \to 4\pi^2 f_\pi^2\,,$$

and obtain in the asymptotic limit:

$$F_\pi^{\mathrm{asymp}}(Q^2) \xrightarrow[Q^2\to\infty]{} \frac{8\pi\alpha_s f_\pi^2}{Q^2}\,. \qquad (10.48)$$

This indeed coincides with (7.28) if we identify the distribution amplitude $\tilde{\phi}_\pi$ introduced there with the asymptotic twist-2 pion DA, so that:

$$\tilde{\phi}_\pi(\alpha) = 6\alpha(1-\alpha)\,, \qquad \int\limits_0^1 d\alpha\, \frac{\tilde{\phi}_\pi(\alpha)}{\bar{\alpha}} = 3\,.$$

At large but finite Q^2, the dominant contribution to the form factor remains the one in (10.42), originating from the LO diagrams without quark-gluon coupling. It is therefore natural to interpret this contribution as a soft-overlap part of the form factor, which is an important result because this mechanism is not accessible in the factorization approach[6] discussed in Chapter 7; hence, the LCSR method offers a unique possibility to take into account both the hard-scattering and soft-overlap mechanisms within one approach, using, as a uniform input, the pion DAs. The price for that is a somewhat indirect way the form factor is obtained: we calculate not the form factor itself but the underlying correlation function and match the result to the dispersion relation combined with quark-hadron duality. A separation of hard-scattering contributions from the soft-overlap ones is therefore not straightforward, as discussed in detail in [114]. There is room for further improvement if the NNLO, two-loop corrections to the correlation function become available, however their calculation is technically very demanding.

To achieve a better accuracy in the LCSR for the pion e.m. form factor, the twist-4 terms and a part of the twist-6 terms were also included in the OPE for the correlation function (10.18). Note that the twist-3, and twist-5 terms are absent in the adopted massless quark

---

[6]Interestingly, in the LCSR asymptotics (10.47) there is also a term proportional to the Sudakov-type double logarithm of Q^2/s. As shown in [114], the influence of this term in LCSR is moderate, hence a strong Sudakov-suppression of the soft overlap part expected in certain factorization schemes is not supported by LCSR.

limit (chiral limit) for this correlation function. In qualitative terms, as already mentioned in Chapter 9, the higher-twist terms of the light-cone expansion take into account two effects: the transverse motion of the quark-antiquark state and the contributions of multiparton states in the pion DAs. A comprehensive discussion can be found in [99] and details of the LCSR calculation are in [114, 115]. For completeness, we quote the contribution of twist-4 DAs to the pion form factor when these DAs are taken in the asymptotic form:

$$F_\pi^{(4)}(Q^2) = \frac{40}{3}\delta_\pi^2(\mu) \int\limits_0^{s_0^\pi} ds\, e^{-s/M^2} \frac{Q^8}{(Q^2+s)^6}\left(1 - \frac{9s}{Q^2} + \frac{9s^2}{Q^4} - \frac{s^3}{Q^6}\right), \qquad (10.49)$$

to be compared with the twist-2 contribution in (10.46). The normalization parameter δ_π^2 of the twist-4 pion DAs is defined in (D.12) via the vacuum-to-pion matrix element of a local quark-antiquark-gluon operator. As can be seen, the twist-4 contribution has an extra suppression factor δ_π^2/M^2 but maintains the same Q^2 dependence as the LO twist-2 term of LCSR. The analytic expressions for the twist-4, and (an approximated) twist-6 contributions, both at the level of correlation function and in the final LCSR can be found in [114, 115]. They contain the pion twist-4 quark-antiquark and quark-antiquark-gluon DAs listed in Appendix D. The diagrams corresponding to these contributions are, respectively, in Figure 10.2(a), where a next-to-leading term of light-cone expansion near $x^2 = 0$ is implied, and in Figure 10.2(b).

The method of LCSRs can be generalized to calculate also the nucleon e.m. form factors at large spacelike momentum transfer. The details and results can be found in [116, 117]. The procedure essentially follows the one we used above for the pion e.m. form factor, with a correlation function in which the nucleon DAs and a quark current with the nucleon quantum numbers replace, respectively, the pion DAs and the axial-vector current.

10.3   THE PHOTON-PION TRANSITION FORM FACTOR

In Chapter 9 we considered the hadronic matrix element (9.4) of the $\gamma^*\gamma^* \to \pi^0$ transition. It was used to study the expansion of a product of two quark currents near the light-cone and led us to the definition of the pion DAs. The leading order answer for this amplitude in (9.58), obtained in terms of the pion twist-2 DA, corresponds to the diagram with a quark propagator between the two current vertices, shown in Figure 9.1. The accuracy of this calculation can be improved further by taking into account the known [118, 119] gluon radiative corrections as well as the next-to-leading terms of the light-cone OPE, including the pion twist-4 DAs – as we have seen, the twist-3 contribution is vanishingly small; hence, an analytical expression for the amplitude (9.4) is available in terms of a few nonperturbative parameters, such as the Gegenbauer moments of the pion DAs, the normalization parameter δ_π^2 of the twist-4 DAs and α_s. It would therefore be very interesting to compare this QCD-based prediction with experiment, measuring the cross section of the process $e^+e^- \to \pi^0 e^+e^-$ in the kinematical region (9.8), where both photon virtualities are spacelike and large. In fact, it is sufficient for both variables $Q^2 = -q^2$ and $|(p-q)^2|$ to be in the ballpark of 1 GeV^2. However, this particular cross section was never measured, because of a strong suppression, caused by the two highly virtual photon propagators. Instead, there are several measurements of the two-photon process with one almost real photon (see, e.g., [120],[121]):

$$\gamma^*(q)\gamma(p-q) \to \pi^0(p)\,. \qquad (10.50)$$

Since we know the amplitude in the case of two virtual photons, it is straightforward to take the leading twist-2 approximation (9.58), and put one of the photon virtualities to zero. At $(p-q)^2 \to 0$, we obtain the photon-pion transition form factor defined in (9.6):

$$F_{\gamma\pi}^{(0)}(Q^2) = \frac{\sqrt{2}}{3} f_\pi \int_0^1 du \frac{\varphi_\pi(u)}{\bar{u}Q^2}. \tag{10.51}$$

In fact, this formula is valid[7], but only at asymptotically large Q^2. Lowering the momentum transfer breaks down the light-cone OPE and the nonperturbative effects, induced by the soft quark-antiquark component of the real photon, become dominant. The question we address is if one can still access the photon-pion transition form factor at moderately large Q^2, taking into account the onset of these effects?

Here we present a method [122] which allows us to calculate this form factor at intermediate momentum transfers, typically, at $Q^2 > 1$ GeV^2, and, simultaneously, provides a smooth transition to the asymptotics (10.51). The method combines a hadronic dispersion relation with the LCSR in terms of the pion DAs.

The idea is to start from the invariant amplitude $F_{\gamma^*\pi}(q^2, (p-q)^2)$ with the two photon virtualities chosen in the region (9.8), and employ the OPE near the light-cone. In what follows, we confine ourselves by the leading twist-2 term $F_{\gamma^*\pi}^{(0)}(Q^2, (p-q)^2)$ given in (9.58). Since our main intention is to demonstrate how the method works, this approximation is sufficient. Radiative corrections and higher-twist contributions can be added afterwards.

The next step is to derive a hadronic dispersion relation for the amplitude $F^{\gamma^*\pi}$, analytically continuing it in the variable $(p-q)^2$ and keeping $q^2 = -Q^2$ fixed. The whole procedure is essentially the same as the one used to derive the LCSR in the previous section. We write down a unitarity relation for the initial hadronic matrix element (9.4), similar to the relation (10.21):

$$2\,\mathrm{Im}_{(p-q)^2}\mathcal{F}_{\mu\nu}(p,q) = \sum_{V=\rho^0,\omega} (2\pi)\delta\left(m_V^2 - (p-q)^2\right)\langle \pi^0|j_\mu^{em}|V(p-q)\rangle\langle V(p-q)|j_\nu^{em}|0\rangle$$

$$+ \underbrace{\sum_h \int d\tau_h \langle \pi^0(p)|j_\mu^{em}|h(p-q)\rangle\langle h(p-q)|j_\nu^{em}|0\rangle}_{2\pi\rho_{\mu\nu}^h(p,q)}, \tag{10.52}$$

where all intermediate hadronic states with the quantum numbers of the e.m. current are taken into account. Since this current has no definite isospin, or more precisely, consists of the two components with $I = 1, 0$, both vector mesons, ρ^0 and ω, contribute to the unitarity relation. In (10.52), their contributions are isolated from the sum over the excited and continuum states with the same quantum numbers.

Identifying the relevant hadronic matrix elements in (10.52), we first define the $\rho \to \pi$ and $\omega \to \pi$ e.m. transition form factors[8]:

$$\frac{1}{3}\langle \pi^0(p) \mid j_\mu^{em} \mid \omega(p-q)\rangle \simeq \langle \pi^0(p) \mid j_\mu^{em} \mid \rho^0(p-q)\rangle = F_{\rho\pi}(Q^2)m_\rho^{-1}\epsilon_{\mu\lambda\alpha\beta}\varepsilon^{(\rho)\lambda}q^\alpha p^\beta; \tag{10.53}$$

where the polarization vector $\varepsilon^{(\rho)}$ is the same for ρ and ω. Here we use isospin symmetry and, in addition, the relation between quark-flow diagrams (see Chapter 4), allowing us to relate the ω and ρ^0 form factors. Within the same approximation, the decay constants of vector mesons are

$$3\langle \omega(p-q) \mid j_\nu^{em} \mid 0\rangle \simeq \langle \rho^0(p-q) \mid j_\nu^{em} \mid 0\rangle = \frac{f_\rho}{\sqrt{2}}m_\rho\varepsilon_\nu^{(\rho)*}. \tag{10.54}$$

---

[7]This form factor was obtained [67] in the same collinear factorization approach as the asymptotic formula 7.29 for the pion e.m. form factor.

[8]Note that in Chapter 4 the $\rho \to \pi$ form factor was already defined in (4.14), with a slightly different normalization, adjusted to the $\rho\pi\gamma$ decay constant, so that $F_{\rho\pi}(Q^2)$ corresponds to $f_{\rho\pi}(q^2)m_\rho$.

In what follows, we neglect the small mass difference between ω and ρ, so that their pole terms in the dispersion relation are merged in one. Finally, we parameterize the hadronic sum in (10.52), introducing the invariant spectral density:

$$\rho^h_{\mu\nu}(p,q) = \rho^h(Q^2,(p-q)^2)\epsilon_{\mu\nu\alpha\beta}p^\alpha q^\beta \theta((p-q)^2 - s_{th}),\tag{10.55}$$

where s_{th} is the lowest threshold, which nominally coincides with the two-pion threshold. At this point we should admit that our assumptions on the hadronic spectrum in the channel of e.m. current, are slightly oversimplified, e.g., we separate the narrow ρ and ω mesons from the continuum of the hadronic states located above s_{th}. Needless to say, the hadronic ansatz can be improved further, e.g., by including the widths of vector mesons and applying a certain resonance formula, as shown in Chapter 5.

Inserting the definitions (10.53), (10.54) and (10.55) in the unitarity relation (10.52), collecting together the vector meson terms and equating the invariant amplitudes which multiply the same Lorentz structure on both parts, we obtain

$$\frac{1}{\pi}\text{Im}_{(p-q)^2}F_{\gamma^*\pi}(q^2,(p-q)^2) = \sqrt{2}f_\rho F_{\rho\pi}(Q^2)\delta(m_V^2 - (p-q)^2)$$
$$+\rho^h(Q^2,(p-q)^2)\theta((p-q)^2 - s_{th}),\tag{10.56}$$

and use it in the dispersion relation

$$F_{\gamma^*\pi}(q^2,(p-q)^2) = \frac{\sqrt{2}f_\rho F_{\rho\pi}(Q^2)}{m_\rho^2 - (p-q)^2} + \int\limits_{s_{th}}^\infty ds\,\frac{\rho^h(Q^2,s)}{s-(p-q)^2}.\tag{10.57}$$

Importantly, no subtractions are needed in this relation, as follows from the QCD asymptotics of the leading-order amplitude (9.58) at $(p-q)^2 \to \infty$.

The next essential step is the quark-hadron duality approximation. Reducing the OPE result to a form of the dispersion integral:

$$F_{\gamma^*\pi}(q^2,(p-q)^2) \simeq F_{\gamma^*\pi}^{(0)}(Q^2,(p-q)^2) = \frac{1}{\pi}\int\limits_0^\infty ds\,\frac{\text{Im}F_{\gamma^*\pi}^{(0)}(Q^2,s)}{s-(p-q)^2},\tag{10.58}$$

we assume that the part of this integral, above a certain effective threshold s_0, is equal to the integral over the spectral density of excited and continuum states:

$$\int\limits_{s_{th}}^\infty ds\,\frac{\rho^h(Q^2,s)}{s-(p-q)^2} = \frac{1}{\pi}\int\limits_{s_0}^\infty ds\,\frac{\text{Im}F_{\gamma^*\pi}^{(0)}(Q^2,s)}{s-(p-q)^2}.\tag{10.59}$$

Equating (10.57) and (10.58), and subtracting on both sides the integrals, according to the above duality relation, we obtain the following sum rule for the $\rho \to \pi$ form factor:

$$\frac{\sqrt{2}f_\rho F_{\rho\pi}(Q^2)}{m_\rho^2 - (p-q)^2} = \frac{1}{\pi}\int\limits_0^{s_0} ds\,\frac{\text{Im}F_{\gamma^*\pi}^{(0)}(Q^2,s)}{s-(p-q)^2}.\tag{10.60}$$

To improve it further, we apply the Borel transformation with respect to the variable $(p-q)^2$ and obtain:

$$\sqrt{2}f_\rho\,F_{\rho\pi}(Q^2) = \frac{1}{\pi}\int\limits_0^{s_0} ds\,e^{(m_\rho^2-s)/M^2}\text{Im}F_{\gamma^*\pi}^{(0)}(Q^2,s).\tag{10.61}$$

Now we are in a position to employ the dispersion relation (10.57), substituting, instead of the form factor $F^{\rho\pi}(Q^2)$, the above LCSR and, instead of the integral over ρ^h, its duality approximation:

$$F_{\gamma^*\pi}(q^2, (p-q)^2) = \frac{\sqrt{2}f_\rho}{m_\rho^2 - (p-q)^2} \frac{1}{\pi} \int_0^{s_0} ds \, \text{Im}F_{\gamma^*\pi}^{(0)}(Q^2, s) \, e^{(m_\rho^2 - s)/M^2}$$

$$+ \frac{1}{\pi} \int_{s_0}^\infty ds \, \frac{\text{Im}F_{\gamma^*\pi}^{(0)}(Q^2, s)}{s - (p-q)^2} . \tag{10.62}$$

The key observation is that the dispersion relation is valid at any $(p-q)^2$, hence, we can safely take the limit $(p-q)^2 \to 0$, obtaining the $\gamma \to \pi^0$ transition form factor:

$$F_{\gamma\pi}(Q^2) = \frac{1}{\pi m_\rho^2} \int_0^{s_0} ds \, \text{Im}F_{\gamma^*\pi}^{(0)}(Q^2, s) \, e^{(m_\rho^2 - s)/M^2} + \frac{1}{\pi} \int_{s_0}^\infty \frac{ds}{s} \text{Im}F_{\gamma^*\pi}^{(0)}(Q^2, s). \tag{10.63}$$

In the adopted twist-2 approximation and at zeroth order in α_s, the OPE result (9.58) for the amplitude $F_{\gamma^*\pi}$ is easily converted into the dispersion integral form, applying the transformation (10.35) of the integration variable:

$$F_{\gamma^*\pi}^{(0)}(Q^2, (p-q)^2) = \frac{\sqrt{2}f_\pi}{3} \int_0^1 \frac{du \, \varphi_\pi(u)}{u \left(\frac{\bar{u}Q^2}{u} - (p-q)^2 \right)}$$

$$= \frac{\sqrt{2}f_\pi}{3} \int_0^\infty \frac{ds}{(s - (p-q)^2)(s + Q^2)} \varphi_\pi(u) \Bigg|_{u = \frac{Q^2}{s + Q^2}}, \tag{10.64}$$

so that the imaginary part is

$$\text{Im}F_{\gamma^*\pi}^{(0)}(Q^2, s) = \frac{1}{s + Q^2} \varphi_\pi(u) \Bigg|_{u = \frac{Q^2}{s + Q^2}}. \tag{10.65}$$

The dispersion representation (10.64) allows us to easily perform the Borel transformation and specify the duality approximation. After that, it is more convenient to return to the initial integration variable u. In this way, we finally obtain the $\gamma \to \pi^0$ transition form factor

$$Q^2 F^{\gamma\pi}(Q^2) = \frac{\sqrt{2}f_\pi}{3} \left(\frac{Q^2}{m_\rho^2} V(Q^2, M^2) + H(Q^2) \right), \tag{10.66}$$

consisting of the two terms:

$$V(Q^2, M^2) = \int_{u_0}^1 \frac{du}{u} \left(\varphi_\pi(u) + \dots \right) \exp\left(-\frac{Q^2(1-u)}{uM^2} + \frac{m_\rho^2}{M^2} \right) \tag{10.67}$$

proportional to the $\rho \to \pi^0$ e.m. transition form factor

$$F^{\rho\pi}(Q^2) = \frac{f_\pi}{3f_\rho} V(Q^2, M^2), \tag{10.68}$$

and

$$H(Q^2) = \int_0^{u_0} \frac{du}{1-u} \left(\varphi_\pi(u) + \dots \right), \tag{10.69}$$

where $u_0 = Q^2/(s_0 + Q^2)$. The dots indicate the power suppressed higher-twist terms and $O(\alpha_s)$ corrections. The most complete expressions for these terms can be found in [123]. Importantly, in the limit $Q^2 \to \infty$, when

$$u_0 \simeq 1 - s_0/Q^2,$$

the term $V(Q^2, M^2)$ is $O(1/Q^2)$ suppressed with respect to $H(Q^2)$ and the form factor $F^{\gamma\pi}(Q^2)$ coincides with the asymptotic formula (10.51). Nevertheless, at moderate Q^2 both components of the form factor are parametrically and numerically important. We emphasize that the formula (10.66) for the photon-pion transition form factor is primarily based on the hadronic dispersion relation and the LCSR plays here a somewhat auxiliary role, providing the residue of the pole term in this relation (and the $\rho \to \pi$ form factor as a useful byproduct).

10.4 VARYING FLAVORS: HEAVY-TO-LIGHT FORM FACTORS

The method presented in the previous sections and applied to the pion e.m. and photon-pion form factors is easily extended to the hadron form factors involved in the flavor-changing transitions. As an example, we consider here in detail the LCSR for the vector form factor of the $B \to \pi$ transition induced by the weak $b \to u$ current. Calculation of this form factor enables to determine the V_{ub} element of the CKM matrix from measurements of the $B \to \pi\ell\nu_\ell$ decay width.

The definition of the $B \to \pi$ form factor is written in a general form in (2.96). Specifying the flavors, we have, e.g., for the $\bar{B}^0 \to \pi^+$ transition:

$$\langle \pi^+(p)|\bar{u}\gamma_\mu b|\bar{B}^0(p+q)\rangle = 2p_\mu f_{B\pi}^+(q^2) + O(q_\mu), \tag{10.70}$$

where the terms proportional to q_μ and containing also the scalar $B \to \pi$ form factor are not shown explicitly.

The starting point is the correlation function

$$\mathcal{F}_\mu(p,q) = i \int d^4x e^{iqx} \langle \pi^+(p)|T\{\bar{u}(x)\gamma_\mu b(x), m_b \bar{b}(0)i\gamma_5 d(0)\}|0\rangle$$

$$= \mathcal{F}(q^2, (p+q)^2)p_\mu + O(q_\mu) \tag{10.71}$$

where the vector part of the weak $b \to u$ current is correlated with a pseudoscalar current which has the B-meson quantum numbers. Here we only isolate the relevant invariant amplitude proportional to the momentum p_μ. Comparing this correlation function with (10.18), we notice that the only essential difference is that a b quark propagates between the points x and 0 instead of a light quark.

To apply the OPE near the light cone to the above correlation function, we choose the two independent invariant variables q^2 and $(p+q)^2$ to vary in the regions

$$q^2 \ll m_b^2, \quad (p+q)^2 \ll m_b^2, \tag{10.72}$$

well below the thresholds at which the currents in the correlation function produce b-flavored hadrons. With this choice, the b-quark emitted and absorbed by the currents is highly virtual and propagates at small average x^2. Thus, also the light quark and antiquark forming the pion are emitted at an almost lightlike separation. It is important that the condition (10.72) does not necessarily demand a spacelike momentum transfer $q^2 < 0$ as it was in the case of the correlation function (10.18). Therefore, the lower (large recoil) part of the kinematical region in the $B \to \pi\ell\nu_\ell$ decays, typically

$$0 < q^2 < m_b^2 - m_b\tau,$$

where $\tau \sim 1$ GeV, belongs to the region accessible with OPE.

In what follows we neglect the u, d-quark and pion masses. Contracting the b-quark fields in (10.71) and inserting the free b-quark propagator, we obtain the LO part of the correlation function:

$$
\begin{aligned}
\mathcal{F}_\mu^{(0)}(p, q) &= im_b \int \frac{d^4x\, d^4k}{(2\pi)^4(m_b^2 - k^2)} e^{i(q-k)x} \Big(m_b \langle \pi(p)|\bar{u}(x)\gamma_\mu\gamma_5 d(0)|0\rangle \\
&\quad + k^\nu \langle \pi^+(p)|\bar{u}(x)\gamma_\mu\gamma_\nu\gamma_5 d(0)|0\rangle \Big) \\
&= im_b \int \frac{d^4x\, d^4k}{(2\pi)^4(m_b^2 - k^2)} e^{i(q-k)x} \Big(m_b[\gamma_\mu\gamma_5]_{\omega\xi} + k^\nu[\gamma_\mu\gamma_\nu\gamma_5]_{\omega\xi} \Big) \\
&\quad \times \langle \pi(p)|\bar{u}_\omega(x)d_\xi(0)|0\rangle .
\end{aligned}
\tag{10.73}
$$

We can now directly insert in the above the decomposition (D.1) of the vacuum-to-pion matrix element in terms of the quark-antiquark DAs. This part corresponds to the diagram which is obtained from Figure 10.2(a) if the virtual u-quark is replaced by the b-quark and the pion is transferred to the final state.

To improve the accuracy, we also include the contribution of the quark-antiquark-gluon DAs originating due to a low-virtuality gluon emitted near the light-cone from the virtual b-quark line. The diagram shown in Figure 10.2(b) describes this contribution. To include this effect in the correlation function, we need the heavy-quark propagator near the light-cone, the massive analog of the propagator derived in the first section of this chapter. The corresponding expression is

$$
S_{(b)}^{(1)}(x, 0) = -g_s \int \frac{d^4k}{(2\pi)^4} e^{-ik\cdot x} \int_0^1 dv\, G^{\mu\nu a}(vx) \frac{\lambda^a}{2} \left(\frac{\slashed{k} + m_b}{2(m_b^2 - k^2)^2} \sigma_{\mu\nu} + \frac{1}{m_b^2 - k^2} vx_\mu \gamma_\nu \right).
\tag{10.74}
$$

At $m_b \to 0$ this expression reduces to (10.16).

Inserting the above propagator instead of contracted b-quark fields in (10.71), we obtain:

$$
\begin{aligned}
\mathcal{F}_\mu^{(G)}(p, q) &= ig_s \int \frac{d^4k\, d^4x\, dv}{(2\pi)^4(m_b^2 - k^2)} e^{i(q-k)x} \langle \pi|\bar{u}(x)\gamma_\mu \Big(vx_\rho G^{\rho\lambda}(vx)\gamma_\lambda \\
&\quad + \frac{1}{2} \frac{\slashed{k} + m_b}{m_b^2 - k^2} G^{\rho\lambda}(vx)\sigma_{\rho\lambda} \Big) \gamma_5 d(0)|0\rangle ,
\end{aligned}
\tag{10.75}
$$

where we can use the decomposition (D.2) in quark-antiquark-gluon DAs.

After lengthy calculation,[9] we obtain for the sum of (10.73) and (10.75):

$$
\mathcal{F}^{(0,G)}(q^2,(p+q)^2) = m_b^2 f_\pi \int_0^1 \frac{du}{m_b^2 - (q+up)^2}\left\{\varphi_\pi(u) + \frac{\mu_\pi}{m_b}u\phi_{3\pi}^p(u)\right.
$$

$$
+ \frac{\mu_\pi}{6m_b}\left[2 + \frac{m_b^2 + q^2}{m_b^2 - (q+up)^2}\right]\phi_{3\pi}^\sigma(u) - \frac{m_b^2\phi_{4\pi}(u)}{2\left(m_b^2 - (q+up)^2\right)^2}
$$

$$
\left. - \frac{u}{m_b^2 - (q+up)^2}\int_0^u dv\,\psi_{4\pi}(v)\right\}
$$

$$
+ \int_0^1 dv \int \frac{\mathcal{D}\alpha}{\left[m_b^2 - \left(q+(\alpha_1+\alpha_3 v)p\right)^2\right]^2}\left\{4m_b f_{3\pi}v(q\cdot p)\Phi_{3\pi}(\alpha_i)\right.
$$

$$
\left.+ m_b^2 f_\pi\left(2\Psi_{4\pi}(\alpha_i) - \Phi_{4\pi}(\alpha_i) + 2\widetilde{\Psi}_{4\pi}(\alpha_i) - \widetilde{\Phi}_{4\pi}(\alpha_i)\right)\right\}, \quad (10.76)
$$

where the nomenclature and expressions for all pion DAs of twist-2,3,4 entering the above equation are presented in Appendix D.

We notice that the twist-3 contribution to this correlation function is parametrically large, being enhanced by the factor μ_π, since $\mu_\pi/m_b \sim 1$. The twist-4 contributions are suppressed by powers of the denominator $1/(m_b^2 - (q+up)^2)$. Altogether, there are two separate hierarchies in this correlation function formed by the even ($t = 2, 4, ..$) and odd ($t = 3, 5, ..$) twist contributions.

The complete OPE result for the invariant amplitude is then represented as a sum of LO and NLO parts:

$$
\mathcal{F}(q^2,(p+q)^2) = \mathcal{F}^{(0,G)}(q^2,(p+q)^2) + \frac{\alpha_s C_F}{4\pi}\mathcal{F}^{(1)}(q^2,(p+q)^2), \quad (10.77)
$$

where the NLO part $\mathcal{F}_1$ has a factorized form of the convolutions:

$$
\mathcal{F}^{(1)}(q^2,(p+q)^2) = f_\pi \int_0^1 du\left\{T_1(q^2,(p+q)^2,u)\varphi_\pi(u)\right.
$$

$$
\left.+ \frac{\mu_\pi}{m_b}\left[T_1^p(q^2,(p+q)^2,u)\phi_{3\pi}^p(u) + T_1^\sigma(q^2,(p+q)^2,u)\phi_{3\pi}^\sigma(u)\right]\right\}, \quad (10.78)
$$

and the hard-scattering amplitudes T_1, $T_1^{p,\sigma}$ result[10] from the calculation of the diagrams similar to the ones in Figure 10.3. The twist-4 NLO terms are not calculated yet.

To obtain the hadronic dispersion relation, we use the expression for the imaginary part which has a similar structure as (10.21) and follows from the unitarity relation, where all intermediate hadronic states with the momentum $p + q$ and quantum numbers of the B meson contribute:

$$
2\,\mathrm{Im}_{(p+q)^2}\mathcal{F}_\mu(p,q) = 2\pi\delta((p+q)^2 - m_B^2)\langle\pi^+(p)|\bar{u}\gamma_\mu d|\bar{B}^0(p+q)\rangle\langle\bar{B}^0(p+q)|m_b\bar{b}i\gamma_5 d|0\rangle
$$

$$
\underbrace{+ \sum_h \int d\tau_h \langle\pi^+(p)|\bar{u}\gamma_\mu d|h(p+q)\rangle\langle h(p+q)|m_b\bar{b}i\gamma_5 d|0\rangle}_{2\pi\rho^h(q^2,(p+q)^2)p_\mu}, \quad (10.79)
$$

---

[9]The details are presented in the online appendix.

[10]Their bulky expressions and the corresponding imaginary parts can be found e.g., in [124].

where the B-meson state is the lowest one and there is a gap between this state and the onset of the spectral density of heavier states starting from the $B^*\pi$ threshold. Note that the $B\pi$ state is not allowed by spin-parity. In fact, the structure of the hadronic spectral density is very similar to the one in the two-point QCD sum rule for the B meson decay constant considered in Chapter 8. From the latter sum rule we also usually take the value of f_B which will appear in the dispersion relation multiplied by the form factor. Using (10.79) and the definitions of the matrix elements entering the B-meson term, we finally obtain, after discarding the equal Lorentz structure p_μ from both sides:

$$\mathcal{F}(q^2, (p+q)^2) = \frac{2m_B^2 f_B f_{B\pi}^+(q^2)}{m_B^2 - (p+q)^2} + \int\limits_{(m_B^* + m_\pi)^2}^{\infty} ds \frac{\rho^h(q^2, s)}{s - (p+q)^2} . \tag{10.80}$$

Quark-hadron duality approximation in this case is given by the approximate equation of the two integrals

$$\int\limits_{(m_B^* + m_\pi)^2}^{\infty} ds \frac{\rho^h(q^2, s)}{s - (p+q)^2} = \frac{1}{\pi} \int\limits_{s_0^B}^{\infty} ds \frac{Im_{(p+q)^2}\mathcal{F}(q^2, s)}{s - (p+q)^2} . \tag{10.81}$$

Applying this approximation and performing the Borel transformation in the variable $(p+q)^2$ we finally obtain the LCSR for the $B \to \pi$ form factor:

$$f_{B\pi}^+(q^2) = \frac{e^{m_B^2/M^2}}{2m_B^2 f_B} \left[F_0(q^2, M^2, s_0^B) + \frac{\alpha_s C_F}{4\pi} F_1(q^2, M^2, s_0^B) \right] , \tag{10.82}$$

where $F_{0(1)}(q^2, M^2, s_0^B)$ originates from the OPE result for the LO (NLO) invariant amplitude $\mathcal{F}^{0(1)}(q^2, (p+q)^2)$ after representing the latter in the dispersion integral form and subtracting the part of the integral according to (10.82).

The LO part of the LCSR has the following expression:

$$F_0(q^2, M^2, s_0^B) = m_b^2 f_\pi \int\limits_{u_0}^{1} du \, e^{-\frac{m_b^2 - q^2 \bar{u}}{u M^2}} \left\{ \frac{\varphi_\pi(u)}{u} \right.$$

$$+ \frac{\mu_\pi}{m_b} \left(\phi_{3\pi}^p(u) + \frac{1}{6} \left[\frac{2\phi_{3\pi}^\sigma(u)}{u} - \left(\frac{m_b^2 + q^2}{m_b^2 - q^2} \right) \frac{d\phi_{3\pi}^\sigma(u)}{du} \right] \right) - 2 \left(\frac{f_{3\pi}}{m_b f_\pi} \right) \frac{I_{3\pi}(u)}{u}$$

$$\left. + \frac{1}{m_b^2 - q^2} \left(-\frac{m_b^2 u}{4(m_b^2 - q^2)} \frac{d^2\phi_{4\pi}(u)}{du^2} + u\psi_{4\pi}(u) + \int\limits_0^u dv\psi_{4\pi}(v) - I_{4\pi}(u) \right) \right\}, \tag{10.83}$$

where

$$u_0 = \frac{m_b^2 - q^2}{s_0^B - q^2} ,$$

and the short-hand notations introduced for the integrals over three-particle DAs are:

$$
I_{3\pi}(u) = \frac{d}{du}\left(\int_0^u d\alpha_1 \int_{(u-\alpha_1)/(1-\alpha_1)}^1 dv\, \Phi_{3\pi}(\alpha_i)\Bigg|_{\substack{\alpha_2 = 1 - \alpha_1 - \alpha_3, \\ \alpha_3 = (u-\alpha_1)/v}}\right),
$$

$$
I_{4\pi}(u) = \frac{d}{du}\left(\int_0^u d\alpha_1 \int_{(u-\alpha_1)/(1-\alpha_1)}^1 \frac{dv}{v}\left[2\Psi_{4\pi}(\alpha_i) - \Phi_{4\pi}(\alpha_i)\right.\right.
$$

$$
\left.\left.+2\widetilde{\Psi}_{4\pi}(\alpha_i) - \widetilde{\Phi}_{4\pi}(\alpha_i)\right]\Bigg|_{\substack{\alpha_2 = 1 - \alpha_1 - \alpha_3, \\ \alpha_3 = (u-\alpha_1)/v}}\right). \qquad (10.84)
$$

The input to LCSR includes the b-quark mass in the $\overline{MS}$ scheme, α_s and parameters of the pion DAs listed in the Appendix D. The numerical values of the latter are estimated either from comparing other LCSRs (e.g., the one for the photon-pion form factor) with experiment, or using dedicated two-point QCD sum rules. The threshold s_0^B is usually estimated by differentiating LCSR over the inverse Borel parameter and dividing the result to the original sum rule. The resulting relation is then equated to the B meson mass squared.

The LCSR results for the form factor are valid in the large recoil region of q^2. The form factor can be parameterized with the z parameterization presented in Appendix C and extrapolated towards large q^2 where it can be compared with the lattice QCD results.

We also mention that the LCSR obtained for the finite b-quark mass allows to investigate the infinitely heavy quark limit of the $B \to \pi$ form factor, applying the same limiting procedure as in (8.176)–(8.178). The result is the nontrivial scaling:

$$
\lim_{m_b \to \infty} f_{B\pi}^+(0) \sim \frac{1}{m_b^{3/2}}
$$

which is sensitive to the end-point behavior of the pion twist-2 DA. A more detailed analysis of this limit is in [125].

A further interesting byproduct of the correlation function (10.71) is the double dispersion relation in both variables $(p+q)^2$ and q^2. Combined with the OPE in pion DAs and quark-hadron duality applied in both channels, this relation results in an LCSR estimate of the $B^*B\pi$ strong coupling [126].

In conclusion we list other important applications of this method. The scalar (tensor) $B \to \pi$ form factor is also accessible by choosing an appropriate structure (changing the Dirac structure of the heavy-to light current). Furthermore, using the kaon DAs instead of the pion ones, it is possible to calculate the $B \to K$ and $B_s \to K$ form factors. Moreover, in LCSRs with the light-meson DAs it is simple to switch to the semileptonic form factors of charmed mesons, by replacing $b \to c$ and $B \to D$ in all formulas and adjusting the scales and duality threshold correspondingly. In this way the $D \to \pi, K$ form factors are calculated with the same accuracy (see, e.g., [100]).

Our last example is the LCSR application to the baryon form factors: the calculation of the $\Lambda_b \to p$ transition form factors [127], which is based on the vacuum-to-nucleon correlation function:

$$
\Pi_a(P,q) = i\int d^4x\, e^{iqx}\langle N(P)|T\left\{\eta^{(\Lambda_b)}(0), j_a(x)\right\}|0\rangle, \qquad (10.85)
$$

where N is a generic (in the isospin symmetry limit) nucleon state with momentum P. This correlation function is calculated in terms of the nucleon DAs, worked out in [107]. The three-quark udb current $\eta^{(\Lambda_b)}$ is similar to the one defined in (3.34) and has the quantum numbers of Λ_b.

10.5 SUM RULES WITH B-MESON DISTRIBUTION AMPLITUDES

The method of LCSRs for heavy-to-light form factors presented in the previous section has one essential limitation. To calculate a form factor of the B meson transition into a certain hadronic state h, one needs to define a set of DAs for this state, to be used in the OPE of the vacuum-to-h correlation function. However, the DAs are well defined and sufficiently elaborated only for a limited number of hadrons, such as the pion, kaon or vector meson in the narrow width limit. Unstable mesons with various spin-parities, dimeson states and charmed mesons represent examples of hadronic states for which the light-cone DAs either have a limited applicability or are not defined at all.

A larger variety of final states in $B \to h$ transition is accessible if one uses another version of LCSRs [128], where the B meson enters as an initial state in the correlation function:

$$\mathcal{F}_{ab}^{(B)}(p,q) = i \int d^4x \, e^{ip \cdot x} \langle 0 | T \left\{ \bar{q}_2(x) \Gamma_a q_1(x), \bar{q}_1(0) \Gamma_b b(0) \right\} | \bar{B}(p_B) \rangle \,. \tag{10.86}$$

Here $p_B = p + q$, $\bar{q}_1 \Gamma_b b$ is a heavy-to-light transition current and $\bar{q}_2 \Gamma_a q_1$ is the interpolating current with the same quantum numbers as the hadronic state h. The advantage is that we can now easily change the flavor content and spin-parity of this state, varying the flavors of $q_{1,2}$ and the Dirac structure Γ_a in the interpolating current. The price for that is a necessity to introduce the B-meson light-cone DAs in order to apply OPE to the correlation function (10.86).

To proceed, we assume that external momenta in the correlation function (10.86) are properly chosen – this will be discussed in detail below – so that a highly virtual q_1 quark is replaced with a propagator $\langle 0 | q_1(x) \bar{q}_1(0) | 0 \rangle$. We encounter the nonlocal hadronic matrix element

$$\langle 0 | \bar{q}_{2\alpha}(x) b_\beta(0) | \bar{B}(p_B) \rangle \,, \tag{10.87}$$

where Dirac indices are left free and the summation over color indices is implied but not shown. This object is expected to generate a set of B-meson light-cone DAs, in a full analogy with the pion DAs discussed in Chapter 9. However, the matrix element (10.87) contains an on-shell B-meson state. Consequently, the presence of a finite and large b-quark mass scale hinders a straightforward expansion of (10.87) near the light-cone $x^2 = 0$[11]. The solution is to exclude the scale m_b, switching to HQET in (10.86) and introducing the B-meson DAs in this effective theory framework.

To realize this transition, we separate the static momentum of the B-meson state from the residual one, as it was done in (1.118):

$$p_B = p + q = m_b v + k \,, \tag{10.88}$$

with v being the four-velocity vector of B, equal to $v = (1,0,0,0)$ in the rest frame. In this frame, $m_B = m_b + \bar{\Lambda}$ and the energy component of the residual momentum is $k_0 \simeq \bar{\Lambda}$. Note that we retain the relativistic normalization of the heavy meson state for convenience, hence

$$|B(P_B)\rangle = |B_v\rangle + O(1/m_b) \,, \tag{10.89}$$

where generic $\sim 1/m_b$ corrections are not specified, being neglected in the adopted approximation. We also replace the b-quark field by an effective heavy quark field h_v, according to (1.130), and redefine the momentum transfer:

$$q = m_b v + \tilde{q} \,, \quad p + \tilde{q} = k \,. \tag{10.90}$$

---

[11]This problem concerns mainly the power expansion in x^2. In particular, the role of the terms of $O(x^2 m_b^2)$ has to be clarified.

The correlation function

$$\mathcal{F}_{ab}^{(B)}(p,q) = i \int d^4x \, e^{ipx} \underbrace{\langle 0|T\{\bar{q}_2(x)\Gamma_a q_1(x), \bar{q}_1(0)\Gamma_b h_v(0)\}|\bar{B}_v\rangle}_{\mathcal{F}_{ab}^{(B_v)}(p,\tilde{q})} + O(1/m_b) \qquad (10.91)$$

is approximated by its counterpart in HQET

$$\mathcal{F}_{ab}^{(B)}(p,q) \simeq \mathcal{F}_{ab}^{(B_v)}(p,\tilde{q}). \qquad (10.92)$$

In the latter function, the two currents consist only of the light-quark and effective fields and the scale m_b is absent. The amplitude $\mathcal{F}_{ab}^{(B_v)}$ depends on the two invariant variables p^2 and $\tilde{q}^2$, and on the intrinsic scale $\bar{\Lambda}$ of the effective state $|\bar{B}_v\rangle$; hence, the situation is very similar to the vacuum-to-pion amplitude considered in Chapter 9. Repeating the same arguments, we convince ourselves that in the deep spacelike region

$$|p^2|, |\tilde{q}^2| \gg \Lambda_{QCD}^2, \bar{\Lambda}^2, \qquad (10.93)$$

if the difference between virtualities is kept large, so that

$$\xi = \frac{|\tilde{q}^2| - |p^2|}{|p^2|} = O(1), \qquad (10.94)$$

the integrand in the correlation function (10.91) is saturated at $x^2 \sim 1/|p^2|$, in the region where the exponent e^{ipx} does not oscillate strongly. Importantly, the scales $\sqrt{|p^2|}$ and $\sqrt{|\tilde{q}^2|}$, being large with respect to nonperturbative QCD scales, remain smaller than the heavy quark mass. Returning to the initial momentum-transfer squared q^2, and estimating the components of momenta similar to the assessment we have done for the vacuum-to-pion amplitude, we find that the conditions (10.93) and (10.94) yield

$$q^2 \simeq m_b^2 + 2m_b\tilde{q}_0 \sim m_b^2 - \frac{m_b|p^2|\xi}{\bar{\Lambda}} \ll m_b^2. \qquad (10.95)$$

Hence, in contrast to the case of a correlation function with light-quark currents, where the condition $q^2 \ll 0$ warrants light-cone dominance, here it is sufficient that q^2 is far below the heavy flavor threshold. At the same time, the momentum transfer can cover the region of large recoil of h in a $B \to h$ transition:

$$0 \leq q^2 \ll m_b^2. \qquad (10.96)$$

We proceed to deriving the LCSR based on the HQET correlation function. The corresponding diagrams are shown in Figure 10.4 and resemble the diagrams in Figure 10.2, except the role of pion is played by the B-meson.

In the following discussion we consider one particular example of the $\bar{B}^0 \to \pi^+$ weak transition. Accordingly, in (10.91) we choose the quark flavors and Dirac matrices to form the $b \to u$ weak vector current and the axial-vector current interpolating the pion. We have, taking into account the approximation (10.92),

$$\mathcal{F}_{\rho\mu}^{(B)}(p,q) = i \int d^4x \, e^{ipx} \langle 0|T\{\bar{d}(x)\gamma_\rho\gamma_5 u(x)\bar{u}(0)\gamma_\mu h_v(0)\}|\bar{B}_v\rangle, \qquad (10.97)$$

and the external momenta squared p^2 and q^2 satisfy, respectively, the conditions (10.93) and (10.96). Contracting the u-quark fields in the free propagator, neglecting m_u, we obtain:

$$\mathcal{F}_{\rho\mu}^{(B)}(p,q) = i \int d^4x \, e^{ip\cdot x} \frac{ix^\lambda}{2\pi^2(x^2)^2} [\gamma_\rho\gamma_5\gamma_\lambda\gamma_\mu]_{\alpha\beta} \langle 0|\bar{d}_\alpha(x)h_{v\beta}(0)|\bar{B}_v\rangle. \qquad (10.98)$$

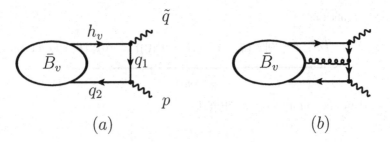

Figure 10.4 Diagrams of the correlation function in HQET: the contribution of (a) quark-antiquark and (b) quark-antiquark-gluon B-meson DAs.

This expression is a convolution of a hard scattering amplitude (the propagator and currents) with the hadronic matrix element

$$\langle 0|\bar{q}_\alpha(x)h_{v\beta}(0)|\bar{B}_v\rangle$$

in HQET, where $q = u, d$ in the isospin symmetry limit; hence, in contrast to (10.87), this matrix element does not contain the heavy quark mass and can be expanded around $x^2 = 0$. In the leading approximation, this expansion, originally suggested in [129] (see also [72] and the review [130]) has the following form:

$$\langle 0|\bar{q}_\alpha(x)[x,0]h_{v\beta}(0)|\bar{B}_v\rangle = -\frac{if_B m_B}{4}\int_0^\infty d\omega\, e^{-i\omega v\cdot x}$$

$$\times\left[(1+\slashed{v})\left\{\phi_+^B(\omega) - \frac{\phi_+^B(\omega) - \phi_-^B(\omega)}{2v\cdot x}\slashed{x}\right\}\gamma_5\right]_{\beta\alpha}. \qquad (10.99)$$

It represents the most general Lorentz decomposition of the vacuum-to-B matrix element in the infinitely heavy quark limit and contains the two B-meson DAs, $\phi_+^B(\omega)$ and $\phi_-^B(\omega)$, normalized to a unit:

$$\int_0^\infty d\omega\, \phi_\pm^B(\omega) = 1\,. \qquad (10.100)$$

The gauge link $[x, 0]$ is inserted to ensure the local gauge invariance of QCD. The variable ω counts the light spectator-antiquark momentum in the effective state B_v. Since this state is formally taken in the infinite mass limit, the upper bound of ω is also infinite[12]. In the local limit, (10.99) correctly reproduces the B-meson decay constant. Since we do not include radiative and $1/m_B$ corrections, f_B is not distinguished from the static decay constant (2.145) defined in HQET.

Inserting (10.99) in (10.98), we obtain

$$\mathcal{F}_{\rho\mu}^{(B)}(p, q) = \frac{f_B m_B}{4}\int_0^\infty d\omega\int d^4x\, e^{i(p-\omega v)\cdot x}\frac{ix^\lambda}{2\pi^2(x^2)^2}$$

$$\times\text{Tr}\left[\gamma_\rho\gamma_5\gamma_\lambda\gamma_\mu(1+\slashed{v})\left\{\phi_+^B(\omega) - \frac{\phi_+^B(\omega) - \phi_-^B(\omega)}{2v\cdot x}\slashed{x}\right\}\gamma_5\right]. \qquad (10.101)$$

---

[12]In fact, this infinite "tail" of the B-meson DAs does not play an important role in the sum rules where the upper limit of integrals over the variable ω is always finite, due to the duality threshold.

Before taking the trace and integrating over the coordinates, we should rearrange the second term in the curly brackets in (10.101) to a more convenient form without $v \cdot x$ in the denominator. This is done by introducing an integrated combination of DAs:

$$\Phi_\pm^B(\omega) \equiv \int_0^\omega d\tau \left(\phi_+^B(\tau) - \phi_-^B(\tau)\right), \tag{10.102}$$

with the following properties:

$$\Phi_\pm^B(0) = \Phi_\pm^B(\infty) = 0, \quad \frac{d\Phi_\pm^B(\omega)}{d\omega} = \phi_+^B(\omega) - \phi_-^B(\omega).$$

Using this definition and partial integration, we transform the integral:

$$\int_0^\infty d\omega\, e^{-i\omega v \cdot x} \frac{\phi_+^B(\omega) - \phi_-^B(\omega)}{2\, v \cdot x} = \frac{i}{2} \int_0^\infty d\omega\, e^{-i\omega v \cdot x}\, \Phi_\pm^B(\omega), \tag{10.103}$$

removing the coordinate factor in the denominator. Substituting the above equation in (10.101) and using the integrals in (A.95) we obtain:

$$\mathcal{F}_{\rho\mu}^{(B)}(p,q) = -i \frac{f_B m_B}{4} \int_0^\infty \frac{d\omega}{(p - \omega v)^2} \left\{ (p - \omega v)_\lambda v_\tau \phi_+^B(\omega) \right.$$

$$\left. + \frac{1}{2}\left(-g_{\lambda\tau} + \frac{2(p - \omega v)_\lambda (p - \omega v)_\tau}{(p - \omega v)^2}\right) \Phi_\pm^B(\omega) \right\} \mathrm{Tr}\left[\gamma_\rho \gamma_\lambda \gamma_\mu \gamma_\tau\right]. \tag{10.104}$$

Since we are mainly interested in the LCSR method, let us avoid cumbersome expressions and choose $q^2 = 0$, the point which corresponds to the maximal recoil of the pion in the $B \to \pi$ transition. To proceed, we express the velocity four-vector in terms of external momenta:

$$v = \frac{p + q}{m_B},$$

and introduce a new variable $s = \omega m_B$, obtaining a useful relation (valid at $q^2 = 0$)

$$(p - \omega v)^2 = \left(1 - \frac{s}{m_B^2}\right)(p^2 - s)$$

which will allow us to simplify (10.104) and to reduce it to a form of dispersion integral.

The expression (10.104) has several Lorentz structures. Since we aim at obtaining the LCSR for the vector $B \to \pi$ form factor $f_{B\pi}^+(0)$, the hadronic dispersion relation for the correlation function has to be investigated first, in order to choose the appropriate kinematical structure for this sum rule. We apply unitarity relation for the correlation function in the variable p^2 at $q^2 = 0$. The imaginary part is similar to (10.21), with the pion ground-state:

$$2\,\mathrm{Im}_{p^2}\mathcal{F}_{\rho\mu}^{(B)}(p,q) = 2\pi\delta(p^2 - m_\pi^2)\langle 0|\bar{d}\gamma_\rho\gamma_5 u|\pi^+(p)\rangle\langle\pi^+(p)|\bar{u}\gamma_\mu b|\bar{B}(p_B)\rangle + 2\pi\tilde{\rho}_{\rho\mu}^h(p,q)$$

$$= 2\pi\delta(p^2 - m_\pi^2) i p_\rho f_\pi \left(2 f_{B\pi}^+(0) p_\mu + \dots\right) + 2\pi\tilde{\rho}^h(0, p^2) p_\rho p_\mu + \dots, \tag{10.105}$$

The second term in the first line of the above equation is the hadronic spectral density of all other hadronic states with the quantum numbers of the pion. As usual, we do not specify this part in anticipation of the duality subtraction. In the second line, we combined the definitions of the pion decay constant and of the $B \to \pi$ hadronic matrix element (see

(2.96)). Putting them together, we see that the kinematical structure $p_\rho p_\mu$ is the most convenient one. In the second line, $\tilde{\rho}^h(q^2, p^2)$ is the relevant invariant spectral density of heavier states and dots denote the other Lorentz structures.

Isolating in the correlation function the invariant amplitude at the structure of our choice:

$$\mathcal{F}_{\rho\mu}^{(B)}(p, q) = i\mathcal{F}^B(q^2, p^2)p_\rho p_\mu + \dots ,$$

we write down the dispersion relation in the variable p^2 at $q^2 = 0$ with the imaginary part given by (10.105):

$$\mathcal{F}^B(0, p^2) = \frac{2f_\pi f_{B\pi}^+(0)}{m_\pi^2 - p^2} + \int_{s_{th}}^\infty ds\, \frac{\tilde{\rho}^h(0, s)}{s - p^2} , \tag{10.106}$$

where $s_{th} = (3m_\pi)^2$. It remains to extract the same structure proportional to $p_\rho p_\mu$ from the OPE expression (10.104). After lengthy algebra, involving traces, Lorentz structures and transformations of the integrals, the result is surprisingly simple. It contains only one of the B-meson DAs and has the necessary form of the dispersion integral:

$$\mathcal{F}^B(0, p^2) = \frac{f_B}{m_B} \int_0^\infty \frac{ds}{s - p^2}\, \phi_-^B(s/m_B). \tag{10.107}$$

Equating this result to the dispersion relation (10.106), we apply the duality approximation:

$$\int_{s_{th}}^\infty ds\, \frac{\tilde{\rho}^h(0, s)}{s - p^2} = \frac{f_B}{m_B} \int_{s_0^\pi}^\infty \frac{ds}{s - p^2}\, \phi_-^B(s/m_B), \tag{10.108}$$

were we introduce the same threshold s_0^π as in the LCSR for the pion form factor. The resulting LCSR for the $B \to \pi$ vector form factor obtained after the Borel transformation is:

$$f_{B\pi}^+(0) = \frac{f_B}{f_\pi m_B} \int_0^{s_0^\pi} ds\, e^{-s/M^2} \phi_-^B(s/m_B), \tag{10.109}$$

where we neglect the pion mass for consistency with the chiral limit used to calculate the correlation function.

Many important details concerning the B meson DAs, especially their perturbative renormalization [131], and their applications to the QCD-factorization description of the photoleptonic B decay considered in Chapter 6 (see, e.g., [132]) remain beyond our scope. Concerning the form of the functions $\phi_B^\pm(\omega)$ there are not so many guiding principles as in the case of pion DAs. For applications one uses mainly phenomenologically based models, such as the exponential model

$$\phi_+^B(\omega) = \frac{\omega}{\omega_0^2} e^{-\omega/\omega_0} , \quad \phi_-^B(\omega) = \frac{1}{\omega_0} e^{-\omega/\omega_0} . \tag{10.110}$$

The key parameter of these DAs is the inverse moment

$$\frac{1}{\lambda_B} = \int_0^\infty d\omega\, \frac{\phi_+^B(\omega)}{\omega} \tag{10.111}$$

which in particular determines the factorization formula for the photoleptonic B decay.

With the sum rule (10.109), having an independent calculation of the $B \to \pi$ form factor, e.g., from the LCSR with pion DAs it is possible to constrain this parameter.

Adjusting the quantum numbers of the interpolating currents in the correlation function (10.86), we can vary the final state and the type of the $B \to h$ form factor. In particular, $B \to D$ and $B \to D^*$ form factors are also accessible [133] in the region of large recoil of charmed mesons where the firm predictions of HQET are not applicable. The accuracy of LCSRs considered in this section can be further enhanced by adding the higher-twist quark-antiquark and quark-antiquark-gluon B-meson DAs. The latter correspond to the diagram in Figure 10.4(b). The most complete classification of B-meson DAs applying a certain twistlike pattern can be found in [134]. The NLO perturbative corrections to the LCSR derived in this section are also already available [135].

Collection of useful formulas

LORENTZ AND DIRAC ALGEBRA

The following definitions are used for

- The coordinate four-vector and partial derivative:

$$x^\mu = (x^0, x^1, x^2, x^3) \equiv (x_0, \vec{x}), \quad \partial_\mu \equiv \frac{\partial}{\partial x^\mu} = \left(\frac{\partial}{\partial x^0}, \vec{\nabla} \right); \tag{A.1}$$

- The metric tensor and ε-tensor (Levi-Civita pseudotensor):

$$g_{\mu\nu} = g^{\mu\nu} = diag(1, -1, -1, -1), \quad \varepsilon_{0123} = -\varepsilon^{0123} = 1; \tag{A.2}$$

- The covariant four-vector and scalar product:

$$x_\mu = g_{\mu\nu} x^\nu = (x^0, -\vec{x}), \quad \partial^\mu \equiv \frac{\partial}{\partial x_\mu} = g^{\mu\nu} \partial_\nu = \left(\frac{\partial}{\partial x^0}, -\vec{\nabla} \right),$$
$$x^2 \equiv x \cdot x = g_{\mu\nu} x^\mu x^\nu = x_\mu x^\mu = x_0^2 - \vec{x}^2; \tag{A.3}$$

- Dirac matrices:

$$\gamma^\mu = (\gamma^0, \gamma^1, \gamma^2, \gamma^3) \equiv (\gamma^0, \vec{\gamma}), \quad \{\gamma^\mu, \gamma^\nu\} \equiv \gamma^\mu \gamma^\nu + \gamma^\nu \gamma^\mu = 2g^{\mu\nu}, \tag{A.4}$$

$$\gamma^0 = \begin{pmatrix} I & 0 \\ 0 & -I \end{pmatrix}, \quad \vec{\gamma} = \begin{pmatrix} 0 & \vec{\sigma} \\ -\vec{\sigma} & 0 \end{pmatrix}, \quad (\gamma^0)^2 = 1, \ (\gamma^i)^2 = -1, \tag{A.5}$$

where I and $\vec{\sigma}$ are the 2×2 unit and Pauli matrices

$$\sigma^1 = \begin{pmatrix} 0 & 1 \\ 1 & 0 \end{pmatrix}, \quad \sigma^2 = \begin{pmatrix} 0 & -i \\ i & 0 \end{pmatrix}, \quad \sigma^3 = \begin{pmatrix} 1 & 0 \\ 0 & -1 \end{pmatrix}; \tag{A.6}$$

$$\gamma^5 = i\gamma^0 \gamma^1 \gamma^2 \gamma^3, \quad \gamma_5 = \gamma^5, \quad \{\gamma^\mu, \gamma^5\} = 0, \tag{A.7}$$

$$\gamma^5 = \begin{pmatrix} 0 & I \\ I & 0 \end{pmatrix}, \quad (\gamma^5)^2 = 1; \tag{A.8}$$

$$\sigma^{\mu\nu} = \frac{i}{2}[\gamma^{\mu}, \gamma^{\nu}] = \frac{1}{2}\varepsilon^{\mu\nu\alpha\rho}\gamma^5\gamma_{\alpha}\gamma_{\rho}\,, \quad \sigma^{\mu\nu}\gamma_5 = -\frac{i}{2}\varepsilon^{\mu\nu\alpha\beta}\sigma_{\alpha\beta}\,. \tag{A.9}$$

Useful relations for the ε-tensor include

- The product:

$$\epsilon^{\alpha\beta\gamma\delta}\epsilon^{\mu\nu\rho\sigma} = -det\begin{pmatrix} g^{\alpha\mu} & g^{\alpha\nu} & g^{\alpha\rho} & g^{\alpha\sigma} \\ g^{\beta\mu} & g^{\beta\nu} & g^{\beta\rho} & g^{\beta\sigma} \\ g^{\gamma\mu} & g^{\gamma\nu} & g^{\gamma\rho} & g^{\gamma\sigma} \\ g^{\delta\mu} & g^{\delta\nu} & g^{\delta\rho} & g^{\delta\sigma} \end{pmatrix} \tag{A.10}$$

- Contractions of the product:

$$g_{\alpha\mu}\varepsilon^{\alpha\beta\gamma\delta}\varepsilon^{\mu\nu\rho\sigma} = -det\begin{pmatrix} g^{\beta\nu} & g^{\beta\rho} & g^{\sigma\sigma} \\ g^{\gamma\nu} & g^{\gamma\rho} & g^{\gamma\sigma} \\ g^{\delta\nu} & g^{\delta\rho} & g^{\delta\sigma} \end{pmatrix},$$

$$\varepsilon^{\alpha\beta\gamma\delta}\varepsilon_{\alpha\beta\rho\sigma} = -2(g_{\rho}^{\gamma}g_{\sigma}^{\delta} - g_{\sigma}^{\gamma}g_{\rho}^{\delta})\,,$$

$$\varepsilon^{\alpha\beta\gamma\delta}\varepsilon_{\alpha\beta\gamma\sigma} = -6g_{\sigma}^{\delta}\,, \quad \varepsilon^{\alpha\beta\gamma\delta}\varepsilon_{\alpha\beta\gamma\delta} = -24\,; \tag{A.11}$$

- The Schouten identity:

$$g_{\alpha\beta}\varepsilon_{\mu\nu\rho\lambda} - g_{\alpha\mu}\varepsilon_{\beta\nu\rho\lambda} + g_{\alpha\nu}\varepsilon_{\beta\mu\rho\lambda} - g_{\alpha\rho}\varepsilon_{\beta\mu\nu\lambda} + g_{\alpha\lambda}\varepsilon_{\beta\mu\nu\rho} = 0\,. \tag{A.12}$$

The operations with Dirac matrices are

- Contractions and traces:

$$\gamma_{\mu}\gamma^{\mu} = 4,\quad \not{x} \equiv x_{\mu}\gamma^{\mu}\,, \quad \mathrm{Tr}(\gamma^{\mu}\gamma^{\nu}) = 4g_{\mu\nu}\,,$$

$$\mathrm{Tr}(\gamma^{\mu}\gamma^{\nu}\gamma^{\alpha}\gamma^{\beta}) = 4(g^{\mu\nu}g^{\alpha\beta} - g^{\mu\alpha}g^{\nu\beta} + g^{\mu\beta}g^{\nu\alpha})\,,$$

$$\mathrm{Tr}(\gamma^{\mu}\gamma^{\nu}\gamma^{\rho}\gamma^{\lambda}\gamma^5) = 4i\epsilon^{\mu\nu\rho\lambda}\,; \tag{A.13}$$

- The products:

$$\gamma_{\mu}\gamma_{\nu} = g_{\mu\nu} - i\sigma_{\mu\nu}\,, \tag{A.14}$$

$$\gamma_{\mu}\gamma_{\rho}\gamma_{\nu} = (g_{\mu\rho}g_{\nu\lambda} + g_{\mu\lambda}g_{\rho\nu} - g_{\mu\nu}g_{\rho\lambda})\gamma^{\lambda} - i\varepsilon_{\mu\rho\nu\lambda}\gamma^{\lambda}\gamma_5\,, \tag{A.15}$$

$$\gamma_{\mu}\sigma_{\rho\nu} = i(g_{\mu\rho}g_{\nu\lambda} - g_{\mu\nu}g_{\rho\lambda})\gamma^{\lambda} + \varepsilon_{\mu\rho\nu\lambda}\gamma^{\lambda}\gamma_5\,,$$

$$\sigma_{\rho\nu}\gamma_{\mu} = i(g_{\mu\nu}g_{\rho\lambda} - g_{\rho\mu}g_{\nu\lambda})\gamma^{\lambda} + \varepsilon_{\rho\nu\mu\delta}\gamma^{\delta}\gamma_5\,, \tag{A.16}$$

$$\gamma_{\mu}\gamma_{\nu}\sigma_{\rho\lambda} = (\sigma_{\mu\lambda}g_{\nu\rho} - \sigma_{\mu\rho}g_{\nu\lambda}) + i(g_{\mu\lambda}g_{\nu\rho} - g_{\mu\rho}g_{\nu\lambda})$$

$$- \varepsilon_{\mu\nu\rho\lambda}\gamma_5 - i\varepsilon_{\nu\rho\lambda\alpha}g^{\alpha\beta}\sigma_{\mu\beta}\gamma_5\,. \tag{A.17}$$

- The contractions in $D \neq 4$ dimensions:

$$\gamma_{\mu}\gamma^{\mu} = D\,, \quad \gamma_{\mu}\gamma_{\alpha}\gamma^{\mu} = (2 - D)\gamma_{\alpha}\,, \quad \gamma_{\mu}\gamma^5\gamma^{\mu} = -D\gamma^5\,,$$

$$\gamma_{\mu}\gamma_{\alpha}\gamma_{\beta}\gamma^{\mu} = 4g_{\alpha\beta} + (D - 4)\gamma_{\alpha}\gamma_{\beta}$$

$$\gamma_{\mu}\gamma_{\alpha}\gamma_{\beta}\gamma_{\delta}\gamma^{\mu} = -2\gamma_{\delta}\gamma_{\beta}\gamma_{\alpha} - (D - 4)\gamma_{\alpha}\gamma_{\beta}\gamma_{\delta}\,. \tag{A.18}$$

- The Hermitian conjugation:

$$(\gamma^\mu)^\dagger = \gamma^0\gamma^\mu\gamma^0, \quad (\gamma^0)^\dagger = \gamma^0, \quad (\gamma^i)^\dagger = -\gamma^i, \quad (\gamma^5)^\dagger = \gamma^5; \tag{A.19}$$

- The charge conjugation defined via the matrix $C = \gamma^2\gamma^0$, so that

$$C^\dagger C = 1, \quad C = -C^T, \quad C^2 = 1, \quad C^{-1}\gamma^5 C = (\gamma^5)^T,$$
$$C^{-1}\gamma^\mu C = -(\gamma^\mu)^T, \quad C^{-1}\gamma^\mu\gamma^5 C = (\gamma^\mu\gamma^5)^T, \quad C^{-1}\sigma^{\mu\nu}C = -(\sigma^{\mu\nu})^T, \tag{A.20}$$

where the index T denotes a transposed matrix.

DIRAC FIELDS

Expansion of the free Dirac field $\psi(x)$ and its conjugate $\bar\psi(x) = \psi^\dagger(x)\gamma^0$ in components with definite momenta, $p = (E, \vec{p})$, $E = \sqrt{|\vec{p}|^2 + m^2}$:

$$\psi(x) = \int \frac{d\vec{p}}{(2\pi)^3\sqrt{2E}} \sum_s \left(a(p,s)\mathrm{u}(p,s)e^{-ipx} + b^\dagger(p,s)\mathrm{v}(p,s)e^{ipx} \right),$$

$$\bar\psi(x) = \int \frac{d\vec{p}}{(2\pi)^3\sqrt{2E}} \sum_s \left(a^\dagger(p,s)\bar{\mathrm{u}}(p,s)e^{ipx} + b(p,s)\bar{\mathrm{v}}(p,s)e^{-ipx} \right), \tag{A.21}$$

where the operators $a(p,s)(a^\dagger(p,s))$ and $b(p,s)(b^\dagger(p,s))$ are, respectively, the annihilation (creation) operators of a free spin-1/2 particle and antiparticle with the momentum p, mass m and polarization s (related to the direction of the spin), e.g., the one-particle state is:

$$|p,s\rangle = \sqrt{2E}a^\dagger(p,s)|0\rangle,$$

with the normalization condition

$$\langle p,s|p',s'\rangle = 2E(2\pi)^3\delta^{(3)}(\vec{p} - \vec{p}')\delta_{ss'}. \tag{A.22}$$

In (A.21) the four-component bispinors $\mathrm{u}(p)$ and $\mathrm{v}(p) \equiv \mathrm{u}(-p)$ obey Dirac equations:

$$(\not{p} - m)\mathrm{u}(p) = 0, \quad (\not{p} + m)\mathrm{v}(p) = 0,$$
$$\bar{\mathrm{u}}(p)(\not{p} - m) = 0, \quad \bar{\mathrm{v}}(p)(\not{p} + m) = 0, \tag{A.23}$$

where $\bar{\mathrm{u}}(p) = \mathrm{u}(p)^\dagger\gamma^0$; the polarization and Dirac indices are not shown explicitly. The normalization conditions and other useful relations involving bispinors are:

$$\bar{\mathrm{u}}(p)\mathrm{u}(p) = 2m, \quad \bar{\mathrm{v}}(p)\mathrm{v}(p) = -2m,$$
$$\bar{\mathrm{u}}(p)\gamma_\mu\mathrm{u}(p) = 2p_\mu, \quad \bar{\mathrm{u}}(p)\gamma_\mu\gamma_\nu\mathrm{u}(p) = 2mg_{\mu\nu},$$
$$\bar{\mathrm{u}}(p)\gamma_\mu\gamma_\alpha\gamma_\nu\mathrm{u}(p) = 2\left[p_\mu g_{\alpha\nu} + p_\nu g_{\alpha\mu} - p_\alpha g_{\mu\nu}\right]. \tag{A.24}$$

To square an amplitude involving a product of two bispinors, we need the complex conjugate of that product, which is obtained using a general relation:

$$\left(\bar{\mathrm{u}}(p_1)\Gamma\mathrm{u}(p_2)\right)^* = \bar{\mathrm{u}}(p_2)\gamma^0\Gamma^\dagger\gamma^0\mathrm{u}(p_1), \tag{A.25}$$

where the Hermitian conjugate $\Gamma^\dagger$ is found from (A.19). The sums over polarizations (density matrices of the bispinors) are:

$$\overline{\mathrm{u}(p)\bar{\mathrm{u}}(p)} \equiv \sum_s \mathrm{u}(p,s)\bar{\mathrm{u}}(p,s) = \not{p} + m,$$

$$\overline{\mathrm{v}(p)\bar{\mathrm{v}}(p)} \equiv \sum_s \mathrm{v}(p,s)\bar{\mathrm{v}}(p,s) = \not{p} - m. \tag{A.26}$$

With the above relations, the sum over polarizations in the squared amplitude takes the form:

$$\overline{|(\bar{\mathrm{u}}(p_1)\Gamma\mathrm{u}(p_2))|^2} = \mathrm{Tr}\{(\not{p}_1 + m_1)\Gamma(\not{p}_2 + m_2)\gamma^0\Gamma^\dagger\gamma^0\}. \tag{A.27}$$

COLOR $SU(3)_C$ ALGEBRA

The $N_c^2 - 1 = 8$ generators of the $SU(3)$-group are given by the eight 3×3 matrices:

$$(t^a)_j^i = \frac{(\lambda^a)_j^i}{2}, \quad (a = 1, ..., 8), \ (i, j = 1, 2, 3)....., \tag{A.28}$$

where the Gell-Mann matrices λ^a are defined as:

$$\lambda^1 = \begin{pmatrix} 0 & 1 & 0 \\ 1 & 0 & 0 \\ 0 & 0 & 0 \end{pmatrix}, \quad \lambda^2 = \begin{pmatrix} 0 & -i & 0 \\ i & 0 & 0 \\ 0 & 0 & 0 \end{pmatrix}, \quad \lambda^3 = \begin{pmatrix} 1 & 0 & 0 \\ 0 & -1 & 0 \\ 0 & 0 & 0 \end{pmatrix},$$

$$\lambda^4 = \begin{pmatrix} 0 & 0 & 1 \\ 0 & 0 & 0 \\ 1 & 0 & 0 \end{pmatrix}, \quad \lambda^5 = \begin{pmatrix} 0 & 0 & -i \\ 0 & 0 & 0 \\ i & 0 & 0 \end{pmatrix}, \quad \lambda^6 = \begin{pmatrix} 0 & 0 & 0 \\ 0 & 0 & 1 \\ 0 & 1 & 0 \end{pmatrix},$$

$$\lambda^7 = \begin{pmatrix} 0 & 0 & 0 \\ 0 & 0 & -i \\ 0 & i & 0 \end{pmatrix}, \quad \lambda^8 = \frac{1}{\sqrt{3}} \begin{pmatrix} 1 & 0 & 0 \\ 0 & 1 & 0 \\ 0 & 0 & -2 \end{pmatrix}. \tag{A.29}$$

Removing the third row and third column in the first three λ-matrices we recognize the Pauli matrices (A.6) determining the three generators $t^i = \sigma^i/2$ of one of the three $SU(2)$ subgroups embedded in the $SU(3)$. The generators of the two remaining subgroups can be identified in analogous way.

The frequently used properties of λ-matrices are[1]:

$$(\lambda^a)^\dagger = \lambda^a, \quad \text{Tr}\lambda^a = 0, \quad \text{Tr}\lambda^a\lambda^b = 2\delta^{ab}, \quad (a, b = 1, ...8). \tag{A.30}$$

The totally antisymmetric tensors f^{abc} (the structure constants of the $SU(3)$ group) emerge in the commutation relation:

$$[\lambda^a, \lambda^b] = 2if^{abc}\lambda^c, \tag{A.31}$$

and obey the following relations:

$$f^{abc} = -\frac{i}{4}\text{Tr}([\lambda^a, \lambda^b]\lambda^c), \quad f^{abc}f^{abd} = 3\delta^{cd}. \tag{A.32}$$

The only nonvanishing elements of f^{abc} (up to antisymmetric permutations) are:

$$f^{123} = 1, \quad f^{147} = -f^{156} = \frac{1}{2},$$

$$f^{246} = f^{257} = f^{345} = -f^{367} = \frac{1}{2}, \quad f^{458} = f^{678} = \frac{\sqrt{3}}{2}. \tag{A.33}$$

The anticommutation relations for λ-matrices contain totally symmetric tensors d^{abc}:

$$\{\lambda^a, \lambda^b\} = \frac{4}{3}\delta^{ab} + 2d^{abc}\lambda^c, \tag{A.34}$$

so that

$$d^{abc} = \frac{1}{4}\text{Tr}(\{\lambda^a, \lambda^b\}\lambda^c), \quad d^{abc}d^{abd} = \frac{5}{3}\delta^{cd}. \tag{A.35}$$

---

[1]In the following relations we replace N_c by 3 and, correspondingly, use the numerical values for the group parameters of $SU(3)$ such as $C_F = (N_c^2 - 1)/(2N_c) = 4/3$, $C_A = N_c = 3$ and $T_F = 1/2$.

For completeness, we quote their nonvanishing elements:

$$d^{118} = d^{228} = d^{338} = -d^{888} = \frac{1}{\sqrt{3}},$$

$$d^{146} = d^{157} = -d^{247} = d^{256} = d^{344} = d^{355} = -d^{366} = -d^{377} = \frac{1}{2},$$

$$d^{448} = d^{558} = d^{668} = d^{778} = -\frac{1}{2\sqrt{3}}. \tag{A.36}$$

Using the antisymmetric and symmetric tensors, the products of two λ-matrices can be expressed as their linear combinations:

$$(\lambda^a \lambda^b)^i_j = i f^{abc}(\lambda^c)^i_j + d^{abc}(\lambda^c)^i_j + \frac{2}{3}\delta^{ab}\delta^i_j, \tag{A.37}$$

In addition, the traces of the products of λ matrices are:

$$\mathrm{Tr}\left(\lambda^a \lambda^b \lambda^c\right) = 2if^{abc} + 2d^{abc},$$

$$\mathrm{Tr}\left(\lambda^a \lambda^b \lambda^c \lambda^d\right) = 2(if^{abe} + d^{abe})(if^{ecd} + d^{ecd}) + \frac{4}{3}\delta^{ab}\delta^{cd}. \tag{A.38}$$

Finally, there is an important analog of Fierz identity for the color matrices:

$$(\lambda^a)^i_j(\lambda^a)^k_\ell = 2\delta^i_\ell \delta^k_j - \frac{2}{3}\delta^i_j \delta^k_\ell. \tag{A.39}$$

QCD EQUATIONS OF MOTION AND OTHER RELATIONS

- Equations of motion for the quark field and its Dirac conjugate:

$$(i\gamma^\mu \overrightarrow{D}_\mu - m_q)q(x) = 0, \quad \bar{q}(x)(i\gamma^\mu \overleftarrow{D}_\mu + m_q) = 0, \tag{A.40}$$

where the covariant derivatives are defined as

$$\overrightarrow{D}_\mu = \overrightarrow{\partial}_\mu - ig_s \frac{\lambda^a}{2} A^a_\mu, \quad \overleftarrow{D}_\mu = \overleftarrow{\partial}_\mu + ig_s \frac{\lambda^a}{2} A^a_\mu. \tag{A.41}$$

For brevity, in (A.40) and in other formulae containing quark fields, we omit the indices of Dirac matrices, unless their explicit display is necessary.

- Equation of motion for the gluon field:

$$D^\rho G^a_{\rho\nu} = -g_s \sum_q \bar{q}\gamma_\nu \frac{\lambda^a}{2} q, \tag{A.42}$$

where $D_\rho \equiv \partial_\rho - ig_s \tilde{t}^a A^a_\mu$, and $\tilde{t}^a$ are the Gell-Mann matrices in adjoint representation.

- The dual gluon field-strength tensor:

$$\tilde{G}^a_{\mu\nu} = \frac{1}{2}\varepsilon_{\mu\nu\alpha\beta} G^{a\alpha\beta}. \tag{A.43}$$

- Commutator of covariant derivatives in QCD:

$$[D_\mu, D_\nu] = -ig_s \frac{\lambda^a}{2} G^a_{\mu\nu}(x). \tag{A.44}$$

- Compact matrix notations for the gluon field, including the coupling:

$$A_\mu(x) \equiv g_s \frac{\lambda^a}{2} A_\mu^a(x), \quad G_{\mu\nu}(x) \equiv g_s \frac{\lambda^a}{2} G_{\mu\nu}^a(x). \tag{A.45}$$

- The gauge link (Wilson line):

$$[x, y] = P \exp \left[i \int_0^1 du(x - y)_\mu A^\mu(ux + \bar{u}y) \right] \tag{A.46}$$

where P is the path-ordering operation.

PROPAGATORS AND VERTICES

Note that we do not use the "standard" Feynman rules of QCD in the momentum representation, they can be found in the textbooks. Instead, we derive the expressions for all necessary diagrams[2], starting from vertices and propagators in the configuration space, presented below.

The propagator of the quark with flavor q is defined as

$$S_{(q)i\alpha\beta}^j(x, y) = -i\langle 0|T\{q_\alpha^j(x)\bar{q}_{\beta\,i}(y)\}|0\rangle, \tag{A.47}$$

where Dirac and color indices are shown explicitly.

The free quark propagator shown in Figure 1.1(a) satisfies the equation for a Green function:

$$(i\partial_\mu\gamma^\mu - m_q)S_{(q)i}^{(0)j}(x - y) = \delta_i^j \delta^{(4)}(x - y), \tag{A.48}$$

and has the following momentum representation

$$S_{(q)i}^{(0)j}(x - y) = \delta_i^j \int \frac{d^4p}{(2\pi)^4} e^{-ip(x-y)} S_{(q)}^0(p), \quad S_{(q)}^0(p) = \frac{(\not{p} + m_q)}{p^2 - m_q^2}. \tag{A.49}$$

For the light-quark propagator, the coordinate representation

$$S_{(q)i}^{(0)j}(x - y) = \frac{\delta_i^j}{2\pi^2} \frac{(\not{x} - \not{y})}{[(x - y)^2]^2} \tag{A.50}$$

is valid in the limit of the massless quarks $m_q = 0$. For a massive quark propagator, the (seldom used) coordinate representation reads:

$$S_{(q)i}^{(0)j}(x - y) = \frac{i\delta_i^j}{16\pi^2} \int_0^\infty \frac{d\alpha}{\alpha^2} \left(m_q + i\frac{\not{x} - \not{y}}{2\alpha} \right) \exp\left[-m_q^2\alpha + \frac{(x - y)^2}{4\alpha} \right]. \tag{A.51}$$

The formula for a differentiated quark propagator:

$$\frac{\partial}{\partial k_\lambda} \left[\frac{(\not{p} \pm \not{k}) + m}{(p \pm k)^2 - m^2} \right]\bigg|_{k=0} = \mp \frac{1}{\not{p} - m} \gamma_\lambda \frac{1}{\not{p} - m}$$

$$= \frac{\mp 2p_\lambda(\not{p} + m) \pm \gamma_\lambda(p^2 - m^2)}{(p^2 - m^2)^2}, \tag{A.52}$$

---

[2]Since no diagrams with gluon loops are calculated explicitly in this book, we omit the elements of diagram technique containing auxiliary Fadeev-Popov fields.

TABLE A.1   Interaction vertices in QCD.

Quark-gluon vertex	$g_s \bar{q}_i(x) \gamma_\mu \frac{(\lambda^a)^i_k}{2} q^k(x) A^{a\,\mu}(x)$
3-gluon vertex	$-\frac{g_s}{2} f^{abc} [\partial_\mu A^a_\nu(x) - \partial_\nu A^a_\mu(x)] A^{b\mu}(x) A^{c\nu}(x)$
4-gluon vertex	$-\frac{g_s^2}{4} f^{abc} f_{ade} A^b_\mu(x) A^c_\nu(x) A^{d\mu}(x) A^{e\nu}(x)$

is useful for deriving correlation functions in external fields, which by default have vanishing four-momentum.

The gluon propagator, defined as

$$D^{ab}_{\mu\nu}(x, y) = i\langle 0|T\{A^a_\mu(x) A^b_\nu(y)\}|0\rangle \,, \tag{A.53}$$

for the free gluon field has the following expression (in the Feynman gauge):

$$D^{(0)ab}_{\mu\nu}(x - y) = i\delta^{ab} \frac{g_{\mu\nu}}{4\pi^2(x - y)^2} = \delta^{ab} \int \frac{d^4k}{(2\pi)^4} e^{-ik(x-y)} D^0_{\mu\nu}(k) \,, \tag{A.54}$$

where

$$D^0_{\mu\nu}(k) = \frac{g_{\mu\nu}}{k^2 - i\epsilon} \,.$$

The QCD vertices in Figure 1.1(b–d) are collected in Table A.1

THE FIXED POINT (FOCK-SCHWINGER) GAUGE

This gauge is used in various QCD calculations involving nonperturbative gluon fields, which enter vacuum condensate densities or serve as a part of the quark-antiquark-gluon DAs.

The gauge condition is defined by introducing a fixed point x_0 in the four-dimensional space and demanding that for an arbitrary point x:

$$(x - x_0)^\mu A^a_\mu(x) = 0 \,. \tag{A.55}$$

For simplicity, $x_0 = 0$ is frequently used. Note that in the fixed point gauge translational invariance can be violated in separate diagrams, but has to be restored in the sum over all diagrams.

Expanding the field $A^a_\mu(x)$ near the fixed point $x = x_0$ and multiplying this expansion by the difference of coordinates:

$$(x - x_0)^\mu A^a_\mu(x) = (x - x_0)^\mu \Big[A^a_\mu(x_0)$$
$$+ \sum_{k=1}^\infty \frac{1}{k!} (x - x_0)^{\alpha_1} ... (x - x_0)^{\alpha_k} \partial_{\alpha_1} ... \partial_{\alpha_k} A^a_\mu(x_0) \Big] \,, \tag{A.56}$$

we notice that, due to the condition (A.55), all terms on the r.h.s. of the above equation are enforced to vanish:

$$(x - x_0)^\mu A^a_\mu(x_0) = (x - x_0)^\mu (x - x_0)^{\alpha_1} ... (x - x_0)^{\alpha_k} \partial_{\alpha_1} ... \partial_{\alpha_k} A^a_\mu(x_0) = 0 \,. \tag{A.57}$$

The main advantage of the fixed-point gauge is that the gluon field is directly related to its strength tensor. To derive this relation, we multiply the definition (1.2) by $(x - x_0)^\mu$:

$$(x - x_0)^\mu G_{\mu\nu}^a(x) = (x - x_0)^\mu \partial_\mu A_\nu^a(x) - (x - x_0)^\mu \partial_\nu A_\mu^a(x), \qquad (A.58)$$

where the term $g_s f_{abc}(x - x_0)^\mu A_\mu^b(x) A_\nu^c(x)$ vanishes due to the gauge condition. Furthermore, combining the partial derivative of (A.55):

$$A_\nu^a(x) + (x - x_0)^\mu \partial_\nu A_\mu^a(x) = 0, \qquad (A.59)$$

with (A.58), we obtain

$$A_\nu^a(x) + (x - x_0)^\mu \partial_\mu A_\nu^a(x) = (x - x_0)^\mu G_{\mu\nu}^a(x). \qquad (A.60)$$

Transforming the coordinate in the above equation:

$$x \to \alpha x + \overline{\alpha} x_0, \quad \partial_\mu \to (1/\alpha) \partial_\mu,$$

where $\alpha \in [0, 1]$ and $\overline{\alpha} = 1 - \alpha$, we then integrate both parts of (A.60) over α:

$$\int_0^1 d\alpha \left[A_\nu^a(\alpha x + \overline{\alpha} x_0) + (x - x_0)^\mu \partial_\mu A_\nu^a(\alpha x + \overline{\alpha} x_0) \right]$$

$$= \int_0^1 d\alpha \, \alpha (x - x_0)^\mu G_{\mu\nu}^a(\alpha x + \overline{\alpha} x_0). \qquad (A.61)$$

After that, employing a general relation for total derivative:

$$\alpha \frac{d}{d\alpha} f(\alpha x) = x^\mu \partial_\mu f(\alpha x), \qquad (A.62)$$

we replace the expression in brackets on the l.h.s. of (A.61) by

$$\frac{d}{d\alpha} \left(\alpha A_\nu^a(\alpha x + \overline{\alpha} x_0) \right)$$

and finally obtain:

$$A_\nu^a(x) = \int_0^1 d\alpha \, \alpha (x - x_0)^\mu G_{\mu\nu}(\alpha x + \overline{\alpha} x_0). \qquad (A.63)$$

Choosing $x_0 = 0$, we arrive at the following expansion of the gauge field in terms of the field strength and its covariant derivatives:

$$A_\nu^a(x) = \frac{1}{2} x^\mu G_{\mu\nu}(0) + \frac{1}{3} x^\mu x^\alpha D_\alpha G_{\mu\nu}(0) + \frac{1}{8} x^\mu x^\alpha x^\beta D_\alpha D_\beta G_{\mu\nu}(0) + \ldots, \qquad (A.64)$$

In this expansion another useful property of the fixed point gauge is employed: it is possible to replace simple derivatives by the covariant ones:

$$f(x) = f(0) + x^\alpha \partial_\alpha f(0) + \ldots = f(0) + x^\alpha (\partial_\alpha - i g_s A_\alpha(0)) f(0) + \ldots$$
$$= f(0) + x^\alpha D_\alpha f(0) + \ldots, \qquad (A.65)$$

where (A.57) is used. Note also that in the Fock-Schwinger gauge the gauge link (A.46) is equal to unit.

QCD VACUUM CONDENSATES

Performing the calculation of condensate contributions to a correlation function, one uses the following vacuum-averaging relations for the vacuum gluon and light-quark ($q = u, d, s$) fields:

$$\langle 0|G_{\alpha\beta}^b G_{\lambda\rho}^c|0\rangle = \frac{\delta^{bc}}{96}\left(g_{\alpha\lambda}g_{\beta\rho} - g_{\alpha\rho}g_{\beta\lambda}\right)\langle 0|G_{\mu\nu}^a G^{a\mu\nu}|0\rangle, \tag{A.66}$$

$$\langle 0|\bar{q}_{i\alpha}D_\sigma q_\beta^k G_{\rho\mu}^a|0\rangle = -\frac{g_s}{3^3 2^6}\langle 0|\bar{q}q|0\rangle^2 (\lambda^a)_i^k \left(g_{\sigma\rho}\gamma_\mu - g_{\sigma\mu}\gamma_\rho + i\varepsilon_{\sigma\rho\mu\nu}\gamma^\nu\gamma_5\right)_{\beta\alpha}, \tag{A.67}$$

$$\langle 0|\bar{q}_{i\alpha}q_\beta^k D_\rho G_{\mu\nu}^a|0\rangle = -\frac{g_s}{3^3 2^5}\langle 0|\bar{q}q|0\rangle^2 (\lambda^a)_i^k \left(g_{\rho\nu}\gamma_\mu - g_{\rho\mu}\gamma_\nu\right)_{\beta\alpha}, \tag{A.68}$$

$$\langle 0|\bar{q}_{i\alpha}\bar{q}_{j\beta}q_\gamma^k q_\delta^l|0\rangle = \frac{1}{144}\langle 0|\bar{q}q|0\rangle^2 (\delta_i^l \delta_j^k \delta_{\alpha\delta}\delta_{\beta\gamma} - \delta_i^k \delta_j^l \delta_{\alpha\gamma}\delta_{\beta\delta}). \tag{A.69}$$

Multiplying both sides of the last relation with generic combinations of Dirac and color matrices and summing over indices we obtain a useful formula:

$$\langle 0|(\bar{q}\Gamma_1 q)(\bar{q}\Gamma_2 q)|0\rangle = \frac{1}{144}\{(\mathrm{Tr}\Gamma_1)(\mathrm{Tr}\Gamma_2) - (\mathrm{Tr}\Gamma_1\Gamma_2)\}\langle 0|\bar{q}q|0\rangle^2, \tag{A.70}$$

where the traces are taken over both Dirac and color indices.

A detailed derivation of these formulas can be found, e.g., in [136].

QUARK CURRENTS AND HADRONIC MATRIX ELEMENTS

• Commutators of quark currents are obtained using the equal time anticommutation relations for free quark field operators:

$$\{q_{\alpha i}^\dagger(x), q_\beta^k(x')\}_{x_0=x_0'} = \delta_{\alpha\beta}\delta_i^k \delta(\vec{x} - \vec{x}'),$$
$$\{q_{\alpha i}^\dagger(x), q_{\beta k}^\dagger(x')\}_{x_0=x_0'} = \{q_\alpha^i(x), q_\beta^k(x')\}_{x_0=x_0'} = 0, \tag{A.71}$$

and the identity

$$[AB, CD] = -AC\{D, B\} + A\{C, B\}D - C\{D, A\}B + \{C, A\}DB. \tag{A.72}$$

The resulting equal-time commutation relation between two local (color-neutral) quark currents reads:

$$\left[\bar{q}_1(x)\Gamma_a q_2(x), \bar{q}_2(x')\Gamma_b q_1(x')\right]_{x_0=x_0'}$$
$$= \delta(\vec{x} - \vec{x}')\left(\bar{q}_1(x)\Gamma_a\gamma_0\Gamma_b q_1(x') - \bar{q}_2(x')\Gamma_b\gamma_0\Gamma_a q_2(x)\right)_{x_0=x_0'}, \tag{A.73}$$

where $\Gamma_{a,b}$ are generic combinations of Dirac matrices.

• Translational invariance allows one to obtain a relation between hadronic matrix elements of the same quark current taken at different points. We employ the momentum operator $\hat{P}_\mu$ in QCD. In the following we actually do not need the explicit form of this operator, but only its main property: the hadronic states with definite momenta are eigenstates of $\hat{P}_\mu$, in particular the QCD vacuum state has zero momentum,

$$\hat{P}_\mu|0\rangle = 0.$$

Applying this operator to a certain initial and final state yields

$$\hat{P}_\mu |h(p_1)\rangle = p_{1\mu} |h(p_1)\rangle , \quad \langle h'(p_2)| \hat{P}_\mu = \langle h'(p_2)| p_{2\mu} . \tag{A.74}$$

The operator $\hat{P}_\mu$ serves as a generator of the four-coordinate translations. The symmetry with respect to this transformation is a part of the Lorentz-Poincare symmetry. For the quark field operators the symmetry leads to an operator equation:

$$i\partial_\mu q(x) = [q(x), \hat{P}_\mu] . \tag{A.75}$$

This differential equation is satisfied by the following relation between the field operators taken at different points,

$$q(x) = e^{i\hat{P}\cdot x} q(0) e^{-i\hat{P}\cdot x} . \tag{A.76}$$

Applying the above relation to the quark field and its conjugate in a local composite operator, such as the quark current

$$\hat{O}_a(x) \equiv \bar{q}(x) \Gamma_a q'(x) , \tag{A.77}$$

where Γ_a is a generic Dirac matrix, we reveal the translation property

$$\hat{O}_a(x) = e^{i\hat{P}\cdot x} \hat{O}_a(0) e^{-i\hat{P}\cdot x} , \tag{A.78}$$

and for a hadronic matrix element we obtain

$$\langle f(p_2)| \hat{O}_a(x) |i(p_1)\rangle = \langle f(p_2)| \hat{O}_a(0) |i(p_1)\rangle e^{-i(p_1 - p_2)x} . \tag{A.79}$$

If no coordinate of the operator is shown explicitly, the origin $x = 0$ is assumed by default.

ISOSPIN SYMMETRY

Here we specify the conventions for isospin symmetry multiplets. In the isospin symmetry limit of QCD (1.52) the u and d quarks build the fundamental doublet of the $SU(2)$ group (see (1.51)):

$$\Psi = \begin{pmatrix} \psi^1 \\ \psi^2 \end{pmatrix} \equiv \begin{pmatrix} u \\ d \end{pmatrix} . \tag{A.80}$$

Instead of using the conjugate (covariant) representation for the antiquarks

$$\bar{\Psi} = (\bar{\psi}_1, \bar{\psi}_2) = (\bar{u}, \bar{d}) ,$$

it is more convenient to define the second fundamental doublet using the unit antisymmetric tensor ε^{ab} ($a = 1, 2$) in the $SU(2)$ space[3]:

$$\bar{\psi}^a = \varepsilon^{ab} \bar{\psi}_b ,$$

so that

$$\bar{\Psi} = \begin{pmatrix} \bar{d} \\ -\bar{u} \end{pmatrix} . \tag{A.81}$$

---

[3]Note that this ansatz does not work for the $SU(3)$ group, where the contravariant and covariant fundamental representations are independent of each other.

We rewrite the components of both doublets in terms of the states $|I, I_3\rangle$ with the isospin I and its third component I_3:

$$
\begin{aligned}
|u\rangle &= |1/2, +1/2\rangle, \quad |d\rangle = |1/2, -1/2\rangle, \\
|\bar{d}\rangle &= |1/2, +1/2\rangle, \quad |\bar{u}\rangle = -|1/2, -1/2\rangle.
\end{aligned} \tag{A.82}
$$

Then it is straightforward to use the table of Clebsch-Gordan coefficients (it can be found e.g., in [1]) to combine quark-antiquark states and obtain the isospin assignment of meson states, e.g., for the pion triplet:

$$
|\pi^+\rangle = |u\bar{d}\rangle = |1, +1\rangle, \quad |\pi^0\rangle = \frac{1}{\sqrt{2}}(|u\bar{u}\rangle - |d\bar{d}\rangle) = -|1, 0\rangle, \quad |\pi^-\rangle = |d\bar{u}\rangle = -|1, -1\rangle. \tag{A.83}
$$

The strange mesons form two mutually conjugated doublets, e.g, the lowest pseudoscalar mesons:

$$
\begin{aligned}
|K^+\rangle &= |u\bar{s}\rangle = |1/2, +1/2\rangle, \quad |K^0\rangle = |d\bar{s}\rangle = |1/2, -1/2\rangle, \\
|\bar{K}^0\rangle &= |s\bar{d}\rangle = |1/2, +1/2\rangle, \quad |K^-\rangle = |s\bar{u}\rangle = -|1/2, -1/2\rangle,
\end{aligned} \tag{A.84}
$$

where the states belonging to the same doublet occupy one line in the above equation. The analogous states for c and b flavored hadrons are:

$$
\begin{aligned}
|\bar{D}^0\rangle &= |u\bar{c}\rangle = |1/2, +1/2\rangle, \quad |D^-\rangle = |d\bar{c}\rangle = |1/2, -1/2\rangle, \\
|D^+\rangle &= |c\bar{d}\rangle = |1/2, +1/2\rangle, \quad |D^0\rangle = |c\bar{u}\rangle = -|1/2, -1/2\rangle,
\end{aligned} \tag{A.85}
$$

$$
\begin{aligned}
|B^+\rangle &= |u\bar{b}\rangle = |1/2, +1/2\rangle, \quad |B^0\rangle = |d\bar{b}\rangle = |1/2, -1/2\rangle, \\
|\bar{B}^0\rangle &= |b\bar{d}\rangle = |1/2, +1/2\rangle, \quad |B^-\rangle = |b\bar{u}\rangle = -|1/2, -1/2\rangle.
\end{aligned} \tag{A.86}
$$

Combining these states, and again using the Tables of Clebsch-Gordan coefficients for 1×1, $1/2 \times 1$ and $1/2 \times 1/2$ direct products, it is easy to obtain the isospin composition of two-meson states: dipion, $K\pi$ and $\bar{K}K$ states, respectively. The latter are needed, in particular, to describe isospin relations between the hadronic matrix elements in $B \to PP\ell\nu_\ell$ decays. A more detailed source on the $SU(2)$ and other Lie groups adapted to particle physics is the book [137].

USEFUL INTEGRALS

• Calculating QCD diagrams, one encounters various four-dimensional integrals in the coordinate or momentum spaces. Before performing the integration, it is often useful to merge the product of factors in the denominator originating from separate propagators into one factor. This is conveniently done, parametrizing the inverse product of arbitrary factors via integration:

$$
\begin{aligned}
\frac{1}{a_1 a_2 ... a_n} = (n-1)! \int_0^1 dx_1 \int_0^{x_1} dx_2 ... \int_0^{x_{n-2}} dx_{n-1} \{a_1 x_{n-1} \\
+ a_2(x_{n-2} - x_{n-1}) + ... + a_n(1 - x_1)\}^{-n},
\end{aligned} \tag{A.87}
$$

a well-known method introduced by Feynman. Differentiating both sides $n - 1, k - 1, ...$ times over $a_1, a_2, ...$, we generalize this parametrization to a product $1/(a_1^n a_2^k ...)$, e.g.:

$$
\frac{1}{a^n b^k} = \frac{(n+k-1)!}{(n-1)!(k-1)!} \int_0^1 dx \frac{x^{n-1}(1-x)^{k-1}}{(ax + b(1-x))^{n+k}}, \tag{A.88}
$$

$$\frac{1}{a^n b^k c^m} = \frac{(n+k+m-1)!}{(n-1)!(k-1)!(m-1)!} \int_0^1 dx (1-x)^{k-1}$$
$$\int_0^x dy \frac{y^{n-1}(x-y)^{m-1}}{(ay + b(1-x) + c(x-y))^{n+k+m}} \, . \tag{A.89}$$

• To isolate divergences, the integration in momentum space is transformed to $D \neq 4$ dimensions:

$$\int d\tau_D \equiv \mu^{4-D} \int \frac{d^D l}{(2\pi)^D} \, . \tag{A.90}$$

For the loop diagrams used in this book, the divergent parts are not relevant, they just have to be correctly separated from the "proper" expressions for the diagrams.

The basic scalar integral in D dimensions is

$$\int d\tau_D \frac{(l^2)^r}{(l^2 - R^2)^m} = i \frac{(-1)^{r-m} \mu^{4-D}}{(16\pi^2)^{D/4}} \frac{\Gamma(r + D/2)\Gamma(m - r - D/2)}{\Gamma(D/2)\Gamma(m)(R^2)^{m-r-D/2}} \, , \tag{A.91}$$

• The integration over coordinate space can also be transformed to $D \neq 4$ dimensions. An example of a scalar integral is:

$$\int \frac{d^D z}{(-x^2 - z^2)^a} = -i\pi^2 \frac{\Gamma(a - D/2)}{\Gamma(a)(-x^2)^{(a-D/2)}} \, . \tag{A.92}$$

Note the opposite signs of Euclidean rotations in the coordinate (x) and momentum (p) spaces: $x_0 = -ix_4$ $(t = -i\tau)$ whereas $p_0 = +ip_4$. The tensor integrals with odd number of four-vectors in the numerator vanish due to the symmetry, the ones with an even number can be reduced to scalar integrals, e.g.:

$$\int d^D z \frac{z_\alpha z_\beta}{(-x^2 - z^2)^a} = \frac{g_{\alpha\beta}}{D} \int \frac{d^D z}{(-x^2 - z^2)^a} \, . \tag{A.93}$$

• To calculate a correlation function starting from the coordinate representation of the massless propagators, one needs to Fourier transform the resulting expression to the momentum space. A general integration formula in D dimensions is (see, e.g.,[79]):

$$\int d^D x e^{ipx} \frac{1}{(x^2)^n} = (-i)(-1)^n 2^{(D-2n)} \pi^{D/2} (-p^2)^{n-D/2} \frac{\Gamma(D/2 - n)}{\Gamma(n)} \, , \tag{A.94}$$

for $n \geq 1$, $p^2 < 0$. Various (tensor) integrals are obtained from this formula directly (by differentiating it over the four-vector p_α):

$$\int d^4 x \, e^{ipx} \frac{x_\alpha}{(x^2)^2} = 2\pi^2 \frac{p_\alpha}{p^2} \, , \quad \int d^4 x \, e^{ipx} \frac{1}{x^2} = -\frac{4i\pi^2}{p^2} \, , \quad \int d^4 x \, e^{ipx} \frac{x_\alpha}{x^2} = 8\pi^2 \frac{p_\alpha}{(p^2)^2} \, ,$$

$$\int d^4 x \, e^{ipx} \frac{x_\alpha x_\beta}{x^2} = \frac{-8i\pi^2}{(p^2)^2} \left(g_{\alpha\beta} - 4\frac{p_\alpha p_\beta}{p^2} \right) \, ,$$

$$\int d^4 x \, e^{ipx} \frac{x_\alpha x_\beta}{(x^2)^2} = -\frac{2i\pi^2}{p^2} \left(g_{\alpha\beta} - \frac{2p_\alpha p_\beta}{p^2} \right) \, ,$$

$$\int d^4 x \, e^{ipx} \frac{x_\alpha x_\beta x_\gamma}{(x^2)^2} = \frac{4\pi^2}{(p^2)^2} \left(g_{\alpha\beta} p_\gamma + g_{\alpha\gamma} p_\beta + g_{\beta\gamma} p_\alpha - 4\frac{p_\alpha p_\beta p_\gamma}{p^2} \right) \, ,$$

$$\int d^4 x \, e^{ipx} \frac{1}{(x^2)^2} = i\pi^2 \ln(-p^2) \, , \int d^4 x \, e^{ipx} \frac{x_\alpha}{(x^2)^3} = -\frac{\pi^2}{4} p_\alpha \ln(-p^2) \, ,$$

$$\int d^4 x \, e^{ipx} \frac{x_\alpha x_\beta}{(x^2)^4} = -\frac{i\pi^2}{48} (p^2 g_{\alpha\beta} + 2p_\alpha p_\beta) \ln(-p^2) \, ,$$

$$\int d^4x \; e^{ipx} \ln(-x^2) = \frac{16i\pi^2}{(p^2)^2} \,, \quad \int d^4x \; e^{ipx} x_\alpha \ln(-x^2) = \frac{-64\pi^2 p_\alpha}{(p^2)^3} \,,$$

$$\int d^4x \; e^{ipx} x_\alpha x_\beta \ln(-x^2) = \frac{64i\pi^2}{(p^2)^3} \left(g_{\alpha\beta} - 6\frac{p_\alpha p_\beta}{p^2} \right) \,, \tag{A.95}$$

where the divergent terms proportional to polynomials in p^2 are omitted, because they vanish after Borel transformation or differentiation.

DISPERSION RELATION, BOREL TRANSFORMATION

• We use the following expression for the dispersion relation:

$$f(x) = \frac{1}{\pi} \int\limits_{x_0} dx' \frac{\mathrm{Im}f(x')}{x' - x - i\epsilon} \,, \tag{A.96}$$

where $\mathrm{Im}f(x) = 0$ at $x < x_0$ and the variables x, x' in case of the hadronic amplitudes are invariant momentum squared. The choice of sign at the infinitesimal element $i\epsilon$ corresponds to the general formula (used under integrals):

$$\frac{1}{x \pm i\epsilon} = PV\left(\frac{1}{x}\right) \mp i\pi\delta(x) \,, \tag{A.97}$$

and the principal value of an integral is defined as:

$$PV \int\limits_a^b dx \, g(x) = \lim_{\epsilon \to 0} \left[\int\limits_a^{c-\epsilon} g(x)dx + \int\limits_{c+\epsilon}^b g(x)dx \right] \,, \tag{A.98}$$

which is essential if $g(c)$ is divergent.

• The analytic continuation of the logarithmic function at $x < 0$ is defined as:

$$\ln(-x) = \ln|x| - i\pi\theta(x) \,. \tag{A.99}$$

Correspondingly, the square of logarithmic function develops an imaginary part at $x < 0$ equal to:

$$\mathrm{Im}\left[\ln(-x)\right]^2 = -2\pi \ln|x|\theta(x) \,. \tag{A.100}$$

The relation between the imaginary part and discontinuity reads:

$$\mathrm{Im}f(x) = \frac{1}{2i} \left[f(x + i\epsilon) - f(x - i\epsilon) \right] \,. \tag{A.101}$$

One particularly important example of (A.97) is used at $x = p^2 - m^2$, that is the analytic continuation of the propagator $1/(p^2 - m^2)$ to the on-shell point $p^2 = m^2$, yielding the Cutkosky rule:

$$\mathrm{Im}\frac{1}{p^2 - m^2} = -\pi\delta(p^2 - m^2)\theta(p_0) \,, \tag{A.102}$$

where for an antiparticle there is $\theta(-p_0)$.

• The Borel transformation of a function $f(-q^2) \equiv f(Q^2)$ at $q^2 < 0$ is defined as:

$$\hat{\mathcal{B}}_{M^2} f(Q^2) = \lim_{\{Q^2,n\} \to \infty, \, Q^2/n = M^2} \frac{(Q^2)^{n+1}}{n!} \left(-\frac{d}{dQ^2} \right)^n f(Q^2) \equiv f(M^2) \,, \tag{A.103}$$

so that any polynomial of Q^2 vanishes after this transformation:

$$\hat{\mathcal{B}}_{M^2}(Q^2)^k = 0, \ k \geq 0 \,. \tag{A.104}$$

The Borel transformations (A.103) of the frequently used functions are:

$$\hat{\mathcal{B}}_{M^2} \frac{1}{(m^2 + Q^2)^k} = \frac{1}{(k-1)!} \left(\frac{1}{M^2}\right)^{k-1} e^{-m^2/M^2} \,, \tag{A.105}$$

$$\hat{\mathcal{B}}_{M^2}(Q^2)^k \ln Q^2 = (-1)^{k+1}\Gamma(k+1)(M^2)^{k+1} \,, \tag{A.106}$$
$$\hat{\mathcal{B}}_{M^2} \exp(-aQ^2) = M^2\delta(1 - aM^2) \,. \tag{A.107}$$

Decay amplitudes and widths

S-MATRIX, UNITARITY AND PROBABILITY

The probability amplitude of a certain process $i \to f$ (decay or scattering), where $|i\rangle$ and $\langle f|$ are, respectively, the initial and final states, is obtained from the S-matrix element:

$$\langle f|\hat{S}|i\rangle \equiv S_{fi}. \tag{B.1}$$

In SM, the matrix $\hat{S}$ is an operator represented as a time-ordered exponent of the (QCD + electroweak) interaction Lagrangian integrated over four-coordinates:

$$\hat{S} = T\Big\{ \exp\Big[i \int d^4x \Big(\mathcal{L}_{QCD}^{int}(x) + \mathcal{L}_{EW}^{int}(x) \Big) \Big] \Big\}. \tag{B.2}$$

In this book we consider e.m. processes with hadrons, for which only the QED part (1.24) of the electroweak Lagrangian contributes:

$$\hat{S} = T\Big\{ \exp\Big[i \int d^4x \Big(\mathcal{L}_{QCD}^{int}(x) + \mathcal{L}_{QED}^{int}(x) \Big) \Big] \Big\}. \tag{B.3}$$

For the weak interaction processes involving hadrons and described by the effective Hamiltonian (1.29), the S-matrix takes the form:

$$\hat{S}_W = T\Big\{ \exp\Big[i \int d^4x \Big(\mathcal{L}_{QCD}^{int}(x) - \mathcal{H}_W(x) \Big) \Big] \Big\}. \tag{B.4}$$

S-matrix obeys the unitarity condition:

$$\hat{S}\hat{S}^\dagger = \hat{S}^\dagger\hat{S} = \mathbb{1}, \tag{B.5}$$

and the hermitian conjugate matrix elements are:

$$\langle f|\hat{S}^\dagger|i\rangle = \langle i|\hat{S}|f\rangle^* = S_{if}^*. \tag{B.6}$$

The $|i\rangle$ and $\langle f|$ belong to the full set of possible physical states. In QCD these are hadronic and vacuum states. They obey the orthonormality condition:

$$\langle n|n'\rangle = \langle n|\mathbb{1}|n'\rangle = \delta_{nn'}. \tag{B.7}$$

The unit operator has the following representation (the completeness condition):

$$\sum_n |n\rangle\langle n| = \mathbb{1}, \tag{B.8}$$

where the sum implies integration (summation) over the momenta (polarizations) of all particles entering each state $|n\rangle$.

The S-matrix is usually decomposed into the unit matrix (no interaction) and the scattering matrix $\hat{T}$:

$$S_{fi} = \delta_{fi} + i(2\pi)^4\delta^{(4)}(P_f - P_i)T_{fi}, \tag{B.9}$$

where $T_{fi} \equiv \langle f|\hat{T}|i\rangle$; $P_i = P_f$ are the momenta of initial and final states and δ-function reflects their conservation. Substituting (B.9) in the first unitarity condition in (B.5), we obtain for the T-matrix elements:

$$T_{fi} - T_{fi}^\dagger = i(2\pi)^4 \sum_n \delta^{(4)}(P_f - P_n)T_{fn}T_{ni}^\dagger. \tag{B.10}$$

Replacing the hermitian conjugates: $T_{fi}^\dagger = T_{if}^*$, $T_{ni}^\dagger = T_{in}^*$ and employing the time-reversal symmetry[1]: $T_{if}^* = T_{fi}^*$, $T_{in}^* = T_{ni}^*$, we obtain the unitarity relation for T-matrix:

$$2\,\mathrm{Im}T_{fi} = (2\pi)^4 \sum_n \delta^{(4)}(P_f - P_n)T_{fn}T_{ni}^* = (2\pi)^4 \sum_n \delta^{(4)}(P_f - P_n)T_{fn}^*T_{ni}, \tag{B.11}$$

where the second equation follows from the second unitarity condition in (B.5), or simply from the fact that $\mathrm{Im}T_{fi}$ is a real quantity. The latter property leads to a nontrivial equality of the complex phases of T_{fn} and T_{ni} (Watson theorem), which is valid when there is only a single intermediate state $|n\rangle\langle n|$ in the unitarity relation. In the special case of the forward scattering when the states $\langle f|$ and $|i\rangle$ coincide, the relation (B.11) simplifies to:

$$2\,\mathrm{Im}T_{ii} = (2\pi)^4 \sum_n \delta^{(4)}(P_i - P_n)|T_{in}|^2, \tag{B.12}$$

known as the optical theorem.

The probability of the process $i \to f$ is given by the modulus square of the S-matrix element. In the presence of interaction:

$$|S_{fi}|^2 = (2\pi)^4\delta^{(4)}(P_f - P_i)|T_{fi}|^2 Vt, \tag{B.13}$$

where the product of the normalization volume V and time interval t is understood as a result of the squaring of δ-function:

$$[\delta^{(4)}(P_f - P_i)]^2 = \delta^{(4)}(P_f - P_i)\frac{1}{(2\pi)^4}\int d^4x e^{i(P_f - P_i)x}$$

$$= \delta^{(4)}(P_f - P_i)\frac{1}{(2\pi)^4}\int d^4x = \delta^{(4)}(P_f - P_i)\frac{Vt}{(2\pi)^4}. \tag{B.14}$$

Most of the processes considered in this book are decays, hence we present a detailed derivation of the decay probability, considering a certain $i \to f$ decay process, where the

---

[1]This symmetry is valid in QCD and QED and for the weak interactions it is correlated with the CP-symmetry. In the processes considered in this book CP is conserved and, therefore, T-symmetry is conserved too.

initial state contains one particle or a single hadron with momentum $P_i = (E_i, \vec{P})$ and the final state consists of N particles and/or hadrons with momenta

$$p_k = (E_k, \vec{p}_k), \quad P_f = \sum_{k=1}^{N} p_k \,. \tag{B.15}$$

The full differential decay probability per unit of time denoted as dw_{fi} is obtained dividing (B.13) by t and multiplying by the number of states per element of the momentum space of all n final particles in the volume V:

$$dw_{fi} = (2\pi)^4 \delta^{(4)}(P_f - P_i \overline{|T_{fi}|}^2 V \prod_{k=1}^{N} \frac{d\vec{p}_i}{(2\pi)^3}(V)^N \tag{B.16}$$

The line over the squared decay amplitude indicates taking average (sum) over polarizations of the initial particle (final particles). Furthermore, each particle field, e.g., quark field, as well as any hadronic state entering the scattering amplitude T_{fi} is normalized with a factor $1/\sqrt{2EV}$ where E is the energy of this particle or state whereas the normalization volume is always implied but usually not shown explicitly (i.e., put to a unity). It is customary to transfer these factors to the phase space, redefining the T-matrix element:

$$T_{fi} = \frac{\mathcal{A}(i \to f)}{[(2E_iV)(2E_1V)....(2E_NV)]^{1/2}} \tag{B.17}$$

in terms of the *decay amplitude* $\mathcal{A}(i \to f)$. After substituting this definition in (B.16) all volume factors cancel out and we obtain the standard expression for the differential decay width which is equivalent to the differential probability per unit of time, $d\Gamma(i \to f) \equiv dw_{fi}$:

$$d\Gamma(i \to f) = \frac{\overline{|\mathcal{A}(i \to f)|}^2}{2E_i} d\tau_N \,, \tag{B.18}$$

where

$$d\tau_N = \prod_{k=1}^{N} \left(\frac{d\vec{p}_k}{(2\pi)^3 2E_k}\right) (2\pi)^4 \delta^{(4)}\left(P_i - \sum_{k=1}^{N} p_k\right) \tag{B.19}$$

is the (relativistically invariant) element of the N-particle phase space. Correspondingly, we adopt a simple convention, relating the S-matrix element at $f \neq i$ directly to the decay amplitude:

$$S_{fi} = i(2\pi)^4 \delta^{(4)}(P_f - P_i)\mathcal{A}(i \to f) \,, \tag{B.20}$$

which implies that we are using the phase space element (B.19).

UNITARITY RELATION FOR THE CORRELATION FUNCTIONS

Here we present a derivation of the unitarity relation for the correlation functions of two currents. We first consider a vacuum-to-hadron correlation function, similar to the ones used for light-cone sum rules in Chapter 10:

$$\Pi_{AB}(p, q) = i \int d^4x \, e^{iqx} \langle h(p)|T\{j_A(x)j_B(0)\}0\rangle \,, \tag{B.21}$$

where the indices A, B distinguish two different current operators j_A, j_B, whose correlation generates transition of the vacuum state to the hadron h.

To proceed, we introduce a fictitious local interaction

$$g\Big(j_A(x)\lambda_A(x) + j_B(x)\lambda_B(x)\Big) \tag{B.22}$$

of the quark currents with the two colorless external fields $\lambda_{A,B}(x)$ not related to QCD. We assume that both couplings are equal and small, $g \ll 1$. In the case when $j_{A(B)}$ is the quark e.m. or weak current, the role of the field $\lambda_{A(B)}$ can be played, respectively, by the photon or W-boson. In the other cases when $j_{A(B)}$ is a generic interpolating or transition current, $\lambda_{A(B)}$ is understood as a hypothetical external field (or source) coupled to that current.

To describe the processes involving the interactions given in (B.22), we should use the corresponding part of the S-matrix:

$$\hat{S} = T\Big\{ \exp\Big[ig \int d^4x \Big(j_A(x)\lambda_A(x) + j_B(x)\lambda_B(x)\Big)\Big]\Big\}, \tag{B.23}$$

expanding it in the perturbative series in g:

$$\hat{S} = \mathbb{1} + ig \int d^4x\, j_A(x)\lambda_A(x) + ig \int d^4y\, j_B(y)\lambda_B(y)$$

$$+ i^2g^2 \int d^4x \int d^4y\, T\Big\{ j_A(x)\lambda_A(x) j_B(y)\lambda_B(y)\Big\} + \dots, \tag{B.24}$$

where the QCD part is implied but not shown.

Let us now consider a process mediated by the currents $j_{A,B}$, when the initial λ_B-particle with momentum p' transits to the final state containing the hadron h and λ_A-particle with momenta p and q, respectively, i.e.:

$$|i\rangle = |\lambda_B(p')\rangle, \quad \langle f| = \langle \lambda_A(q)h(p)|. \tag{B.25}$$

To realize this transition, one needs a superposition of both interactions in (B.22), that is, the $O(g^2)$ term in the S-matrix expansion (B.24), so that the transition matrix element is:

$$S_{fi}^{(2)} = i^2g^2 \int d^4x \int d^4y \langle \lambda_A(q)h(p)|T\Big\{ j_A(x)\lambda_A(x) j_B(y)\lambda_B(y)\Big\}|\lambda_B(p')\rangle. \tag{B.26}$$

To this order of g the states with $\lambda_{A,B}$ can be factorized from the QCD states:

$$|\lambda_B(p')\rangle = |\lambda_B(p')\rangle \otimes |0\rangle, \quad \langle \lambda_A(q)h(p)| = \langle \lambda_A(q)| \otimes \langle h(p)|, \tag{B.27}$$

bringing the matrix element in (B.26) to the form:

$$\langle \lambda_A(q)h(p)|T\Big\{ j_A(x)\lambda_A(x) j_B(y)\lambda_B(y)\Big\}|\lambda_B(p')\rangle$$

$$= \langle \lambda_A(q)|\lambda_A(x)\Big(\langle h(p)|T\Big\{ j_A(x) j_B(y)\Big\}|0\rangle \Big)\lambda_B(y)|\lambda_B(p')\rangle. \tag{B.28}$$

After that we expand the quantum fields $\lambda_{A,B}(x)$ in momentum components multiplied by the creation and annihilation operators, leaving only those components which match the initial and final states:

$$\langle\lambda_A(q)|\lambda_A(x)\left(\ldots\right)\lambda_B(y)|\lambda_B(p')\rangle = \langle\lambda_A(q)|\left(\sum_k \epsilon_A(k)a_A^\dagger(k)e^{ikx} + \ldots\right)$$

$$\times\left(\ldots\right)\left(\ldots + \sum_{k'}\epsilon_B(k')a_B(k')e^{-ik'y}\right)|\lambda_B(p')\rangle$$

$$= \sum_{k,k'}\epsilon_A(k)\epsilon_B(k')e^{ikx-k'y}\langle\lambda_A(q)|a_A^\dagger(k)\left(\ldots\right)a_B(k')|\lambda_B(p')\rangle$$

$$= \sum_{k,k'}\epsilon_A(k)\epsilon_B(k')e^{ikx-k'y}\delta_{kq}\delta_{k'p'}\left(\ldots\right) = \epsilon_A(q)\epsilon_B(p')e^{iqx-p'y}\left(\ldots\right), \qquad (B.29)$$

where $a_A^\dagger(k)$ and $a_B(k')$ are, respectively the creation and annihilation operators in the k, k' momentum expansion of the $\lambda_{A,B}$ fields and $\epsilon_{A,B}$ are the corresponding polarization vectors which are chosen to be real quantities. Substituting (B.29) in (B.26), we obtain

$$S_{fi}^{(2)} = i^2 g^2 \epsilon_A(q)\epsilon_B(p')\left[\int d^4x\, e^{iqx}\int d^4y\, e^{-ip'y}\langle h(p)|T\{j_A(x)j_B(y)\}|0\rangle\right]. \qquad (B.30)$$

The expression in brackets can be transformed further, applying translations of the currents according to (A.78) and a transformation of the integration variable:

$$\left[\ldots\right] = \int d^4x e^{iqx}\int d^4y e^{-ip'y}\langle h(p)|\underbrace{e^{ipy}e^{-i\hat{P}y}}_{=1}\{T\{j_A(x)e^{i\hat{P}y}j_B(0)e^{-i\hat{P}y}\}|0\rangle$$

$$= \int d^4y e^{-ip'y+ipy+iqy}\int d^4x e^{iqx-iqy}\langle h(p)|T\{e^{-i\hat{P}y}j_A(x)e^{i\hat{P}y}j_B(0)\}|0\rangle$$

$$= \int d^4y e^{-i(p'-p-q)y}\int d^4x e^{iq(x-y)}\langle h(p)|T\{j_A(x-y)j_B(0)\}|0\rangle$$

$$= (2\pi)^4\delta^{(4)}(p'-p-q)\int d^4x\, e^{iqx}\langle h(p)|T\{j_A(x)j_B(0)\}|0\rangle \qquad (B.31)$$

so that finally:

$$S_{fi}^{(2)} = i(2\pi)^4\delta^{(4)}(p'-p-q)g^2\epsilon_A(q)\epsilon_B(p')\,i\int d^4x e^{iqx}\langle h(p)|T\{j_A(x)j_B(0)\}|0\rangle. \qquad (B.32)$$

The corresponding T-matrix element (see (B.9), where the unity part does not contribute in the presence of interaction) is proportional to the correlation function (B.21):

$$T_{fi}^{(2)} = g^2\epsilon_A(q)\epsilon_B(p+q)\Pi_{AB}(q). \qquad (B.33)$$

The unitarity relation (B.11) for the above scattering amplitude reads:

$$2\,\mathrm{Im}T_{fi}^{(2)} = (2\pi)^4\sum_n\delta^{(4)}(P_f - P_n)T_{fn}^{(1)}T_{ni}^{(1)*}. \qquad (B.34)$$

The most important point is that, as indicated by the indices, in order to match the powers of coupling g on both sides of this relation, we include on the r.h.s. only the matrix elements of the first order in g transitions between initial, intermediate and final states. The set of intermediate hadronic states $|n\rangle$ includes all possible states which simultaneously match the transitions $|0\rangle \to \langle n|$ via j_B current and $|n\rangle \to \langle h|$ via j_A current. Starting again from the S-matrix element, this time using the $O(g)$ terms in the expansion (B.24), and repeating the steps similar to the ones in the above derivation of (B.33), we obtain

$$
\begin{aligned}
S_{ni}^{(1)} &= ig \int d^4 y \langle n|j_B(y)\lambda_B(y)|\lambda_B(p')\rangle = ig\epsilon_B(p') \int d^4 y\, e^{-ip'y} \langle n|j_B(y)|0\rangle \\
&= ig\epsilon_B(p') \int d^4 y\, e^{-ip'y} \langle n|e^{i\hat{P}y} j_B(0) e^{-i\hat{P}y}|0\rangle \\
&= ig\epsilon_B(p') \int d^4 y\, e^{-ip'y+ip_n y} \langle n|j_B|0\rangle \\
&= i(2\pi)^4 \delta^{(4)}(p'-p_n) g\epsilon_B(p') \langle n|j_B|0\rangle
\end{aligned}
$$

(B.35)

with $p' = p + q$, and

$$
T_{ni}^{(1)} = g\epsilon_B(p+q)\langle n|j_B|0\rangle.
$$

(B.36)

Analogously,

$$
T_{fn}^{(1)} = g\epsilon_A(q)\langle h(p)|j_A|n\rangle.
$$

(B.37)

Taking the imaginary part of (B.33) and using (B.36) and (B.37) we obtain from (B.34):

$$
\begin{aligned}
2g^2 \epsilon_A(q)\epsilon_B(p+q)\mathrm{Im}\Pi_{AB}(q) &= g^2 \epsilon_A(q)\epsilon_B(p+q) \\
&\times (2\pi)^4 \sum_n \delta^{(4)}(P_f - P_n)\langle h(p)|j_A|n\rangle\langle n|j_B|0\rangle^*,
\end{aligned}
$$

(B.38)

and, after cancelling the equal factors on both sides, we finally obtain the unitarity relation for the correlation function:

$$
2\,\mathrm{Im}\Pi_{AB}(q) = \sum_n \!\!\!\!\!\int d\tau_{N(n)} \langle h(p)|j_A\overline{|n\rangle\langle n|} j_B|0\rangle^*.
$$

(B.39)

Here we denoted by overline the sum over polarizations and made explicit the integration over the phase space of each intermediate state n in the sum, assuming it has $N(n)$ particles and using the phase space integration element $d\tau_{N(n)}$ as defined in (B.19).

Finally, we notice that the a similar derivation is valid also for a vacuum-to-vacuum correlation function (8.1) used in Chapter 8 for QCD sum rules. One simply has to replace in (B.21) the state $\langle h(p)|$ by $\langle 0|$ and put $p \to 0$. Repeating step by step the procedure described above, we arrive at the unitarity relation (8.5).

KINEMATICS OF DECAY WIDTHS

Here we present useful formulae for decay widths, starting from the simplest two-body decay:

$$
M(P) \to m_1(p_1) + m(p_2),
$$

(B.40)

where particles are denoted by their masses, with momenta in parentheses, so that

$$
P = p_1 + p_2.
$$

In two-body decays all kinematical invariants are fixed by the particle masses:

$$
P^2 = M^2, \quad p_{1,2}^2 = m_{1,2}^2, \quad p_1 \cdot p_2 = \frac{1}{2}(M^2 - m_1^2 - m_2^2).
$$

(B.41)

Hence, the amplitude squared is a constant and can be taken out of the phase-space integral, so that the general formula (B.18) in case of the two-body decay can be written as

$$\Gamma_2 = \frac{\overline{|\mathcal{A}|^2}}{2M} \int d\tau_2 \,. \tag{B.42}$$

We choose the rest frame of the decaying particle, in which

$$P = (M, \vec{0}) \,, \quad \vec{p}_1 = -\vec{p}_2 \,, \quad |\vec{p}_1| = |\vec{p}_2| \equiv p \,,$$

and the energy conservation yields:

$$E_1 + E_2 = \sqrt{p^2 + m_1^2} + \sqrt{p^2 + m_2^2} = M.$$

Solving the last equation, we obtain the 3-momentum of decay products:

$$p = \frac{\lambda^{1/2}(M^2, m_1^2, m_2^2)}{2M} \,, \tag{B.43}$$

where the standard Källen function is defined:

$$\lambda(a, b, c) = a^2 + b^2 + c^2 - 2ab - 2ac - 2bc \,. \tag{B.44}$$

In the phase space factor written according to (B.19), the integral over $\vec{p}_2$ removes the three-dimensional δ-function:

$$\int d\tau_2 = \frac{1}{(2\pi)^2} \int \frac{d\vec{p}_1}{2E_1} \frac{d\vec{p}_2}{2E_2} \delta^{(4)} (P - p_1 - p_2) = \frac{1}{16\pi^2} \int \frac{d\vec{p}_1}{E_1 E_2} \delta(M - E_1 - E_2) \,. \tag{B.45}$$

The integration over $\vec{p}_1$ is performed in spherical coordinates:

$$\int d\tau_2 = \frac{1}{16\pi^2} \int_0^\infty p^2 dp \int d\Omega \frac{\delta\left(M - \sqrt{p^2 + m_1^2} - \sqrt{p^2 + m_2^2}\right)}{\sqrt{p^2 + m_1^2}\sqrt{p^2 + m_2^2}} = \frac{\lambda^{1/2}(M^2, m_1^2, m_2^2)}{8\pi M^2} \,, \tag{B.46}$$

where $d\Omega = d(\cos\theta)d\phi$ and for the integral over p we used the property of δ-function:

$$\delta\big(f(x) - f(a)\big) = \left(\left|\frac{df}{dx}\right|\right)^{-1} \delta(x - a).$$

Finally, we obtain for the width:

$$\Gamma_2 = \overline{|\mathcal{A}|^2} \frac{\lambda^{1/2}(M^2, m_1^2, m_2^2)}{16\pi M^3} \,. \tag{B.47}$$

Turning to the three-body decay

$$M(P) \to m_1(p_1) + m_2(p_2) + m_3(p_3) \,, \tag{B.48}$$

where $P = p_1 + p_2 + p_3$, we encounter a richer kinematics with two independent variables[2] formed by the invariant masses of the particle pairs, e.g., $(p_1+p_2)^2$ and $(p_2+p_3)^2$. A standard

---

[2]They are analogous to the variables s, t of the $2 \to 2$ scattering which is actually a crossing channel of the three-body decay.

way to describe the kinematics of three-body decay is to introduce the two-dimensional Dalitz plot of these variables (see, e.g., the minireview on particle kinematics in [1]).

Here we use a slightly different approach which directly leads to the differential width over one of the invariant masses. We denote

$$p_2 + p_3 \equiv q, \quad p_2 - p_3 \equiv r,$$

so that $P = p_1 + q$, and choose q^2 and $(r \cdot p_1)$ to be the independent variables on which the three-body decay amplitude, in general, depends. Calculating the differential width:

$$d\Gamma_3 = \frac{1}{2M} \overline{|\mathcal{A}|^2} \, d\tau_3, \tag{B.49}$$

it is convenient to single out from $d\tau_3$ the two-body phase-space element $d\tau_2$ of the particles m_2 and m_3. To this end, we start from the definition of the phase space element (B.19) for $n = 3$:

$$d\tau_3 = \frac{d\vec{p}_1}{(2\pi)^3 2E_1} \frac{d\vec{p}_2}{(2\pi)^3 2E_2} \frac{d\vec{p}_3}{(2\pi)^3 2E_3} (2\pi)^4 \delta^{(4)} (P - p_1 - q), \tag{B.50}$$

and use the following identity:

$$1 = \int dq^2 \delta(q^2 - (p_2 + p_3)^2) \int d\vec{q}\, \delta^{(3)}(\vec{q} - \vec{p}_2 - \vec{p}_3)$$

$$= \int dq^2 \int \frac{d\vec{q}}{2q_0} \delta^{(4)}(q - p_2 - p_3). \tag{B.51}$$

In the above, in order to obtain the second equation, the integral over q^2 was taken in the c.m. frame of the particles m_2 and m_3, so that

$$\vec{p}_2 + \vec{p}_3 = 0, \quad q_0 = \sqrt{q^2} = E_2 + E_3, \tag{B.52}$$

and

$$\delta(q^2 - (p_2 + p_3)^2) = \frac{1}{2q_0} \delta(q_0 - (E_2 + E_3)).$$

We are then able to rearrange (B.50) as

$$d\tau_3 = \frac{1}{2\pi} dq^2 \frac{d\vec{p}_1}{(2\pi)^3 2E_1} \frac{d\vec{q}}{(2\pi)^3 2q_0} (2\pi)^4 \delta^{(4)} (P - q - p_1) d\tau_2 = \frac{1}{2\pi} dq^2 d\tau_2' d\tau_2, \tag{B.53}$$

where the phase space element $d\tau_2'$ corresponds to the two-body final state consisting of the particle m_1 and (composite) particle with the mass $\sqrt{q^2}$. Factorization of the phase space element in (B.53) is convenient when the decay amplitude squared has a form

$$\overline{|\mathcal{A}|^2} = \sum_n f_n(q^2)(r \cdot p_1)^n, \tag{B.54}$$

with a polynomial dependence on the variable $(r \cdot p_1)$. Then it is possible to integrate over $d\tau_2'$:

$$\int d\tau_2' = \frac{\lambda^{1/2}(M^2, m_1^2, q^2)}{8\pi M^2}, \tag{B.55}$$

and reduce the differential width to a form:

$$\frac{d\Gamma_3}{dq^2} = \frac{\lambda^{1/2}(M^2, m_1^2, q^2) \lambda^{1/2}(q^2, m_2^2, m_3^2)}{2^8 \pi^3 M^3 q^2} \sum_n f_n(q^2) I_n(M^2, m_{1,2,3}^2, q^2) \tag{B.56}$$

with the universal kinematical coefficients

$$I_n(M^2, m_{1,2,3}^2, q^2) \equiv \frac{\int (rp_1)^n d\tau_2}{\int d\tau_2}, \tag{B.57}$$

where

$$\int d\tau_2 = \frac{\lambda^{1/2}(q^2, m_2^2, m_3^2)}{8\pi q^2}. \tag{B.58}$$

The total width of the three-body decay is obtained integrating (B.56) over the kinematically allowed region of q^2:

$$\Gamma_3 = \int_{(m_2+m_3)^2}^{(M-m_1)^2} dq^2 \frac{d\Gamma_3}{dq^2}. \tag{B.59}$$

The coefficients I_n are conveniently calculated in the c.m. frame (B.52) of the particles m_2 and m_3 in which $\vec{q} = 0$, and we choose the direction of $\vec{P} = \vec{p_1}$ along the z-axis. The energy-momentum conservation yields:

$$|\vec{p_1}| = \frac{\lambda^{1/2}(M^2, m_1^2, q^2)}{2\sqrt{q^2}}, \quad E_1 = \frac{M^2 - m_1^2 - q^2}{2\sqrt{q^2}},$$

$$|\vec{p_2}| = |\vec{p_3}| = \frac{\lambda^{1/2}(q^2, m_2^2, m_3^2)}{2\sqrt{q^2}}, \quad E_2 = \frac{q^2 + m_2^2 - m_3^2}{2\sqrt{q^2}}, \quad E_3 = \sqrt{q^2} - E_2.$$

In the chosen frame, we can express the invariant variable (rp_1) in terms of the (polar) angle θ between $\vec{p_2}$ and $\vec{p_1}$:

$$\begin{aligned}(rp_1) &= (E_2 - E_3)E_1 - (\vec{p_2} - \vec{p_3}) \cdot \vec{p_1} = \frac{1}{2q^2}\Big[(m_2^2 - m_3^2)(M^2 - m_1^2 - q^2) \\ &- \lambda^{1/2}(q^2, m_2^2, m_3^2)\lambda^{1/2}(M^2, m_1^2, q^2)\cos\theta\Big].\end{aligned} \tag{B.60}$$

After that we insert in (B.57) the function $(rp_1)^n$ transforming the phase-space integral similar to the one in (B.45). The next steps are to integrate over $p \equiv |\vec{p_2}|$ with δ-function and over the azimuthal angle ϕ, retaining the integration over the polar angle θ and using (B.58):

$$\begin{aligned}I_n &= \left(\frac{1}{\int d\tau_2}\right) \int_0^\infty p^2 dp \int_{-1}^1 d\cos\theta \int_0^{2\pi} d\phi \frac{\delta\left(\sqrt{q^2} - \sqrt{p^2 + m_2^2} - \sqrt{p^2 + m_3^2}\right)}{16\pi^2 \sqrt{p^2 + m_2^2}\sqrt{p^2 + m_3^2}}(rp_1)^n \\ &= \frac{1}{2(2q^2)^n} \int_{-1}^1 d\cos\theta \Big[(m_2^2 - m_3^2)(M^2 - m_1^2 - q^2) \\ &- \lambda^{1/2}(q^2, m_2^2, m_3^2)\lambda^{1/2}(M^2, m_1^2, q^2)\cos\theta\Big]^n.\end{aligned} \tag{B.61}$$

The remaining integration, to which only the even powers of $\cos\theta$ contribute, yields e.g., for $n = 1, 2$:

$$I_1 = \frac{1}{2q^2}(m_2^2 - m_3^2)(M^2 - m_1^2 - q^2), \tag{B.62}$$

$$\begin{aligned}I_2 &= \frac{1}{12(q^2)^2}\Big[3(m_2^2 - m_3^2)^2(M^2 - m_1^2 - q^2)^2 \\ &+ \lambda(M^2, m_1^2, q^2)\lambda(m_2^2, m_3^2, q^2)\Big].\end{aligned} \tag{B.63}$$

Parameterization and z-expansion of form factors

Consider a generic form factor $F(q^2)$ determining the hadronic matrix element of a certain $H \to h$ transition generated by the current j_a:

$$\langle h(p_2)|j_a|H(p_1)\rangle = L_a(q, p_2)F(q^2) + ... , \qquad (\mathrm{C.1})$$

where $q = p_1 - p_2$, L_a is the Lorentz structure and the ellipsis indicates other possible form factors. The meson in the initial (final) state has a mass m_H (m_h), so that $p_1^2 = m_H^2$ ($p_2^2 = m_h^2$). There are two essential combinations of the hadron masses:

$$t_\pm = (m_H \pm m_h)^2 . \qquad (\mathrm{C.2})$$

The parameter t_+ is the threshold of the region $q^2 > t_+$, where the timelike form factor $F(q^2)$ is defined via the crossing-transformed matrix element:

$$\langle h(p_2)\bar{H}(p_1)|j_a|0\rangle = L_a(q, p_2)F(q^2) + ... \qquad (\mathrm{C.3})$$

with $q = p_1 + p_2$. In this region the form factor develops an imaginary part generated by the intermediate states and given by the sum of contributions to the unitarity relation.

Below t_+, the form factor $F(q^2)$ is real valued, except if there is a hadron H^* with the quantum numbers of the current j_a and a mass $m_{H^*} < \sqrt{t_+}$. Assuming for simplicity that there is only one subthreshold hadron, we encounter an isolated pole of the form factor at $q^2 = m_{H^*}^2$. A prominent example is the $B \to \pi$ vector form factor with a B^*-meson pole located at $q^2 = m_{B^*}^2 < (m_B + m_\pi)^2$. Depending on the flavor combination in the $H \to h$ transition, the lowest pole of the form factor $F(q^2)$ can also be located above the threshold t_+. This is the case e.g., for the pion e.m. form factor and for the $K \to \pi$, $D \to \pi$ vector form factors in which, respectively, ρ and K^*, D^* are located above corresponding thresholds t_+ and decay strongly to 2π and $K\pi$, $D\pi$.

The parameter t_- determines the upper boundary of the physical region for a semileptonic decay $H \to h + leptons$:

$$0 < q^2 < t_- , \qquad (\mathrm{C.4})$$

where we neglect for simplicity the lepton masses.

Suppose the form factor $F(q^2)$ has been calculated or measured in a certain interval of the semileptonic region or at spacelike $q^2 < 0$. The task is to compare the results with another calculation or measurement available in a different interval of q^2 in the same region[1]; hence,

---

[1]We do not discuss here an extrapolation from spacelike to a timelike region, which is generally not possible in a model-independent way.

one has to extrapolate (interpolate) the form factor $F(q^2)$ from one interval of q^2 to another (between the intervals). To this end, one needs a rigorous and improvable *parameterization* of the form factor. This problem is less trivial than a simple fit restricted to the region of validity of a calculation or measurement, in which case any function approximating numerical results for the form factor is sufficient.

One possibility is to use, as a starting point, the hadronic dispersion relation for $F(q^2)$. After isolating the H^* pole contribution, we have

$$F(q^2) = \frac{f_{H^*} g_{H^* H h}}{m_{H^*}^2 - q^2} + \frac{1}{\pi} \int_{t^+}^{\infty} ds \frac{\mathrm{Im} F(s)}{s - q^2 - i\epsilon}, \tag{C.5}$$

where f_{H^*} and $g_{H^* H h}$ are, respectively, the decay constant of H^* and the $H^* H h$ strong coupling. The integral over the imaginary part in which all other hadronic states contribute via unitarity relation, starts from the threshold t_+. Suppose that $F(s)$ vanishes at $s \to \infty$ sufficiently fast and the dispersion integral converges. In this case the dispersion relation (C.5) itself represents a parameterization of the form factor which is valid at any q^2 including the whole semileptonic region (C.4). However, while the H^* pole term is a simple function of q^2, even an approximate form of the function $\mathrm{Im} F(s)$ is largely unknown, making (C.5) not convenient for a practical use. Since the integral converges, one possible remedy is to replace it by an additional effective pole that accumulates all excited states with the same quantum numbers as H^*:

$$F(q^2) = \frac{f_{H^*} g_{H^* H h}}{m_{H^*}^2 - q^2} + \frac{r}{m_{eff}^2 - q^2}. \tag{C.6}$$

This parameterization being used at $q^2 < t_-$, contains two additional parameters – the residue r and the mass m_{eff} of the effective resonance. Alternatively, one can explicitly replace the effective pole by resonances corresponding to the radially excited H^* states. Such an ansatz resembles multiresonance models used for the pion form factors and discussed in Chapter 5.5. In the case of B and D meson form factors, the problem is that radially excited B^* or D^* states are still not well established. Moreover, an extrapolation of the form factor using different versions of effective-pole or multiresonance models towards the $q^2 \sim t_-$ region is not unique and in practice turns out to be sensitive to the parameters of the states above H^*.

The Omnés representation (5.70), which is also implicitly based on a hadronic dispersion relation, offers another possibility to parameterize a form factor in an explicit form. Usually a few subtractions are applied to make the integral more convergent and, thus, less dependent on the asymptotics of the phase, adding subtraction constants to the set of inputs. Most importantly, the phase of the elastic $Hh \to Hh$ scattering used in this representation, is basically known only for the pion and kaon scattering. In order to obtain it for heavy H and light h mesons, one needs to perform a dedicated analysis (see, e.g., [138]).

A model-independent parameterization of the form factors widely used nowadays is based on a certain mapping of the complex q^2 plane. This method is very general. One essentially uses the conjecture (based on the unitarity relation) that the form factor $F(q^2)$ is free from singularities in the whole complex q^2 plane, apart from the positive real axis[2]. Below we briefly present one practically useful version of this parameterization suggested in [140]; another frequently used version can be found in [141].

---

[2] A detailed description of this method, with its origins and underlying mathematics can be found in the recent book [139].

The main idea is to map the complex q^2 plane onto another complex plane of the new variable z. This new variable is related to q^2 in the following way (conformal mapping)

$$z(q^2, t_0) = \frac{\sqrt{t_+ - q^2} - \sqrt{t_+ - t_0}}{\sqrt{t_+ - q^2} + \sqrt{t_+ - t_0}}, \tag{C.7}$$

where t_0 is an additional real valued parameter, for which a convenient choice is:

$$t_0 = t_+ \left(1 - \sqrt{1 - \frac{t_-}{t_+}} \right) = (m_H + m_h)(\sqrt{m_H} - \sqrt{m_h})^2, \tag{C.8}$$

so that $t_0 \leq t_-$, but other choices are also possible. From (C.7) it follows that at $q^2 > t_+$ the variable z is complex-valued. The specific points and intervals on the q^2-plane map to the points on the z-plane as follows:

$$z(t_+, t_0) = -1, \quad z(0, t_0) = -z(t_-, t_0), \quad z(t_0, t_0) = 0. \tag{C.9}$$

The semileptonic region (C.4) is mapped onto a relatively small interval of the z-variable with $z \ll 1$. Isolating the only singularity, we define a function

$$f(z) = (1 - q^2/m_{H*}^2) F(q^2)$$

which can be expanded in Taylor series around $z = 0$ and this expansion is usually reliable due to the smallness of z. The parametrization up to $O(z^N)$ based on this expansion is

$$F(q^2) = \frac{1}{1 - q^2/m_{H*}^2} \sum_{k=0}^{N} \widetilde{b}_k \left[z(q^2, t_0) \right]^k, \tag{C.10}$$

where, for the vector form factor, taking into account certain details concerning the near-threshold behavior, an additional condition is imposed [140]

$$\widetilde{b}_N = -\frac{(-1)^N}{N} \sum_{k=0}^{N-1} (-1)^k \, k \, \widetilde{b}_k. \tag{C.11}$$

It is convenient to choose the form factor at zero momentum transfer $F(0)$ as one of the independent parameters, correspondingly rescaling the coefficients in the series expansion: $\widetilde{b}_k = F(0) b_k$, so that

$$b_0 = 1 - \sum_{k=1}^{N-1} b_k \left[z(0, t_0)^k - (-1)^{k-N} \frac{k}{N} z(0, t_0)^N \right].$$

and finally,

$$F(q^2) = \frac{F(0)}{1 - q^2/m_{H*}^2} \left\{ 1 + \sum_{k=1}^{N-1} b_k \left(z(q^2, t_0)^k - z(0, t_0)^k \right. \right.$$

$$\left. \left. -(-1)^{k-N} \frac{k}{N} \left[z(q^2, t_0)^N - z(0, t_0)^N \right] \right) \right\}. \tag{C.12}$$

The pion light-cone distribution amplitudes

Here we collect the set of the pion DAs and their parameters introduced and used in Chapters 9 and 10.

- The quark-antiquark DAs of the pion are represented in a form of a decomposition of the vacuum-to-pion matrix element (for definiteness, we assume the π^+ in the final state):

$$\langle \pi^+(p)|\bar{u}^i_\omega(x_1)d^j_\xi(x_2)|0\rangle_{x^2\to 0} = \frac{i\delta^{ij}}{12}f_\pi \int_0^1 du \ e^{iup\cdot x_1 + i\bar{u}p\cdot x_2}$$

$$\times \left([\not{p}\gamma_5]_{\xi\omega}\varphi_\pi(u) - [\gamma_5]_{\xi\omega}\mu_\pi\phi^p_{3\pi}(u) + \frac{1}{6}[\sigma_{\beta\tau}\gamma_5]_{\xi\omega}p_\beta(x_1-x_2)_\tau\mu_\pi\phi^\sigma_{3\pi}(u) \right.$$

$$\left. + \frac{1}{16}[\not{p}\gamma_5]_{\xi\omega}(x_1-x_2)^2\phi_{4\pi}(u) - \frac{i}{2}[(\not{x}_1-\not{x}_2)\gamma_5]_{\xi\omega}\int_0^u \psi_{4\pi}(v)dv \right). \qquad (\mathrm{D}.1)$$

- The analogous decomposition in the quark-antiquark-gluon DAs reads:

$$\langle \pi^+(p)|\bar{u}^i_\omega(x_1)g_s G^a_{\mu\nu}(x_3)d^j_\xi(x_2)|0\rangle_{x^2\to 0} = \frac{\lambda^a_{ji}}{32}\int \mathcal{D}\alpha e^{ip(\alpha_1 x_1 + \alpha_2 x_2 + \alpha_3 x_3)}$$

$$\times \left[if_{3\pi}(\sigma_{\lambda\rho}\gamma_5)_{\xi\omega}(p_\mu p_\lambda g_{\nu\rho} - p_\nu p_\lambda g_{\mu\rho})\Phi_{3\pi}(\alpha_i) \right.$$

$$-f_\pi(\gamma_\lambda\gamma_5)_{\xi\omega}\left\{ (p_\nu g_{\mu\lambda} - p_\mu g_{\nu\lambda})\Psi_{4\pi}(\alpha_i) + \frac{p_\lambda(p_\mu x_\nu - p_\nu x_\mu)}{(p\cdot x)}(\Phi_{4\pi}(\alpha_i) + \Psi_{4\pi}(\alpha_i)) \right\}$$

$$\left. -\frac{if_\pi}{2}\epsilon_{\mu\nu\delta\rho}(\gamma_\lambda)_{\xi\omega}\left\{ (p^\rho g^{\delta\lambda} - p^\delta g^{\rho\lambda})\widetilde{\Psi}_{4\pi}(\alpha_i) + \frac{p_\lambda(p^\delta x^\rho - p^\rho x^\delta)}{(p\cdot x)}(\widetilde{\Phi}_{4\pi}(\alpha_i) + \widetilde{\Psi}_{4\pi}(\alpha_i)) \right\} \right].$$

$$(\mathrm{D}.2)$$

In the above, $x_i = \xi_i x$, where ξ_i are arbitrary numbers, $x^2 = 0$; $\bar{u} = 1 - u$ and

$$\mathcal{D}\alpha = d\alpha_1 \, d\alpha_2 \, d\alpha_3 \, \delta\left(\sum_{i=1}^3 \alpha_i - 1 \right).$$

The path-ordered gauge links are omitted assuming the fixed-point gauge for the gluons.

• The decompositions (D.1) and (D.2) include the following DAs, ordered according to their twist:

t	quark-antiquark DAs	quark-antiquark-gluon DAs
2	φ_π	–
3	$\phi_{3\pi}^p,\ \phi_{3\pi}^\sigma$	$\Phi_{3\pi}$
4	$\phi_{4\pi},\ \psi_{4\pi}$	$\Phi_{4\pi},\ \Psi_{4\pi},\ \widetilde{\Phi}_{4\pi},\ \widetilde{\Psi}_{4\pi}$

The definitions of separate DAs are obtained, multiplying both parts of (D.1) and (D.2) by the corresponding combinations of γ matrices and taking the traces over the Dirac and color indices. In particular, multiplying (D.1) by $(\gamma_\alpha\gamma_5)_{\omega\xi}$, the definition (9.44) of the twist-2 DA is obtained. The notation for twist-3 and twist-4 DAs is taken (with a slight modification) from [99].

• The following expressions for the DAs are used in LCSRs. The adopted accuracy is at least to the next-to-leading order in the conformal spin expansion:

– twist-2 DA:

$$\varphi_\pi(u,\mu) = 6u\bar{u}\left(1 + a_2(\mu)C_2^{3/2}(u - \bar{u}) + a_4(\mu)C_4^{3/2}(u - \bar{u})\right), \qquad (D.3)$$

where the first two Gegenbauer polynomials are included in the nonasymptotic part, with the coefficients having the following scale dependence (see (9.52), (9.53)):

$$a_2(\mu) = \left(\frac{\alpha_s(\mu)}{\alpha_s(\mu_0)}\right)^{\frac{25C_F}{6\beta_0}} a_2(\mu_0), \quad a_4(\mu) = \left(\frac{\alpha_s(\mu)}{\alpha_s(\mu_0)}\right)^{\frac{91C_F}{15\beta_0}} a_4(\mu_0), \qquad (D.4)$$

– twist-3 DAs:

$$\phi_{3\pi}^p(u) = 1 + 30\frac{f_{3\pi}}{\mu_\pi f_\pi}C_2^{1/2}(u - \bar{u}) - 3\frac{f_{3\pi}\omega_{3\pi}}{\mu_\pi f_\pi}C_4^{1/2}(u - \bar{u}),$$

$$\phi_{3\pi}^\sigma(u) = 6u(1 - u)\left(1 + 5\frac{f_{3\pi}}{\mu_\pi f_\pi}\left(1 - \frac{\omega_{3\pi}}{10}\right)C_2^{3/2}(u - \bar{u})\right), \qquad (D.5)$$

with μ_π defined in (9.60), and

$$\Phi_{3\pi}(\alpha_i) = 360\alpha_1\alpha_2\alpha_3^2\left[1 + \frac{\omega_{3\pi}}{2}(7\alpha_3 - 3)\right]. \qquad (D.6)$$

The parameters $f_{3\pi}$ and $\omega_{3\pi}$ are defined via the matrix elements of the following local operators:

$$\langle\pi^+(p)|\bar{u}\sigma_{\mu\nu}\gamma_5 G_{\alpha\beta}d|0\rangle = if_{3\pi}\left[(p_\alpha p_\mu g_{\beta\nu} - p_\beta p_\mu g_{\alpha\nu}) - (p_\alpha p_\nu g_{\beta\mu} - p_\beta p_\nu g_{\alpha\mu})\right], \quad (D.7)$$

and

$$\langle\pi^+(p)|\bar{u}\sigma_{\mu\lambda}\gamma_5[D_\beta, G_{\alpha\lambda}]d - \frac{3}{7}\partial_\beta\bar{u}\sigma_{\mu\lambda}\gamma_5 G_{\alpha\lambda}d|0\rangle = -\frac{3}{14}f_{3\pi}\omega_{3\pi}p_\alpha p_\beta p_\mu. \qquad (D.8)$$

The scale dependence of the twist-3 parameters is given by:

$$\mu_\pi(\mu_1) = \left(\frac{\alpha_s(\mu)}{\alpha_s(\mu_0)}\right)^{-\frac{4}{\beta_0}} \mu_\pi(\mu_0)\,,$$

$$f_{3\pi}(\mu) = \left(\frac{\alpha_s(\mu)}{\alpha_s(\mu_0)}\right)^{\frac{1}{\beta_0}\left(\frac{7C_F}{3}+3\right)} f_{3\pi}(\mu_0)\,,$$

$$(f_{3\pi}\omega_{3\pi})(\mu) = \left(\frac{\alpha_s(\mu)}{\alpha_s(\mu_0)}\right)^{\frac{1}{\beta_0}\left(\frac{7C_F}{6}+10\right)} (f_{3\pi}\omega_{3\pi})(\mu_0)\,. \tag{D.9}$$

– twist-4 DAs:

$$\phi_{4\pi}(u) = \frac{200}{3}\delta_\pi^2 u^2\bar{u}^2 + 8\delta_\pi^2\epsilon_\pi\Big\{u\bar{u}(2+13u\bar{u})$$

$$+2u^3(10-15u+6u^2)\ln u + 2\bar{u}^3(10-15\bar{u}+6\bar{u}^2)\ln\bar{u}\Big\}\,,$$

$$\psi_{4\pi}(u) = \frac{20}{3}\delta_\pi^2 C_2^{1/2}(2u-1)\,, \tag{D.10}$$

$$\begin{aligned}
\Phi_{4\pi}(\alpha_i) &= 120\delta_\pi^2\varepsilon_\pi(\alpha_1-\alpha_2)\alpha_1\alpha_2\alpha_3\,, \\
\Psi_{4\pi}(\alpha_i) &= 30\delta_\pi^2(\mu)(\alpha_1-\alpha_2)\alpha_3^2[\tfrac{1}{3}+2\varepsilon_\pi(1-2\alpha_3)]\,, \\
\widetilde{\Phi}_{4\pi}(\alpha_i) &= -120\delta_\pi^2\alpha_1\alpha_2\alpha_3[\tfrac{1}{3}+\varepsilon_\pi(1-3\alpha_3)]\,, \\
\widetilde{\Psi}_{4\pi}(\alpha_i) &= 30\delta_\pi^2\alpha_3^2(1-\alpha_3)[\tfrac{1}{3}+2\varepsilon_\pi(1-2\alpha_3)]\,.
\end{aligned} \tag{D.11}$$

The parameters δ_π^2 and ϵ_π are defined as

$$\langle\pi^+(p)|\bar{u}\widetilde{G}_{\alpha\mu}\gamma^\alpha d|0\rangle = i\delta_\pi^2 f_\pi p_\mu\,, \tag{D.12}$$

and (up to twist 5 corrections):

$$\langle\pi^+(p)|\bar{u}[D_\mu,\widetilde{G}_{\nu\xi}]\gamma^\xi d - \frac{4}{9}\partial_\mu\bar{u}\widetilde{G}_{\nu\xi}\gamma^\xi d|0\rangle = -\frac{8}{21}f_\pi\delta_\pi^2\epsilon_\pi p_\mu p_\nu\,, \tag{D.13}$$

with the scale-dependence:

$$\delta_\pi^2(\mu) = \left(\frac{\alpha_s(\mu)}{\alpha_s(\mu_0)}\right)^{\frac{8C_F}{3\beta_0}} \delta_\pi^2(\mu_0)\,, \quad (\delta_\pi^2\epsilon_\pi)(\mu) = \left(\frac{\alpha_s(\mu)}{\alpha_s(\mu_0)}\right)^{\frac{10}{\beta_0}} (\delta_\pi^2\epsilon_\pi)(\mu_0)\,. \tag{D.14}$$

• For convenience, we present the expressions for the relevant Gegenbauer polynomials:

$$C_0^{3/2}(x) = 1\,, \quad C_1^{3/2}(x) = 3x\,, \quad C_2^{3/2}(x) = \frac{3}{2}(5x^2-1)\,,$$

$$C_3^{3/2}(x) = \frac{5}{2}(7x^3-3x)\,, \quad C_4^{3/2}(x) = \frac{15}{8}(21x^4-14x^2+1)\,,$$

$$C_0^{1/2}(x) = 1\,, \quad C_1^{1/2}(x) = x\,, \quad C_2^{1/2}(x) = \frac{1}{2}(3x^2-1)\,,$$

$$C_3^{1/2}(x) = \frac{1}{2}(5x^3-3x)\,, \quad C_4^{1/2}(x) = \frac{1}{8}(35x^4-30x^2+3)\,, \tag{D.15}$$

and the orthogonality relation

$$\int_0^1 du\, u\bar{u}\, C_n^{3/2}(u-\bar{u})C_m^{3/2}(u-\bar{u}) = \delta_{nm}\frac{(n+1)(n+2)}{4(2n+3)}\,. \tag{D.16}$$

Bibliography

[1] M. Tanabashi et al. Review of particle physics. *Phys. Rev.*, D98(3):030001, 2018.

[2] L. B. Okun. *Weak interactions of elementary particles*. Nauka, Moscow, (in Russian), 1962.

[3] L. B. Okun. *Leptons and quarks*. North-Holland, Amsterdam, Netherlands, 1982.

[4] J. J. Sakurai and J. Napolitano. *Modern quantum mechanics*. Cambridge University Press, Cambridge, 2017.

[5] V. B. Berestetsky. The dynamical properties of elementary particles and the theory of the scattering matrix. *Sov. Phys. Usp.*, 5:7–36, 1962.

[6] M. E. Peskin and D. V. Schroeder. *An introduction to quantum field theory*. Addison-Wesley, Reading, USA, 1995.

[7] T. Muta. *Foundations of quantum chromodynamics: An introduction to perturbative methods in gauge theories, (3rd ed.)*. World Scientific, Hackensack, N.J., 2010.

[8] B. L. Ioffe, V.S. Fadin, and L. N. Lipatov. *Quantum chromodynamics: Perturbative and nonperturbative aspects*. Cambridge Univ. Press, 2010.

[9] A. I. Vainshtein, V. I. Zakharov, V. A. Novikov, and M. A. Shifman. ABC's of instantons. *Sov. Phys. Usp.*, 25:195, 1982.

[10] S. R. Sharpe. Phenomenology from the lattice. In *Theoretical Advanced Study Institute in elementary particle physics (TASI 94): CP Violation and the limits of the Standard Model, Boulder, Colorado*, pages 0377–444, 1994.

[11] Ch. Gattringer and Ch. B. Lang. *Quantum chromodynamics on the lattice*. Lect. Notes Phys., Springer, Berlin, Heidelberg, 2010.

[12] A. S. Kronfeld. Twenty-first century lattice gauge theory: Results from the QCD Lagrangian. *Ann. Rev. Nucl. Part. Sci.*, 62:265–284, 2012.

[13] K. G. Chetyrkin, J. H. Kuhn, and M. Steinhauser. RunDec: A Mathematica package for running and decoupling of the strong coupling and quark masses. *Comput. Phys. Commun.*, 133:43–65, 2000.

[14] G. Buchalla, A. J. Buras, and M. E. Lautenbacher. Weak decays beyond leading logarithms. *Rev. Mod. Phys.*, 68:1125–1144, 1996.

[15] B. L. Ioffe. Axial anomaly: The modern status. *Int. J. Mod. Phys.*, A21:6249–6266, 2006.

[16] J. Gasser and H. Leutwyler. Quark masses. *Phys. Rept.*, 87:77–169, 1982.

[17] H. Leutwyler. Chiral dynamics. In *At the frontier of particle physics, vol. 1* 271-316, Shifman, M. (ed.) World Scientific*, pages 271–316, 2000.

[18] B. L. Ioffe. Chiral effective theory of strong interactions. *Phys. Usp.*, 44:1211–1227, 2001.

[19] T. Feldmann, P. Kroll, and B. Stech. Mixing and decay constants of pseudoscalar mesons. *Phys. Rev.*, D58:114006, 1998.

[20] S. Nussinov and M. A. Lampert. QCD inequalities. *Phys. Rept.*, 362:193–301, 2002.

[21] M. Gell-Mann, R. J. Oakes, and B. Renner. Behavior of current divergences under SU(3) x SU(3). *Phys. Rev.*, 175:2195–2199, 1968.

[22] V. A. Novikov, M. A. Shifman, A. I. Vainshtein, and V. I. Zakharov. Are all hadrons alike? *Nucl. Phys.*, B191:301–369, 1981.

[23] B. V. Geshkenbein and B. L. Ioffe. The role of instantons in generation of mesonic mass spectrum. *Nucl. Phys.*, B166:340–364, 1980.

[24] V. A. Novikov, L. B. Okun, M. A. Shifman, A. I. Vainshtein, M. B. Voloshin, and V. I. Zakharov. Charmonium and gluons: Basic experimental facts and theoretical introduction. *Phys. Rept.*, 41:1–133, 1978.

[25] M. Neubert. Heavy quark symmetry. *Phys. Rept.*, 245:259–396, 1994.

[26] M. A. Shifman. Lectures on heavy quarks in quantum chromodynamics. In *ITEP lectures on particle physics and field theory. Vol. 1, 2, World Scientific*, pages 409–514, 1995.

[27] Th. Mannel. Effective theory for heavy quarks. In *Proceedings, 35. Internationale Universitatswochen fur Kern- und Teilchenphysik: Schladming*, volume 479, pages 387–428, 1997.

[28] N. Uraltsev. Topics in the heavy quark expansion. In *At the frontier of particle physics, vol. 3* 1577-1670, Shifman, M. (ed.) World Scientific*, pages 1577–1670, 2000.

[29] A. V. Manohar and M. B. Wise. Heavy quark physics. *Camb. Monogr. Part. Phys. Nucl. Phys. Cosmol.*, 10:1–191, 2000.

[30] Th. Mannel. *Effective field theories in flavor physics*. Springer Tracts in Modern Physics. 203, Springer, Berlin, 2004.

[31] S. R. Amendolia et al. A measurement of the space - like pion electromagnetic form-factor. *Nucl. Phys.*, B277:168, 1986.

[32] G. M. Huber et al. Charged pion form-factor between $Q^2 = 0.60$ GeV^2 and 2.45 GeV^2. II. Determination of, and results for the pion form-factor. *Phys. Rev.*, C78:045203, 2008.

[33] J. Bijnens, G. Colangelo, G. Ecker, and J. Gasser. Semileptonic kaon decays. In *2nd DAPHNE physics handbook*, pages 315–389, hep-ph/9411311,1994.

[34] S. Aoki et al. FLAG Review 2019, [arXiv:1902.08191[hep-lat]]. 2019.

[35] A. Ali, P. Ball, L. T. Handoko, and G. Hiller. A comparative study of the decays $B \to (K, K^*)\ell^+\ell^-$ in Standard Model and supersymmetric theories. *Phys. Rev.*, D61:074024, 2000.

[36] W. Altmannshofer, P. Ball, A. Bharucha, A. J. Buras, D. M. Straub, and M. Wick. Symmetries and asymmetries of $B \to K^*\mu^+\mu^-$ decays in the Standard Model and beyond. *JHEP*, 01:019, 2009.

[37] G. Burdman, Z. Ligeti, M. Neubert, and Y. Nir. The decay $B \to \pi$ lepton neutrino in heavy quark effective theory. *Phys. Rev.*, D49:2331–2345, 1994.

[38] G. F. Sterman and P. Stoler. Hadronic form-factors and perturbative QCD. *Ann. Rev. Nucl. Part. Sci.*, 47:193–233, 1997.

[39] S. Pacetti, R. Baldini Ferroli, and E. Tomasi-Gustafsson. Proton electromagnetic form factors: Basic notions, present achievements and future perspectives. *Phys. Rept.*, 550-551:1–103, 2015.

[40] C.-H. Chen and C. Q. Geng. Baryonic rare decays of $\Lambda_b \to \Lambda \ell^+ \ell^-$. *Phys. Rev.*, D64:074001, 2001.

[41] Th. Feldmann and M. W. Y. Yip. Form factors for $\Lambda_b \to \Lambda$ transitions in SCET. *Phys. Rev.*, D85:014035, 2012. [Erratum: *Phys. Rev.* D86,079901(2012)].

[42] E. V. Shuryak. Hadrons containing a heavy quark and QCD sum rules. *Nucl. Phys.*, B198:83–101, 1982.

[43] M. Gell-Mann, D. Sharp, and W. G. Wagner. Decay rates of neutral mesons. *Phys. Rev. Lett.*, 8:261, 1962.

[44] E. Eichten, K. Gottfried, T. Kinoshita, K. D. Lane, and T.-M. Yan. Charmonium: The model. *Phys. Rev.*, D17:3090, 1978. [Erratum: *Phys. Rev.* D21,313(1980)].

[45] A. H. Hoang. Heavy quarkonium dynamics. In *At the frontier of particle physics, vol. 4* 2215-2331, Shifman, M. (ed.) World Sci.*, 2002.

[46] R. J. Eden, P. V. Landshoff, D. I. Olive, and J. Ch. Polkinghorne. *The analytic S-matrix*. Cambridge Univ. Press, Cambridge, 1966.

[47] G. Breit and E. Wigner. Capture of slow neutrons. *Phys. Rev.*, 49:519–531, 1936.

[48] J. H. Kuhn and A. Santamaria. Tau decays to pions. *Z. Phys.*, C48:445–452, 1990.

[49] G. J. Gounaris and J. J. Sakurai. Finite width corrections to the vector meson dominance prediction for $\rho \to e^+ e^-$. *Phys. Rev. Lett.*, 21:244–247, 1968.

[50] N. N. Achasov and A. A. Kozhevnikov. Electromagnetic form factor of pion in the field theory inspired approach. *Phys. Rev.*, D83:113005, 2011. [Erratum: *Phys. Rev.* D85,019901(2012)].

[51] C. A. Dominguez. Pion form-factor in large N(c) QCD. *Phys. Lett.*, B512:331–334, 2001.

[52] R. F. Lebed. Phenomenology of large N(c) QCD. *Czech. J. Phys.*, 49:1273–1306, 1999.

[53] Ch. Bruch, A. Khodjamirian, and J. H. Kuhn. Modeling the pion and kaon form factors in the timelike region. *Eur. Phys. J.*, C39:41–54, 2005.

[54] S. Binner, J. H. Kuhn, and K. Melnikov. Measuring $\sigma(e^+ e^- \to hadrons)$ using tagged photon. *Phys. Lett.*, B459:279–287, 1999.

[55] X.-W. Kang, B. Kubis, Ch. Hanhart, and U-G. Meißner. B_{l4} decays and the extraction of $|V_{ub}|$. *Phys. Rev.*, D89:053015, 2014.

[56] S. Faller, Th. Feldmann, A. Khodjamirian, Th. Mannel, and D. van Dyk. Disentangling the decay observables in $B^- \to \pi^+\pi^-\ell^-\bar{\nu}_\ell$. *Phys. Rev.*, D89(1):014015, 2014.

[57] A. Khodjamirian and D. Wyler. Counting contact terms in $B \to V\gamma$ decays. In *From integrable models to gauge theories, Gurzadyan, V.G. (ed.) World Scientific, 227-241*, 2001.

[58] A. Khodjamirian, Th. Mannel, A. A. Pivovarov, and Y. M. Wang. Charm-loop effect in $B \to K^{(*)}\ell^+\ell^-$ and $B \to K^*\gamma$. *JHEP*, 09:089, 2010.

[59] A. Khodjamirian, Th. Mannel, and Y. M. Wang. $B \to K\ell^+\ell^-$ decay at large hadronic recoil. *JHEP*, 02:010, 2013.

[60] J. Koponen, A. Zimermmane-Santos, Ch. Davies, G. P. Lepage, and A. Lytle. Light meson form factors at high Q^2 from lattice QCD. *EPJ Web Conf.*, 175:06015, 2018.

[61] R. P. Feynman. *Photon-hadron interactions*. W.A. Benjamin, Inc. Reading, Massachusetts, 1972.

[62] V. A. Matveev, R. M. Muradyan, and A. N. Tavkhelidze. Automodality in strong interactions. *Lett. Nuovo Cim.*, 5S2:907–912, 1972.

[63] S. J. Brodsky and G. R. Farrar. Scaling laws at large transverse momentum. *Phys. Rev. Lett.*, 31:1153–1156, 1973.

[64] V. L. Chernyak, A. R. Zhitnitsky, and V. G. Serbo. Asymptotic hadronic form-factors in quantum chromodynamics. *JETP Lett.*, 26:594–597, 1977. [Pisma Zh. Eksp. Teor. Fiz.26,760(1977)].

[65] G. R. Farrar and D. R. Jackson. The pion form-factor. *Phys. Rev. Lett.*, 43:246, 1979.

[66] A. V. Efremov and A. V. Radyushkin. Factorization and asymptotical behavior of pion form-factor in QCD. *Phys. Lett.*, 94B:245–250, 1980.

[67] G. P. Lepage and S. J. Brodsky. Exclusive processes in perturbative quantum chromodynamics. *Phys. Rev.*, D22:2157, 1980.

[68] I. G. Aznaurian, S. V. Esaibegian, and N. L. Ter-Isaakian. On the asymptotics of the nucleon form factors in the quark-gluon model. *Phys. Lett.*, 90B:151, 1980. [Erratum: *Phys. Lett.* 92B,371(1980)].

[69] B. V. Geshkenbein and M. V. Terentiev. The enhanced power corrections to the asymptotics of the pion form factor. *Phys. Lett.*, 117B:243–246, 1982.

[70] H.-N. Li and G. F. Sterman. The perturbative pion form-factor with Sudakov suppression. *Nucl. Phys.*, B381:129–140, 1992.

[71] M. Beneke, G. Buchalla, M. Neubert, and C. T. Sachrajda. QCD factorization for $B \to \pi\pi$ decays: Strong phases and CP violation in the heavy quark limit. *Phys. Rev. Lett.*, 83:1914–1917, 1999.

[72] M. Beneke and T. Feldmann. Symmetry breaking corrections to heavy to light B meson form-factors at large recoil. *Nucl. Phys.*, B592:3–34, 2001.

[73] M. A. Shifman, A. I. Vainshtein, and V. I. Zakharov. QCD and resonance physics. Theoretical foundations. *Nucl. Phys.*, B147:385–447, 1979.

[74] M. A. Shifman, A. I. Vainshtein, and V. I. Zakharov. QCD and resonance physics: Applications. *Nucl. Phys.*, B147:448–518, 1979.

[75] V. A. Novikov, M. A. Shifman, A. I. Vainshtein, and V. I. Zakharov. Wilson's operator expansion: Can it fail? *Nucl. Phys.*, B249:445–471, 1985.

[76] M. A. Shifman. QCD sum rules: The second decade. In *QCD 20 Years Later: Proceedings, Workshop, Aachen, Germany, June 9-13, 1992*, pages 775–794, 1993.

[77] K. G. Chetyrkin and A. Khodjamirian. Strange quark mass from pseudoscalar sum rule with $O(\alpha_s^4)$ accuracy. *Eur. Phys. J.*, C46:721–728, 2006.

[78] P. A. Baikov, K. G. Chetyrkin, and J. H. Kuhn. Scalar correlator at $O(\alpha_s^4)$, Higgs decay into b-quarks and bounds on the light quark masses. *Phys. Rev. Lett.*, 96:012003, 2006.

[79] V. A. Novikov, M. A. Shifman, A. I. Vainshtein, and V. I. Zakharov. Calculations in external fields in quantum chromodynamics. Technical Review. *Fortsch. Phys.*, 32:585, 1984.

[80] K. Maltman and J. Kambor. Decay constants, light quark masses and quark mass bounds from light quark pseudoscalar sum rules. *Phys. Rev.*, D65:074013, 2002.

[81] B. L. Ioffe. Condensates in quantum chromodynamics. *Phys. Atom. Nucl.*, 66:30–43, 2003. [Yad. Fiz.66,32(2003)].

[82] L. J. Reinders, H. Rubinstein, and S. Yazaki. Hadron properties from QCD sum rules. *Phys. Rept.*, 127:1, 1985.

[83] S. Narison. QCD spectral sum rules. *World Sci. Lect. Notes Phys.*, 26:1–527, 1989.

[84] P. Colangelo and A. Khodjamirian. QCD sum rules, a modern perspective. In *At the frontier of particle physics, vol. 3* 1495-1576, Shifman, M. (ed.) World Scientific*, 2000.

[85] K. G. Chetyrkin, J. H. Kuhn, A. Maier, P. Maierhofer, P. Marquard, M. Steinhauser, and C. Sturm. Charm and bottom quark masses: An update. *Phys. Rev.*, D80:074010, 2009.

[86] M. B. Voloshin. Heavy quarkoinum outside the potential model. *Sov. Phys. Usp.*, 28:88–89, 1985.

[87] T. M. Aliev and V. L. Eletsky. On leptonic decay constants of pseudoscalar D and B mesons. *Sov. J. Nucl. Phys.*, 38:936, 1983.

[88] P. Gelhausen, A. Khodjamirian, A. A. Pivovarov, and D. Rosenthal. Decay constants of heavy-light vector mesons from QCD sum rules. *Phys. Rev.*, D88:014015, 2013. [Erratum: *Phys. Rev.* D91,099901(2015)].

[89] A. Khodjamirian. Dispersion sum rules for the amplitudes of radiative transitions in quarkonium. *Phys. Lett.*, 90B:460–464, 1980.

[90] V. A. Beilin and A. V. Radyushkin. Quantum chromodynamic sum rules and $J/\psi \rightarrow \eta_c \gamma$ decay. *Nucl. Phys.*, B260:61–78, 1985.

[91] A. Khodjamirian. On the calculation of $J/\psi \to \eta_c \gamma$ width in QCD. *Sov. J. Nucl. Phys.*, 39:614, 1984. [Yad. Fiz.39,970(1984)].

[92] B. L. Ioffe and A. V. Smilga. Pion form-factor at intermediate momentum transfer in QCD. *Phys. Lett.*, 114B:353–358, 1982.

[93] V. A. Nesterenko and A. V. Radyushkin. Sum rules and pion form-factor in QCD. *Phys. Lett.*, 115B:410, 1982.

[94] P. Ball, V.M. Braun, and H.G. Dosch. Form-factors of semileptonic D decays from QCD sum rules. *Phys. Rev.*, D44:3567–3581, 1991.

[95] P. Ball, V.M. Braun, Y. Koike, and K. Tanaka. Higher twist distribution amplitudes of vector mesons in QCD: Formalism and twist-three distributions. *Nucl. Phys.*, B529:323–382, 1998.

[96] P. Ball and V. M. Braun. Handbook of higher twist distribution amplitudes of vector mesons in QCD. In *Continuous advances in QCD. Proceedings, 3rd Workshop, QCD'98, Minneapolis, USA, April 16-19, 1998*, pages 125–141, 1998.

[97] V. M. Braun, G. P. Korchemsky, and D. Müller. The uses of conformal symmetry in QCD. *Prog. Part. Nucl. Phys.*, 51:311–398, 2003.

[98] V. L. Chernyak and A. R. Zhitnitsky. Asymptotic behavior of exclusive processes in QCD. *Phys. Rept.*, 112:173, 1984.

[99] P. Ball, V. M. Braun, and A. Lenz. Higher-twist distribution amplitudes of the K meson in QCD. *JHEP*, 05:004, 2006.

[100] A. Khodjamirian, Ch. Klein, Th. Mannel, and N. Offen. Semileptonic charm decays $D \to \pi \ell \nu_\ell$ and $D \to \pi \ell \nu_\ell$ from QCD light-cone sum rules. *Phys. Rev.*, D80:114005, 2009.

[101] V. M. Braun, S. Collins, M. Gockeler, P. Perez-Rubio, A. Schafer, R. W. Schiel, and A. Sternbeck. Second moment of the pion light-cone distribution amplitude from lattice QCD. *Phys. Rev.*, D92(1):014504, 2015.

[102] K. G. Chetyrkin, A. Khodjamirian, and A. A. Pivovarov. Towards NNLO accuracy in the QCD sum rule for the kaon distribution amplitude. *Phys. Lett.*, B661:250–258, 2008.

[103] P. Ball, V. M. Braun, and N. Kivel. Photon distribution amplitudes in QCD. *Nucl. Phys.*, B649:263–296, 2003.

[104] M. Diehl, T. Gousset, B. Pire, and O. Teryaev. Probing partonic structure in $\gamma^* \gamma \to \pi\pi$ near threshold. *Phys. Rev. Lett.*, 81:1782–1785, 1998.

[105] A. G. Grozin. On wave functions of mesonic pairs and mesonic resonances. *Sov. J. Nucl. Phys.*, 38:289–292, 1983.

[106] M. V. Polyakov. Hard exclusive electroproduction of two pions and their resonances. *Nucl. Phys.*, B555:231, 1999.

[107] V. Braun, R. J. Fries, N. Mahnke, and E. Stein. Higher twist distribution amplitudes of the nucleon in QCD. *Nucl. Phys.*, B589:381–409, 2000. [Erratum: *Nucl. Phys.* B607,433(2001)].

[108] I. I. Balitsky. String operator expansion of the T-product of two currents near the light-cone. *Phys. Lett.*, 124B:230–236, 1983.

[109] I. I. Balitsky and V. M. Braun. Evolution equations for QCD string operators. *Nucl. Phys.*, B311:541–584, 1989.

[110] I. I. Balitsky, V.M. Braun, and A. V. Kolesnichenko. The decay $\Sigma^+ \to p\gamma$ in QCD: Bilocal corrections in a variable magnetic field and the photon wave functions. *Sov. J. Nucl. Phys.*, 48:348–357, 1988.

[111] I. I. Balitsky, V.M. Braun, and A. V. Kolesnichenko. Radiative decay $\Sigma^+ \to p\gamma$ in quantum chromodynamics. *Nucl. Phys.*, B312:509–550, 1989.

[112] V. L. Chernyak and I. R. Zhitnitsky. *B* meson exclusive decays into baryons. *Nucl. Phys.*, B345:137–172, 1990.

[113] V. M. Braun and I. E. Halperin. Soft contribution to the pion form-factor from light cone QCD sum rules. *Phys. Lett.*, B328:457–465, 1994.

[114] V. M. Braun, A. Khodjamirian, and M. Maul. Pion form-factor in QCD at intermediate momentum transfers. *Phys. Rev.*, D61:073004, 2000.

[115] J. Bijnens and A. Khodjamirian. Exploring light cone sum rules for pion and kaon form-factors. *Eur. Phys. J.*, C26:67–79, 2002.

[116] V. M. Braun, A. Lenz, N. Mahnke, and E. Stein. Light cone sum rules for the nucleon form-factors. *Phys. Rev.*, D65:074011, 2002.

[117] V. M. Braun, A. Lenz, and M. Wittmann. Nucleon form factors in QCD. *Phys. Rev.*, D73:094019, 2006.

[118] F. del Aguila and M. K. Chase. Higher order QCD corrections to exclusive two photon processes. *Nucl. Phys.*, B193:517–528, 1981.

[119] E. Braaten. QCD corrections to meson-photon transition form-factors. *Phys. Rev.*, D28:524, 1983.

[120] J. Gronberg et al. Measurements of the meson-photon transition form-factors of light pseudoscalar mesons at large momentum transfer. *Phys. Rev.*, D57:33–54, 1998.

[121] S. Uehara et al. Measurement of $\gamma\gamma^* \to \pi^0$ transition form factor at Belle. *Phys. Rev.*, D86:092007, 2012.

[122] A. Khodjamirian. Form-factors of $\gamma^*\rho \to \pi$ and $\gamma^*\gamma \to \pi^0$ transitions and light cone sum rules. *Eur. Phys. J.*, C6:477–484, 1999.

[123] S. S. Agaev, V. M. Braun, N. Offen, and F. A. Porkert. Light cone sum rules for the $\pi^0\gamma^*\gamma$ form factor revisited. *Phys. Rev.*, D83:054020, 2011.

[124] G. Duplancic, A. Khodjamirian, Th. Mannel, B. Melic, and N. Offen. Light-cone sum rules for $B \to \pi$ form factors revisited. *JHEP*, 04:014, 2008.

[125] A. Khodjamirian, R. Ruckl, and C. W. Winhart. The scalar $B \to \pi$ and $D \to \pi$ form-factors in QCD. *Phys. Rev.*, D58:054013, 1998.

[126] V. M. Belyaev, V. M. Braun, A. Khodjamirian, and R. Ruckl. $D^*D\pi$ and $B^*B\pi$ couplings in QCD. *Phys. Rev.*, D51:6177–6195, 1995.

[127] A. Khodjamirian, Ch. Klein, Th. Mannel, and Y. M. Wang. Form factors and strong couplings of heavy baryons from QCD light-cone sum rules. *JHEP*, 09:106, 2011.

[128] A. Khodjamirian, Th. Mannel, and N. Offen. Form-factors from light-cone sum rules with B-meson distribution amplitudes. *Phys. Rev.*, D75:054013, 2007.

[129] A. G. Grozin and M. Neubert. Asymptotics of heavy meson form-factors. *Phys. Rev.*, D55:272–290, 1997.

[130] A. G. Grozin. B-meson distribution amplitudes. *Int. J. Mod. Phys.*, A20.

[131] B.O. Lange and M. Neubert. Renormalization group evolution of the B meson light cone distribution amplitude. *Phys. Rev. Lett.*, 91:102001, 2003.

[132] M. Beneke and J. Rohrwild. B meson distribution amplitude from $B \to \gamma \ell \nu$. *Eur. Phys. J.*, C71:1818, 2011.

[133] S. Faller, A. Khodjamirian, Ch. Klein, and Th. Mannel. $B \to D^{(*)}$ form factors from QCD light-cone sum rules. *Eur. Phys. J.*, C60:603–615, 2009.

[134] V. M. Braun, Y. Ji, and A. N. Manashov. Higher-twist B-meson distribution amplitudes in HQET. *JHEP*, 05:022, 2017.

[135] Y.-M. Wang and Y.-L. Shen. QCD corrections to $B \to \pi$ form factors from light-cone sum rules. *Nucl. Phys.*, B898:563–604, 2015.

[136] B. L. Ioffe and A. V. Smilga. Meson widths and form-factors at intermediate momentum transfer in nonperturbative QCD. *Nucl. Phys.*, B216:373–407, 1983.

[137] H. Georgi. Lie algebras in particle physics. *Front. Phys.*, 54:1–320, 1999.

[138] J. M. Flynn and J. Nieves. Form-factors for semileptonic B $\to \pi$ and D $\to \pi$ decays from the Omnes representation. *Phys. Lett.*, B505:82–88, 2001. [Erratum: *Phys. Lett.* B644,384(2007)].

[139] I. Caprini. *Functional analysis and optimization methods in hadron physics*, volume 9783030189488 of *Springer Briefs in Physics*. Springer, 2019.

[140] C. Bourrely, I. Caprini, and L. Lellouch. Model-independent description of $B \to \pi \ell \nu_\ell$ decays and a determination of $|V_{ub}|$. *Phys. Rev.*, D79:013008, 2009. [Erratum: *Phys. Rev.* D82,099902(2010)].

[141] C. G. Boyd, B. Grinstein, and R. F. Lebed. Model independent extraction of $|V_{cb}|$ using dispersion relations. *Phys. Lett.*, B353:306–312, 1995.

Index

Printed in the United States
by Baker & Taylor Publisher Services